AF806643

Scientific Objectivity and Its Contexts

Evandro Agazzi

Scientific Objectivity and Its Contexts

Springer

Evandro Agazzi
School of Medicine, Bioethics Department
Universidad Panamericana
Mexico City
Mexico

ISBN 978-3-319-04659-4 ISBN 978-3-319-04660-0 (eBook)
DOI 10.1007/978-3-319-04660-0
Springer Cham Heidelberg New York Dordrecht London

Library of Congress Control Number: 2014931790

© Springer International Publishing Switzerland 2014
This work is subject to copyright. All rights are reserved by the Publisher, whether the whole or part of the material is concerned, specifically the rights of translation, reprinting, reuse of illustrations, recitation, broadcasting, reproduction on microfilms or in any other physical way, and transmission or information storage and retrieval, electronic adaptation, computer software, or by similar or dissimilar methodology now known or hereafter developed. Exempted from this legal reservation are brief excerpts in connection with reviews or scholarly analysis or material supplied specifically for the purpose of being entered and executed on a computer system, for exclusive use by the purchaser of the work. Duplication of this publication or parts thereof is permitted only under the provisions of the Copyright Law of the Publisher's location, in its current version, and permission for use must always be obtained from Springer. Permissions for use may be obtained through RightsLink at the Copyright Clearance Center. Violations are liable to prosecution under the respective Copyright Law.
The use of general descriptive names, registered names, trademarks, service marks, etc. in this publication does not imply, even in the absence of a specific statement, that such names are exempt from the relevant protective laws and regulations and therefore free for general use.
While the advice and information in this book are believed to be true and accurate at the date of publication, neither the authors nor the editors nor the publisher can accept any legal responsibility for any errors or omissions that may be made. The publisher makes no warranty, express or implied, with respect to the material contained herein.

Printed on acid-free paper

Springer is part of Springer Science+Business Media (www.springer.com)

To Lulú whose loving insistence has urged me to conclude this long-awaited work

Foreword

This work has a long history, so long that it well warrants being considered my life's work. In my book in the philosophy of physics in Italian, *Temi e problemi di filosofia della fisica*, which was published in 1969, I presented a theory of scientific objectivity which succeeded in attracting the attention of a number of scholars at that time, including, for example, Marian Przelecky, Riszard Wóycicki and Marja Kokoszynska in Poland and Carl Hempel in the United States. That book, however, was born under an unlucky star, its publisher becoming insolvent soon after its publication. And, though the unsold copies were acquired by another publisher and put on the market as a second edition in 1974, this new house discontinued its philosophical collections after a short while. The result of all this was that the book appeared hardly at all in bookstores, even in Italy, and for many years has only been available in libraries or by direct order from the publisher via a rather complicated procedure.

On the positive side, however, the Spanish translation of the book, which came out in 1978, has had a broad circulation in Spain and Latin America thanks to the solid marketing policy of the Spanish publisher. And over the years, I have had the opportunity to present its central ideas at conferences and in lectures, and to increase their circulation, particularly among philosophers of science. The pleasant consequence of this was that I could note their wide acceptance and even see them inspiring other scholars. Less pleasant, however, was the fact that I seldom received credit for being the source of these ideas. This had nothing to do with a lack of professional honesty, but simply with the fact that no other work of mine existed (particularly none in English) to which reference could be made, except for a few scattered papers.

The only way to correct this situation seemed to me to be to write a book in English, which could enjoy the possibility of the broad readership offered by this language, especially in the philosophy of science. I discarded the idea of a translation of my original book because it contained several parts that were specifically relevant to physics while not being directly related to my original position regarding scientific objectivity, and furthermore contained only partial elaborations of more general issues that deserved greater attention. Therefore, I decided to prepare a new book, in which the view of scientific objectivity already proposed in *Temi e problemi* would be presented and further developed, along with much broader references and a discussion connected to past and present authors whose work was relevant to the topic. The result is the present work, in which certain parts of *Temi e problemi* have been omitted, while, at the same time, chapters and sections have been included whose content was at most only hinted at in the earlier book.

I was able to begin realising my project when I had the opportunity to present and discuss my ideas in detail while teaching a seminar for graduate students at the University of Pittsburgh in 1977 and working as a research fellow at the Center for Philosophy of Science of that University. A similar opportunity presented itself in 1978, when I was visiting the University of Dusseldorf. It was there that I began to work out the general structure of the book and organise my numerous notes. The fulfilment of the project, however, would have required a year's concentrated effort, and the opportunity to exert such effort did not offer itself for a long while. On the contrary, my academic activities and my numerous international responsibilities increased between 1978 and 1993, leaving me only a couple of occasions for uninterrupted work (twice in Oxford during summer vacations, one term in Stanford in 1981 and one term again at the Center for Philosophy of Science in Pittsburgh in 1992).

This does not mean that, during the intervals between these fortunate opportunities, this work remained in a state of hibernation; indeed it has constantly been in the forefront of my mind, and several parts of it have been redacted, and even published from time to time as self-contained papers, which have been incorporated with a few adaptations as sections of this book. Their listing in the references of this work is, therefore, a documentation of its progressive construction over many years.

The long history of the writing of this book explains how I have come to be indebted to many people for inspiration and suggestions, though this may not always be clear from its content. For example, the fundamental orientation of the whole of my thought, due to my having been a disciple of Gustavo Bontadini, only occasionally surfaces in my constant reference to and criticism of 'epistemological dualism', and in my way of conceiving of the cognitive status of metaphysics (two fundamental aspects of his teaching); and the same may be said regarding the essential contribution made to the final shaping of my theory of objectivity by the reflections of Vittorio Mathieu on this topic. During my stay in Pittsburgh in 1977, I had the privilege of spending many hours in philosophical discussion with Wilfrid Sellars, sharing with him many points of view; however, only our major point of disagreement is what appears in the book, namely my not accepting his opposing of the 'manifest' and the 'scientific' images of the world. In a similar way, the conversations I had with Karl Popper are also reflected in this work mainly through criticisms I express concerning several of his doctrines.

In spite of all this, let me mention at least a few philosophers with whom I had especially fruitful exchanges of views: Larry Laudan and Nicholas Rescher during my first stay at Pittsburgh, again Rescher and Peter Machamer during my second stay, Patrick Suppes, Edward Zalta and John Etchemendy during my stay at Stanford, Alwin Diemer and Wolfram Hogrebe during my stay in Düsseldorf, and Kurt Hübner and Hans-Georg Gadamer on several scattered occasions.

More substantial have been the suggestions I received from those who have accepted to read and discuss parts of this work during its elaboration. Jonathan Cohen, under whose supervision I had spent a year as a research postgraduate in Oxford in 1960, read my developing work on the occasion of several stays I spent at Oxford much later. Also Rescher had the kindness of doing the same in Pittsburgh in 1992. I also received valuable comments from Marco Buzzoni, a former student of mine who (having become a respected colleague in the meanwhile) has helped me in a thorough revision of a first draft of this work. I have received equally valuable comments and suggestions from Mario Alai regarding certain central parts of this book. The scholar who has most directly assisted me in the redaction of the book, however, is Craig Dilworth, whom I first met in 1977 on the occasion of a lecture I gave in Uppsala, where he was working on his doctoral dissertation (later published as the book *Scientific Progress* in 1981). We discovered a fundamental affinity between the 'perspectivist' view of scientific theories he was advocating and my own theory of scientific objectivity that was also perspectivist, and began a collaboration that has lasted right up to the present. His critical appraisal of my writings, including the present work, has been precious to me, as will be clear from my several references to his work, references that do not imply either a direct influence of his views on mine, nor the reverse, but rather a fruitful convergence of often different paths. The frequent references to his work, however, are also intended to compensate for the little attention that mainstream philosophers of science have paid to his very valuable production.

A few words now regarding certain features of this work. Its general spirit is in keeping with the *analytic* approach that has characterized philosophy of science during the twentieth century, and this is a natural consequence of the fact that my training in philosophy of science has been based on a detailed study of this tradition, from logical empiricism to the subsequent developments within the Anglo-American world. From the beginning, however, I did not share certain elements of this tradition, that is, its radical empiricism, syntacticism, linguistic exclusivism and lack of historical sensitivity. It is true that such features have been gradually overcome during the evolution of the said tradition, but the fact of having been free from them from the beginning has offered to my perspective, I believe, the advantage of anticipating several of such developments, and also of avoiding certain excessive reactions they contained. For example, the awareness of the limitations implied in the purely linguistic view of scientific theories has often led people to discard completely the so-called 'statement view' of theories and the nomological-deductive model of scientific explanation. According to my view, theories are not *just* systems of statements, but they are *also* this, because they are linguistic explications of the content of a particular *Gestalt* proposed for the understanding and explanation of a given domain of objects. Therefore, the sentential view and the nomological-deductive model can be preserved as a *partial* characterization of scientific theories, whose more adequate characterization needs the introduction of *hermeneutic* tools. Similarly, the appreciation of the dependence of the meaning of a concept on its linguistic context has come as a development of the linguistic approach to theories, and has prompted the ideas of meaning variance, incommensurability, incomparability, and so on. I had arrived at a similar conclusion, instead, as a consequence of my studies on formal systems and axiomatic method, in which I had stressed a genuine *semantic function* of the axiomatic method (in addition to its commonly recognized syntactic function); this means that the axiomatic context contributes to the shaping of the *sense* of the concepts occurring in a theory (therefore, meaning variance is a real fact). However, not having remained prisoner of the 'linguistic turn,' I always maintained that sense cannot produce or ensure *reference*, for which an extralinguistic source must be provided, and this source consists in *operations* that are not reducible to the *observations* that radical empiricism requires, since they are essentially related to *praxis* and can be connected to sense thanks to its *intensional* nature. This position, in turn, has led me to vindicate a fundamental role for *truth* in science (something that had been almost banned from philosophy of science) and to study how truth can be attained, either by direct reference, or by argument, and this offers a foundation for admitting also the truth of non-observationally testable statements. Finally, the referential commitment of truth justifies a (carefully and duly specified) *realist* view of science. In my perspective, scientific objectivity is not context-dependent in a purely linguistic sense, but in a *historical* sense (of which the linguistic dependence is only a very particular aspect). The exploration of such a historical contextualization (that does not amount to *relativism*) opens the way to a due appreciation of all the right points stressed by the sociological interpretation of science, without falling into its excessive conclusions, and at the same time it justifies the consideration of those problems (for example, problems of an ethical and metaphysical nature) that cannot be treated in a consideration of science as a closed system of concepts and procedures. This approach has also provided a more comprehensive framework for the treatment of the relations between theories and models, and the strongly 'structuralist' conception of models, already explicitly presented in my book of 1969.

Owing to all this, several important works that were published during the long elaboration of my book did not appear all that new and original to me, since their basic views had already been anticipated in certain sections of this work or, sometimes, even published in papers of mine. Nevertheless, I am indebted to them for having pushed me to better formulations, or for further deepening certain views that I had conceived of independently. By the way, it was because of the publication of such works that I have been obliged to resume and revise from time to time my work in order to keep it up to date with the pertinent literature, a fact that has obviously slowed down its redaction. Yet this should not be understood as a pretension of 'completeness' and, in particular, it does not mean that I underestimate the importance of authors whom I do not mention. I do not 'ignore' them, but I simply had some particular reason for not mentioning them in the book. (In particular, in spite of having lectured for 19 years in French and German at the University of Fribourg, and of being well acquainted with twentieth century French and German philosophy of science, I preferred not to mention the relevant French and German authors, rather than make a few occasional references in footnotes to works hardly accessible to the English-speaking readers to whom this work is particularly addressed.)

I want to conclude by mentioning the favourable conditions that have significantly helped the realisation of this work. Regarding the stimulating intellectual atmosphere and the availability of research facilities, I must stress the importance of my repeated stays in Oxford and the great opportunities offered me by my two stays at the Center for the Philosophy of Science of the University of Pittsburgh (where the concrete redaction of the greatest part of this book was completed). In order to terminate my work, however, I needed a long period of time to devote almost exclusively to this enterprise, and this has been granted me by a research appointment of the Accademia Nazionale dei Lincei. It is thanks to this that this longstanding life's work of mine could be 'practically' concluded. By saying 'practically', however, I mention an innate hypercritical attitude of mine that has imposed on me the obligation of a 'final global revision', via which repetitions and redundancies would be eliminated; but the leisure required for this revision has hardly occurred, so I finally decided to close this enterprise after a 'normal' careful control. Owing to such a long elaboration I must say that this book is like Theseus' ship that, after many repairs and replacements of its parts, was no longer the original one (and I gladly admit that several 'layers' can be found in the book's structure which are, however, systematically connected), but at the same time I am satisfied that it did not result, after all, in a Penelope's web.

Contents

Chapter 1
Historical and Philosophical Background

1.1 Objectivity as a Replacement for Truth in Modern Science

The attitude towards science today is rather prudent compared to how it has been in the past, or, as some might say, it is not as naive as it was, for it no longer considers science the proper guardian of the *truth* concerning nature or, more generally, concerning the different fields into which the domain of scientific investigation has expanded. We shall examine some of the reasons explaining the older attitude. In place of this once widely accepted point of view, one now finds another no less deeply-rooted perspective—among professional scientists as well as various cultivated people—namely, the belief that the assertions of science, though not deserving simply to be called *true*, must nevertheless be considered *objective*.[1]

This kind of contraposition between truth and objectivity is not easy to understand, and we shall be specifically concerned with it later in this book; but for the moment let us stress the fact that such a reconceptualisation of scientific knowledge neither corresponds to nor implies a devaluation of science itself. Quite the contrary, we can easily show that science has practically become the very paradigm of rigorous knowledge, or that, at least in the opinion of many scholars, the only knowledge deserving of the name is scientific knowledge, when knowledge is considered as something distinct from, for example, emotional, artistic, religious and other such attitudes towards reality. In this way, while at the beginning of the

[1] Of course, *certain* philosophers of science deny that science deserves the qualification of providing *objective* knowledge, and even claim that the concepts of scientific method and scientific rigour are fictitious. According to them, science is simply a social practice, with no special features making it superior to or even clearly distinct from other social practices. We shall explore some of these claims later in this book, and show how they are unjustified and misleading. However we shall not begin this discussion here, for two reasons: first, because we would like to investigate the much sounder and much more widely accepted view (especially among scientists) that science actually provides objective knowledge. Second, because the critical evaluation of the opposite thesis will be more precise after the clarification of what we can really mean by scientific objectivity. After this clarification it will be possible to accept certain claims of the socially-oriented position, without giving up the requirement of scientific objectivity.

E. Agazzi, *Scientific Objectivity and Its Contexts*, DOI: 10.1007/978-3-319-04660-0_1,
© Springer International Publishing Switzerland 2014

twentieth century only such areas of research as mathematics, physics, chemistry, biology and, perhaps, psychology, were considered to constitute sciences, we now speak rather unproblematically of the sciences of economics, political science, philology, history, and so on. In other words, *every* field of research is today admitted as a possible branch of scientific knowledge, provided that it is pursued in accordance with certain standards of rigour; and this implies that science is no longer characterised by *what* it investigates, but by *how* it investigates.

This transition from an emphasis on content to an emphasis on method has surely complicated the problem of obtaining a correct understanding of scientific objectivity, for this neglect of the subject-matter of science has gradually induced many scholars to maintain that science is not properly concerned with *objects*, but that it is simply *objective* in the sense that it remains faithful to certain formal or methodological requirements. We are going to discuss whether this way of understanding objectivity is sound, but for the moment we simply take this as a factual portrayal of how scientific objectivity is often understood. But, once this step is taken, the conflict between objectivity and truth must soon emerge, for there is no room for truth proper if no reference to something which is in some way external to or distinct from the discourse is envisaged.

To be more precise, we note that two distinct features have been indicated above as characterising the present-day conception of science: objectivity *and* rigour, and it is by no means evident that these two requirements are synonymous, or that objectivity 'reduces' to rigour. Indeed, the traditional conception (going back to Plato and Aristotle, and continuing in the West until the Renaissance) characterised science through two distinct features: *truth* and rigour. The idea of science as providing knowledge of a kind that deserves the highest qualification emerged gradually in ancient Greek philosophy by requiring that this knowledge not only be knowledge of what is the case, but also of the *reason* or *reasons* for its being the case. This 'giving of the reason' (*logon didonai*) soon led to the traditional ideal of science as a *demonstrative* discourse—as a discourse where logically cogent proofs are provided of what is claimed—and we can correctly see in this requirement a first qualification of the notion of the rigour which must accompany truth in science (leaving aside, for the moment, what these reasons should be). This is why purely empirical or descriptive knowledge (even if true) was not qualified as *science*, but as *history* in a very broad sense (in which the concept of 'natural history' could easily be accommodated). Given this fact, we see that the requirement of rigour has constantly characterised the idea of science during the history of Western civilisation, and has expanded from the original proposal of logically *proving* the truth of certain assertions (starting from more basic truths which could provide reasons for them), to the proposal of reliably *ascertaining* the said truth. In this sense the development of the 'empirical' methods in the natural sciences, no less than in the fields of philological, historical, sociological, and psychological research, belongs to this enlargement of the concept of rigour (and of science). However, the refinements in the pursuit of mathematical rigour—which have been characteristic of the history of mathematics in the last two centuries—equally express the development of the requirement of rigour that has constantly marked science.

As a consequence, we must recognise that, while science is now characterised as providing knowledge endowed with both objectivity and rigour, it is only the first aspect, i.e. objectivity, that constitutes a radical novelty vis-à-vis the past. But, at least partly because of decreasing confidence in the requirement of truth, the new notion of objectivity has tended to become identical with that of rigour. This tendency has been particularly strong in the field of mathematics through the dominance of the formalistic outlook, and it has not gained a comparable strength in the field of the empirical sciences. In the present book we shall concentrate almost entirely on the problem of scientific *objectivity*: not because we underestimate the central role of rigour in science, nor because we are unaware that rigour and objectivity are so strictly interwoven that it would be impossible to separate them, but simply because certain problems which do not coincide with the question of rigour are related to the notion of objectivity. We could try to sketch this distinction by saying that scientific rigour still corresponds to the fulfilment of the requirement of 'giving the reasons' for whatever one claims in science (i.e., it consists in spelling out *how* and *why* one came to a particular claim), while objectivity corresponds rather to the clarification of *what* one is claiming—the meaning and content of one's claim. From what we have said it also becomes clear why, when speaking of science generally, we shall mean *empirical science*; objectivity in mathematics (as something distinct from pure rigour) would require a much more elaborate discussion.[2]

In conformity with the explanations just outlined, we shall try to understand what kind of change has occurred in our way of conceiving of *what* science says, can know, and is about; and to do this we require a brief historical analysis.

One must first recognise that modern science, as it emerged from the 'Galilean revolution,' had certain anti-essentialist and anti-substantialist features that distinguished it sharply from the philosophical attitude current at that time. Galileo's proposal, not to be concerned with the 'essence' of physical realities ("natural substances," as he says), but simply with certain of the 'affections' that they show, and not to worry about the ultimate causes of empirical phenomena, but only about discovering a simple mathematical description of them, not only suggested an attitude of intellectual modesty, but also precisely indicated a particular methodology, which was followed by Newton (at least in the *Principia)* and by the creators of the 'new science' of mechanics (as Galileo called it).[3]

[2] I have sketched these discussions elsewhere, e.g. in Agazzi (1961), (1966), (1978c), (1978d), (1994) and (1997), maintaining that a purely formalistic approach does not capture the nature of the formal sciences, in which certain 'contents' are also present. More recently I have collected the most significant of my papers on this and related topics in Agazzi (2012).

[3] To be more precise, Boyle, Hooke and almost all of the new experimentalists theorised about the causes underlying the phenomena they uncovered, but without concern for the 'ultimate' causes. This can be understood by becoming aware that the search for causes is the core of any *explanation*, and even Galileo (as we shall see later) did not dispense with this basic condition of knowledge. The question, then, reduces to 'what kind of causes' one is ready to admit, and here the meaning of 'ultimate' causes may receive a certain precision. For example, J. S. Mill (who on this point is in fundamental agreement with the tradition of British empiricism) says: "I make no

The science which actually developed proved so powerful in predicting and explaining facts in so many branches of natural investigation (from those concerning the most common events of everyday experience up to the cosmological perspectives of celestial mechanics) that—after less than a hundred years of such impressive achievements—people became inclined to believe, towards the end of the eighteenth century, that this science had provided mankind with the only authentic instrument for investigating the *true* structure of physical reality, thus attaining the goal which the 'traditional' philosophy of nature had unsuccessfully striven after for centuries. As a consequence, a kind of metaphysics of science was born, which found its clearest expression in what is frequently called the mechanistic worldview of the nineteenth century.[4] Mechanics was held not only to be the fundamental branch of physics, but also the interpretation key to every natural phenomenon, since its principles had received the widest application and were expected to have an *unlimited* application. Thus the attempt was begun to reduce all the traditional parts of physics, such as acoustics, optics, and the theory of heat, to mechanics, that is, to investigate these fields assuming as fundamental concepts only those of the *motion* of *material* particles, or that of a *material* ether, under the influence of certain *forces*. The same effort was made when electrodynamics was created, by designing various complicated 'mechanical models' for the 'electromagnetic ether.' This part of history is so well known that we feel we can dispense with giving details.[5] We have

(Footnote 3 continued)
research into the ultimate or ontological causes of anything… The only notion of cause which the theory of induction requires, is such a notion as can be gained from experience." (Mill 1881, p. 326). Mill's assertion mirrors his *positivistic* general attitude that implied the refusal of *any* ontology and reduced causality to mere *uniformity* (a reduction, however, that implicitly presupposes the 'ontological' assumption of the principle of uniformity of nature). In the case of Galileo and other scientists of the seventeenth century, the causes were endowed with an ontological status, but, as we shall see, a status that was 'delimited' by those 'mechanical affections' that they thought to be the only ones relevant to natural science.

[4] A mechanistic philosophy of nature was influential and widespread already in the seventeenth century. However, this philosophy constituted a metaphysical view which was not *generated* or *suggested* by modern physical science, but which rather (at least to a certain extent) *promoted its birth*. Moreover, it could even be excessive to call it a metaphysics in a proper sense, since many of the scientists that adopted it did not pretend that the mechanistic approach was adequate for interpreting *the whole* of the natural phenomena. Indeed, Boyle and Newton, for example, explicitly affirmed that the phenomena of life overstepped the framework of mechanical explanation, and also within the domain of inanimate nature, electricity and magnetism were not expected to be necessarily encompassed in that framework. The mechanistic worldview of the nineteenth century, on the other hand, was a genuine metaphysical view that resulted from the widespread success of the science of mechanics in gradually absorbing different branches of physics, which produced the conviction that *the whole* of natural phenomena could be explained mechanically. For more details on this point, see Agazzi (1969), pp. 23–26.

[5] Simply as examples, let us quote a few very significant statements of some leading scientists of that time.

In his famous paper 'On the Conservation of Force,' Helmholtz says:

> The task of the physical sciences may therefore be determined, ultimately, as that of reconducting natural phenomena to immutable forces, attractive or repulsive, whose

briefly recalled these facts in order to stress that, during the nineteenth century, science was regarded as an inquiry that was completely committed ontologically, and as the guardian of eternal *truth* concerning physical reality.

It is well known that this basic conception of reality experienced a radical crisis in the last decades of the nineteenth, and in the first years of the twentieth century, as a consequence of difficulties which first appeared in electrodynamics and thermodynamics, and then reappeared as a consequence of the growth of relativity and quantum theory. This crisis was very shocking for many scientists, and eventually led most of them to a profound change in their attitude towards scientific knowledge as such. For, not only had the categories of every physical world-picture (such as those of space and time) been altered, but practically every concept of the old, mechanistic physics underwent a radical revision. These changes had two main consequences. First, these mechanistic concepts seemed not to be as ontologically faithful and reliable as the older scientists had believed them to be; and second, extremely serious difficulties arose regarding the possibility of constructing an intelligible picture of the microworld. Physicists thus soon became accustomed to

---

(Footnote 5 continued)

> intensity depends on distance. The possibility that this task be fulfilled constitutes at the same time the condition of the full intelligibility of nature (Helmholtz 1847, p. 16).

Maxwell, at the end of his *Treatise on Electricity and Magnetism*, in spite of having given as the most significant result of his research the famous equations of the electromagnetic field, which do not presuppose anything about the specific 'substance' of this field itself, feels almost obliged to propose as a programme for future research the exact determination of the structure of this field in the form of a mechanical medium:

> It is true that at one time those who speculated as to the causes of physical phenomena were in the habit of accounting for each kind of action at a distance by means of a special ethereal fluid, whose function and property it was to produce this action. They filled all space three and four times over with ethers of different kinds, the properties of which were invented merely to 'save the appearances'... In fact, whenever energy is transmitted from one body to another in time, there must be a medium or substance in which the energy exists after it leaves one body and before it reaches the other... Hence all these theories lead to the conception of a medium in which the propagation takes place, and if we admit this medium as a hypothesis, I think it ought to occupy a prominent place in our investigations, and that we ought to endeavour to construct a mental representation of all the details of its action, and this has been my constant aim in this treatise (Maxwell 1881, II, pp. 865–866).

As regards our final example, Lord Kelvin, it is well known that the inability to realise a mechanical model of the electromagnetic field led him even to refuse his consent to the Maxwellian electromagnetic theory of light:

> I never satisfy myself until I can make a mechanical model of a thing. If I can make a mechanical model I can understand it. As long as I cannot make a mechanical model all the way through I cannot understand and that is why I cannot get the electromagnetic theory... I want to understand light as well as I can, without introducing things that we understand even less of (Thomson 1884, pp. 270–271).

the idea that it was not their task to formulate an intuitive ontological representation or picture of their domain of research; and this led to a mistrust of the conception of physics as a description of the real world, for there was apparently no room for speaking of a conformity between the inner structure of the microworld and the scientific picture of it, since such a picture could not even be proposed.

As a consequence, first physics, then other sciences following its example, became ontologically uncommitted, more or less in the sense of Kant's denial of the possibility of knowing the *noumenon*, the "thing-in-itself"; and a certain variety of epistemological positions replaced the old trust in the ability of science to provide knowledge of reality, in the sense that this ability was rather unproblematically admitted for the empirical part of this knowledge but not for its theoretical part. In order to present these issues with full clarity we should analyse the distinction between scientific empiricism and scientific realism, as well as other related questions, but we do not feel obliged to do this at this point, since we are going to debate these questions in the sequel of this work. Therefore we shall be content with a few general remarks.

Some scholars (whom we can indicate as defenders of a form of empiricism/positivism) showed a purely pragmatist or 'instrumentalist' attitude; that is, they considered science to be charged with the much more modest aim of providing people with useful instructions about how to behave successfully in their relations with nature, how to make reliable predictions about some interesting facts, how to organise the different pieces of information that we are able to obtain from experience, from empirical evidence, and so on.[6]

Others wanted to remain faithful to the idea that science is a 'cognitive' enterprise; but they inclined towards a phenomenalistic epistemology. They did not pretend that science must have to do with reality, but accepted rather that it could limit its concerns to a specific world of *phenomena*. Phenomena are here to be understood as constituting a realm of intellectual constructions which are linked to experience by particular internal and external conditions, conditions which also keep them from becoming speculative. This line of thought, aside from including patent elements of Kant's philosophy, also had some important predecessors among nineteenth-century scientists (such as Helmholtz and Hertz), but it received new force from the state of affairs that quantum mechanics brought about.[7]

[6] The most developed and best known doctrine along these lines is probably that of Ernst Mach, who at an earlier stage in his thinking had been a supporter of the mechanistic worldview (e.g. when he wrote his *Treatise of Physics for Physicians*, 1863), only later to become one of the most influential critics of this outlook and generally of every 'metaphysical' commitment of science.

[7] As a very interesting example of this position we may mention that of Max Planck. See in particular his essay of 1930, "Positivismus und reale Aussenwelt" (in Planck 1933, pp. 208–232). Here he sharply distinguishes the inaccessible 'real world' from the 'world of sense perceptions,' and sees the task of science as that of constructing a physical world-picture which should depict an objective relationship between these two worlds. These conceptions are expressed in a less detailed way at the beginning of a more accessible paper: "The Scientist's Picture of the Physical Universe," which is available in English in Planck (1932).

One of the most characteristic points of quantum mechanics is, of course, that it is not possible to conceive of an object which could remain unaffected by the very process of observing it, or by the measurement procedures which are applied to it. It has become usual in the literature to say that in microphysics there is definitely no possibility of separating the 'subject' from the 'object' so as to be able to attribute particular separate measured quantities to each of them at the same time. This situation was frequently described by saying that we never have to do with nature, but always with the interrelationship between nature and man. In such a perspective, it was very tempting to identify this new structure, this indiscernible unity of subject and object, as the 'phenomenon' that the new science had to handle as its proper subject-matter.[8]

Clearly, in both of the positions sketched above, science no longer seems to be concerned with describing *reality* (for, according to the one view, it has not a descriptive, but a pragmatic aim, while according to the other it retains its descriptive intention, but is unable to reach reality proper). Even less does it seem to have the task of explaining phenomena in terms of an underlying reality (which is a central point in the empiricist/realist controversy). As a consequence science could no longer be conceived as a form of *true* knowledge (at least in the familiar sense of "true" according to which it consists in matching reality).

On the other hand, the fact that science had given up the task of obtaining truth about reality did not imply that it provided an *arbitrary* form of knowledge. Even some conventionalist perspectives that were proposed in the context of the pragmatist or instrumentalist viewpoint (Mach, Duhem, Poincaré) did not attribute arbitrariness to science.[9] The result was an effort to preserve the non-arbitrary and

---

[8] To limit ourselves once again to but a single author, we could mention Werner Heisenberg, who expressed a view of this kind in several of his philosophically oriented writings. Consider his saying, for example: "As a final consequence, the natural laws formulated mathematically in quantum theory no longer deal with the elementary particles themselves but with our knowledge of them" (Heisenberg 1958a, p. 15).

[9] Historical accuracy would certainly require making distinctions among these positions. For example, in the case of Duhem, conventionalism is simply a methodological scheme for analysing the structure of scientific theories, and does not exclude a pretension to scientific truth ("the image of an order and organisation of reality," as he says), as is also clear from his polemics against the British physics of model construction and from his 'realist' interpretation of even the theoretical concepts of science. For him the conciliation of these two aspects is possible by considering the historical development of science. Poincaré, as is well known, not only strongly criticised the extreme conventionalism of Le Roy—especially in the last chapter of *The Value of Science* (Poincaré 1904)—but also moderated his own conventionalism by claiming that the practical success of scientific knowledge in applications and predictions is a witness of its coping—at least to some extent—with 'reality.' However, this claim does not make Poincaré an instrumentalist, since he explicitly admits the existence of scientific truth (not reducible to the said practical success) in a sense which is rather close to the already quoted position of Planck, as may be seen, for example, from this passage from *Science and Hypothesis*:

> And one could not say that in such a way we are reducing physical theories to the simple role of practical recipes. These equations express relations and, if the equations remain true, it is because these relations retain their reality. The equations teach us, after

rigorous character of scientific knowledge by founding it on the basis of certain structural criteria, some of them being of a pragmatic, some being of a cognitive character, according to the different outlooks. All this gave rise to the conception of science with which we started our considerations: science provides *objective* knowledge which is nevertheless not *true* knowledge since it does not concern reality. This lack of ontological commitment appeared as the main identifying mark of the new characterisation of science.

At this point it is useful to explain what we mean by *knowledge* in this book. The notion of knowledge is linguistically expressed in different ways according to different languages. In English a unique term is available ("knowledge," with the related verb "to know") and this fact has led to the necessity of distinguishing *knowledge by acquaintance* from *propositional knowledge*, that is, 'knowing *p*' from 'knowing *that p*': e.g., "I know this red pencil" (by acquaintance) from "I know *that* this is a red pencil" (propositional knowledge). This distinction does not occur in other languages where there are two distinct terms for these two kinds of knowledge. Knowledge by acquaintance and its related verb are rendered, for instance, by: *conoscenza-conoscere* (Italian), *connaissance-connaître* (French), *conocimiento-conocer* (Spanish), *Erkenntnis-kennen* (German). In such languages propositional knowledge is denoted by a term that serves at the same time as a substantive and as a verb, for instance by: *sapere* (Italian), *savoir* (French), *saber* (Spanish), *Wissen* (German). Therefore, on the one hand, one does not say, for

(Footnote 9 continued)

> (the change of interpretation) as before, that a certain relation exists between something and something else; only, this something which we called *motion* before, is now called *electric current*. But these denominations were nothing but images put at the place of real objects that nature will eternally hide from us. The true relations between these real objects are the only reality which we can attain, and the only condition is that there exist between these objects the same relations existing between the images which we are obliged to put in place of them. If these relations are known to us, what does it matter if we consider it useful to replace one image with another? (Poincaré 1902, p. 190).

In concluding these remarks let us only note that Poincaré's speaking in this way is reminiscent of Henrich Hertz' way of expressing himself in the Introduction to his *Principles of Mechanics*:

> We form for ourselves images or symbols of external objects; and the form which we give them is such that the necessary consequents of the images in thought are always the images of the necessary consequents in nature of the things pictured. In order that this requirement may be satisfied, there must be a certain conformity between nature and our thought. Experience teaches us that the requirement can be satisfied, and hence that such a conformity does in fact exist.... The images which we here speak of are our conceptions of things. With the things themselves they are in conformity in *one* important respect, namely, in satisfying the above-mentioned requirement. For our purpose it is not necessary that they should be in conformity with the things in any other respect whatever. As a matter of fact, we do not know, nor have any means of knowing, whether our conceptions of things are in conformity with them in any other than this *one* fundamental respect (Hertz 1894, pp. 1–2).

example, "je *connais* que Paris est la capitale de la France", but "je *sais* que Paris…," and, on the other hand, one does not say, for example, "je *sais* New York" but "je *connais* New York."

At least in the Latin languages, however, another distinction must be introduced in the meaning of "sapere," "savoir," "saber," and so on, since these verbs are also used to indicate what could be called *performative knowledge* indicating a certain *competence* or *ability*; when it is said, for example, "egli *sa* parlare inglese," "il *sait* parler anglais," "el *sabe* hablar inglés," the meaning of these sentences is rendered in English by "he *can* speak English" (likewise in German. "er *kann* Englisch sprechen."

The preceding are not just philological digressions since the distinction between the two kinds of knowledge (by acquaintance, and propositional) is really relevant to several issues and, in particular, regards the fact that knowledge by acquaintance is intrinsically private while propositional knowledge can be communicated and become 'intersubjective.' This is a capital point in the treatment of *scientific knowledge* which is obviously *public.* Therefore, when we speak of *knowledge* in the present work we intend it to be *propositional knowledge* (unless we explicitly indicate that we are referring to knowledge by acquaintance).

Now a question arises: has knowledge something to do with *truth?* A satisfactory answer to this question needs a clarification of the notion of truth that will be offered later in this book. However, at least a few hints must be given here. Since the intuitive idea of truth implies some kind of 'correspondence' between a representation and the thing represented, one might spontaneously say that knowledge by acquaintance is always true. A more refined analysis, however (that will be carried out in due time), shows that truth is much more appropriately said of *propositions*. Therefore, it is more expedient to say that knowledge by acquaintance is not true or false, but, perhaps, *right*, while propositional knowledge has to do with truth. But what does it mean "has to do?" Could propositional knowledge really be knowledge without being true? Apparently not; in fact it is admitted in general epistemology that, in order to affirm "I know that *p*," the first necessary (though not sufficient) condition is that *p* be *true*. Therefore, to require from (propositional) knowledge that it be true seems little pertinent since knowledge 'includes' truth and it would be self-contradictory, for instance, to speak of 'false knowledge.' Despite all this, we are confronted with an historical fact, scientific *knowledge* (which is a form of propositional knowledge) is considered by many as being a form of knowledge without truth which still deserves to be considered knowledge. This is why we must, for methodological reasons, recognise this situation, look for its reasons, and eventually see whether and how it should be corrected by retaining the rights of truth also in the field of *scientific knowledge* (as will be done in this work).

A last remark. In the tradition it was usually required (starting with Plato and Aristotle) that genuine knowledge requires *understanding and explanation* of what is directly known. Positivists (as we have seen) rejected these additional requirements but this was only the consequence of their *radical empiricism* and the implicit dogmatic *presupposition* that *reason* has no *cognitive* import. Since we do not share such a tenet (and shall give reasons for this rejection) we shall include understanding and explanation in the structure of *knowledge*. This will justify our claim that science (also in its *theoretical* part) aims at *truth* as a consequence of the fact that it strives towards the fullest acquisition of (propositional) knowledge. Not all authors share this thesis,[10] but we are confirmed in maintaining it also by the fact that *contemporary* epistemology (and not just traditional epistemology) requires that, in order to affirm correctly "I know that *p*," not only must *p* be true (a necessary but not sufficient condition), but must also be 'justified'; and understanding and explaining are certainly very relevant requirements for justification.

1.2 The Scientific Revolution Revisited

An interesting question that might now be asked is whether such a disengagement from ontology, such a reconceptualisation of science as affording an objective form of knowledge rather than a true one, as took place at the beginning of the twentieth century, must be conceived of as a 'lowering' that science was compelled to accept as a consequence of the recognised impossibility of its being anything more than that, or whether such an epistemological condition was already 'structurally' embodied in science itself. If the latter were the case, we should easily be led to the conclusion that such an outcome of the crisis of modern science actually meant a recovering of its original purity, rather than a resignation from its previous tasks.

In order to investigate this question we must analyse the conceptual meaning of the Scientific Revolution (that is, the intellectual revolution that gave rise to modern science at the time of the Renaissance), and especially the very turning point in it which coincides with Galileo's position. In fact, many of the aspects that are often considered as most characteristic of and decisive for that revolution

[10] For example, Craig Dilworth maintains that truth *and* explanation are the basic goals of science, and considers truth to be the fundamental property of *scientific laws* (that are located at an empirical level, though not in an empiricist/positivistic sense). They provide *knowledge* while theories are put forth in order to provide *understanding* and *explanation*, and do not constitute knowledge (Dilworth 2007). Though Dilworth defends scientific *realism,* in the sense that he convincingly argues that science cannot avoid trying to (causally) explain, by means of *theories,* the empirical by postulating that the observed (or rather measured) features of reality are produced by certain underlying not empirically accessible entities, he does not clearly maintain that such entities really *exist,* and in such a way his realism is a 'weak' one, in comparison with the one that defends the 'reality of the unobservables,' to put it briefly. On the contrary, we are going to subscribe to this 'stronger' form of realism, and this is probably the most important difference of our position with regard to that of an author with whose conceptions we largely agree.

appear not to be such once they are submitted to a closer scrutiny.[11] The idea of attributing an essential role to empirical support in statements concerning nature, for example, was not really unique to the Renaissance, for it is to be found in the whole tradition of the late Scholastics, especially of the Oxford Scholastics. The same can be said regarding the criticism of Aristotelian physics, which was surely one of the most obvious features that accompanied the Scientific Revolution. Here too we can assert that such criticism was not novel, there having been detailed and critical discussions of Aristotelian physics amongst the late Scholastics (not to speak of the rejection of several of its aspects for philosophical or theological reasons already in the thirteenth century). Also, the use of mathematics in the description of nature had a very long tradition, both in the sense of conceiving of mathematics as a particularly well-established world of rational certainty and eternal truth, as well as in the sense of conceiving of the natural world as mathematically structured.[12]

If such was the situation, we can correctly ask why these conditions did not act simultaneously to give rise to modern science some centuries before Galileo. The answer to this question seems to be the following: despite every appearance, these patterns did not determine a form of knowledge which necessarily was *at variance* with the Aristotelian model of knowledge of nature. The criticism of Aristotelian physics, for example, was promoted within the Aristotelian framework itself, so as to improve or perhaps correct it, but not to reject it. This can be said because the fundamental points of view, the conceptual tools, the categories applied and, more particularly, the aims of natural investigations were the same as those of Aristotle. This can also be said of the importance given to empirical evidence or to the

[11] The reader will forgive us if, owing to the specific nature of this book, the short discussion of the scientific revolution which follows will lack the scholarly backing that would be necessary if our presentation had a specifically historical aim. In particular we shall take for granted that a 'scientific revolution' actually took place, a thesis that not all historians share today, at least in the almost literal sense attached to this expression when Alexandre Koyré introduced it in 1939 (and which is mirrored, for example, in the title *Scientific Revolution* of Rupert Hall 1954). A more nuanced position has been defended by more recent authors (see, e.g., Shapin 1996) who stressed that we cannot identify a single well-delimited and historically coherent event (chronologically located between the end of the sixteenth and the beginning of the seventeenth century) so revolutionary and crucial that it changed in a radical and irrevocable way the knowledge humans had of nature and the methods of attaining this knowledge. Much more continuity with respect to the past, and much less homogeneity in the way of considering the 'novelty' of their own approach is to be found among the very protagonists of this complex process. It is undeniable, however, that those people (such as Galileo, Bacon, Descartes, Boyle, Pascal) even in the titles of their works wanted to stress the 'novelty' of their approaches, methods and contributions, and explicitly opposed them to the heritage of the past. Therefore the real problem is that of understanding in what this novelty consists and to what extent it is a novelty, without pretending that the solution will result in the indication of just a few clear-cut 'factors'. As a consequence we too shall focus only upon *certain aspects* which are of special importance to the *philosophical* analysis we are interested in, while leaving out of consideration several other aspects which are amply studied in the very rich literature concerning this great historical phenomenon.

[12] Clear documentation of what we are saying is provided in the monumental work of Pierre Duhem (Duhem 1913).

mathematical interpretation of natural phenomena. If such attitudes are adopted within one particular conceptual perspective they could provide a great impulse to what was subsequently called modern science, while, if adopted within an older perspective, they might simply contribute to improving a received metaphysical conception of reality. Let us only think of the Pythagorean and Platonic world-pictures, in each of which mathematics played a central role, but which were not scientific in the modern sense of this term.

We can now ask: what was the perspective or conceptual framework that had to be superseded in order for modern science to come to light? We can briefly answer that it was the substantialist viewpoint, and we shall now proceed to explain what we mean by this.

1.3 The Essentialist and Substantialist Points of View

Both "essentialism" and "substantialism" are terms which occur regularly in contemporary philosophy and, in particular, in discussions concerned with the nature of science. Think only of the polemics against essentialism led by Popper, or of the attacks against substantialism led by Cassirer. These are surely not the only thinkers to share this critical attitude. We are not interested here, however, in exploring these various positions. In the present context we would like to stress rather how the different authors agree in tracing the origins of these doctrines back to early Greek philosophy, and especially to Aristotle. According to Cassirer, for instance, the foundation of the Aristotelian doctrine of concept resides in the Aristotelian ontology of substance in such a way that "the complete system of scientific definitions would also be a complete expression of the substantial forces which control reality."[13] As for what substance *is*, Cassirer puts special emphasis on the Aristotelian doctrine according to which substance was conceived of as a kind of substratum in which the different features of being are inherent. This distinction between a substratum and features of being not only led to the later well-established distinction between a thing and its properties, but moreover produced a subordination of relations to essences; for a relation, in order to exist, must presuppose that which is being put into relation and cannot modify its essence.

Popper's definition of essentialism presupposes this conception both in its conceptual features and in its historical reconstruction. Actually, he characterises essentialism in a broad sense as "the doctrine that science aims at ultimate explanation."[14] Then, when he wants to provide a more detailed depiction of its nature, he indicates two aspects, one being an epistemic attitude, namely, the belief that "the scientist can succeed in finally establishing the truth of… theories beyond all reasonable doubt," and the other having an implicit ontological grounding, in

[13] Cassirer (1923), pp. 7–8.

[14] Popper (1963), p. 105.

that it claims that "the truly scientific theories describe the 'essence' or the 'essential nature,' of things—the realities which lie behind the appearances."[15] He, like Cassirer, maintains that the historical origin of this doctrine goes back to Aristotle. Leaving aside for the moment the negative judgement that both Cassirer and Popper express concerning the essentialist or substantialist doctrines, one must first note that they are correct in recognising the main lines of these doctrines in Aristotle. But one should perhaps also note that the respective features of these two doctrines are not so strictly connected as they seem to be on first consideration. In other words, while "substance" and "essence" are very often used synonymously in colloquial language, certain features allow us to distinguish them and to give them a different role as far as the cognitive status of science is concerned. We would now like to outline the reasons which have led to a certain confusion of these two concepts, and the reasons which recommend keeping them distinct; both are already to be found in Aristotle.

Let us first see how Aristotle conceives of the *essence*. According to him, a very general meaning of essence can be seen to correspond to an answer to the question: "what is?" (*ti esti*). Such a question, however, can be answered in different ways. For example, we can answer: "Socrates is a philosopher," "Sugar is a white and sweet powder," "Man is a rational animal." All of them are correct answers to the respective question: "what is?" but the first simply mentions a particular characteristic that Socrates could have failed to posses, the second mentions certain qualities that sugar might perhaps not always show, while the third mentions features that are considered *necessary* for something to be a man. As a conclusion, Aristotle restricts the genuine meaning of essence to this last case: a thing's essence must be conceived of as the whole of its constitutive *properties*, that is, as the complex of characteristics which necessarily makes it what it is and not some other thing (essence as the *to ti en einai*). It is of great significance that he equates this notion of the essence (which we could call 'necessary essence') with one of the fundamental meanings he gives to the concept of *substance* (*ousia*), and for this reason we can call it also 'substantial essence.' It is also true that this fundamental ontological structure fully determines (as Cassirer has stressed) the cornerstone of the Aristotelian scientific discourse, that is, the predicative judgements in which various properties are attributed to the substance. In particular, this doctrine accounts for the foundational role attributed by Aristotle to *definitions* (which seems rather difficult for modern scholars to understand). Actually, we are accustomed to giving only an *intra-linguistic* role to definitions, as tools for 'establishing the meaning' of terms, while Aristotle considered them, in addition, as endowed with the task of 'expressing the essence'; and this is at least a partially *extra-linguistic* role.

In this sense (which was later to be characterised as that of being a *real* as compared to a *nominal* definition) definitions must be *true*, and indeed must constitute the starting point of every discourse aimed at being true not accidentally

[15] Ibid., pp. 103–104.

but by necessity. All these are well known theses, which are familiar to everyone who has had the opportunity of studying, for example, the second book of the *Posterior Analytics*. It is on the basis of this general conception that Aristotle is led to his axiomatic-deductivistic model of scientific knowledge, which has as its foundation the 'knowledge of principles,' provided by a set of definitions able adequately to express the essence of the reality upon which the discourse is being developed. Only if such a sound starting point is provided can the subsequent rigorous deduction, secured by syllogistic inference, lead to the construction of a satisfactory scientific edifice.

If we look at the history of Greek science, we can see that the adoption of the Aristotelian scheme was in a way responsible both for its splendours and for its shortcomings. As a matter of fact, Euclid's *Elements* is a masterpiece in mathematics which is structured according to the model established in the *Posterior Analytics*. In general one can say that that model was very well suited to the mathematical or formal disciplines in which the definitions (in the sense of *real* definitions) of the entities involved play an accessory role, while the axioms are the fountainhead of actual knowledge. But the application of the same scheme failed to prove equally fertile in the case of the empirical sciences, for there the alleged necessity of first establishing some essence-expressing definitions, in order to proceed from them via a rigorous deduction, often led to sterile aprioristic speculations, rather than to a substantial empirical science. If this explains, at least to some extent, why the classical tradition was able to produce a glorious mathematics, but only a meagre empirical science, it also raises the question whether this shortcoming is to be imputed to the Aristotelian doctrine of the essence rather than to other aspects of the above scheme.[16]

In order to answer this question we must consider that the term "*ousia*" has a multiplicity of meanings in Aristotle, and in particular that it is used by him not only to denote essence (as we have just seen) but also substance. Indeed, one must say that 'substance' is its primary and fundamental meaning, and that 'essence' appears as a kind of particularisation of that meaning. This can be clearly seen when we ask: what do we mean when we speak of the substantial elements or features of a particular thing? Among the possible answers to this question that Aristotle considers, two are of particular interest. One identifies those substantial features with the *essence* (*to ti en einai*), that is, with the complex of qualities which characterise the thing and which cannot be omitted without it ceasing to be

[16] We do not wish the foregoing statements to be taken too literally. Indeed Aristotle's biology is often considered with respect by modern scholars, and his physics contains many deep insights, careful analyses and profound discussions. However, it is a fact that Aristotle himself did not consider those parts of his natural investigation which are eminently descriptive (as is the case with most of his biology) to be *scientific*—in *his* sense. It is also a fact, on the other hand, that the doctrines of his physics are usually more significant from a metaphysical point of view (let us only mention the doctrine of change, or the analysis of causes) than from a point of view that we would call scientific in *our* sense. Therefore, even if we do not pretend that his physics should be judged from the point of view of our physics, it is nevertheless certain that it does not show the features of an *empirical* science proper.

that thing. These are the features that must be caught and expressed in the definition, which fixes the class to which the thing belongs.

Another way of conceiving of substance is to think of it as the *substratum* (*to hypokeimenon*) to which all of a thing's qualities must be related, to the extent that they are qualities 'of something.' Clearly this second way of conceiving of substance is suggested with particular force by the logical form of judgement, in which something is predicated 'of something else,' so that the idea of the substratum appears as the ontological counterpart of the subject in the subject-predicate linguistic structure. This distinction has been gradually abandoned in the course of the development of modern philosophy, in particular as a consequence of the development of modern science (as Cassirer has tried to show in his important work), but we shall not consider this historical point in our study. For the moment let us simply note that, according to Aristotle, the qualities of its substantial essence are the *causes* of the different characteristics and behaviours exhibited by an entity, and this is why he maintains that an adequate definition of the essence should allow for a satisfactory *deduction* of such features.[17]

It falls outside the aim of this book to enter into a detailed discussion of this Aristotelian doctrine, which one finds developed especially (but not only) in Book Z of the *Metaphysics*.[18] But from the few remarks we have made here it appears on the one hand that essentialism and substantialism could in a way coincide, as there is at least one classical sense (already present in Aristotle) according to which substance is essence. On the other hand, however, there are reasons for not putting these two notions on the same footing. Actually, there does not seem to be any reasonable objection to speaking of the essence of a thing as the sum of the features which distinguish it from other things, and assimilate it to other things in one and the same grouping (call it genus, species, class or what you will). This admission is indeed compatible with several conceptions of the ontological status of these features, about the naturalness or conventionality of their being put together (i.e. cohering in the same thing, or being conceived of as existing on the same ontological level), and so on. And, moreover, it expresses a condition without which no intellectual grasping of reality appears possible (for, in order to comprehend reality, we must be able both to *distinguish* entities on the basis of their *disparate* features, and to recognise them as potentially being of the *same kind* on the basis of their *common* features).

---

[17] We note, by the way, that a similar conception was shared also by the initiators of modern physics and lasted until the end of the nineteenth century in this science. The main difference was represented by the 'restriction' of the substantial essence they adopted, and by the adoption of a different kind of causality: whereas Aristotle had considered this causality as being essentially a *final* one (in the case of 'natural' events), modern physics considered it as an *efficient* causality, which for Aristotle was rather typical of artificially produced events (For a discussion of these points, see Dilworth 2007).

[18] A still very useful discussion of the different logical and ontological interconnections of these meanings of "substance" in Aristotle is provided in Trendelenburg (1846). A valuable recent work regarding the aspects of our issue which are treated in Book Z of Aristotle's *Metaphysics* is Frede-Patzig (1988).

The idea of a substratum, within this approach, appearing as a kind of ontological counterpart to a linguistic structure, seems to have no special reason for being retained on its own merits. In fact (and this criticism is present to some extent in Aristotle himself), it remains mysterious how one could even conceive of this 'entity,' which is hidden under a crust of qualities. It must be in principle indeterminable, because to determine it would be tantamount to specifying *its* qualities, which in turn would be incompatible with its alleged function to be only the *bearer* of qualities.

This kind of reasoning explains why the admission of the essence (under a variety of forms) could be maintained also by several philosophical schools that eliminated the doctrine of substance, beginning with the ancient Stoics. From this point of view, the copula "is" was simply meant to express not a relation of *inherence* of a property in an essence, but a *factual* relation that happens to hold between that which is signified by the subject and that which is signified by the predicate, in the sense that the predicate 'occurs' in the individual that is signified by the subject. This way of thinking became common in the so-called "terministic logic" of late Scholasticism, where the contraposition of the theory of the *suppositio* against the theory of the *inherence* was actually but an aspect of the contraposition of the theory of the essence against the theory of the substance. The central feature of this disentanglement was the thesis that no characteristic of an entity can be deemed to be superior to any other, since no one of them is *necessary*. This led to the elaboration of a particular doctrine of the essence. Initiated by Hobbes, developed by Locke, and followed by several scholars in the empiricist tradition up to Mill, and by various contemporary authors interested in the debate of 'natural kinds,' this doctrine is characterised by the elaboration of the notion of *nominal essence*, where the adjective "nominal" indicates that the essence is nothing more than that characteristic (or set of characteristics) that we use in order to "give a name" to an object. The principal element of this doctrine is that the essence ceases to be something dependent on the object to which it is attributed, but rather depends on the contingent fact that man selects certain properties of the object in order to *identify* it and give it a name as a kind of identification tag. Expressing this in a different way, we could say that, within this line of thought, essence migrates from ontology to the theory of meaning: indeed, as Quine pointed out,[19] the meaning is what essence results in after divorcing the object from the reference and uniting it with the word. This, however, would not be a faithful portrayal of the situation for it only mirrors the outcome of the doctrine of the essence for those authors who adhere to the 'linguistic turn' of twentieth century philosophy. For Locke and his followers (including several contemporary scholars) the nominal essence does not exhaust the whole meaning of essence, and does not even constitute the most genuine one: surprising as it might sound, Locke explicitly accepts a second meaning of essence, which is totally in keeping with the Aristotelian conception, with only a terminological difference: he calls it *real*

---

[19] Quine (1963), II, 1.

essence instead of "substance" (but this is understandable owing to his well-known criticism of the idea of substance). Actually, after having defined the nominal essence as "the abstract idea to which the name (of a class or species, that he calls "sort") is annexed; so that everything contained in this idea is essential to that sort," he continues by saying: "This, though it be all the essence of the natural substances that *we* know, or by which we distinguish them into sorts, yet I call it by a peculiar name, the *nominal essence*, to distinguish it from the real constitution of substances, upon which depends this nominal essence, and all the properties of that sort; which, therefore, as has been said, may be called the *real essence*."[20] We have said that this admission of a real essence is surprising because one does not see why, after such an admission, one had to introduce in addition the "peculiar" notion of nominal essence. This surprise, however, vanishes if we consider that, for Locke, the real essence cannot be known by humans, and in this assertion we have clear evidence that Locke was prisoner of that 'epistemological dualism' we will duly discuss later.[21]

The moral of this story could be expressed by saying that while the concept of essence is hardly eliminable from a discourse concerning knowledge, efforts can be made in order to dispense with the concept of substance in so far as substance is conceived as a substratum of properties that, for the very reason of being a substratum or a pure bearer, escapes any possibility of being known, since our knowledge cannot be anything other than knowledge of certain properties. One could discuss whether this was indeed the genuine notion of a substratum 'in which' properties inhere according to the Aristotelian tradition, but we are not interested in this discussion here. We want rather to analyse a second meaning of substance that plays a crucial *ontological* role in Aristotle's doctrine and is also related to his notion of *substratum*. This role has to do with the intelligibility of *change* or *becoming*: in order to say that 'something' has changed or become different from what *it* was, we must admit, on the one hand, that it 'remained the

---

[20] Locke (1690), III, 6, 2.

[21] On the contrary, this distinction is used as a tool for supporting a *realist* view of science by those contemporary authors who do not share this epistemological dualism. See, e.g., Mackie (1976) where a refinement of Locke's distinction is proposed, and Dilworth (2007). It may be interesting to note that the distinction between 'nominal essence' and 'real essence' is a kind of reformulation of the scholastic distinction between 'nominal definition' and 'real definition.' The first expressed those characteristics that we include in the *meaning* of a term and that may help us even in the determination of the existence of an entity endowed with such characteristics. This, however, does not entail that we have an adequate knowledge of the essence of such an entity, which can remain largely inscrutable. For example, Thomas Aquinas, responding to an objection according to which in order to prove the existence of God we should know in advance his essence, says that we actually use in our proof the nominal definition of God (that contains certain characteristics we mean God should have), and then we prove that such an entity must exist, without pretending to know his essence in depth: "For in order to prove that something exists, it is necessary to use as a medium term *that which the name signifies* but not *what this is,* because the question *what is* comes after the question *whether it is* (*Summa Theologiae*, I, q. 2, a.2). The advantage of the traditional terminology is that it did not introduce the rather strange notion of two essences, but the more reasonable distinction between the essence and a definition.

same thing' and, on the other hand, that it has become a different thing. This is possible if we distinguish between a *permanent* core of this entity, which Aristotle calls *substance,* and a changeable way of the entity's presenting itself. Since this core remains unaltered 'under' the different mutations, it can be equated with a *sub-stratum.* Not, however, a substratum with regard to properties or qualities, but with regard to existence. Note, however, that this ontological substratum cannot be something undetermined; therefore its stable characteristics are called by Aristotle and later classical ontologists *properties* (that constitute the thing's *essence*) while the thing's changeable features are often called *qualities*.[22] The properties of a substance are constant and present in every instance of the substance, while its qualities can change.[23] One must note, however, that this ontological role of substance has a clear *metaphysical* character, and was not appreciated by those 'new scientists' who were much concerned not to become involved in metaphysical speculations. Therefore, it was only *implicitly* at work in their theoretical constructions.[24] As a consequence, when these authors spoke of substance and essence they usually had in mind the picture of a mysterious substratum that remains 'behind' the perceptible qualities as an alleged 'bearer' of them; and one can see here the first germs of that 'epistemological dualism' (with all its gratuity and methodological difficulties) that was destined to play such an important role in the history of Western philosophy—especially from Descartes to Kant. But we shall return to this point later.

As a consequence of the said dualism one could say that the most reasonable move would have been to retain the genuine spirit of essentialism (which appears to be sound and rather unproblematic) while rejecting the spurious notion of substance (for the retention of which there exists no convincing evidence or argument, unless one explicitly enters the domain of ontology). This separation of the two seems particularly justified if one shares the analysis provided, for instance, by Cassirer, which shows how the 'independence' of substance from qualities and relations was meant to be one of its characteristic features. Therefore, since our knowledge of the world consists in the determination of qualities and relations, we can safely avoid any reference to substance in its ontologically committed sense and at most use this term only in one of the four senses

---

[22] We are using this distinction rather loosely, for reasons of brevity and clarity. In particular we are not giving to "property" the technical meaning of "proprium" that was explicitly defined by Aristotle and taken up again in more sophisticated ways by the tradition. The *proprium* is a characteristic that pertains to a *whole* class of objects and pertains to its members *always and solely* but is not part of the substantial essence, though being strictly dependent on it (the example given by Aristotle is the capability of learning grammar in the case of man).

[23] What we have said regards what in traditional ontology was called "accidental change" (in which a substance is permanent and only its accidents change). Also a "substantial change" was considered: in this case one substance must disappear in order for another substance to come to be, and this is again possible because something remains permanent, i.e., *matter* (by the way, this is why matter is also considered as one possible meaning of substance by Aristotle himself).

[24] We shall come to this issue in Chap. 10, devoted to the topic "Science and Metaphysics." See Dilworth (2007) for a detailed treatment of this question.

mentioned by Aristotle, that is, as indicating any concretely existing individual (in his terminology: the *first substance*). Unfortunately, however, this solution was, at least historically, not so easy to implement. Indeed, for Aristotle, essence was conceived of as a set of properties; but not all properties that can be attributed to an entity were said to belong to its essence. Therefore, the essence itself did not appear as something patent and simply waiting to be determined, but rather as something *hidden*, which was to be uncovered and dug out from 'under' or 'behind' the crust of inessential and even defective properties which can distract our attention and put us on the wrong path. This was tantamount to powerfully suggesting a significant reidentification of the notions of essence and cognitively-hidden substratum. The essence was more and more imagined to be a kind of core, a *receptaculum* of 'hidden qualities' which served at the same time as the obscure *substratum* of the other properties, and which therefore challenged our ability to bypass the curtain of appearances. But if we remember that the distinguishing feature of substance, from an ontological point of view, was its role as *substratum* (in the comprehension of change), we easily understand that a confluence of the meanings of "substance" and "essence" was occurring again.

This is why we cannot easily separate essentialism from substantialism. We must admit that Popper reflects a widespread way of conceiving of essence when he characterises it, as in the passage quoted earlier, as "the 'essence' or the 'essential nature' of things—the realities which lie behind the appearances." This, however, was not in keeping with Aristotle: though Aristotle distinguishes between properties and qualities, he still puts them on the same level of reality; nothing is hidden, but there are simply two different 'ways of existence': the substance exists "in itself," while properties and qualities exist only "in a substance." But this was an *ontological* distinction, not very palatable to people who were exclusively sensitive to *epistemological* requirements.[25]

What resulted from the above process was, in a way, a necessary conclusion. If we start a cognitive endeavour concerning something, our goal can be no more (but also no less) than to establish 'what it is,' besides having ascertained 'that it is,' and also having described 'how it is.' But to establish, with regard to a certain entity, 'what it is' obviously entails determining its *essence* (in the genuine and uncompromised sense we mentioned earlier). We shall henceforth refer to this as the 'correct' notion of essence. It is therefore no wonder that the general aim of every full-fledged cognitive enterprise, or striving for 'scientific' knowledge as it used to be called, had to be, for Aristotle and his followers, that of *knowing the essence* (not in the sense of knowledge by acquaintance, but of propositional knowledge). This was indeed the ideal of 'traditional' science. But this programme became involved in a good deal of historically understandable, but logically

[25] Let us note that, according to Aristotle, we proceed in our knowledge from what is "prior for us" (i.e., the immediately known qualities of things) to what is "prior in nature" (i.e., the essential properties), and this is a progression simply implying the use of more complex capabilities of our knowing apparatus (i.e., senses and intellect). This doctrine has been widely adopted in the philosophical tradition.

unnecessary, complications as a consequence of the additional burden with which the knowledge of the essence was charged. One was to strip from a thing all the 'external' properties that were simply involved in the contingent fact 'that' it was there, and also in the equally contingent ways 'how' it exhibited itself, in order to uncover, 'behind' this contingent façade, 'what' it really was. Clearly the enterprise of grasping the essence has much of a guessing quality about it, and its attempts to obtain results had to lead to frustration.[26]

This is precisely the intellectual situation that Galileo was no longer willing to accept.[27]

1.4 The Core of the Galilean Revolution

If we consider Galileo's attitude we easily see that it was truly revolutionary because it disregarded precisely that which was the very core of scientific knowledge according to the traditional doctrine, that is, the ability to capture the real *essence* of things. Galileo, not only in his practical way of investigating nature but also in his conscious theoretical reflection, explicitly refused to 'attempt the essence,' as is revealed in the following excerpt from his third letter to Mark Welser on sunspots:

> In our speculating we either seek to penetrate the true and internal essence of natural substances, or content ourselves with a knowledge of some of their affections. Attempting the essence I hold to be as impossible an undertaking with regard to closest elemental substances as with more remote celestial things… But if what we wish to fix in our minds is the apprehension of some affections of things, then it seems to me that we need not despair of our ability to acquire this respecting distant bodies just as well as those close at hand—and perhaps in some cases even more precisely in the former case than in the latter.[28]

---

[26] By saying this we are far from intending to trivialise these efforts. A study of such non-trivial methodological discussions as those regarding *composition* and *division*, or *analysis* and *synthesis*, would show how skilfully these issues were often treated. However, it seems undeniable that a much more radical step was needed in order to progress, and this is what we are now trying to explore.

[27] We are fully aware that, in this brief discussion, we have omitted any mention of the *intellectual intuition* operated by the *noûs* which, according to Aristotle, is the tool for uncovering the essence (and which constitutes the ground for that *induction* or *epagogé* that is the path to the essence in a way very different from that of modern post-Baconian induction). It would lead us too far afield to consider these doctrines here. This issue will be considered to some extent later, in our discussion of scientific realism. Let us simply note that a recovering of the role of intellectual intuition is implicit in what may be considered the most convincing revival of the doctrine of the essence in contemporary philosophy, i.e. in Husserl's phenomenology, with its notion of W*esenschau.*

[28] Galileo, *Opere* V, pp. 187–188; translated in Drake (1957), pp. 123–124. I have slightly modified this translation by using "affections" instead of "properties," not only in order to be more faithful to the letter of the Galilean text (where the Italian word "*affezioni*" occurs), but especially because "affection" was at that time a technical term in philosophy, and this fact—as

Here we find a clear distinction between the internal 'essence' and the 'affections' of natural entities, plus the declaration that we can hope to gain some knowledge of such entities only if we confine our attention to their affections. If we remember that the imperative of knowing the essence had been the characteristic mark of *philosophy* since Socrates' celebrated "*ti esti*," we can conclude that Galileo's proposal was, at least in part, that of abandoning the strictly *philosophical* viewpoint in investigating nature. We shall return later to a closer investigation of the question of essence in Galileo's conception of science, but we can already say that this was the conceptual feature which makes Galileo's step the very heart of the Scientific Revolution; it really meant a transition from *philosophy* to *science* in the modern sense of this word. Natural science was being understood as *non-philosophical* knowledge (despite the fact that it continued to be called "natural philosophy" for a couple of centuries) in the sense that it had abandoned the investigation of ultimate grounds and reasons that has been the typical attitude of philosophy throughout its history.[29]

This fact becomes even clearer if we take into consideration some scholars who are sometimes regarded as forerunners or even as pioneers of the scientific revolution. Let us consider, for example, Francis Bacon. The reason why, in all fairness, he cannot be regarded as a founder of modern science (although he gave a very clear picture of the inductive method as something different from simple enumeration, which proved extremely fruitful for scientific research) is not so much the fact—often underscored—that he was unable to recognise the proper role of mathematics in natural science, nor that he cannot be credited with any scientific discovery proper, but rather that he still claimed the specific *task* of natural investigation to be that of uncovering the *form* of things. And, although he devoted much effort to trying to distinguish his form from Aristotle's, he was not actually able to show any appreciable difference, for his form meant, exactly as did that of Aristotle, the ultimate and deepest 'essence' of things. He writes, indeed (§ 4 of the Second Book of the *Novum Organon*):

> We will lay this down, therefore, as the genuine and perfect rule of practice, that it should be certain, free, and preparatory, or having relation to practice. And this is the same thing as the discovery of a true form; for the form of any nature is such, that when it is assigned the particular nature infallibly follows. It is therefore, always present when that nature is present, and universally attests to such presence, and is inherent in the whole of it. The same form is of such a character, that if it be removed the particular nature infallibly vanishes. It is, therefore, absent, whenever that nature is absent, and perpetually testifies such absence, and exists in no other nature. Lastly, the true form is such, that it deduces the nature from some source of essence existing in many subjects, and more known (as they term it) to

(Footnote 28 continued)
we shall see in the sequel—gives great significance to Galileo's approach. I have also reintroduced Galileo's significant expression "attempting the essence."

[29] We do not maintain that this step was sufficient to characterize the whole spirit of modern science. Indeed it is not sufficient to account for scientific theorizing and we shall see in the sequel that additional elements entered the very Galilean epistemology, elements which cannot be reduced to this preliminary step of a quasi-positivistic flavour.

> nature, than the form itself. Such, then, is our determination and rule with regard to a genuine and perfect theoretical axiom, that a nature be found convertible with a given nature, and yet such as to limit the more known nature, in the manner of a real genus.[30]

Bacon's ideal of knowledge, in other words, was still that of philosophical and not of scientific knowledge in the modern sense. His "new organon" was intended to be the elaboration of a stringent methodology capable of attaining the 'necessary essence' of natural substances (in the same sense as the Aristotelian *to ti en einai*) through a systematic and articulated study of *empirical evidence* rather than through *intellectual intuition*. Therefore his position can be characterised as an explicit *empiricism*, which is, again, a particular philosophical doctrine having no specific 'scientific' connotation in the modern sense.

The same can be said (though for different reasons) of Renaissance Italian philosophers such as Telesio, Bruno and Campanella (who were practically contemporaries of Galileo). The naturalistic flavour of their philosophy may be noted, as well as the fact that these philosophers shortened the distance between natural facts and the metaphysical principles capable of making them understandable. But they nevertheless remained faithful to the metaphysical-essentialist viewpoint, even when they looked for new principles in nature itself. The title of Telesio's main work, *De rerum natura juxta propria principia* (1565–1585), is in a way self-explanatory and paradigmatic. Nature must be explained by recourse to 'its own' principles; but they are still 'principles,' in the sense of metaphysical ultimate patterns, which were thought to correspond to the deepest essence of natural reality (they are, for example, heat and cold, condensation and rarefaction, that is, such things that, even in their sources, are reminiscent of the ancient naturalistic Pre-Socratic philosophy). Similar considerations may also be brought to bear on Bruno's animistic or monistic cosmology, or on Campanella's panpsychistic world outlook, according to which the "sense of things" was accessible only through a kind of mystical identification with the divine world order, and the mastering of nature was possible through magic (*De* sensu *rerum et magia*, 1604). If we compare such doctrines with Galileo's viewpoint, a difference becomes immediately apparent. The above authors believed that a better understanding of nature could result from *changing philosophy* (i.e., from finding *new* ultimate 'essential' principles of Nature), while Galileo maintained such an understanding to be obtainable only through a *non-philosophical* investigation, in the sense of disregarding the research of *any* such ultimate principles. This fact, by the way, is confirmed by the circumstance that the scholars who promoted the rise of the 'new science' of nature fought with equal force against the Aristotelian physics and against this widespread 'naturalism' of their contemporaries.[31]

---

[30] Bacon (1620), p. 138.

[31] We find in the distinction mentioned here what may be a more significant reason for the separation of modern natural science from the magic and the occult arts which were still flourishing at that time, as well as for the more general antagonism between the scientific spirit and magical or occultist approaches that has become dominant since then. Those who claim that magic and occultism, being directed towards mastering the powers of nature, aided the birth of

The extent to which Galileo's view differed from the 'philosophical' perspective may be further appreciated if we consider Galileo's attitude towards the problem of looking for the *causes* of phenomena. An explicit definition of science which frequently occurs in the classical tradition is the following: *scientia est per causas scire*. The search for a cause, or better for *the* causes, was considered of major importance simply because the notion of *cause* was intimately connected with that of *essence*. This may pose difficulty for today's philosophers who are accustomed to practically one simple kind of cause, i.e., that which 'produces' its effect. But if we go back to ancient philosophy, we see that the concept of cause had a much wider meaning. Consider, for example, Aristotle's doctrine of the 'four causes.' One of these causes (later labelled the "efficient cause") is comparable to the present-day notion of cause; but the doctrine also admits a "material," a "formal," and a "final" cause, none of which is to be conceived of as something external to a thing, acting upon it and thereby producing a particular effect. (The notion of a final cause has been recovered rather recently, and not without resistance, in considering human action.) Each of these is, rather, an 'internal' principle, strictly bound to the essence of the thing so as to express its way of acting or behaving. A closer scrutiny of the efficient cause itself reveals that this cause too is to have a direct connection with the essence of a thing, so that, in the last analysis, the search for causes largely coincides with the task of investigating the essence. (Recall the pervasive Scholastic metaphysical principle *operari sequitur esse*, which implicitly stressed such an interdependence between the essence and the way it reveals itself as a 'cause,' that is, as an active principle.)[32]

Once such a strong link between essence and causes is appreciated, one has no difficulty in understanding why Galileo, who was so diffident towards the notion of essence, had to be equally diffident with regard to causes.[33] Here we quote some

(Footnote 31 continued)
modern science are taking an historically contingent fact as evidence for an actually unproved interaction. The active presence of these two different trends in Renaissance culture (magic on the one side, incipient science on the other), and their having a certain common aim, simply express one of the many contradictions of this fascinating epoch, but do not allow one to overlook that the trend represented by science was at variance, and not in keeping, with the other more traditional trend. This remains true even despite the fact that certain outstanding representatives of the incipient modern natural science incarnated in their work both of these contradictory attitudes (the most impressive example is probably that of Kepler).

[32] In Greek philosophy the terms *aition*, *aitia* and *arche* were practically synonymous. The Latin translations of the first two is *causa* (cause), and that of the last is *principium* (principle). Taking this fact into account, it is easy to understand why the classical ideal of knowledge was that of determining 'causes.' This simply meant looking for *reasons* (as we should say), which make reality understandable and which may be—according to cases—efficient causes in our sense, but also final causes, general principles and ultimate essential properties of things.

[33] Actually, Galileo's attitude towards causes was much more nuanced than that which, for the sake of brevity, we shall consider here. What Galileo disregards is rather the investigation of efficient causes, while he is not insensitive to problems which in the classical terminology would fall in the realm of other causes (especially formal, but even final causes). These types of cause are often concealed under a slightly different terminology (such as that which speaks of *reasons*),

lines from his more 'scientifically' conceived and composed work, that is, from the *Dialogues Concerning Two New Sciences.* When the moment comes for him to discuss the accelerated motion of falling bodies, he says:

> The present does not seem to me to be an opportune time to enter into the investigation of the cause of the acceleration of natural motion, concerning which various philosophers have produced various opinions, some of them reducing this to approach to the centre; others to the presence of successively less parts of the medium (remaining) to be divided; and others to a certain extrusion by the surrounding medium which, in rejoining itself behind the moveable, goes pressing and continually pushing it out. Such fantasies, and others like them, would have to be examined and resolved, with little gain. For the present, it suffices our Author that we understand him to want us to investigate and demonstrate some attributes (*passiones*) of a motion so accelerated (whatever be the cause of its acceleration) that the momenta of its speed go on increasing, after its departure from rest, in that simple ratio with which the continuation of time increases, which is the same as to say that in equal times, equal additions of speed are made. And if it will be found that the events that then will have been demonstrated are verified in the motion of naturally falling and accelerated heavy bodies, we may deem that the definition assumed includes that motion of heavy things, and that it is true that their acceleration goes increasing as the time and the duration of motion increases.[34]

As one can easily see, the whole problem is reduced here to that of establishing the correct description of a very limited and particular 'affection' of

(Footnote 33 continued)
and may be found in the very application of mathematical reasoning to the study of physical questions. This is not strange, and will be a clear result of the sequel of our discussion, where the 'realist' meaning of Galileo's appeal to mathematics in physical questions will be discussed. A valuable analysis of this complex issue is provided in Machamer (1978).

[34] Galileo (1638), *Opere* VIII, pp. 202–203; English translation, pp. 158–159. It is not accidental that Galileo says that it "does not seem to be an opportune time" (at this point of the *Dialogues*) to investigate the cause of (gravitational) acceleration. Indeed this not only leaves open the possibility that some other time might be opportune, but it also could indicate that at other times Galileo himself had investigated that cause. In fact, a careful survey of the progression of Galileo's studies on motion shows that he only gradually came to consider the accelerated motion of falling bodies, and that he had been looking for a causal explanation of this acceleration without obtaining a satisfactory result (as no one, including Newton, has since). In addition, it would be strange to consider Galileo's statement at this point as a rejection of the investigation of causes as such, since the second day of the *Dialogues* had just been devoted to the investigation of the cause of the cohesion of solid bodies. However, this investigation had hardly been conclusive. As a consequence, it seems correct to say that Galileo's attitude towards causes is parallel to his attitude towards essences. Having experienced frustration in trying to find causes, he came to consider it an "impossible undertaking," and restricted himself to the achievable task of "demonstrating some of the affections of accelerated motion." We could note, however, that this problem has to do with *gravitation* in particular, and that, even in the case of Newton, it led to his "*hypotheses non fingo*." Both Galileo and Newton adopted the notion of contiguous efficient causes (i.e., mechanical causes) for explaining physical phenomena, and while such causes could easily be determined in many areas, they could not when it came to gravitation. As with the problem of essences, Galileo does not declare the problem of causes to be absurd or uninteresting; it is simply bracketed and left for some more "opportune time" (that of philosophical speculation). We shall see that something similar may also be said regarding Newton's investigation of the cause of gravitational attraction.

physical bodies—that of falling with an accelerated motion—without asking the traditional question about the 'essence' of this motion, which inevitably would bring along with it the very intriguing question as to the 'causes' that produce it. It is also of interest to consider the way Galileo proposes to achieve his goal. He explicitly indicates that we have to proceed by formulating some reasonable conjectures, starting obviously with the simplest ones, and develop them with regard to their logical consequences. If it turns out that such deduced 'testable consequences' (as we should call them today) coincide with a faithful description of the observed 'affections,' we could retain them as a result of their being well established.[35] Galileo even says, on another occasion, that if this should happen not to be the case, we ought not feel obliged to consider such a conjecture as intrinsically untenable. Indeed its internal correctness would not be affected by its empirical inadequacy, and it would still remain a good description of a 'possible' motion, although not of the motion we originally wanted to describe. In order to describe this motion we ought to go on and try new conjectures, and put these to the test, until we eventually reach that conjecture which appears to be in agreement with the observed facts with which we were concerned.[36]

If one considers the epistemological line of this discussion, one can find in it an early sketch of what Popper refers to as the method of conjectures and refutations,

[35] Consider this quotation from Galileo's *Two New Sciences*:

> And first, it is appropriate to seek out and clarify the definition that best agrees with that (accelerated motion) which nature employs. Not that there is anything wrong with inventing at pleasure some kind of motion and theorising about its consequent properties, in the way that some men have derived spiral and conchoidal lines from certain motions, though nature makes no use of these (paths); and by pretending these, men have laudably demonstrated their essentials from assumptions (*ex suppositione*). But since nature does employ a certain kind of acceleration for descending heavy things, we decided to look into their properties so that we might be sure that the definition of accelerated motion which we are about to adduce agrees with the essence of naturally accelerated motion. And at length, after continual agitation of mind, we are confident that this has been found, chiefly for the very powerful reason that the essentials succesively demonstrated by us correspond to, and are seen to be in agreement with, that which physical experiments (*naturalia experimenta*) show forth to the senses (Galileo 1638, *Opere* VIII, p. 197; English translation, p. 153).

[36] Consider for example this passage from a letter of 7 January 1639 from Galileo to G. B. Baliani:

> I argue *ex suppositione* about motion, so that even though the consequences should not correspond to the events of the natural motion of falling heavy bodies, it would little matter to me, just as it derogates nothing from the demonstrations of Archimedes that no moveable is found in nature that moves along spiral lines. But in this I have been, as I will say, lucky: for the motion of heavy bodies and its events correspond punctually to the events demonstrated by me from the motion I defined (Galileo, *Opere*, XVIII, pp. 12–13; translated in Drake 1975, p. 156).

with all its anti-inductionist flavour; and it is interesting to find it formulated at the very beginning of modern science, together with the first conscious characterisation of science itself as something distinct from philosophy.

1.5 The Question of the Essence and Epistemological Dualism

According to our interpretation, that which best characterises the Galilean revolution, and which at the same time marks the most distinctive feature of 'modern' science, was Galileo's renunciation of the task of attempting to know the essence of things (the problem of causes being somewhat less important and subordinate to that of the essence). This seems to imply, first, that already with Galileo we find a clear opposition to 'essentialism.'[37] But to claim this cannot mean anything precise if one does not rely upon a precise definition of essentialism. This is why we must try to directly evaluate the Galilean attitude; and to do so we must try to understand what is really meant by the 'problem of the essence,' continuing in greater depth the discussion started in Sect. 1.3.

There is a kind of natural and irrefutable meaning of 'essence,' which can be expressed by saying that no existing thing, in the widest sense of the concept of existence, can be conceived of as being general or undetermined but, in order to be 'something,' has to be *definite*, with features which distinguish it from everything else. In other words, the idea of the essence is the conceptual counterpart of an ontological principle, that of the 'determinateness' of being, which in Medieval philosophy led to the formulation of one of the famous 'transcendental' features of reality, that of *unum*. This principle was formulated in the Scholastic texts as follows: *omne ens est indivisum in se et divisum a quolibet alio* (every being is a unity in itself and is distinguished from every other being). Such a principle is in itself so clear and evident that one would hardly contest it today, even if its ancient formulation is somewhat old-fashioned. (Note that an entity's constituting a *unity* in itself does not prevent it from having *parts*). In fact, if I say, for instance, that there is a book on my desk at this moment, I must rely on some features of this entity which enable me to distinguish it, for example, from a cat or from a pipe and allow me to say that it is a book. Moreover I can also say that I saw this book in a dream last night, or that I saw it on my desk yesterday. In these cases the book had different kinds of 'existence' (it existed not as a 'perceived entity' but as a

[37] This, for example, is in contrast to what Popper says when (in defining essentialism in his *Logic of Scientific Discovery*) he explicitly claims "Galilean philosophy of science" to be essentialist. This, however, is not an important question, for it depends basically on two factors: the particular meaning which one gives to the notion of essentialism, and the accuracy of Popper's portrayal of the doctrines of Galileo. Both of these factors might well be investigated, but we are not interested in this issue here.

'dreamed entity' or as a 'remembered entity'), but it still had the same 'essence,' given that it was recognisable as *the same* book each time.

But here certain complications arise, for if we really identify the essence with the system of properties which qualify a being, and therefore also distinguish it from everything else, we are obliged to ascribe to the essence really *all* the features an individual entity possesses. As a matter of fact, what distinguishes *this* book from other things (including other books) might well be some feature we normally consider 'inessential' to it. Due to this type of difficulty, philosophers were led to conceive of the essence as something *general* or *universal*, that is, as suitable for the identification of genus and species rather than individuals.[38] In this way, they found a kind of natural relationship between the 'essence' and the 'substance.' As a matter of fact, the celebrated distinction between 'substance' and 'accidents' was originally conceived of as a purely ontological distinction between an existence 'in itself,' and an existence 'in something else.' (For example, a human being is a substance because it is something which exists in itself, while having blue eyes is an accident because it does not exist in itself but only as a 'way of being' of a substance, that is, of a human being.) But it soon became customary to carry on this ontological distinction on another plane in which one came to speak of 'substantial' and 'accidental' features of things. In such a way *substantial features* became synonymous with *essential features*; and essence, after having been considered as the complex of features which place an individual in a certain species, became the substance itself.[39]

---

[38] In fact the Aristotelian essence characterises *species*, and it is in order to deal with the sort of problems hinted at here that later philosophers elaborated more sophisticated concepts, such as those of *quidditas* and *hecceitas*.

[39] In the above statements we deliberately adopted the half-colloquial way of using terms such as *substantial* and *accidental*. This use does not correspond to the original delineation of 'predicables' expressed by Aristotle in the *Topics*, where he first says that "of what is peculiar to anything, part signifies its essence, while part does not" (A4, 101^b17–23). He then proceeds to characterise a *definition* as "a phrase signifying a thing's essence"; a *property* (Latin *proprium*) as "a predicate which does not indicate the essence of a thing but yet belongs to that thing alone and is predicated convertibly of it"; a *genus* as "what is predicated in the category of essence of a number of things exhibiting differences in kind"; an *accident* as "something which though it is none of the foregoing—i.e. neither a definition nor a property nor a genus—yet belongs to the thing: something which may possibly either belong or not belong to any one and the self-same thing, as (e.g.) 'sitting posture' may belong or not belong to some self-same thing" (A5, 101^b37–102^b8, passim).

In these passages we have a summary of most of Aristotle's semiotic analysis of terms, where such basic notions are characterised as those of *definition* (*oros*), property in the technical sense of *proprium* (*idiom*), genus (*genos*), *kind* or *species* (*eidos*), (specific) *difference* (*diaphora*), and *accident* (*symbebekos*). All these notions, as is clear from the quotations, are more or less closely connected with the essence, and essence is here intended as what makes a thing be what it necessarily is (*to ti en einai*). This analysis, which is performed on a linguistic level owing to the fact that the question at issue is that of classifying different kinds of predicables, becomes more complex and also less clear as soon as other levels of meaning are taken into consideration, and especially when questions of reference become involved. For example, already in the *Topics* the celebrated list of ten categories is presented (A9, 103^b20–37), and Aristotle remarks that each of

(Footnote 39 continued)
the four predicables will fall, in every actual statement, into one or another of the ten categories. Here, that which is usually translated as the category of *essence* (the first on the list) is designated with another name: *ti esti*, instead of *to ti en einai* as before. But if we consider this list as it appears in a parallel text of the *Categories*, we can see that the category of essence is termed *ousia*, while all the other categories are listed under exactly the same names and even in the same order as in the *Topics*.

There are obviously reasons for this shifting of terminology, which mainly consist in the fact that the referential aspect of the discourse is being taken into consideration. Actually, the more usual way of translating *ousia* is as "substance" and not "essence," and this differentiation becomes more apparent in Aristotle himself where he remarks that essence, in the sense of intrinsic nature, is in every category, not only in that of substance; hence, "substance" and "essence" or "*ousia*" and "*ti esti*" are not synonymous (*Topics* A9, 192^b27–38). However, at this point a line is being cast towards ontology, as the term "*ousia*" receives a double meaning through the distinction between primary substance (*prote ousia*) and secondary substance (*deutera ousia*), the first referring to any actually existing individual, and the second to what may be predicated of a subject or may exist in a subject (*Cat.* 5, 2a 11–19).

From the above brief presentation one can see how many distinction criteria are implied (e.g. between necessary and non-necessary and between convertible and non-convertible predication, and between independent existence, existence in a subject, and predicability of a subject). These criteria, on the other hand, are not always parallel, but may often interfere with one another; and this gives rise to difficulties in the interpretation of the Aristotelian texts, difficulties which are partly due to the lack of certain technical devices "which later logicians and philosophers have found indispensable in making their points clear, inverted commas and the free invention of abstract norms," as William and Mary Kneale correctly point out (Kneale 1962, p. 27). But they are also partly due to the depth and difficulty of the philosophical issues involved. This is why, in particular, Aristotle himself oscillates on some important points, and why his Latin translators of the Middle Ages had much difficulty in creating a terminology capable of expressing his subtle distinctions. But they continued to discuss the core of the matter, and developed many ingenuous and subtle theories in order to tackle the most debatable questions.

It lies outside the scope of this study to explore these developments. Let us simply mention the fact that, as far as our problem of *substance* and *essence* is concerned, the very fact that the Scholastics conceived of logic as a theory of what they termed "second intentions" (being in thought and not in nature) is of relevance. Indeed they were led to a considerable amount of semiotic analysis which directly concerns our issue, though their way of conceiving of the *intentiones secundae* was by no means constant, as we can see by comparing, for example, Thomas Aquinas (thirteenth century), and Ockham and Albert of Saxony (early and late fourteenth century). Scholastic logic consists essentially of two parts, the doctrine of the properties of terms (*proprietates terminorum*), and the doctrine of consequences (*consequentiae*). The first is replete with interesting discussions relevant to our problem, such as those connected with the different theories concerning supposition, appellation, ampliation and such properties of terms, with all their sub-distinctions, that have their direct impact in the celebrated great debate about *universals* which divided the schools for about four centuries. Works such as the *Dialectica* of Peter Abelard (1079–1142), the *Introductiones in logicam* of William of Shyres (c. 1200–1270), the *Summulae logicales* of Peter of Spain (c. 1210–1277), the *Summa totius logicae* of William Ockham (c. 1285–1349), the *De puritate artis logicae* of Walter Burleigh (1275–1343), the *Perutilis logica* of Albert of Saxony (c. 1316–1390), and the *Logica magna* of Paul of Venice (c. 1372–1429) contributed to such a development of this theme as has not been equalled in subsequent centuries, including ours. But, as often happens, the enlargement and specialisation of the inquiry did not lead to a unification and standardisation of terminology and classification. In addition, logicians of the sixteenth and seventeenth centuries took different directions in their

The effects of this development were not all positive, for it projected upon the essence a number of ambiguities which were already resident in the notion of substance. As a matter of fact, the very distinction between substance and accident was already such as to suggest the idea that substance is a kind of hidden substratum that 'carries' the accidents and is concealed behind them. Through the identification of essence with substance, essence came to be thought of as containing in itself all the 'substantial' features of a thing, due to its position as 'the intimate core' of every individual entity, while the 'accidental' features were to constitute the 'veil of appearances' which had to be penetrated in order to reach the essence. In this way an ontological dualism gradually took shape which held that there is a surface and a core of every reality, and that our knowledge is always challenged to reach the core by penetrating through the crust or superficial stratum of accidental patterns. Such a conception has become so familiar that we have retained many references to it even in everyday language. We commonly speak, for example, of superficial knowledge as contrasted with deep or profound knowledge; we speak of an investigation which 'goes to the bottom of things,' and so on (the position of Locke, that we have already considered, may be seen as a mature expression of this view, that was historically prepared through a slow development).

We must now take into consideration a second kind of dualism, which appeared explicitly in the history of philosophy only in the seventeenth century, and which we shall here call *epistemological dualism*.[40] According to this conception—which seldom became an explicit doctrine but acted as a tacit presupposition under the doctrines of many philosophers from, say, Descartes to Kant—what we really know when trying to consider reality is our representation ('ideas') of it, but not reality itself. Here one immediately faces the problem of how we can be sure of a correspondence between our ideas and the reality to which they are to correspond. This is the famous question of the 'bridge' between ideas and reality that was so ingeniously but so unsatisfactorily looked for by the rationalist and empiricist

(Footnote 39 continued)
investigations, so that the question of essence and affections, which we have met in speaking about Galileo, became much less precise and more open to ambiguities, as we have tried to explain. (For an exposition of the foregoing historical development, the reader is directed to such books as Bochenski 1956, Kneale 1962, Moody 1953, and Boehner 1952). The 'correct' notion of essence has surfaced again in contemporary philosophy, not only with Husserl's phenomenology, but also in the philosophy of science, particularly since the works of Kripke, and has given rise to many discussions and claims concerning (modal) logic no less than ontology that are reminiscent, even in their terminology, of several medieval distinctions. In saying this we do not mean to imply that we subscribe to Kripke's form of essentialism; we only intend to point out the intrinsic importance of the issue we have been discussing here.

[40] This expression (translated into English from the Italian "*dualismo gnoseologico*") is due to Gustavo Bontadini, who has analysed this philosophical phenomenon with particular care and acumen, showing it to be the most characteristic feature of modern philosophy from Descartes to Kant. See in particular Bontadini (1947) and (1952). One must say, however, that this conception is often referred to in contemporary philosophical literature under the name of *representationalism*.

philosophers of the seventeenth and eighteenth centuries. The transition from what we have called the *ontological* dualism between the surface and the core of reality, to this *epistemological* dualism, is easily made by considering the superficial, accidental aspects of reality not as being features of reality proper, but rather as belonging to our representation of it. In such a way, instead of considering a splitting of reality into two sides or parts, and admitting in such a way a kind of first class and second class reality (such as in the form of substances and accidents), one is led to separate the content of our *knowledge* from *reality* (though continuing to admit that the aim of our knowledge be that of attaining reality itself). Through these precisions we want to discard a meaning of "epistemological dualism" that could be suggested by the use of the term "dualism." This expression must not be understood in the sense that we have two kinds, or forms, or steps of knowledge (such as sensible knowledge and intellectual knowledge), which is a perfectly legitimate position, but in the sense that the essential goal of knowledge (that is, to know reality *as it is*), cannot be attained directly, but (hopefully) by passing through an intermediate diaphragm: according to this view, what we immediately know are our *representations* or *ideas*, and not reality. Therefore, the chief question became that of determining whether or not, starting from our ideas, we can indirectly obtain knowledge of reality.

The transition from the first to the second kind of dualism is also to be found in Galileo's works. In a celebrated passage in his *Saggiatore* (*The Assayer*) he introduced that which later became the famous distinction between primary and secondary qualities of things, primary qualities being those which may be conceived of as belonging to things in themselves, while secondary qualities were simply the effect of our knowing activity, of our coming in contact with the thing in question via our sense organs.[41] They correspond in this way to what we have called our *representations* of the thing[42]:

> Now I say that whenever I conceive any material or corporeal substance, I immediately feel the need to think of it as bounded, and as having this or that shape: as being large or small in relation to other things, and in some specific place at any given time; as being in motion or at rest; as touching or not touching some other body; and as being one in number, or few, or many. From these conditions I cannot separate such a substance by any stretch of my imagination. But that it must be white or red, bitter or sweet, noisy or silent, and of sweet or foul odour, my mind does not feel compelled to bring in as necessary accompaniments. Without the senses as our guides, reason and imagination unaided would probably never arrive at qualities like these. Hence I think that tastes, odours, colours, and so on are no more than mere names so far as the object in which we place them is concerned, and that they reside only in the sensitive body so that, once the animal is removed, they are all removed and annihilated as well. But since we have imposed upon

[41] Actually, already in ancient philosophy the core of this doctrine had already been proposed by Democritus and later atomist philosophers, and this is in keeping with the revival of atomism that was taking place in seventeenth century Europe.

[42] For a detailed analysis of this important work of Galileo, see, e.g., Agazzi (1967).

> them special names, distinct from those of the other primitive and real accidents, we wish to believe that they really exist as actually different from those.[43]

As one easily sees, some qualities are supposed to belong to reality in itself (such as figure, movement, size, position, and so on) all of which are 'quantitative' in character, while others are supposed to have no existence either in themselves or in things, but simply to be the effect of the action through which the 'animal' comes in contact with the thing. We could safely say that, according to Galileo, the first (which are significantly called "real accidents") are 'essential' qualities, while the others are pure appearance, and do not concern science.

This view is supported by another well-known passage of the *Assayer* in which Galileo states, in an argument with his interlocutor, what the 'true characters' of nature are:

> Philosophy is written in this grand book, the universe, which stands continually open to our gaze. But the book cannot be understood unless one first learns to comprehend the language and read the letters in which it is composed. It is written in the language of mathematics, and its characters are triangles, circles, and other geometric figures without which it is humanly impossible to understand a single word of it.[44]

After all this, one could easily say: Galileo maintained that qualitative features do not belong to the essence of physical reality and, moreover, he relegated them to the status of simple subjective patterns of our knowing activity. He reached in such a way the conclusion that only quantitative and mathematisable features constitute the essence of physical reality, and was thus led to the step of applying mathematics in the description and explanation of natural phenomena, a step which proved so decisive for the development of modern science. If we wanted to tease, we might say that, in so doing, he tacitly and unconsciously accepted the making of his science into something concerned with the *essence*, despite his explicitly stated proposals, because he had simply changed the traditional perspective concerning what the essence of physical reality actually was (he had, so to speak, *externalised* the essence).

All this may well be true, but for the moment let us remark that such a change in the idea of the essence of (physical) reality was so subtle and implicit that Galileo himself might not have been fully aware of it, since it could hardly be distinguished from his purely *methodological* proposal, that is, from the adoption of a new method of inquiry in which our attention has no longer to be directed towards the goal of grasping the essence but simply towards that of describing certain 'affections' of natural substances.

In attempting to understand these proposals, we have been led to consider how the essence was actually conceived of by the Scholastic or Aristotelian philosophers of that time, namely as a kind of hidden core of reality which had to be

---

[43] Galileo (1623), *Opere* VI, pp. 347–348; translated in Drake (1957), p. 274. We have modified the last lines of Drake's translation slightly, in order to remain more faithful to the Galilean formulation.

[44] Galileo (1623), *Opere* VI, p. 232; translated in Drake (1957), pp. 237–238.

uncovered and grasped through an effort of philosophical intuition capable of penetrating the surface of accidental properties. Against such a programme, Galileo (who was still affected by this historically transmitted dualistic conception of the essence) proposed the conception of a new pattern of knowledge consisting precisely in the scrutiny of the neglected surface, of the accidental features of reality. Once this is understood, one may even maintain that Galileo's privileging of the quantitative or mathematisable qualities does not imply in itself an essentialistic conception, but may simply be interpreted as the drawing of a distinction *inside* the realm of the accidental features or 'affections' of reality, according to which only some of them actually belong to reality (the "real accidents," or the mathematisable ones) while the others are purely subjective. Which of these two possible interpretations of Galileo's conception is correct is neither easy nor even possible to decide since neither was explicitly formulated by him; and one should probably say that both of them were at work in his thought.

But now, having seen that two ways of conceiving of the essence had been developed by the tradition (the 'correct,' according to which essence is 'what something is,' and the 'incorrect,' according to which a thing's essence is hidden), we can ask which of the two was the target of Galileo's attacks. Clearly it was the incorrect one. We have actually shown, by discussing at length some of Galileo's most typical utterances, that he intended to abandon the programme of attempting to grasp the essence, conceived of as a hidden core of reality. This means that though he had an incorrect, dualistic notion of essence, he decided not to bother pursuing it. However, Galileo does sometimes speak of essence in a non-dualistic sense, for example when he designates as essence the real features of some of the 'affections' which are the target of his investigation. For example, he says that "the definition which we will give of our accelerated motion would correspond to the essence (*essentia*) of the naturally accelerated motion."[45] One must say that science has preserved such an attitude up to now, except during the brief mechanistic infatuation of the nineteenth century.[46]

---

45 "*... eam, quam allaturi sumus de nostro motu accelerato definitionem, cum essentia motus naturaliter accelerati congruere contigerit*" (Galileo 1638, *Opere* VIII, p. 197).

46 This explains why authors such as Krajewski (1977) and Nowak (1980) interpret Galileo's conception in a strongly essentialist manner. It is exactly those features which correspond to the Galilean primary qualities that they consider essential, features which Galileo was able to identify thanks to his extraordinary capacity for idealisation. But we must say that in such a way the other aspect of Galileo's revolution is neglected, that which he expressed through his explicit refusal to grasp the essence and remain content with the knowledge of certain affections. These authors are probably inclined to disregard this point because they are both dualist in the sense explained above (as they state themselves). A more subtle position is defended in Harré (1964), where a very interesting analysis is devoted to Galileo's distinction of primary and secondary qualities (pp. 85–93). The conclusion of this analysis is that Galileo's primary qualities actually amount to a new (physicalistic/atomistic) expression of the essence of matter. This claim can be accepted, because it does not express so much the meaning and the intention of Galileo's *scientific* work and attitude, but rather the philosophical and metaphysical framework which developed in connection with his scientific approach.

1.6 Science and the Non-dualistic Meaning of Essence: What are "Affections"

We would like briefly to discuss whether the 'correct' conception of essence as being 'what a thing is' was also rejected by Galileo. Although Galileo himself did not distinguish between this notion of essence and that according to which essence is hidden, we can say that he would not have had anything against making it, as he could not conceivably have been against establishing with regard to a certain thing 'what it is,' and how it is not to be confused with other things, independently of the epistemological question of the extent to which the knowledge of this essence can be attained. Moreover, this correct idea of the essence is independent of the distinction between 'substance' and 'accidents' (which, as we have seen, Galileo also adopts), because accidents too have their essences (as already stressed by Aristotle) if they are at all to be recognisable and identifiable. So, for example, we might declare not to concern ourselves with the 'essence' of water, but choose to investigate instead some of its easily observable 'affections,' such as its freezing and boiling points. Still, in order to perform such an investigation we need to know 'what it is' to freeze or to boil, that is, we have to know the essence (in the non dualistic sense) of these processes without which we could not even begin to speak about them.[47]

For Galileo the question appears to be even more complicated, for in certain passages he expresses his confidence in being able to reach, by means of the methods he proposes, the 'true essence of things.'[48] However, it is by trying to

---

[47] Note that, in the Lockean sense, these features are part of the 'nominal essence' of water. However, this does not avoid the fact that, at least for them, we must know their essence in a full sense (or, if we wanted to speak of nominal and real essence also in their case, we could not go on indefinitely with this strategy, because we are to stop at a stage at which the knowledge of the real essence of something is attained). This is why (though recognizing that the Lockean terminology may be useful in certain discussions) we prefer not to adopt it systematically, owing to its possible 'dualistic' interpretation (that is present in Locke himself, who considers unknowable the real essence). The analytic role of the notion of nominal essence will be preserved in later parts of this work when we come to speak of the 'referential features' of an object (features which, in addition, will not be conceived in a purely empiricist sense).

[48] Galileo's *realist* attitude may be found in several short passages of his work, but the most extensive discussion is perhaps in a letter he wrote to P. Dini (23 March, 1615) in which he opposes the view that the Copernican theory should be accepted simply as a suitable tool for "saving the appearances (phenomena)," and not as a description of what is really happening in nature. He stresses that Copernicus had already fulfilled the task of computationally saving the appearances (in his earlier writings) according to the traditional Ptolemaic view, but then

> ... wearing the philosopher's dress, and considering whether such a constitution of the parts of the universe could really exist in *rerum natura*, and having seen that this was not the case, and also estimating that the problem of this true constitution was worth being investigated, he engaged himself in the investigation of such a constitution, recognising that, if a disposition of the parts of the universe was able to satisfy the appearances in spite of being fictitious and not true, much better would this result be obtained from the true and

understand and solve this difficulty that we can clarify the real issues involved (both historically and conceptually) in the question of essentialism and substantialism. In order to do this we must briefly consider the notion of 'affection,' which plays a strategic role—as we have seen—in certain of Galileo's epistemological claims.

As used today, the term "affection" normally has a very different meaning from what it had in the past, especially at the time of Galileo. The primary current meaning of this term relates it to the sphere of *affectivity*, and generally expresses a positive emotional attitude towards a person (or a thing, a situation, a way of life). Another meaning is that related to medicine, where "affection" is sometimes used as synonymous with "disease," especially when the intention is that of specifying which organ or function is 'affected' by a given malady. Now this is the sense which is residual in contemporary language from the much more general meaning of "affection" in the past, when the term was used to indicate whatever feature a certain being could be said to be 'affected' by. In this way affections were in the last analysis properties or states of a being. This explains why modern translators of ancient texts usually translate "*affectio*" by "property," a reasonable solution from a practical point of view, but one which may conceal certain important issues in a scholarly discussion.[49]

(Footnote 48 continued)
real disposition; at the same time one would have gained in philosophy a knowledge as eminent as that which consists in knowing the true disposition of the parts of the world. (Galileo, *Opere* V, pp. 297–298)

In the first of the three famous letters to M. Welser on sunspots, he had already written:

> The philosopher-astronomers, besides trying to save at any rate the appearances, try to investigate—as the greatest and most marvellous problem—the true constitution of the universe, since such a constitution exists, and it exists in a way which is unique, true, real, and impossible to be otherwise, and worth being put before any other knowable question by the speculative mind, owing to its greatness and nobility. (Galileo, *Opere* V, p.102)

In these passages Galileo is referring to a distinction that was rather customary in the astronomy of the Middle Ages, that between "geometers-astronomers" and "philosophers-astronomers". The firsts were those who (to use a modern way of speaking) proposed skillful mathematical models in which the celestial phenomena could be suitably accommodated (this is the sense of "saving the appearances"), without the pretention that they mirrored the real structure of the universe. For this reason it was admitted that they could be very different. The philosophers-astronomers, instead, where those scholars who intended to propose a real picture of the universe, that is, a philosophical cosmology with pretention of truth. Galileo considers Copernicus as having been both, and he himself intended to be at the same time a 'geometer' capable of proposing working mathematical models of physical phenomena, and at the same time a natural 'philosopher' aiming at providing a true description of certain natural processes. A good presentation of this double aspect of Galileo's work is offered in Minazzi (1994).

[49] This is confirmed by the fact that—in particular in the English translations of Galileo quoted above—the translators have used "property" instead of "affection"; and now it is clear why we have changed their translation slightly by reintroducing "affection" here. However it must be said that the old meaning does not seem to have been totally discarded in contemporary use, if we

However, this linguistic explanation is still of little interest. What must be added is that in late Scholastic philosophy, which had developed a whole theory of the *affectiones entis*, "affection" had a philosophically *technical* meaning. Without providing any details of this theory, it is sufficient for us to quote the two short definitions of "*affectio externa*" and "*affectio interna*" which are given in an authoritative philosophical lexicon published in 1613, that is, more or less in the same year that Galileo was writing the words we have quoted. *Affectio externa* is "*quae subiecto advenit ob externam causam*" (that which comes to the subject because of an external cause), while *affectio interna* is "*quae manat a subiecti principiis intimis*" (that which emanates from the intimate principles of the subject).[50] It is more than likely that Galileo did not read this lexicon. Lexica, however, do not create meanings but rather record, clarify and perhaps 'codify' existing meanings. Therefore we can safely say that Galileo was using a current technical term when speaking of 'affections,' and this is easily confirmed by a simple inspection of the passages quoted. In the third letter to Welser he is clearly referring to 'external affections' when he says that he intends to content himself with knowledge of some of the affections of natural substances, without investigating "the true and internal essence" of things (and this essence, as we have seen, was considered at that time to be something containing the 'intimate principles' from which the said affections 'emanate,' to use the eloquent terminology of the quoted lexicon). In the passage taken from the *Dialogues Concerning Two New Sciences*, we have seen that Galileo refuses to take into consideration the *causes* of the accelerated motion of falling bodies, but rather "merely to investigate and to demonstrate some of the affections of accelerated motion (whatever the cause of this acceleration may be)," and this clearly shows that he has in mind the "external affections" according to the then current distinction.

It is interesting to see that this terminology—and the conceptual features it was intended to express—continued to play an important role for a long while after Galileo, and at the same time was impregnated with some of the Galilean preferences we already encountered when we considered the privilege conceded by

(Footnote 49 continued)
consider the *Concise Oxford Dictionary*, where the last meanings indicated for "affection" are: "mode of being; property, quality, attribute".

[50] See Goclenius (1613), p. 78. It is perhaps not completely superfluous to recall that in the philosophical tradition preceding the eighteenth century the term "subject" indicates not a knowing subject or person (as is most current now), but an individual entity in general (more or less with the same meaning as is preserved in our concept of *subject-matter*). In particular, a problem typical of the 'epistemological dualism' mentioned in the preceding section was that of knowing whether the *subiectum* (i.e., the real ontological thing in itself one tries to know) faithfully corresponds to the *obiectum* (i.e., to the 'representation' of it which is 'put before' our act of knowing). This way of using the terminology (which is standard, e.g., in Descartes) may create difficulties for the contemporary reader since we are used to considering the 'object' as the thing in itself, and the 'subject' as the knowing subject (or mind). These remarks are useful here because the *subiectum* occurring in the given definitions is the *ens* to which the affections come either from the outside or the inside, and is not a knowing mind.

him to the "real accidents" related to the quantifiable properties. The conjunction of these features appears in the notion of *mechanical affections* which, though not occurring explicitly in the pages of Galileo himself, perfectly expresses his programme of what should constitute the concern of science. Let us only quote two passages, one from Boyle and one from Locke, and make a few comments on them. Boyle writes:

> That which I chiefly aim at, is to make it probable to you by experiments, that almost all sorts of qualities, most of which have been by the schools either left unexplicated, or generally referred to I know not what incomprehensible substantial forms, may be produced mechanically, I mean by such corporeal agents as do not appear either to work otherwise than by virtue of the motion, size, figure, and contrivance of their own parts (which attributes I call the mechanical affections of matter).[51]

This statement is an expression of the mechanistic world outlook which accompanied the development of the new science of mechanics, and at the same time it is clear that the "attributes" mentioned here, and the whole flavour of the discourse, are directly reminiscent of the famous passage from Galileo's *Assayer* concerning primary and secondary qualities, as well as of the Galilean proposal to leave aside knowledge of the intimate essence of things, which is here termed their

---

[51] See Boyle (1672), vol. III, p. 13. Boyle presents a concise, but very incisive, defence of his mechanistic worldview in an essay published in 1674, 'The Excellency and Grounds of the Corpuscular or Mechanical Philosophy,' where the expression *mechanical affections* often occurs and indicates the privileged qualities of the Galilean *Assayer*. Boyle says, for example, that "the phenomena of the world thus constituted are physically produced by the mechanical affections of the parts of matter, and that they operate upon one another according to mechanical laws" (p. 189); and elsewhere: "both the mechanical affections of matter are to be found, and the laws of motion take place, not only in the great masses, and the middle size lumps, but in the smallest fragments of matter" (p. 194). Besides the notion of mechanical affections, we find in this essay also those of mechanical laws and mechanical principles. This is not surprising, since these pages were written after the publication of Newton's *Principia*, in which mechanics had emerged as a fully fledged physical doctrine. (Boyle's way of speaking is very Newtonian when he says, e.g., that "these principles do afford such clear accounts of those things, that are rightly *deduced* from them only," p. 190; my italics.) Furthermore, this mechanistic view is already becoming a metaphysical doctrine with all the features of exclusivity and reductionism that this implies. This is clear throughout the whole essay; in corroboration we quote only a passage from its Recapitulation:

> The parts of matter endowed with these catholick affections are, by various associations, reduced to natural bodies of several kinds, according to the plenty of that matter, and the various compositions and decompositions of the principles; which all presuppose the common matter they diversify; and these several kinds of bodies, by virtue of their motion, rest, and other mechanical affections, which fit them to act on, and suffer from one another, become endowed with several kinds of qualities, (whereof some are called manifest, and some occult,) and those, that act upon the peculiarly framed organs of sense, whole perceptions, by the animadversive faculty of the soul, are sensations (p. 208).

This is almost literally Democritean atomism, already resumed in the passage of Galileo's *Assayer* mentioned above.

The quotation is taken from Boas Hall (1965), where Boyle's essay is fully reproduced.

substantial forms. Finally, it is very significant that those attributes (i.e., primary qualities) are qualified, in the very spirit of Galileo, as the *mechanical affections* of matter.

Now let us consider the following passage from Book IV, Chapter III, § 25 of Locke's *Essay Concerning Human Understanding*:

> But whilst we are destitute of senses acute enough to discover the minute particles of bodies, and to give us *Ideas* of their mechanical affections, we must be content to be ignorant of their properties and ways of operation; nor can we be assured about them any further, than some few trials we make, are able to reach.[52]

Here, while Locke on the one hand explicitly mentions mechanical affections, on the other he has no confidence in being able to reach them at the level of the atomic constitution of matter through a solid knowledge of things (which was proclaimed by Galileo and still shared by Boyle). Why is this? Not because such mechanical affections are necessarily impossible to ascertain in and of themselves, but because, according to Locke, they ought to be attributed also to the "minute particles" of bodies, particles which we cannot observe. At least to a certain extent, Locke is saying that these affections, though not mysterious in themselves, may remain hidden as far as their actual applicability to the (alleged) invisible microstructure of bodies is concerned. In such a way, even the mechanical affections appear to be subject to the same kind of criticism (though to a less drastic extent) as Locke had expressed against the notion of substance, and which we find clearly summarised, for example, in Book II, Ch. XIII, § 19 of the *Essay*, where he first says that:

> They who first run out the notion of accidents, as a sort of real beings, that needed something to inhere in, were forced to find out the word Substance, to support them…

and then criticises this doctrine which claims that

> *Substance*, without knowing what it is, is that which supports *Accidents*. So that of *Substance* we have no *Idea* of what it is, but only a confused obscure one of what it does.[53]

In other words: the mechanical affections should be part of the properties of the 'real essence' of material bodies, in order to fulfil their task of producing the properties (including the mechanical affections themselves), of the 'nominal essence' of these bodies. But this is said to be impossible, owing to Locke's 'dualistic' presupposition regarding the unknowability of the real essence.

One could say that here Locke is expressing himself more as a Newtonian than a Galilean, for his chief criterion for admitting the legitimacy of a cognitive claim has not so much to do with the *kind* of properties, attributes, or affections involved in the claim, but rather with what the claim is about being more or less remote from immediate experience. Let us not forget that Newton, in his General Scholium to the *Principia*, included after his famous "*hypotheses non fingo*"

[52] Cf. Locke (1690), p. 556.

[53] Cf. Locke (1690), p. 175.

mechanical 'hypotheses' among those he wanted not to invent, and that he had qualified as hypotheses all claims which could not be "deduced from the phenomena and generalised by induction."[54]

The significance of this change on the part of Locke resides in the fact that an *epistemological criterion* is clearly being introduced here for deciding questions that had been of an *ontological* nature before. This is apparent from the very fact that in both of the above-cited passages (and in the whole of the Lockean approach) the pivotal notion is that of the *ideas*. But it is also confirmed by the fact that the difference between substance and accidents is presented here not as that between something which can exist in itself and something which can exist only "*in alio*" (an ontological difference), but in the rougher and more pictorial way as that between something which supports and something which is supported, which—as we have seen—is again an expression of the epistemological dualism (we know the supported but not the support). In order to overstep the limitations of this phenomenalist approach it was necessary to recognise the indispensable role and legitimacy of *theorising*: something Galileo had already done and subsequent science was to continue doing, making the progress of natural knowledge depend on the appropriate choice of the 'affections' to be investigated (an ontological requirement) and not on their sensory accessibility (epistemological requirement), as Galileo had already pointed out by noting that our knowledge of such affections can sometimes be easier in the case of distant physical entities (that are much removed from sensory accessibility) than in the case of things "close at hand."

It would lead us too far afield to follow the path which led to this change of perspective from the indication of ontological requirements to the prescription of epistemological criteria for the successful pursuit of the study of nature. We would simply like to conclude this survey of historical points by indicating how this doctrine of 'affections,' which from one point of view was a development of the more traditional doctrine of accidents ("affections" and "accidents" being

[54] Let us stress that what is being said here is not intended to be an interpretation of the *whole* attitude Newton adopted as to the ontological commitments of his physics, but only a rather literal interpretation of this famous passage. In particular the admission of particles and of the *vis insita* are already obvious violations of this precept in the very core of Newton's physics; and in the decades following the publication of the *Principia* Newton devoted serious effort to attempting to provide an ontology capable of sustaining his pivotal theoretical construct, that is, attraction, and, more generally, force. After having dismissed 'internal' or occult properties of bodies, he felt obliged to look for some 'active principles' which could operate in some way outside of bodies and provide a medium for the transmission of force, so that he could avoid the conceptual difficulties involved in the notion of action at a distance. In these endeavours he even went so far as to admit the existence of an ether equipped with an exceptional combination of properties; but he never found a satisfactory solution to his problem (nor has anyone else). From this point of view, Newton's fruitless efforts are reminiscent of the fruitless efforts spent by Galileo in searching for the causes of the acceleration of falling bodies (which is indeed the same problem), and testify to the same intellectual attitude: a realist aspiration towards a full ontological understanding (in terms of *mechanical affections*) of nature, reinforced by the consideration that they are mathematically expressible 'affections,' since nothing more can be attained. For an excellent presentation of this story, see McMullin (1978).

practically synonymous in the citations from Galileo), gradually developed into more complex philosophical doctrines. A very compact passage of Spinoza gives us an excellent capsule-conception of this convergence. In Book I, Chapter 3 of his *Cogitata metaphysica* he defines the "*affectiones entis*" as "*quaedam attributa, sub quibus unuscuiusque essentiam vel existentiam intelligimus, a qua tamen non nisi ratione distinguuntur*" (certain attributes, under which we understand the essence or existence of anything, but which can be distinguished from it only by our reason).[55] In this definition of "affection" we find mention of "attributes," "existence" and "essence," while it is well known that in the general system of the Spinozian ontology *modes* are also introduced, which are said to be affections of the substance.

Practically all the ingredients of ontology as it was discussed at the time of Galileo are displayed here; and the topic of the 'modes' was going to become one of the most elaborated. (It occupies an important position, e.g., in the researches of Descartes, Spinoza, Locke and Hume.) 'Modes' retain the character of being strictly related to substance, and yet not identifiable with it. Also Locke expresses this conception (though in his own language of 'ideas') when he says in Book II, Chapter XII, § 4 of the *Essay*:

> *Modes* I will call complex *Ideas*, which however compounded, contain not in them the supposition of subsisting by themselves, but are considered as Dependencies on or Affections of Substances; such are the *Ideas* signified by the Words *Triangle, Gratitude, Murther,* etc.[56]

All this clearly shows us two things: that the decision to restrict oneself to 'affections' did not mean a lack of ontological commitment, implying what we

[55] See Spinoza (1663), p. 124. In fact, the term *affectio* has a very wide circulation in Spinoza's writings, as may easily be seen, for example, from the *Lexicon Spinozianum* of E. Giancotti Boscherini (see Giancotti Boscherini 1970). In particular, Spinoza tells us that his meaning for "affection" coincides with that of "attribute" as it is used by Descartes in his *Principles of Philosophy*; and in this way we are led to see how this theme of affections (sometimes somewhat disguised) dominated the epistemological and ontological discussions of the time. Descartes' passage hinted at by Spinoza deserves consideration, since it represents a bridge linking the rich Cartesian discussion on modes and attributes to our present historical remarks:

> And indeed here we are understanding by *modes* exactly the same thing as we understand elsewhere by *attributes* or *qualities*. But when we consider that the substance is affected or altered by these things, we call them *modes*; when the kind of this substance can be named from this alteration, we call them *qualities*; and finally, when we more generally consider these only as being inherent in a substance, we call them *attributes* (Descartes, *Oeuvres* VIII, p. 28. Quoted from the translation by V. R. and R. Miller, Reidel, Dordrecht/Boston, 1984, pp. 24–25).

The fact that modes are what affects or alters a substance clearly indicates their close relationship with affections. This terminological evolution also explains why the discussion of modes on the part of many seventeenth and eighteenth century authors must be considered a development of the more general discussion of affections.

[56] Cf. Locke (1690), p. 165.

termed 'epistemological dualism'; and that affections are attributes which cannot be separated from things, but only distinguished from their essences and existence by an act of logical analysis (therefore, they are 'real' and at the same time are intelligible only as related to a 'reality' of which they are affections). Therefore neither the negation of the ontological existence (in some appropriate sense) of the affections, nor their being alien to the essence, is to be found in these doctrines; and this is also the case with Galileo. As we shall see later, the suitable organization of specific attributes amounted for the new natural sciences to the construction of their domain of objects, whose essence was not 'hidden,' but was characterised precisely through such attributes.

Leaving Galileo aside, for contemporary science too the question of a correct use of the notion of essence appears not to be negligible, especially because the Galilean prescription not to 'attempt the essence' has been popularised as though it meant the rejection of *any* investigation of the essence. As a matter of fact, we have been told since the time we were children that, for example, modern physics does not pretend to know 'what light is,' but simply describes and explains certain 'phenomena' connected with light, such as reflection and refraction. In a similar way it is said that science does not pretend to say what electricity is or what atoms are, and so on, but simply states a set of laws regulating the so-called electrical or atomic 'phenomena' and nothing else. It is not difficult to recognise that such affirmations are the expression of a positivistic conception of science, but the question is to know whether they are right or wrong.

Even granting, for the sake of the argument, that science is not interested in knowing 'what is' light, electricity or atoms, one must say that many essences (in the correct sense) must be known, for it is certain that in order to distinguish reflection from refraction, the magnetic from the thermal effects of an electric current, atomic reactions from atomic decay, and so on, one has to know the 'essence' of these phenomena, one has to know 'what they are,' quite apart from the fact that, in order to *understand* and *explain* them some proposals regarding the nature or essence of the 'things' of which they are phenomena must be provided (but this discourse will be taken up much later in this work).

From what has been said above it follows that, although in order for modern science to come into being it was important that it drop the programme of 'attempting the essence,' the 'essence' it was rejecting was only fictitious in any case. This rejection was important and decisive, but we shall later find it useful to return to what we have termed the 'correct' notion of essence in order to avoid its only effective meaning being incorrect, in which case it would reproduce the mistaken position of epistemological dualism for contemporary science, interpreting it as a simple phenomenal knowledge incapable of describing reality. In particular, many of the issues involved in the discussion concerning scientific *objectivity* are confused because of this dualistic position, and a non-dualistic theory of scientific objectivity allows one to maintain a responsible and correct form of essentialism.

Moreover, even such technical notions as those of affection, attribute and mode, which many contemporary philosophers are inclined to ridicule as though they were simply archaic curiosities, will surface again under new names in discussions which are held to be very modern. Finally, we would like to note that the rather detailed presentation of general metaphysical and ontological notions, and of their discussion by several authors in connection with the birth of modern natural science, will serve as a useful historical backing when we come to analyse the issue of the relations between science and metaphysics in the Chap. 10 of this work.

1.7 The Maturation of the Model of Science Between Galileo and Kant

From our preceding discourse one might have received the impression that the Galilean proposal had been such a decisive and clear breakthrough that no one, in order to investigate nature successfully, could dispense with it. As a matter of fact, however, things were not that simple, because the Galilean proposal could also be considered an invitation to avoid difficult and engaging investigations, and to regress to the level of a merely superficial knowledge of the accidental features of reality, deprived of any necessity and rigour. This explains why many philosophers of nature, including those who shared a worldview quite close to the perspectives of the 'new science' of mechanics (the most representative of them is certainly Descartes) preferred to invent new metaphysical systems for interpreting the world and what was in it (e.g. animals and man) rather than follow the methodological prescriptions of Galileo.[57] It was only the *concrete success* rapidly encountered by

[57] In a letter addressed to Mersenne on 11 October, 1638, Descartes makes many critical comments about the recently published Galilean *Dialogues Concerning Two New Sciences*. Without entering into Descartes' rather detailed examination of the book, it will be sufficient for us to quote the first lines of this letter, which contain a general appreciation of Galileo's work:

> I shall begin this letter with my remarks about Galileo's book. I find, generally speaking, that he philosophises much better than ordinary people, for he rids himself to the extent he can of the errors of the Schools, and tries to examine physical matters by means of mathematical reasoning. In this I feel completely in agreement with him, and maintain that there is no other way to find the truth. But he seems to me to be very defective in that he makes digressions all the time, and he does not stop to explain any matter completely. This shows that he did not discuss his questions systematically and that, having left the causes of nature out of consideration, he has looked for the reasons of some particular effects and in such a way his construction has no foundation (Descartes, *Oeuvres*, II, p. 380).

Descartes' last remarks are especially illuminating, as they clearly characterise as imperfections or even as major defects exactly those features of Galileo's methodology (such as disregarding first causes and strictly delimiting the domain of inquiry) which we have been led to recognise as being the most pioneering among his intuitions. What Descartes is advocating here is in fact a new *philosophy of nature* which is still of a fully metaphysical character, that is, that pretends to grasp the intrinsic essence of the material world 'as such' and *deduce* from it the particular features of physical events (as he has tried to do and pretended to have done in his

the new physical science, especially as developed by Newton along the Galilean pattern, which eventually led to the general acceptance of this pattern as constituting the new model of science.

When we speak of 'concrete success' we are not referring to technological advances that the new science of mechanics was able to produce (indeed, these technological applications originated mainly in the nineteenth century). We intend rather to call attention to the impressive amount of systematic and uncontroversial

---

(Footnote 57 continued)
'physical' works). It is in a way only accidental that the principles of this philosophy were mechanical rather than otherwise since they were the consequence of his well known partition of *the whole* of reality into two fundamental *substances*, the *res cogitans* (the spiritual reality) and the *res extensa* (the material reality). This is why, coming to the material world, he could maintain that its essence reduces to extension and that, therefore, the science of extension (i.e., geometry) was sufficient for investigating it ("the whole of my physics—he said—is nothing but geometry") and for this reason he did not care about supporting his physics by means of experiments and did actually claim that *all* the properties of physical entities (including living beings) can be adequately accounted for mechanically, in an often ingenious but always aprioristic way.

One might be inclined to think that this was due to the 'rationalistic' style of Cartesian philosophy, but this is not true. A philosopher such as Hobbes, for instance, who is often classified as an 'empiricist,' always maintained the traditional thesis that "philosophy is the science of causes" and could not feel satisfied with a kind of inquiry (such as that promoted by Boyle and the members of the Royal Society) that was certainly interested in the causal structure of nature, but was believed to uncover it through a accumulation of careful empirical records. For Hobbes, a rationally founded knowledge of the causes *from which* knowledge of the effects could be deduced was the condition for making a science and not a simple history of natural phenomena. This is the reason why, in spite of being a 'mechanistic philosopher' and having spent most of his life in England, he constantly and polemically refused the experimental method and was never admitted as a member of the Royal Society. As with Descartes, we can say that he did not accept that transition to a 'non philosophical' investigation of nature that (in the sense already explained) constituted the core of the Galilean revolution.

This is not the case with regard to the 'mechanical philosophy' of Boyle and several other thinkers at that time. The main difference between the often-conflicting positions among these thinkers is perhaps that according to some of them the mechanical principles were rather of an aprioristic nature, while for others they were 'deduced' from experience. For the latter it was obviously much easier to take experience seriously into account, to be in a better position for making their principles better tailored to the concrete features exhibited by the investigation, and in such a way to confine the 'metaphysical' flavour of these principles to the role of general regulative frameworks rather than prescriptive tenets. In particular, they were not taken as expressing the essence of matter, but only certain widespread characteristics of natural phenomena. Therefore, they had a limited scope in two different senses. First, in the sense of not encompassing *the whole* of natural phenomena: Boyle, for instance, in the passage quoted above, declares that "almost all sorts of qualities" (but not *all* qualities) can be causally explained mechanically and, in other passages of his works, explicitly admitted that even an investigation of final causes is legitimate in the domain of living beings and perhaps also elsewhere; and Newton was open to the consideration of "active principles" different from the mechanical principles also in certain domains of physics. Second, in the sense of refraining from giving to their mechanical principles a 'substantialist' purport for they limited their scope to the domain of the "mechanical affections" (as we have seen), that is, to a specific *aspect* of nature that was not claimed to exhaust *the whole* even of single natural entities.

knowledge this science was able to amass within less than a century, affording an interpretation of nature in which a few principles were able to explain equally well the acceleration of a falling body, the oscillations of a pendulum, the elliptic orbits of the planets, and several other features of terrestrial and celestial motions. All of this, moreover, was systematised with perfect mathematical rigour, and showed a degree of universality and necessity which, far from reminding us of the fragmentary character of accidental knowledge, was endowed with what Aristotle would have considered to be the best marks of an authentic apodictic science. In brief, it rapidly became the case that, thanks to this new science, humankind had come to know *much more and much better* about nature than in all its past history; this is what we mean by concrete success.

This completed the revolution started by Galileo. He had promoted a form of non-philosophical knowledge which, as such, was felt by many to be something (or even much) less than *scientific* knowledge, the paradigm of science still being considered at that time to be philosophy and, more particularly, metaphysics. After the creation of Newtonian mechanics and its development in the eighteenth century, the model or paradigm of science had changed. This kind of knowledge, and no longer that afforded by philosophy, began to be considered to constitute science in a proper sense, while at the same time it even became possible to ask whether *metaphysics* is possible as a science. As is clear and well known, the terminus of this maturation is to be found in Kant, whose *Critique of Pure Reason* clearly indicates this change of paradigm, and is witness to the final victory that the Galilean programme had scored over its rival.[58]

A question which may now arise is: how far did the basic philosophical views of Galileo remain unchanged in Kant's conception of science, and which new elements revealed themselves only with the advent of the transcendental philosophy? This is not a peripheral question owing to the great influence Kant's philosophy had on the way of conceiving of science. Here we could answer that the Galilean prescription not to 'attempt the essence' is fully retained in Kant's doctrine of the unknowability of the noumenon. Indeed, the very notion of the noumenon expresses in the most significant way that idea of the essence which we

---

[58] We are not going to provide more details on this point, which has been analysed in Agazzi (1978). However, regarding the quite inadequate attention long paid by philosophers to the methodological and epistemological claims of the 'new science,' we should like to quote, as a confirmation of our view, the following words of Rom Harré:

> Not only have the arguments of Galileo been neglected, but there are also unique forms of argument used by Newton which have neither been repeated nor criticised in the works of professional philosophers since. Even Berkeley does not attack Newton's philosophy but Locke's. This can hardly be because the arguments of Galileo and Newton lack merit, but may be due to the fact, remarkable if true, that empiricism was advocated, condemned and disputed by generations of philosophers who seldom, so far as one can judge, made themselves thoroughly acquainted with the work of empirical science for which that philosophy was the ultimate justification (Harrè 1964, p.87).

qualified as 'incorrect' in the preceding pages, for the noumenon is conceived as the 'thing-in-itself' which lies behind the phenomena and cannot be reached by any inquiry. The notion of the noumenon is therefore a misleading notion (just as it was misleading to conceive of the essence as the 'core' of reality to which accidental properties are simply attached in a kind of extrinsic relationship). In addition, the claim that the essence *cannot* be known is a purely *dogmatic presupposition* which is even less well grounded than 'attempting the essence,' since Galileo represented such an attempt as a 'desperate enterprise' (i.e., as a *practically* unsolvable problem), while Kant clearly states, with no arguments to support this claim, that the noumenon is unknowable *in principle*. This makes his presupposition not only dogmatic but rather close to a flat contradiction, since to affirm the existence of something undoubtedly means to include it inside the domain of knowledge (otherwise how could we claim that it exists?). But this fact forbids our saying that it lies outside our knowledge at the same time. This criticism is of course the essence of the rejection of the idea of the noumenon made by the idealist philosophers coming after Kant (especially Fichte and Hegel, and their followers in the early twentieth century).[59]

---

[59] It lies outside the scope of this book to enter into the discussion of such a difficult and controversial issue as that concerning the genuine doctrine of the noumenon in Kant's *Critique of Pure Reason*. Also, the difficulty of such an analysis is increased by the fact that Kant himself made substantial modifications to the doctrine in the second edition of his work without, however, arriving at a coherent picture. Let us only mention, for example, that in the first edition the theme of the noumenon is strictly connected with the doctrine of the "transcendental object which is not in itself an object of knowledge, but only the representation of appearances under the concept of an object in general, viewed as determinable through the manifold of those appearances" (A 250) (hence it cannot even be thought of apart from the sense-data which are referred to it). But this doctrine was completely eliminated, for some very intrinsic reasons, from those main sections which were reformulated in the second edition, though it remained in some other sections of less central importance. However the concept of the "transcendental object = x" transforms itself into the notion of the noumenon as far as it plays the role of a limiting concept (*Grenzbegriff*) which is indispensably involved in the constitution of human experience:

> The concept of a noumenon is thus a merely *limiting concept*, the function of which is to curb the pretensions of sensibility; and it is therefore only of negative employment. At the same time it is no arbitrary invention, it is bound up with the limitation of sensibility, though it cannot affirm anything positive beyond the field of sensibility (A 255, B 310–311).

This statement, appearing in both the first and second editions of the *Critique*, prepares the way for the well-known distinction between the negative and positive meanings of the term *noumenon* that is especially stressed in the second edition. Taken positively, the term means "an object of a non-sensuous intuition" (which is an illegitimate meaning according to the "critical philosophy"); taken negatively it only means "a thing so far as it is not an object of our sensuous intuition," and in this sense it tends to become indistinguishable from the notion of the unknown "thing-in-itself." This is already the case in the first edition, and is preserved in the second: "But in so doing it [the understanding] at the same time sets limits to itself, recognising that it cannot know these noumena through any of the categories, and that it must therefore think them only under the title of an unknown something" (A 256, B 312). It is clear that, in such a way, Kant is substituting the concept of a noumenon for the less definite concept of the "thing-in-itself."

But now we must try to see what the aspects of reality are that one's knowledge *can* reach. In the case of Galileo we already saw that they are certain special *affections* of things and, more precisely, those *real accidents* which correspond, roughly speaking, to quantitative or at least quantifiable features of reality. In the case of Kant, we know that phenomena as he understands them are the only objects of proper knowledge; and we can say that, at least to some extent, they correspond to the Galilean 'affections,'[60] as contrasted with the 'essence' (though in a more

(Footnote 59 continued)
But this has far-reaching consequences, because it opens the way to admitting the existence of unknown and *unknowable* "things in themselves" behind these "appearances." Let us simply quote a single passage (that appears in both editions) in which the object in itself, the transcendental object, and the concept of appearance all coalesce within a few lines:

> The understanding, in limiting sensibility, thinks for itself an *object in itself*, but only as *transcendental object*, which is the cause of *appearance* and therefore *not itself appearance*, and which can be thought neither as quantity nor as reality nor as substance, etc. (A 288, B 344; my italics).

In this way the assumption that things in themselves exist becomes explicit, despite Kant's greater insistence upon the impossibility of applying any of the categories to them. We can see a double reason for this assumption (or admission). One is the natural conviction that the reference of representations to objects must be their reference to things in themselves; the other is the view (which goes back to Descartes) that it is by a *causal inference* that we advance from a representation to its 'external' ground. Of course, in developing his critical teaching, Kant was obliged to realise the serious difficulties involved in his more or less implicit application of the categories of substance and causality outside the realm of the *empirical* objects which seemed to be bound to the difference between appearance and reality. It is only in the *Dialectics* that this distinction could be seen as something different from that between experience and the non-experienced. However, we shall refrain from following this further development. That which we have said thus far should be sufficient to explain in which sense Kant remains within the closed circle of 'epistemological dualism,' and to understand why his own work already contained the intellectual requirements for an overcoming of his position. For a deeper analysis of this issue we might suggest the very excellent discussion contained in Kemp Smith (1918), or the classical work Adickes (1924).

[60] It would be interesting (though lead us too far afield) to see how Kant preserves and at the same time modifies the vocabulary of 'affections.' We shall content ourselves simply with sketching some lines concerning this issue, without giving quotations or references. The main difference consists in the fact that it is not the object, but the knowing subject that is said to be 'affected,' so that the traditional doctrine of the *affectiones entis* becomes a doctrine of the *affectiones cognoscentis*. This affection regarding the knowing subject is sometimes expressed as things in themselves affecting the I in itself, and sometimes as external things affecting, under the form of appearances, the subject's sensibility (which is therefore characterised by its 'receptivity'). This latter is certainly the most stable doctrine in Kant, since he says that our cognitive capacity is awakened by objects which "affect our senses," and that the object is given to us only as far as it affects our mind (*Gemüt*) by giving rise to perceptions. In other words, it is central to Kant's mature critical philosophy that intuition be bound to the senses and be based on 'affections,' while concepts are bound to the understanding and are based on 'functions.' These features of the affections correspond in a way to the scholastic doctrine of the 'external affection.' But we find in Kant also a counterpart of the 'internal affection,' since he says that the *Gemüt* may

substantial sense this correspondence does not hold since Galileo's 'affections' are properties of things in themselves; they are "real accidents" in the sense of accidents belonging to reality; Kant's phenomena are not). We can therefore conclude that, within the limits of an acceptable degree of approximation, Galileo and Kant agree on their fundamental points: the essence of things (conceived of in a 'dualistic' sense) remains fully outside the domain of science (i.e., of proper knowledge); but this on the other hand does not prevent science from having a domain of sound, fully significant, authentic and even universal and necessary knowledge, which is the domain of the 'phenomena.' (We can, without real ambiguity, adopt this specifically Kantian term in the context of the more flexible Galilean terminology.)

However, it is at this point that the Kantian and the Galilean conceptions can no longer be assimilated, for they are incomparable as far as the *foundation* of the legitimacy, adequacy and soundness of (phenomenal) scientific knowledge is concerned. Actually, we can say that in the case of Galileo phenomenal knowledge is reliable for two reasons: first, because both subjective and objective phenomenal features (affections, accidents, and so on) are epistemologically accessible, while essence is not; and second, because, as far as science is concerned, knowledge can be had of objective 'phenomenal' features (the *real accidents* or the mathematisable properties of real things)—which Kant considers impossible. In this sense (as we have explicitly noted and stressed in the foregoing section) we can say that there is an element of non-dualistic essentialism in Galileo which enables him to distinguish between reality and appearance, and to say that at least some of the accidents (those corresponding to the so-called 'primary qualities') do not belong to appearance. If we confine our investigations to objective accidents we do not reach the 'essence' in the most engaging (and mistaken) sense; but we do know the essence of things, at least to some extent, in a correct sense. It is because of this peculiar feature of certain privileged aspects of reality that, according to Galileo, a natural science which limits itself to their investigation may be expected to attain

(Footnote 60 continued)
"affect itself," and that reason may "affect our internal sense" in a way which is similar to that according to which "something, which lies at the ground of external appearances" affects the "external sense."

This change of perspective is perfectly accounted for through the adoption of the epistemological dualism of which we have already spoken, and which led to the displacing of the core from the object to the subject. However, since in such a way the dominant paradigm has become that of the 'external affection' (which was explicitly defined as that "which comes to the subject because of an external cause"), it was unavoidable that this external cause be surreptitiously or tacitly implied. This is why Kant sometimes also maintains that the "thing-in-itself" affects the subject (as has been pointed out by several scholars such as Riehl, Vaihinger and Adickes). But since this creates difficulties with other parts of his doctrine (especially with his doctrine of causality), already several of his contemporaries and immediate followers (e.g. Maimon, Jacobi, Schulze, Fichte) rejected this possibility of an affection coming from the thing-in-itself, and in such a way the very rejection of the thing-in-itself was prepared. This was performed by classical German Idealism, and actually amounted to an overcoming of the epistemological dualism.

the highest level of certainty, universality and necessity. These are characteristics emanating *from the subject-matter itself*.[61]

The position maintained by Kant is very different. He agrees with Galileo in ascribing the first reason for the legitimacy of phenomenal knowledge to the 'accessibility' of phenomena as compared with noumena (though having a notion of phenomena categorially different from that of Galileo, who never actually uses this term); but then the justification of the most positive aspects of scientific knowledge follows a new path. First of all, according to Kant, there is no way of splitting phenomena into reality and appearance as Galileo had done, because both are, in a way, appearance. The extremely engaging task for Kant was therefore that of divesting appearance of all the negative connotations it had gathered during the entire history of Western thought, during which it was very often identified with error, illusion, or unreliable or mistaken belief. Moreover, the task was actually that of showing that the realm of appearance was able to give rise to a kind of knowledge in which universality and necessity were not only possible but guaranteed.

Such properties, not being provided by particular features of the subject-matter itself, were secured in Kant by means of the genial idea of the synthetic a priori. It is the legislative action of our knowing power which unifies the phenomenal appearances under certain structural patterns, such as those of space and time, substantial unity, causal relation, and so on. In such a way, the real accidents of which Galileo had spoken became instantiations and articulations of a particular transcendental a priori function, either of our sensibility or of our understanding, and this provided the basis for the universality and necessity of our knowledge. In this sense it is correct to say that Kant's transcendental philosophy accomplished the task of providing the philosophical foundation of modern science—which had been started by Galileo and developed by Newton—within the *framework of the dualistic epistemological presupposition*, that is, the position that we aim at knowing reality, but can only know our representations.

The most interesting feature in all this process might perhaps be seen to reside in the fact that Kant was able to accept the claim that our knowledge is only referred to appearances, without drawing what had been for centuries the obvious conclusion of this assumption, namely that our knowledge would then necessarily be subjective. (Let us remember that Galileo himself had discarded 'appearances' because they are subjective and disappear as soon as "the animal is removed.") Kant's conclusion is in fact the opposite, for he clearly characterises such knowledge, under certain conditions, as "objective" (*objektiv*).

We are led in such a way to the following point: in the case of Galileo, we can say that the kind of knowledge proposed is objective *because* it has a pertinent *reference to objects*, by pointing at some suitable privileged features of them. In

[61] It is precisely for this reason that the intuitions of Galileo and his contemporaries could actually develop into a mechanical philosophy endowed with a metaphysical flavour of exclusivity and reductionism, as we have already remarked.

the case of Kant we have, on the contrary, an 'objectivity' *without dependence on* any *object*, but which is based on the (transcendental) conditions of the cognitive process itself. This has the collateral implication that while for Galileo the object was *given*, and the *construction* of our mind concerned only (by means of conjectures and tests) the theory *about it*, in the case of Kant it is the *object* itself which is constructed prior to every theoretical or even empirical investigation concerning it. This is why, though Kant qualifies knowledge as "objective" because it is knowledge "of objects," this claim has a new meaning, since it can no longer denote something 'existing in itself' to which our knowledge refers.[62]

We are now in a position to appreciate why the rather broad historical discussion we have developed in the preceding sections is by no means a digression, but rather has much to do with the central topic of this book, that is, with the structure of scientific objectivity. As a matter of fact, we began by remarking that modern science has given up the pretence of affording true knowledge, by restricting itself to providing simply an 'objective' form of knowledge; and we traced this attitude back to the deception suffered by contemporary science at the beginning of the twentieth century, when a radical crisis in what we now call classical science occurred. If we now try to express the significance of that crisis we could begin by saying that, owing to the euphoric optimism aroused by the theoretical and practical successes of modern science, the intellectual attitude in the nineteenth century had resulted in the adoption of the Galilean conception, that is, in the assumption that through science we adhere to certain privileged features of reality (the mathematisable or measurable ones), and that this enables us to know some part of the *real structure* of the existing world. It was, to use the earlier locution, *objectivity as reference to objects*.[63]

---

[62] The various considerations we have devoted to Galileo in the preceding parts of this volume have been put together (and slightly enlarged) in an article we devoted to the realist nature of Galileo's science. See Agazzi (1994).

[63] We could express this view by saying that the tremendous cognitive performance of Newtonian mechanics during the eighteenth century had led scientists to the firm belief that *natural laws* are the really existing objective patterns of reality, and that science is able to reach this stable core of nature. This core cannot be reached by either the deceptive knowledge of common sense, which is unable to master the riddle of superficial phenomena, or by the vacuous efforts of speculative philosophy. Hegel himself refers to this "quiescent kingdom of laws" which, however, he considers as a still unconscious manifestation of the Idea:

> The difference is expressed in the *law*, which is the stable presentment or picture of unstable appearance. The *suprasensible* world is in this way a *quiescent kingdom of laws*, no doubt beyond the world of perception—for this exhibits the law only through incessant change—but likewise *present* in it, and its direct immovable copy or image.This kingdom of laws is indeed the truth for understanding; and that truth finds its *content* in the distinction which lies in the law. At the same time, however, this kingdom of laws is only the *preliminary truth* and does not give all the fullness of the world of appearance (Hegel 1807; English translation, p. 195).

The above-mentioned crisis implied the elimination of this kind of objectivity in the sense that *no object* seemed to be specifiable or even thinkable, and for a while a kind of scepticism seemed to prevail, marked by the features of instrumentalism and conventionalism. Science appeared to have been brought back to mere 'appearances' in the most radical and subjective sense of this term. (Think of Mach's analysis of perceptions and of his idea of the reduction of the content of knowledge to this sole basis.) But, as we have seen, after an initial period of discomfort, science became once again aware of its cognitive tasks, and tried to justify its status as the provider of 'objective knowledge.'

Now we can ask: what kind of objectivity was this meant to imply? Was it an objectivity with reference to objects, or an objectivity without objects? It is all too natural to answer that it could be nothing but an objectivity without objects (given the situation which had just been left behind), but this leaves us with many other questions. For we know of this kind of objectivity as it had been proposed by Kant, but it does not seem (in spite of neo-Kantian philosophers' being rather influential at that time) that scientists were interested in subscribing to Kant's doctrine when qualifying their science as objective. Some Kantian touches are certainly to be found, especially in the case of certain more philosophically minded scientists, but it is by no means possible to maintain that the actual core of Kant's view regarding objective knowledge, that is, the transcendental function of the a priori, was the basis accepted by scientists for expressing the meaning of scientific objectivity. For this reason we would now like to explain in greater detail this contemporary idea of scientific 'objectivity without objects.'

Chapter 2
The Characterisation of Objectivity

2.1 Objectivity Without Objects? The Strong and the Weak Senses of Objectivity

Before considering science in particular, let us note that a certain disengagement of the notion of objectivity from the idea of object may already be found in ordinary language. The meaning of the term "objectivity" seems primarily to be characterised through an (indirect) reference to the *subject* rather than through reference to the object. When one says, for example, that a certain judgement is objective, that a certain inquiry has been led in an objective manner, or that a certain quality is objectively possessed by something or by someone, one usually means that the judgement, inquiry, or quality does not depend on the subject or subjects who express the judgement, make the inquiry, or attribute the quality. In other words, while subjectivity seems to be the first mark of our knowledge, it is also considered as its worst defect, a defect with which humankind has struggled for centuries, our ideal being a form of knowledge which, though inevitably acquired by various subjects, is nevertheless independent of them in its validity.

But why should we care so much to have a body of knowledge independent of subjects? The answer to this question gives us the key to understanding the meaning of the phrase in the title: "objectivity without objects." Indeed, it seems to be constitutionally embodied in our minds (or at least in Western civilisation's mind) that the only way we have to check whether our efforts to know reality are successful is to verify that the picture of reality we arrive at is 'independent of the subject,' that is, that other subjects agree with us regarding the veracity of this picture. Note how artificial, in a way, this view is. The natural task of our knowing is indeed that of 'grasping' reality; and, abstractly speaking, we should say that such a goal is reached with the obtaining of 'objective knowledge,' that is, knowledge which matches that portion of reality that it is its purpose to match. But, on the other hand, man seems always to be afraid of not being able to complete such a task; and doubts regarding this matter come from the fact that very frequently different persons, confronted with the same portion of reality, describe it in different ways. The conclusion is easy: if different pictures are

E. Agazzi, *Scientific Objectivity and Its Contexts*, DOI: 10.1007/978-3-319-04660-0_2,
© Springer International Publishing Switzerland 2014

proposed concerning the same reality, none of them (or possibly just one) can be 'objective,' that is, can 'correspond to the object,' whereas all of them (with one possible exception) must be considered as purely 'subjective'—as expressing a certain way of envisaging objective reality which is typical of some single subject.[1]

All this is so simple that it seems even trivial, but it actually explains some fundamental features of the concept of objectivity. As we have just seen, the presence of different subjective pictures is meant to be *sufficient* to exclude any one of them constituting objective knowledge. It follows then that a *necessary* (though perhaps not sufficient) *condition* for objective knowledge is its being 'independent of the knowing subject.' It is not at all easy to say which additional condition should be added to this necessary one, and this question has been controversial for several centuries.[2] But the turning point should now be clear: even if one accepts defining a piece of knowledge as 'objective' if it is a faithful representation of its intended *object*, the main problem becomes that of having a tool for ensuring *whether* some instance of knowledge has that precious property; and the safest mark of its actually having it seems to be its being independent of the subject. This explains the seemingly curious fact that objectivity has preserved a kind of indirect characterisation, that is, characterisation through reference to the subject, which, in principle, should have nothing to do with the notion of an object.

If we keep this indirect characterisation in mind, we may also obtain a better understanding of certain features which, through the entire history of Western philosophy, have been maintained as indispensable for any instance of purported knowledge to be genuine: *universality* and *necessity*. We are faced here, once more, with something unnatural, for every concrete act of knowledge has primarily

---

[1] We do not take into consideration here the special case of introspection, that would deserve an *ad hoc* analysis.

[2] The problem of being able to establish such additional requirements has been present in the history of philosophy in the guise of the search for 'criteria of truth,' and has constituted the core of the debate about *scepticism*. Indeed, scepticism cannot be trivialised as being the doctrine which claims that "truth does not exist," or that "we are always wrong" (a claim which may even seem self-contradictory, if it is proposed as a valid assertion). It is rather the doctrine which does not believe in the possibility of finding criteria for overcoming the subjective aspect of 'appearances.' This view was already advanced by the Greek Sophists and later by the sceptics of the Hellenistic age (e.g. the Pyrrhonists such as Sextus Empiricus and the sceptical Academics such as Arcesilaus and Carneades), who rejected the Stoic 'foundationalist' view of epistemology according to which it is possible to identify those appearances (*phantasia*) which deserve our assent with certainty. With the rediscovery in the sixteenth century of the writings of Sextus Empiricus, the arguments of the Greek sceptics found wide circulation in the philosophical community, and determined the so-called *crise pyrrhonienne* of the early seventeenth century. This crisis was of concern to people such as Montaigne, Descartes and their influence eventually reached Hume through the mediation of Bayle and others. But scepticism, understood in this sense, did not disappear from philosophy after this. We may note that we have spoken of *objective* knowledge rather than *true* knowledge, since truth is more properly considered a property of judgements and propositions. Knowledge may never be false in a proper sense, although truth has much to do with objective knowledge, as we shall see in the sequel.

to do with *single* experiences, with *individually* localised facts and things. Moreover, if we think of an 'object' in general, it always appears to be an individual, strictly bound to its particular features and to its location in space and time; it appears, in a way, as exactly the opposite of something which might be conceived of as perpetually established, as unaffected by change in space and time.

Where does this strange idea arise from, i.e. that an instance of knowledge, in order to be objective, must have the properties of universality and necessity? Let us limit ourselves to two main lines along which this doctrine has developed. The first concerns ontology: the Eleatic discovery that being as such cannot be affected or limited by non-being seemed to entail that every single determination of reality, every portion of it, was obliged to share this fundamental characteristic and, as a consequence, not to undergo the fate of mutation (which would be its turning from 'being' to 'non-being'). So every 'object,' since it shares the general features of being, must be conceived of as something that has always been and always will be exactly as it is in any arbitrary moment of its existence (which means, to put it differently, that every objective determination is necessary and universal, as we have said). But, against this a priori established immutability of reality (which seems an indispensable condition for thinking of reality in a consistent way), *experience* shows change to be a general feature of the world. The solution of this difficulty is well known: the Eleatics confined change to the illusory 'opinion' (*dóxa*) of sense perception, while considering pure reason as being able to attain the 'truth' (*alétheia*) concerning immutable reality. (It may be noted that the previously discussed doctrine of essence as an immutable substratum lying beneath a layer of mutable 'accidents' is directly related to this picture of reality; note also that a first solution to this problem, though set by the Eleatics, was proposed by the atomists).

This solution would not be of particular interest for us if it were not for the linking of this ontological way of thinking with the second line mentioned above. This line leans towards the side of knowledge: an object, we could say, is admittedly something individual and sharply localised in space and time. But there is some justification for considering our 'knowledge' of it actually to be knowledge only if that 'knowledge' appears to be 'independent of the subject,' that is, if it is valid for every knowing subject ('universal') and ('necessarily') the same for all such subjects.

It is easy to understand that "universality" and "necessity" are not meant in the same way according to the two different lines of thought. The first meaning, which refers to the permanence of the ontological structure, gives to the idea of universality a connotation of spatio-temporal immutability, whereas the second meaning simply refers to a uniformity of appreciation by different knowing subjects, and no spatio-temporal feature is involved.

Although these conceptions of universality and necessity were, and are, distinct, a practical confluence of both took place in the history of philosophy, and they helped one another to attain the status of being the distinguishing marks of objectivity. To express this fact synthetically, we could say that the ontological structure of the object, as well as the warranties of our having sound knowledge of it, have stressed the two characteristics of universality and necessity as the outstanding and fundamental marks of objectivity. And all this has happened in spite of the fact that objects manifest themselves as individuals, and that knowledge develops primarily as a result of unique experience.[3]

Let us now consider the knowing activity as such of any person. It is certainly undeniable that such an activity is intrinsically characterised by the aim of being *objective* in the sense of being able to 'capture' the real features of objects. As a result of the foregoing discussion, we must say that, *if* it is successful in this enterprise, *then* it must result in something universal and necessary; and this is tantamount to saying that universality and necessity conjointly appear as a *necessary condition* in order that a form of knowledge be objective.

We have here a situation which is strictly parallel to the one we recognised earlier concerning the 'independence of the subject.' There too we noted that such a feature appeared, first, as a *consequence*, that is, as a *necessary* but perhaps not *sufficient* condition for objectivity. The interesting fact is now that these three features (independence of the subject, universality and necessity) not only appear to be consequences of objectivity understood as reference to objects, but are also logically interrelated. This consideration will prove fruitful in our subsequent investigations.

Should we wish to label these three characteristics in a different way, we might say that they belong to the *formal* aspect of objectivity, the *substantial* aspect being represented by the 'reference to the object.' From this it follows that whenever one is in the position to claim that a certain feature is objective in the *stronger* and much more engaging substantial sense, one may easily obtain as a consequence that this feature is also objective in the *weaker* and less engaging formal sense of being independent of the subject, necessary, and universal. The reverse is, strictly speaking, not true, as simple formal logic immediately reveals. If a certain feature happens to enjoy the formal characteristics of universality, necessity and independence of the subject, one is not entitled to claim on this basis

---

[3] A significant step in this direction was already taken by Plato . While the Eleatics had separated opinion from truth, Plato recognised that we can have "true opinions" (or, as he more often says, "right opinions"), and the question becomes that of distinguishing *opinion*—including true or right opinion (i.e. *alethe doxa* or *orthe doxa*)—from authentic *knowledge* (*episteme*). The answer given in the passage of the *Meno* where he debates this issue is that authentic knowledge consists in stabilising the fugitive truth of right opinions by the solid links of arguments capable of providing its reasons or "causes" (Meno, 97–98a). In such a way universality and necessity appear to be the distinguishing marks of authentic *knowledge*, and are in a way more decisive than truth itself, since truth alone (which may be understood as a correspondence to the object) may also be possessed accidentally by simple (right) opinion (which, from a *practical* point of view, is equivalent to knowledge). A powerful development of this Platonic doctrine (to which Plato himself has abundantly contributed, especially in the *Theaetetus*) was realised by Aristotle (especially in the *Posterior Analytics*), for whom authentic knowledge has to be not only true, but also endowed with universality and necessity, which are granted by suitable forms of rational argument. Since then, in many streams of Western thought, universality and necessity (in different forms and with different degrees of strength) have never ceased to characterise the notion of authentic knowledge. (For the *Meno* passage mentioned above, see Plato, pp. 380–382).

that this state of affairs is a consequence of the fact that this feature is substantially inherent in the object, nor that it entails such a fact.

After having labelled as formal (or weak) and substantial (or strong) these two senses of objectivity, we must note that the weak sense gradually became the only one to be taken into consideration within the theory of knowledge. This has meant that, after a time, the formal characteristics came to be thought of as *sufficient* warranties for objectivity or, if one prefers, that the formal characteristics became, in a way, substantial as well. This process was already completed in Kant's philosophy. For him, 'objective' simply means universal and necessary and, hence, independent of individual subjects; and he gives no particular reason for this identification, which seems to him obvious. But for us it is not difficult to understand the historical reasons for the identification. Kant appears at the end of a long process in Western philosophy, during which the problem of knowing reality in the strong sense of objectivity showed itself more and more to be hopelessly unsolvable. With Kant, any hope in this direction is explicitly relinquished, and he denies that the object in the ontological sense may be known. It is just a thinkable 'noumenon' which our knowledge does not reach. By this point, it was no longer reasonable to keep the strong sense of objectivity alive, for all its meaning was confined to the weak sense.

As is clear, the reduction of the entire sense of objectivity to its 'weak' component was simply a consequence of the 'dualistic epistemology' we have discussed in earlier sections. It was thanks to this prejudice that this reduction did not appear to be a simple logical mistake (namely, of taking a necessary condition as also being sufficient) but simply to be the consequence of a matter of fact (i.e., the impossibility of ever fulfilling the requirement of the 'strong' sense of objectivity).

If we now consider science in particular, we might say that one can see in the history of modern science a kind of summary of the general diversity we have sketched for the concept of 'objective knowledge.' Starting with Galileo, science has been considered as providing objective knowledge in the *strong* sense because, as we have seen, it was supposed to be directly concerned with certain *intrinsic* (even if no longer *essential*) properties of things. With very few exceptions, this conviction remained deeply rooted in the minds of working scientists, as well as in the general outlook of common sense, until the end of the nineteenth century. Meanwhile philosophy, on the other hand, had made a transition from the ancient conception of strong objectivity to the new conception of *weak* objectivity. Towards the end of the nineteenth century, and more substantially with the beginning of the twentieth century, something analogous to the loss of confidence in the possibility of 'reaching the object' took place in science as well, reproducing in a way the situation which, in philosophy, had occurred during the period between Galileo and Kant.[4]

[4] Rather than relativity theory, it was quantum mechanics that involved the said doubt regarding the capability of science to attain its intended objects. The reason is that while in classical physics it was not only ideally, but also concretely, possible to introduce a separation between the object

The acme of this crisis paralleled the Kantian assumption of the unknowability of the noumenon, and was expressed in terms of a widespread scepticism concerning the authentic cognitive power of science. The overcoming of the crisis finally consisted in the shaping of a new criterion of scientific objectivity, which was a *weak* one, as it no longer presupposed a reference to objects (it is the

(Footnote 4 continued)
of study and the instrument used to investigate it, this was no longer possible in quantum physics, as was clearly stated, for example, by Niels Bohr. He recognised, on the one hand, that:

> The discussion… thus emphasized once more the necessity of distinguishing, in the study of atomic phenomena, between the proper measuring instruments which serve to define the reference frame and those parts which are to be regarded as objects under investigation and in the account of which quantum effects cannot be disregarded (Bohr 1958, p. 228),

but at the same time he stressed:

> *The impossibility of any sharp separation between the behavior of atomic objects, and the interaction with the measuring instruments which serve to define the conditions under which the phenomena appear* (ibid., p. 210).

From these remarks, only a short step was needed to come to the idea of the impossibility of speaking, at least in the new science, of an 'independent reality.' To begin with, this may be expressed as the conviction that 'ordinary perception' becomes useless at the atomic and subatomic level:

> The very recognition of the limited divisibility of physical processes, symbolised by the quantum of action, has justified the old doubt as to the range of our ordinary forms of perception when applied to atomic phenomena. Since, in the observation of these phenomena, we cannot neglect the interaction between the object and the instrument of observation, the question of the possibilities of observation again comes to the foreground. Thus, we meet here, in a new light, the problem of the objectivity of phenomena which has always attracted so much attention in philosophical discussion (Bohr 1934, p. 93).

But then this becomes a much more general epistemological thesis in which the possibility of objectivity intended as the grasping of an ontologically given object seems to be precluded not by the constitution of our minds, but "by nature herself":

> The discovery of the quantum of action shows us, in fact, not only the natural limitation of classical physics, but, by throwing new light upon the old philosophical problem of the objective existence of phenomena independently of our observation, confronts us with a situation hitherto unknown in natural science. As we have seen, any observation necessitates an interference with the course of the phenomena, which is of such a nature that it deprives us of the foundation underlying the causal mode of description. The limit, which nature herself has thus imposed upon us, of the possibility of speaking about phenomena as existing objectively, finds its expression, as far as we can judge, just in the formulation of quantum mechanics (ibid., p. 115).

We have here some of the earliest and clearest formulations of the 'Copenhagen interpretation' of quantum mechanics, which has dominated the philosophical interpretation of the discipline for decades. It is interesting, however, to note that no flavour of subjectivism is implicit in the passages quoted from Bohr. This means that the problem of an 'objectivity without objects' was perceived as a kind of logical necessity exactly because it was a way of rejecting subjectivistic conclusions. For a valuable discussion of Bohr's philosophical position, studied in connection with the intellectual environment of Bohr himself, see Faye (1991).

'objectivity without objects' about which we have spoken in this section).[5] The qualification of this as a weak form of objectivity implies that certain alternative or formal features were proposed for characterising it. As a matter of fact, this actually occurred, and we might even say that these features were similar to the characterising marks of independence of the subject, universality, and necessity which we have considered in the general case of objective knowledge. It is rather obvious, however, that these marks should themselves receive some more specific and technical characterisation in the case of science, as they actually did. It will therefore be our next task to indicate some of the most significant of these current interpretations of scientific objectivity, especially those which appear in the thought of scientists.[6]

2.2 Some Ways of Qualifying Scientific Objectivity

We shall not attempt any kind of complete presentation of the different meanings of "scientific objectivity" one can find in the specialised literature, for our purpose is not a documentary or historical one, but simply that of presenting a theoretical discussion and analysis, which is much better served by the consideration of a few significant examples. For the same reason, we are not going to become involved in the much-debated question of the *subjectivist* interpretation of modern science, which has developed in a non-negligible manner in physics due to certain authoritative subjectivist interpretations of quantum theory and of the role played in it by the 'observer.' We shall not deny, in other words, that there have been and still are subjectivist interpretations of modern science. Rather, we point to the fact that the majority of the interpretations favour an *objectivist* view of science, and we shall try to determine what is usually and most significantly meant by this.

The most widespread sense of objectivity is undoubtedly that which identifies it with *intersubjectivity*.[7] One could say that this is the sense prevalent among

5 It is not without interest that this distinction between 'weak' and 'strong' objectivity, which the author of this work has being proposing for several decades, has been adopted by certain physicists interested in the epistemology of their science, such as, for example, Bernard d'Espagnat.

6 The consideration of the specifically scientific ways of characterising the nature of objectivity is of great significance for the general philosophical treatment of this complex issue. Indeed, it is the 'specialisation' of the concept of an object that is taking place in science which has produced several consequences in modern and contemporary philosophy. As V. Mathieu remarks, "Science has specialized the concept of an object more and more, and has given such good reasons for its way of proceeding that philosophy would have never been allowed to ignore the new situation, even if it had not been able to find these reasons by itself (what it has done instead, especially in the case of Kant)" (Mathieu 1960, p. 15).

7 In order to avoid misunderstandings, let us explicitly state that the term "intersubjectivity" will be used throughout this work as meaning a *property* of propositions, judgements, theories and

working scientists—scientists who are constantly in the position of considering and experiencing science as a *public* discourse. For them, the need of mutual information, the practice of international co-operation, the exchanges between specialists of related fields, the usefulness of reciprocally testing experiments and computations and of comparing theoretical viewpoints are to such an extent the very essence of their activity that the objective character of scientific statements must appear to coincide with the features of intersubjectivity. They see further requirements, connected with the question of the actual correspondence of scientific statements with the intrinsic structure of an underlying reality, as never offering themselves for real consideration (except, perhaps, at a theoretical level, but even in this case intersubjectivity should be the criterion for evaluating the soundness of the theoretical arguments). Only if one were to cease being engaged in the proper work of research, and to start considering it 'from the outside,' might one feel interested in problems of that kind. But even then one would probably see no reason for looking for further requirements that objectivity should meet in order for it to be suitable *for science*. This means that even if other kinds of objectivity can be envisaged from a more general philosophical point of view, these alternatives are often considered to be of little interest with regard to the sort of objectivity that matters in science (though, of course, they cannot be considered irrelevant for a 'philosophy of science' whose aim is, among others, that of qualifying the special nature of scientific objectivity).

This attitude (which we have briefly sketched as expressing the view of science as a form of 'public' discourse) may be manifest in a variety of forms whose important feature for us is the identification of objectivity with intersubjectivity, a feature which all of them share. Here we can easily recognise the requirement of being 'independent of the subject' (which we have already suggested to be one of the most typical ways of characterising objectivity). Such a prerequisite is, of course, presented in a much more sophisticated way, in the sense that several criteria are now offered to secure this independence, but the substance remains essentially unchanged.[8]

(Footnote 7 continued)
knowledge in general. Therefore it must not be confused with the *situation* in which people are confronted with the problem of communicating among themselves. Intersubjectivity in this second sense is a much-investigated problem in contemporary philosophy, and it clearly has a flavour which we could call *existential*, while our sense of "objectivity" is more abstract, and has a specifically epistemological connotation. Just as an example of an approach to intersubjectivity of this second kind let us mention Husserl (1973).

[8] We shall limit ourselves to a couple of outstanding examples of the characterisation of scientific knowledge as intersubjective knowledge that have been given by *scientists*. Indeed, it would not be difficult to find a good deal of evidence for such a characterisation in several *philosophical* doctrines. These include Frege's uninterrupted fight against "psychologism" so as to vindicate the "objectivity" of the contents of thought and of logic, Husserl's efforts to overcome the privacy of the individual subject's knowledge through phenomenological reduction, Carnap's efforts to grant an intersubjective status to science by means of a strictly syntactic and formal interpretation of science itself, capable of counterbalancing "methodological

(Footnote 8 continued)
solipsism," and Popper's explicit reduction of objectivity to intersubjectivity. The presence of this preoccupation among philosophers of such different persuasions indicates how contemporary philosophy has been compelled to find remedies for the consequences of modern philosophy's *discovery of subjectivity*. However, since these philosophical doctrines are well known (and since we shall have the opportunity to refer to some of them in the sequel), we prefer to quote, instead, a few significant testimonies provided by scientists.

The first example is that of Henri Poincaré, who indicates intersubjectivity as the characterising mark of what he has already called the "objectivity of science." What is particularly interesting is that he develops this view in a section of his work which has "The Objectivity of Science" as its title, which is included in a chapter of his book *The Value of Science* (Poincaré 1904) entitled 'Science and Reality,' which in turn belongs to the concluding part of the book, "The objective value of science." Therefore we can say that for such an outstanding 'working scientist' (no less than a serious philosopher of science) as Poincaré, the *value* of science was identified with its objectivity; and this was itself identified with the existence of an intersubjective agreement (this, of course, does not eliminate the fact that Poincaré was a positivist and, as such, inclined to underestimate other reasons for which science is 'valuable'; in particular its contribution to the *understanding* of reality). "Such, therefore," he says "is the first condition of objectivity; what is objective must be common to many minds and consequently transmissible from one to the other, and as this transmission can only come about by 'discourse' … we are even forced to conclude: no discourse, no objectivity" (Poincaré 1904, p. 136) . He then notes that the privacy of our sensations makes them non-transmissible, so that "all that is objective is devoid of all quality and is only pure relation" (p. 136). This is why, the aim of science being that of obtaining objective knowledge, the enterprise must be confined to the investigation of relations: "Science, in other words, is a system of relations" (p. 137). This restriction is far from diminishing the value of science, since it is the foundation of scientific objectivity: "To say that science cannot have objective value since it teaches us only relations is to reason backwards, since it is relations alone which can be regarded as objective" (p. 137). Developing these ideas, Poincaré maintains that the intrinsic nature of things cannot be objectively determined, and in this sense he partially denies what we have called the 'strong' sense of objectivity, for he accepts that science concerns primary qualities (measurable properties); what he is against is the postulation on the part of theories that there exists a deeper reality responsible for that of the primary qualities. He accepts realism (wittingly or unwittingly) on the empirical level, but not on the theoretical. However, he is a 'realist,' for he argues that, while we must say that science cannot inform us of "the true nature of things," it allows us know "the true relations of things" (p. 138), and these relations are grasped in a way that reveals their cogency and *permanence*, and this is a characteristic which we commonly advocate when we claim that certain objects of ordinary experience are real. Therefore, the characteristic of being real cannot be denied to those systems of relations which are the object of science. In fact—he notes—scientific change concerns 'theories' which aim at expressing in some way the intrinsic nature of reality, but does not affect the scientific laws expressing the said relations (pp. 138–140).

Our second example is that of N. R. Campbell who in Chap. 1 of his *Physics: The Elements* (1920) explicitly distinguishes science from other kinds of knowledge through the characteristic of "universal assent." He does not claim that this criterion is "ultimate," but says that "the truly ultimate is one that is scarcely capable of precise expression" (p. 15). In particular he does not dogmatically reject the legitimacy of "metaphysical" investigations concerning the foundations of science (which he actually approaches, for instance, in Chap. 9), but explicitly recognises that they are of a different nature with regard to the rest of his methodological investigation. Therefore, despite the fact that he does not use the term "objectivity" or "objective knowledge," one must say that the whole of his book clearly expresses a view of science as providing intersubjectively valid knowledge (as well as 'intellectual satisfaction,' the 'ontological' side of which remains rather an open question).

As noted earlier, the requirement of being independent of the subject was demanded of objectivity for certain epistemological reasons. But we also noted that some other characteristics (although still of a formal nature) were proposed on the basis of reasons which are of a more ontological flavour, namely those of universality and necessity. Does one find such requirements still attached to the modern concept of scientific objectivity?

At first sight, one might be inclined to believe that modern science has abandoned any claim to such engaging characteristics, for—as many philosophers of science have claimed—science has accepted that each of its statements might be found wrong, that every determination, even of scientific data, is always subject to revision, that the scope of every law is limited, and so on. But if one does not stop at the most superficial meaning of these statements, one can find that the question is not that easy to answer. One could observe, for example, that scientific laws are conceived of as being universally valid within their domain of application, though this may be a very restricted one (as we shall closer consider in the sequel).

Though we do not wish to discuss this very general point now, we would like to stress, for the moment, that there exists a feature, advocated by many to be a genuine mark of objectivity, which is directly reminiscent of the old requirements of universality and necessity: *invariance*. Among the scientists who have insisted on this principle, Max Born may be particularly prominent.[9] According to this point of view, the main feature of our coming in contact with the objects of our experience is that we can describe them in different ways, depending on the various frames of reference we adopt for recording our observations. All these descriptions are indeed different from one another, but it turns out that these different 'projections' of the same object can be submitted to certain transformation rules which constitute *groups* in the mathematical sense of this word; and these groups admit of *invariants*. Now, while it would not be reasonable to pretend that all different projections are objective (because they are different) it seems very reasonable to reduce objectivity to this core of invariants which are preserved under the various points of view. One may remark that this meaning of objectivity is at least implicitly understood, in a way, in the theory of relativity. It is true that no 'privileged observer' is admitted by this theory, and there are no physical measurements which may be considered independent of the system of co-ordinates to which they are referred. But, on the other hand, this 'relativity' is not at all the 'final stage' of physical investigation, but rather the point of departure which must in a way be overcome. Actually, the aim of the theory of relativity is to find a

(Footnote 8 continued)

While we shall only occasionally and briefly hint at Poincaré's thoughts in the sequel (because he presents them very sketchily, in spite of their being particularly sound), we shall give a much more detailed account of Campbell's position in the notes of the Sect. 2.3.

[9] Max Born is perhaps the most convinced supporter of the identification of objectivity with invariance. Some of his considerations in this regard may be found, for example, in Born (1956) (see especially the essay 'Physical Reality') and Born (1964) (particularly p. 725 and Appendix 3: 'Symbol and Reality').

formulation of the basic physical laws which is invariant with respect to all reference frames in which magnitudes are measured.

When we discussed the conceptual foundations of the traditional requirements of universality and necessity, we found that they were grounded on the basic permanence of the structure of reality. It seems that something of this kind is at work here as well, for invariance seems to be the characteristic closest to the idea of 'not undergoing mutation' which used to be advocated by traditional ontology. However, the requirement of invariance (that, strictly speaking, must be articulated as an invariance of form and invariance of substance, both of them being actually at work in science) is very often reduced to something formal that does not necessarily entail ontological commitment.

In order to see this, it suffices to consider that invariance (as formulated in *explicit* and exact terms by modern physicists) is a property of the mathematical formulation of the observed phenomena rather than of the phenomena themselves. In mentioning this objection we are actually implicitly posing the question of whether the identification of objectivity with what we have called 'weak' objectivity can be maintained as completely satisfactory, and this does not seem always to be the case. In order to see this, we should remember that weak objectivity primarily expresses the epistemological side of objectivity, and therefore stresses those characteristics which are least reminiscent of any 'reference to objects,' that is, universality and necessity, the uncoloured neutrality of which made them the most natural candidates for summarising such an ontologically uncommitted point of view. But, if we look at further efforts to qualify scientific objectivity which have been made by certain scholars, we can see that a much wider spectrum of characteristics, which come close to those of intersubjectivity and invariance, have been suggested.

A list of such characteristics, which an author such as Margenau qualifies as 'metaphysical requirements,' is the following: logical fertility, multiple connections, stability, extensibility, causality, simplicity and elegance. They come into play when the question is not so much that of ensuring the objectivity of some single empirical determination (for which intersubjectivity and invariance, interpreted as uniformity of recording using standard instruments, suffice), but rather the objectivity of certain theoretical or intellectual constructs. In these cases the simple 'verification' of the theoretical construct, which might seem at first sight to provide the most adequate tool for discriminating objective patterns from subjective imagination, proves insufficient, and a suitable combination of some or all of the above listed 'metaphysical requirements' must be used in order to make a choice between conflicting views.[10]

Let us focus on a general feature that these 'metaphysical requirements' have in common. These requirements are, in a way, still 'formal,' or at least

---

[10] Just as we did not present the details of the conception which identifies objectivity with invariance, we shall not illustrate these further criteria here, preferring to refer the reader to the literature where they are presented in an excellent and detailed form. See especially Margenau (1950), Chap. 5.

'methodological,' since they concern features that must be possessed by statements, or by systems of statements, considered in themselves. Yet at least some of them are also clearly endowed with an ontological flavour: multiplicity of connections, stability, extensibility and causality, for instance, are patently some of the most usual criteria indicating the presence of some sort of concrete 'substratum.' This could thus mean that, despite the fact that the scholars who give special credit to these requirements explicitly avoid giving an ontological meaning to scientific objectivity, they are nevertheless expressing a certain shift towards a particular conception of objectivity in the 'strong' sense, that is, towards objectivity conceived as reference to an object proper.

If one tries to understand why these scholars incline towards recovering the strong sense of objectivity, but actually only move half way in this direction, one will find that 'epistemological dualism' or 'representationalism' is once again involved. A few lines from a paper by Margenau and Park afford a good picture of the situation: "Not many scientists," they say "let alone quantum physicists, are *naive* realists. For if one seeks the objective, understood as the cause of sensations, in the things that appear in sensation, one's research is at once led beyond appearances, since even the simplest scientific observations show that things are not as they are perceived."[11] Here the idea of the object as something which lies hidden behind the appearances, an idea which we have already seen to be typical of epistemological dualism, is clearly recognisable. After an analysis of the difficulties involved in the effort of digging out the objects from 'behind the appearances,' the two authors conclude: "We therefore dismiss ontological objectivity from further consideration."[12]

One could, however, raise the question whether a more effective way of avoiding the serious drawbacks connected with the dualistic presupposition would not be that of dropping it altogether. Thus, if one were *not* to conceive of the object as 'something' unknown lying 'behind' our experiences—an entity which only reveals certain indications of its presence—it might prove possible to rescue a correct and satisfactory ontological interpretation of scientific objectivity. But this is a question which will be our concern later in this work.

However, even without giving special importance to this dualistic presupposition, which actually operates on an implicit rather than explicit level, we may note that the epistemological and the ontological sides of the problem of objectivity are intrinsically involved in the general attitude contemporary science has adopted towards its subject-matter. In fact, for contemporary science, the object is not so much something that must *exist* as something that must be *known*. This point certainly concerns one of the central methodological choices of modern physics. When Einstein started his analysis of simultaneity, which was to lead him to the relativisation of this notion and, more generally, of time itself, he discarded the common-sense view (which was also the view shared by professional

[11] Margenau and Park (1967b), p. 101.

[12] Op. cit., pp. 163–164.

physicists up to his time) according to which events *are* or *are not* simultaneous, and went on to examine how we can *know* whether they are simultaneous. The relativity of simultaneity which he found did not concern the illusory simultaneity of two events 'as such,' but the simultaneity that we can physically ascertain (i.e., by means of physical measurements).

Quantum theory, as is well known, adopted a similar attitude. Only *measurable* quantities are admitted, so that even when two situations are conceptually distinct, they are considered to be one and the same situation if they cannot be distinguished by the tools admitted in the theory. This, for instance, is evident in the case of probabilities concerning the location of particles. Also, if one considers operationalism one can see that the same requirement was at the root of some of its strongest claims. The denial of any scientific concept's having a permanent and universal character was explicitly bound to the postulate that we must confine ourselves to affirming only what we really know; and, in the case of the exact sciences, this reduces to what we are able to *measure*; hence the programme of identifying concepts with *measuring operations*.

This concentration on the idea of *being known* inevitably leads to attributing a kind of privilege to the epistemological aspect of scientific objectivity, for the locution is clearly elliptical, and its easily understood complement may be expressed as 'being known to or by someone.' In such a way the reference to the knowing subject seems unavoidable and, as a knowing activity is necessarily a first-person activity, the risks of subjectivity become immediately apparent. Thus we are led to the efforts considered in the preceding pages to avoid this danger by overcoming subjectivity through such tools as intersubjectivity and invariance.

Are these tools completely satisfactory? The answer seems to be in the negative; and if we wish to investigate the reasons for this dissatisfaction we might find a first indication by considering another sense in which the expression given above is elliptical. Indeed the notion of *being known* is elliptical not only, so to speak, on its right hand side, but also on its left. To state what is meant completely, we should say: "the being known of something by someone." It turns out from this remark that, although the object must be primarily conceived of as something that must be *known*, it must still *be* something, i.e. it must exist. This explains why, in the effort to better characterise objectivity, certain requirements had to be set which clearly show traces of this ontological side of the issue.

At this point one might say that we are confronted with a small puzzle; and, actually, we cannot hope to go on without analysing a concept which, strangely enough, we have not really taken into consideration yet. As a matter of fact, we began our considerations by observing that the concept of objectivity is always characterised in an indirect way, that is, by reference to the subject, instead of by direct reference to the object. But, after having followed the line indicated by this consideration, we find at last (which is rather sensible after all) that we cannot reasonably go on speaking of objectivity without a better determination of the concept of *object*.

We have said that our having spoken for so many pages about objectivity without concerning ourselves with making the notion of object precise was somewhat curious. This is true, but we must not forget that in our discourse we have tried to follow the line of thought which is largely prevalent among scientists and philosophers of science today, who usually are partisans of a view involving an 'objectivity without objects.' In following this line, we actually discovered many features that have been put forward for qualifying objectivity, which are hardly justifiable without accepting a kind of hidden nostalgia for the object. It is therefore all too natural to ask the question whether this notion of the object is simply a skeleton in the closet which must be removed at least for reasons of intellectual clarity, or whether there are more explicit or clearer indications that this concept has an accepted circulation in the vocabulary of science.

Brief reflection shows that it indeed has such an accepted circulation, namely when scientists, philosophers of science, and even laypersons say that every science is characterised by its own *domain of objects*. Moreover, it is customary to recognise that one of the main features which distinguish science from everyday discourse is precisely the fact that science always envisages only a restricted and specific domain of objects. But it is also clear that, in such a way, a characteristically 'referential' way of speaking is being adopted in the case of science, and this means that scientific language is considered not only to be a tool which is common to certain people, but also as being *about* something. This is perhaps the most direct symptom we have of the permanence of an ontological side in the notion of scientific objectivity, and this is why it is at any rate sensible, useful, and perhaps necessary, to devote some attention to the concept of a scientific object.

2.3 An Analysis of the Concept of Object

We shall begin to explore the concept of object by considering the use of the term "object" in everyday language. Easily recognisable within that context is the idea of an object as that of an individual *being*, of an ontological entity, of a portion of reality with which we are confronted. One might say that this is just a naive common-sense notion which is uncritical or misleading. However, this notion is fundamental in an important sense, namely, in that we cannot help starting from it, if for no other reason than because we too necessarily share that meaning in all our everyday conversations with other people, and because this remains, after all, the most immediate meaning we feel inclined to attribute to the concept. This fact, of course, does not prevent us from developing the analysis of this notion far beyond the simple connotations it receives within the context of common sense.

In any case, an important feature is already implied in this original conception of the object as an *existent*, namely that it must be the same for *all* subjects who know it by acquaintance. This is a consequence of the fundamental character of reality, which was already stressed by the Eleatic School when Parmenides recognised that the only way of specifying the notion of being is to understand that it simply means the opposite of non-being. It is therefore intrinsically impossible that something real, something existing, can under certain conditions be non-real or non-existing.

Applied to knowledge, this general principle leads to the conclusion that something real cannot exist under certain circumstances and for certain subjects while not existing under other circumstances or for other subjects. Of course, this does not mean that every portion of reality is always in the cognitive presence of every possible subject, but simply that, as far as something real is put in the cognitive presence of different subjects, they cannot help knowing it[13]; thought or cognition cannot make being non-existent, they cannot annihilate reality. In this way we have found an explicit justification for the already mentioned conviction according to which weak objectivity *follows* from strong objectivity. What is *intrinsic to the object*, and therefore real, must also be *known by every subject* who is in a position suitable for knowing it. This conclusion, if knowledge is understood simply to be 'by acquaintance,' is shared by positivism and verificationism; if it is extended to propositional knowledge as well, it can be shared only by people who are ready to attribute a cognitive power also to reason; in this case the 'cognitive presence' includes also several theoretical requirements.[14] Therefore what is intrinsic to the object must be intersubjective. This is simply the epistemological counterpart of the 'principle of Parmenides,' which states as the fundamental law of ontology the impossibility, under *whatever circumstances* (and hence also under the specific circumstances of the cognitive activity), of denying the existence of being. All the same, the experience individuals have of their commerce with other people violates this claim.

This is so since, among the different qualities which we happen to attribute to objects, some are perhaps admitted by all subjects, but many are surely not. And this is the case not only in the sense that two different persons may not agree that an object has these or those qualities, but also in the sense that one and the same person may attribute one such property to an object at a certain moment, and deny it at another (or even deny its existence). It was already noted by the ancient Sophists that, for example, wine tastes agreeable to a healthy man and disagreeable to the same man if he is sick. One could remark of course that, strictly speaking, the feeling of pleasure that the healthy man enjoys when drinking wine and the opposite feeling of disgust experienced by the sick man are both equally real. Such a remark is unobjectionable, but it cannot eliminate the fact that, within the domain of qualities or properties that we can consider as real, as really perceived by the subject, some are considered to be bound to the subject's perceptions, while others

[13] By "cognitive presence" we mean a situation in which the subject is equipped with the ability, and finds himself in the conditions necessary and sufficient, for knowing (by acquaintance) a particular object.

[14] It is obvious, but perhaps not superfluous, to make explicit that we are not maintaining that whatever exists is knowable 'in general,' but only that it cannot help being known if it has the characteristics enabling it to come into a particular 'cognitive presence' in the sense explained above.

seem at least to be independent of these perceptions and to be equally well perceived by all subjects.[15]

The very presence of this distinction easily induces one to express it in a form which is seemingly only slightly different, by saying that the qualities that change their status with a change of subjects are not really possessed by the object but are simply *subjective* (in the sense of being a result of the perceiving activity of the subject), whereas the other qualities are really inherent in the object, and as such deserve to be called *objective*. This distinction is the root of the classical partition between primary and secondary qualities that was so widespread (under different forms) in the philosophy of the seventeenth and eighteenth centuries, and which found one of its first and most typical expressions in the celebrated pages of Galileo's *Assayer*, as noted in Chap. 1. This way of introducing the distinction is potentially dangerous, for it suggests the admissibility of such ambiguous statements as that which says that subjective qualities are not really present in the object, leaving the possibility open for us to imagine that they actually are not real. From our present point of view, to say that they are not real would certainly be a mistake, for here we are equating reality and existence: what we should correctly say—using, for instance, the distinction between the primary and secondary qualities of things—is that they have a 'different sort of reality.'[16] This, in any case, would not imply that the secondary qualities reside *uniquely* in the subject, but that they depend on a particular *relation* between certain features of the object and certain cognitive capabilities of the subject; this, however, is a question that we shall address later; and, at the same time, we must recognise that secondary qualities (for reasons to become clear later) are not taken into consideration by science.

Still, this interpretation is not inevitable; and we can adhere to the above distinction provided that we do not take the step of qualifying the subjective qualities as illusory or as *pure* appearance having no relation to objective reality, which would introduce a dualistic prejudice into the discourse. Thus far we have

---

[15] In fact this line of reasoning was not inaugurated by the Sophists but rather by Democritus, who termed knowledge derived via the senses "bastard" cognition. "Legitimate" cognition, on the other hand, is knowledge obtained by reasoning concerning the properties of the ultimate constituents of matter (the atoms). Here we find the first drawing of the distinction between primary and secondary qualities that is essentially the same as that expressed in the passage of Galileo's *Assayer*. This is not accidental, since a revival of Greek atomism took place contemporaneously with the birth of modern science, as is well known (and Galileo expresses his acceptance of atomism elsewhere in the *Assayer*).

[16] We have consciously said "from our present point of view" in order not to disregard the fact that, in many ontological discussions, it is useful to introduce certain technical distinctions between reality and existence. However, we do not need such technical distinctions here, and shall at most make a modest use of them when discussing the general problem of realism. Just to give an indication of how distinguishing between reality and existence might be profitable in the case of our example, we note that subjective qualities do in fact exist in a particular domain of reality (i.e. consciousness) different from the domain of physical reality. It is this fact that entitles us to claim that they are real, since reality admits of different spheres of existence.

simply admitted a splitting of *reality* into two fields, that of subjectivity (containing all those features of reality that are real only for an individual subject) and that of objectivity (containing all those features that are real for all knowing subjects).

This situation is, nevertheless, not a stable one. Objectivity must, indeed, rapidly become prevalent. This has to do with the fact that, although anything which is different from nothing may be said to be real, this notion has an *analogical* character, that is, it is attributed with different degrees of pertinence or force to different kinds of entities. This amounts to saying that 'reality' is usually meant to correspond, in the most proper or strongest sense, to a given category of being, while other categories deserve to be called real only in a weaker sense. In everyday language, for example, reality in the most proper and strongest sense is meant to be what philosophers call the external world (see also the difference between *Wirklichkeit* and *Realität* in German). According to certain philosophies, on the contrary, reality may be primarily the sphere of our internal self-consciousness (e.g., for Cartesianism).

Being aware of this situation, we can easily understand why and how objectivity may obtain a privileged status. Indeed it is very often considered to be the only warrant one has that something is *real* in the 'strongest' sense of this word. If I should like to convince someone of the reality of something which is present to me (i.e. known by acquaintance), the only way I have at my disposal is to try to make it present to him as well, that is, to transform this reality from a subjective into an objective one (examples will be provided later). But, even for the single subject, objectivity turns out to appear more important than subjectivity. Indeed, everyone is inclined to say, for example, that the bad taste he attributed to wine when he was ill had not *really* to do with the wine, but with his illness; similarly, one usually says that the things he experienced during a dream were not really there, though he really had the relevant perceptions in the dream. But why does one deny *proper* reality to such things? Simply because one sees that other subjects do not say that they experience the same things, and because one does not oneself experience them after the illness or after the dream. Here one can see the implicit force of the 'principle of Parmenides' (the permanence of reality is the basic feature of reality itself), combined with the fact that objective reality ranks higher than subjective reality.

Clearly, certain ponderous presuppositions lie behind this way of thinking, such as that there is a kind of normal status in which what we perceive is (strongly) real (e.g. health and good eyesight are supposed to contribute to such a status, as illness and dreaming do not). But rather than discuss the legitimacy of these tenets here, let us simply state the fact that, due to them, subjectivity becomes progressively excluded from one's considerations of reality. As a matter of fact, objectivity, which at the beginning was taken only as an *indication* of reality, has become identical with reality itself, in the sense that aspects of reality which cannot be considered objective seem to be condemned to remain altogether negligible.[17]

---

[17] Concerning science, a particular insistence on the fact that scientific statements are characterised by their being intersubjective is expressed by Campbell in the first chapter of his (1920), 'The Subject Matter of Science':

> Judgements do not form part of the proper subject matter of science until they are free from the smallest taint of personality, unless they are wholly independent of volition and unless universal assent can be obtained for them. In practice the last criterion is applied almost exclusively; the subject matter of science may be defined as those immediate judgements concerning which universal agreement can be obtained (p. 21).

Campbell's discussion also resembles our analysis of 'subjective' and 'objective' features of reality, which is mixed up, however, with the very problematic idea of the 'external world':

> Evidence for an external world. What this criterion is will be easily seen if we ask ourselves why the particular class of immediate judgements, of which those that are the basis of science form part, are described as judgements of the material or of the external world. (For our present purpose we may regard "material" and "external" as identical, for our knowledge of any external objects which are not material is based on our knowledge of those which are material). We all realize instinctively that the judgements which compose our conscious life can be divided into two classes, those which represent events happening within ourselves and those which represent events in the external world. The first class includes our judgements concerning our tastes and our desires and our purely logical judgements: the second class includes the judgements we associate with sense-perceptions. The distinction between these two classes arises from a difference in the extent to which the judgements are common to persons other than ourselves. In respect to the first class of judgements we find that other persons often dissent entirely from us; in respect of the second class we find there is something common between ourselves and any other person with whom we can enter into communication. It is the community of our judgements of the second class with those of others that leads us to attribute them to some agency which is neither we nor they, but something external to all; it is the divergence of our judgements of the first class from those of others which leads us to attribute them to something inherent in our own personality (p. 19).

About the actual possibility of obtaining universal certain judgements, Campbell limits himself to mentioning three classes of judgement for which this seems actually to be the case, without excluding other possibilities:

> Is it possible to find any judgements of sensation concerning which all sentient beings whose opinion can be ascertained are always and absolutely in agreement? The best answer that can be given is to state at once what judgements appear to be absolutely free from contradictions such as we have been considering. I believe there are at least three groups of such judgements:
>
> (1) Judgements of simultaneity, consecutiveness and "betweenness" in time. I believe that it is possible to obtain absolutely universal agreement for judgements such as, the event A happened at the same time as B, or A happened between B and C.
>
> (2) Judgements of coincidence and "betweenness" in space.
>
> (3) Judgements of number, such as, The number of the group A is equal to, greater than or less than, the number of the group B.
>
> These three groups will be termed respectively time-, space- and number- judgements (p. 29).

The above discussion had two important tasks to fulfil. The first was that of showing how the concept of object, at its origin, is bound to that of reality in a very correct way, not as a counterpart of reality but as a specific subdomain of it. (We could say, perhaps, that the domain of objectivity appears to be a subset of reality and not its complementary set. This image is not simply pictorial, for if objectivity were to be the complement of reality, one would be obliged to specify the general domain with respect to which these two sets are complementary, and this set would be either the 'non-real' or the 'suprareal'—in either case, something enigmatic).[18] Second, we discovered the reason why subjectivity has obtained such a bad reputation. As a matter of fact, we are used to thinking that one must avoid subjective judgements, appreciations, and so on. However, no clear reason is usually advanced for this view. Now, if the above analysis is correct, we find that the implicit reason is that subjectivity is considered too weak a warranty for reality (in its 'proper' sense).

We are now in the position to move closer to the core of our question. First, we have found in this analysis that the two familiar characteristics of *intersubjectivity* and *invariance* are the basic structural marks of objectivity, since we have qualified as objective only those features of reality which are the same (invariant) for different subjects (intersubjectivity). Yet there is much more in these two notions as they appear now than there was at the beginning of our discussion, when they were simply the *consequence* of the 'principle of Parmenides.'

(Footnote 17 continued)

Note that he explicitly avoids any identification of this form of objectivity with "truth," and this confirms what we have stated at the beginning of this book, that is, that the notion of objectivity was meant to be a replacement for that of truth:

> Is the criterion of universal assent ultimate? A few further remarks are necessary to avoid misconception. It must be insisted again that our object in this discussion is merely to ascertain what is the criterion which science applies in the selection of its subject matter; we are not concerned to ask why it applies its criterion. If these judgements which are selected are indeed ultimate and fundamental, to ask such a question would be to trespass beyond the province of science; ultimate judgements are those for the acceptance of which no reason can be alleged. We must be extremely careful not to assert that universal assent is a test of "truth" or that our fundamental judgements are "true" because they are universally accepted. If such an assertion were made, the door would be opened to all kinds of objections which might appear very trivial to students of science, but yet would have to be faced and answered. (p. 34).

[18] We could express this idea more precisely by saying that the notion of objectivity is eminently epistemological and reflects itself on ontology in the sense that certain parts or aspects of reality are characterised by the fact of being 'subjective,' and others by the fact of being 'objective.' This is mirrored by the fact that (as we have seen), the notion of *object* presupposes that of *subject* and vice versa since an object is what is referred to by a subject, and a subject is what refers to an object.

The first novelty is that they are no longer *only* necessary conditions for objectivity, as was the case when the state of affairs of 'being such for all subjects' was seen only as a consequence of the immutability of being. Now we must say that this condition (being the *defining condition* for that subdomain of reality which we shall *call* the realm of objectivity) is indeed a necessary *and sufficient* condition for it, as is the case with definitions in general. This is, moreover, much more than a pure consequence of a formal feature, such as that implied in the structure of a definition; for we can say that whereas in the case of weak objectivity, as contrasted with strong, the features of intersubjectivity and invariance seemed to 'emanate' from reality as reliable criteria for it, in the present case this privileged status is not attributed to them. As a matter of fact, subjectivity also refers to reality (because there cannot be perception, even in the most subjectivist sense, of the non-existent), so that intersubjectivity and invariance are no longer considered to be 'characteristic marks' of reality, but rather 'specific characteristics' of a *particular* sector of reality, the sector of objectivity.

A second feature deserving note is that the notion of intersubjectivity itself has now received a more profound qualification, for we have here characterised as intersubjective those features of reality which are knowable by many subjects or, equivalently, by one and the same subject under different conditions. Therefore, from the point of view of the requirement of intersubjectivity, one individual subject 'splits' into a plurality of subjects, so that the condition of invariance has to be understood not as an invariance with respect to different evaluating persons, but with respect to different acts of knowing, no matter whether they are performed by different subjects or by the same subject on different occasions. This gives to the notion of objectivity, understood as intersubjectivity and invariance with respect to subjects, an abstract character which appears to be susceptible of rigorous treatment. (Let us also note that in this way our reasoning does not *depend* on the presupposition of the existence of a multiplicity of subjects; however, we have no difficulty in assuming this multiplicity to be the case.)

From what we have said, a further important consequence also emerges regarding the specific field of *scientific* objectivity. The consequence is that since this notion of objectivity implies a reference to a *plurality* of subjects, these subjects cannot be conceived of as minds, consciousnesses, or anything of the kind. In fact, consciousness necessarily represents, in every act of knowledge, that very part of knowledge that is irremediably private, for the only thing that two subjects cannot hope to be able to share, to make intersubjective, is their respective consciousness of reality. Thus the interesting result ensues that the subjects about which we have often spoken thus far cannot be conceived of as *egos*, but simply as *detectors* or *recorders* of different aspects of reality. Such a conclusion, however, should not be surprising if one only considers how an observer or a subject is actually conceived of in the exact sciences. Despite every appearance, the subject is considered equivalent to an observation instrument, and this may be seen as the main reason for which quantum mechanics is not subjectivistic, notwithstanding that the 'observer' is mentioned in it, since such an observer is never an individual subject in the usual epistemological or psychological sense of this term.[19]

When intersubjectivity is conceived of as an agreement among impersonal measuring instruments, clearly all that is required is simply that those properties that are meant to be objective in this sense must be *invariant* with respect to all these instruments, or, speaking more generally, with respect to all these systems or 'frames of reference.' To put it differently, since the subject is reduced to a precisely delineated 'viewpoint' (in the most general sense of this term, which could be even better rendered by the notion of a generalised frame of reference with respect to which not only space and time, but a broad variety of 'co-ordinates' may be considered), intersubjectivity coincides with invariance with respect to such viewpoints. This not only allows us to find our second mark of objectivity (i.e., invariance) confirmed once again, but also to see its essential interchangeability with the general features of intersubjectivity. For invariance, as we have seen, is most properly defined with respect to frames of reference, and it therefore applies literally if subjects are considered to be generalised frames of reference. Under such a condition, intersubjectivity also turns out to be nothing more than this invariance.

The above discussion of the relationship between intersubjectivity and invariance should not divert our attention from a delicate point indicated earlier. We stated that a subject cannot share with other subjects his or her consciousness, his or her awareness of reality. Now the question arises: what then can a subject share with other subjects? To examine the situation more closely we may recall that the object is something that is known in a way which is equally valid for every subject. But, on the other hand, something's being known necessarily implies the existence of a subject conceived of as a consciousness. Thus it would appear that there is a problem of rendering 'public' something that is intrinsically and essentially 'private.' We shall discuss this question in the Sect. 2.4. For the moment let us only note that, if we put the problem in the above terms, it is clearly unsolvable. But it turns out that we are not compelled to pay such an impossible price to obtain objectivity, for objectivity does not require, for instance, one's being *aware* (as a knowing subject, in the full sense of the word) of the *awareness* of one's interlocutor when speaking about a certain property of a thing, but simply to be aware of an *agreement* with him regarding that property. In other words: I cannot know

[19] For a rather detailed discussion of this issue, which also takes the relevant literature into consideration, see Agazzi (1969), Sect. 48. Furthermore Heisenberg, in discussing the famous problem of the intervention of the subject in quantum mechanics, escapes subjectivism precisely by putting the human subject on an equal footing with an instrument: "Of course the introduction of the observer must not be misunderstood to imply that some kind of subjective features are to be brought into the description of nature. The observer has, rather, only the function of registering decisions, that is, processes in space and time, and it does not matter whether the observer is an apparatus or a human being" (Heisenberg 1958b, p. 137). In connection with this discussion, see also Popper : 'Quantum Mechanics without "The Observer",' in Bunge (1967c), pp. 7–44.

the private *knowledge* that my interlocutor has of a red surface; but we both can know, can be aware of, the agreement we have reached in qualifying such a surface as red. As said, we shall later investigate to a certain extent how such an agreement can be obtained. For the moment it suffices to stress that the agreement can be made publicly, and this is all that matters with regard to objectivity.

The elements we now have at our disposal enable us to derive some consequences regarding the general methodology of science. The first is the requirement of the *replicability* of those situations which are expected to reveal objective features of reality. In fact, if an objective feature must be valid for every subject, it follows that no subject could in principle be excluded from the possibility of knowing it; and this amounts to its being the case that, whenever certain precisely stated conditions are satisfied, the same feature must be observed by any subject whatever.[20] Of course, practical difficulties may be of considerable importance, but they cannot completely eliminate the possibility of repeating the observation in question (the question is one of principle, not practice). No exception is represented, from this point of view, by so-called irrepeatable events. A stellar explosion, for example, is a fact that cannot be observed a second time. But what we claim is not to be able to repeat the observation of *that* explosion, but simply that any particular astrophysical theory presupposes the possibility of observing the general phenomenon of stellar explosion with a probability which, though very small, must nevertheless differ from zero; and that such a theory indicates the conditions under which such an event could again be observed.

Replicability also plays a central role in one of the most fundamental procedures of scientific method, that is, in testing.[21] *Testability* is equally well suited to a verificationist as to a falsificationist approach to science. Moreover it is not limited to the pure testing of hypotheses, but may also include the checking of experimental conditions and empirical data, so that the majority of epistemologists make it the determining feature of scientific knowledge. However, we might still wish to know why testability has this privileged role; and the answer to this question might be that testability is bound to the empirical character of science, to its obligation to be something different from pure invention or fantasy. Answers of this kind may be acceptable, but they miss the important point that reference to testability is necessary for objectivity (understood as intersubjectivity) to receive its philosophically most satisfactory characterisation.

In order to see this, let us start by formulating a rather intriguing question about objectivity as defined thus far. We have said that an object is something that can be known to be such by *many* subjects. Put in this way, objectivity sounds very much like nothing other than an 'enlarged subjectivity,' which is not very satisfactory.

[20] It should be noted that this affirmation implicitly presupposes an *ontological* principle that is tacitly assumed by common sense no less than by science, i.e., the principle of the uniformity of nature, since the condition that every subject should in principle be capable of knowing something objective does not imply that the conditions giving rise to that something on one occasion will give rise to it on another.

[21] Which implies, again, that testing, too, presupposes the principle of the uniformity of nature.

Even if we should modify our statement by saying that an object is something which can be known to be such by *all* subjects, we still might feel dissatisfied since one could still imagine that all subjects might simply agree by chance (knowledge cannot depend purely on agreement). The requirement one would like to be able to set instead might be formulated more or less like this: an object is something that *must* be such as can be known to be such by *all* subjects.

Now, how is it possible to state something that *must* be agreed upon by *all* subjects? The answer may be suggested by considering the way according to which one establishes the existence of all-properties (i.e., properties concerning a given totality) in the realm of the exact sciences. Two cases must be distinguished: either the property is predicated of the totality itself (*collective universal*) but not of its single members, such as when we say that 'the days of the week' are seven in number; or the property is predicated of each individual belonging to the totality (*distributive universal*), as when we say that all men are mortal. When we have to do with a property which is meant to be universal in a distributive sense, what we do is simply establish its existence in the case of a *generic* individual belonging to the envisaged totality. When we wish to prove, for example, that *all* the points of a certain line have a particular property, we simply select at random a *single* point and, without endowing it with any further special features, we prove that it has that property. The same happens everywhere. When we have a collection, no matter whether finite or infinite, and we wish to prove that all its members have a certain property, we simply try to prove that such a property is possessed by a *generic* member of the collection (i.e., by a member to which no other properties are attributed than those that define the collection). In other words, *every* is considered equivalent to *whatever*, and the reason for this probably resides in the principle of the identity of indiscernibles. (If we have no means by which to distinguish a 'generic' element of a collection from the others, there is no reason why they should not have the same properties that it does).

Coming now to our problem of establishing that the existence of a certain property is agreed upon by all subjects, we may think of showing that it is agreed upon by *whatever* subject we may select. This amounts to saying that, whenever an arbitrary subject may wish to *test* the hypothesis that this particular property exists, it *must* (in principle) be possible for him to do so, and he (in principle) *must* obtain the same result as any other subject would who carries out the same test. In this light, testability is nothing less than the tool through which intersubjectivity can be conceived of as something more engaging than simply a broadened form of subjectivity; and, as such, it deserves to be the defining characteristic of objectivity. We could also express this fact by saying that it is only via testability that we can give to intersubjectivity the character of a *quaestio iuris*, and not simply that of a *quaestio facti*, of a pure matter of fact which does not involve any kind of necessity or normativity in itself.

However, one must be careful to understand what the real matter at issue is here. We are referring to the *meaning* of "intersubjectivity," and we are saying that a conspicuous aspect of its meaning is that intersubjectivity is not simply a broadened subjectivity. But this implies that, intersubjective agreement being reached by testing, *any* subject must *in principle* obtain the same results when performing the same tests.

Let us now leave the question of principle (represented by the analysis of the meaning of intersubjectivity), and address the question of fact (represented by the actual performance of a specific test designed to ascertain the objective nature of some proposed statement). Clearly, even if a great number of subjects were able to test this statement and obtain a positive result, one could still theoretically retain the doubt that they were possibly not ‘generic’ subjects, and that all of them shared a certain peculiar feature which (perhaps unconsciously) led them to agree with regard to that property, while other more generic (i.e. not biased) subjects might well not corroborate the statement (i.e., might find that it is not confirmed by the test). As is clear, the situation is here similar to that with which we are familiar in the critical analysis of ‘verification.’ Even after a hundred positive tests the possibility always exists that the next test be negative, thus admitting a *practical certainty*, but not an *absolute certainty* (in other words, one could not exclude the possibility that all the tests were favoured by fortuitous circumstances). In addition, let us note the possible lack of ‘generality’ of the subjects or, rather, the suspicion that those who performed the test were all equally biased such that the results were uniformly fallacious (a question which, especially in the case of what are termed the human sciences, is by no means Byzantine). In any case, we can conclude that the question of fact has only a ‘practical’ relevance, and as such it can be handled by means of those standard patterns of accuracy and critical prudence which in current experimental practice lead to reliable results within some determinable degrees of confidence.

This resorting to replication and testing has the additional non-negligible advantage of removing any psychological hidden meaning from the notion of independence of the subject. In fact, it is very easy to understand such an expression as meaning ‘independent of the subject’s will.’ For instance, when one says “whether or not I perceive what is before me does not depend on me,” one often means that whether or not one has such a perception is not dependent on one’s will. The inadequacy of this characterisation becomes immediately evident, however, if one thinks of the perceptions one experiences in dreams or hallucinations, which are indeed independent of one’s will, but which are nevertheless not objective, as has been explained earlier.

Even worse are perhaps those characterisations of the independence of the subject which lean on the idea that the object is ‘external’ to the subject. Apart from the rather naive picture of the subject as something circumscribed by his bodily sense organs, by his skin, or even by his cranium, it is probably due to this obscure tenet that psychic facts (being internal to the subject) are still thought by many to lie outside what can be treated by any objective study.

On the other hand, if we carefully consider the reasons which have led us to see the independence of the subject as a distinguishing feature of objectivity, we note that they were purely epistemological in character, and involved no reference to an individual’s free will, or to localisations in space. Correspondingly, the features of intersubjectivity and invariance, specified through the requirements of replicability and testing, are completely free of such spurious mixtures, though they have the above-mentioned psychological and spatial requirements among their corollaries. In fact, if something is objective in the sense of being testable by other subjects as well as by me, it is also necessarily independent of my will; and, moreover, if this testing must be such that it can be done by any subject whatever, this implies that it must be 'external' to me both in space and in time.

2.4 How to Overcome the Privacy of the Subject

We can now return to the question regarding the crucial point in the transition from the subjective situation (which is necessarily implied in knowing something) to the intersubjective one (which is characteristic of objectivity) that seems to be the requirement of rendering 'public' something which is essentially 'private.'

Here again let us clarify the question at issue. What is not in question is *the fact that* our knowledge has this kind of public status, since the phenomenological evidence of our usual commerce with other persons indicates *that* we are able to exchange information and communicate with them. This is why such much-debated questions as those regarding solipsism and other minds are at least to a certain extent ill-raised. They start from an *imagined* problematic situation (exactly as in the case of the imagination lying behind the presupposed epistemological dualism) and, taking it as given, try to overcome it.[22] The correct starting point, on the contrary, is to begin with the phenomenological evidence, and then, by analysing it accurately, to try to understand *how* it is possible that a certain kind of (propositional) knowledge may become public while being private in its origin. With regard to this point we have already stressed that what matters is not that my 'knowledge by acquaintance' of red is the same as that of other

[22] Here are a few brief considerations explaining in which sense we say that intersubjective communication is given as 'phenomenological evidence' (without any pretension of providing in a couple of lines a full foundation for such an engaging claim). It would perhaps be too hasty and superficial to say that this evidence is granted by the simple fact that individuals are able to exchange views, opinions, orders, instructions and information, and to understand each other in an effective way (although this fact would be almost impossible to explain without admitting at least a certain measure of intersubjective communication). What seems more significant in this respect is the fact that in speaking with other persons we usually *experience* that what we are told by them corresponds exactly to what we *expected* to hear e.g. as an answer to our question. For example, if I ask someone, "What time is it?" and she answers, "Snow is white," I feel that she has not understood me, that a communication between us did not obtain. If she answers, "It is noon," I feel that she has understood me since her answer is *of the kind* I expected, even if it should be wrong. Of course, I might sceptically raise the doubt that her answer was such just by chance, but then I could try to check the soundness of this doubt by further questions and, if they continued to be answered in the way that corresponds to my expectations, I should have really *no reason* for claiming that we do not understand one another. That is to say that this kind of evidence has the same degree of soundness as ordinary evidence (for, since Descartes, we know that it is always possible to raise *artificial* doubts about any sensory evidence). Hence we may conclude that intersubjective communication constitutes phenomenological evidence, and the question is only that of explaining *how* this is possible.

subjects, or is shared by all the subjects with whom I should like to establish an intersubjective dialogue. As has already been pointed out, this is surely impossible, for I could never imagine myself 'looking inside' other people's minds in order to perceive their perceptions and compare them with my own. The solution is offered, we have said, by the fact that in order for objectivity to be granted we simply need to be in *agreement* with other subjects, for example, about the intersubjectively manifest application of what each of us takes to be our notion of red, without knowing whether what others call red is what we call red. In other words, what we need, and are able to attain, is not an impossible agreement—for instance—of our subjective 'knowledge by acquaintance' of red, but our agreement on the 'propositional knowledge' *that* this pencil is red.

Is such an agreement possible? It is, through *operations*. This fact is very general and is not limited to scientific practice. When we wish to test whether we agree with someone else about a certain notion (that is, about any content of knowledge) the only means at our disposal is to see whether we both make the same *use* of that notion. It is not *apprehending the same thing* in applying the notion that can demonstrate agreement about the notion, but *applying the notion in the same way* in what are otherwise the same circumstances.[23] If I have certain reasons to be doubtful about my interlocutor's having the same notion of red as mine, I could, for example, invite her to select from a bundle of pencils a red one. If the person's way of *operating* is the same as that which I should have adopted in all circumstances of this kind, I am fully justified in concluding that 'red' is an intersubjective notion for us, even if, for example, the other person sees what I should call green in situations where I see what I should call red (but such a difference would remain forever a private affair, not communicable between us).

The same can be said, of course, for more complex notions, and even for those which are much more abstract in character. For instance, if a teacher wants to know whether her pupil has acquired the 'correct' notion (that is, the notion determined by her science) of, say, a logarithm, she cannot rest content with his pupil's being able to repeat certain definitions. Rather, in order to be sure that the pupil has really grasped the notion and has not simply learned some suitably connected words, the teacher will invite him to *operate* with logarithms, to solve some problems where these are involved and so on, until it is possible to ascertain that the pupil *uses* the notion of logarithm correctly.

The epistemological relevance of what we have said has been stressed in a doctrine which is often overly emphasised by its supporters and overly discredited by its opponents, that is, operationalism. When this doctrine emphasised the overall importance of operations, it actually insisted on a capital point, that operations constitute the basic condition for determining objectivity, in that they allow public agreement with regard to particular properties, which allows those properties to become objective. This is true as a matter of fact; however, the notion

[23] "Not the way of apprehending, but the way of using a thing may reveal if we agree about it." (Mathieu 1960, p. 31).

of an operation is not always so clearly presented by operationalists as to show this essential function in a proper light. Evidence that they usually did not recognise this function may be found in some explicit statements by Bridgman himself who, strangely enough, was a supporter of a subjectivist attitude towards science. "There is no such thing as a public or mass consciousness," he says:

> In the last analysis science is only my private science, art is my private art, religion my private religion, etc. The fact that in deciding what will be my private science I find it profitable to consider only those aspects of my direct experience in which my fellow beings act in a particular way, cannot obscure the essential fact that it is mine and naught else. 'Public Science' is a particular kind of science of private individuals.[24]

These claims clearly show that Bridgman was unable to reconcile the fact that knowledge quite generally is something which may be stated only in the 'first person' with the fact that scientific knowledge in particular must be something independent of the subject. This latter state of affairs is, of course, not one that a science begins with, but is rather a goal that must be accepted when attempting to create a science, a goal ultimately to be arrived at via a long and complex journey.

This point was correctly stated by Born, among others, who once described how science is arrived at at the end of a process involving the progressive elimination of the individual subject:

> 'Natural science' is placed at the end of this series, at the point where the I, the subject, represents but an insignificant part; every progress in the modelling of the concepts of physics, of astronomy, of chemistry, indicates a further step towards the goal of excluding the I. This, of course, does not concern the act of knowing, which is bound to the subject, but the final picture of nature, the basis of which is the idea that the ordinary world exists in a way independent of and not influenced by the process of knowing.[25]

The only thing missing in this passage is an indication of the way in which this goal may be reached. If the 'act of knowing is bound to the subject,' how is it possible to 'exclude the I'? The answer we have tried to give seems reasonable: if *knowing* is necessarily bound to the subject, it is not on the ground of knowledge that we may hope to discard the subject. The alternative ground we have, however, which allows this discarding of the subject, is that of *doing*.[26]

---

[24] Bridgman (1936), pp. 13–14.

[25] Born (1956), p. 2.

[26] These considerations also apply to the doctrine presented by van Fraassen (2008). He repeatedly stresses—on the one hand—the "indexicality" of any scientific "representation," which is always made by someone from his/her private vantage point, and—on the other hand—he affirms that the 'public' status of representations (that he obviously requires for scientific representations) obtains through their *use*. For this reason he points out that this problem does not concern semantics proper, but rather *pragmatics*: "The notion of *use*, the emphasis on the pragmatics rather than syntax or semantics of representation in general, I will give pride of place in the understanding of scientific representation" (p. 25). It is clear that the mention of use refers to the familiar partition of semiotics into syntax, semantics and pragmatics that remains essentially within the framework of an analysis of language and gives a primacy to the problem of communication. We could say that this approach has clear affinities with the Wittgesteinian

To express the above in a more pictorial way, we could say that, while the subject cannot reveal his mind to others, he can show them how he does things and what he has done. As a consequence, while two or more subjects could never check whether they have the same thought, they can always check whether they are performing the same operations, for these are perceivable by both of them. (What each directly perceives is different; therefore abstraction is required, as in all instances of knowledge *that*, i.e., of propositional knowledge.) Thus when we say that a notion cannot itself be public, while agreement concerning it can be, we mean that such an agreement is concretely expressed by a coincidence of the relevant operations and of their results.

After giving these specifications one does not want to be misunderstood and, for example, be thought of as an unsophisticated pragmatist who has abandoned the ground of ideas and the mind in favour of a return to the narrow perspective of science as simply a way of operationally mastering the physical world. Even if we were not to add in the sequel explicit considerations concerning the cognitive side of objectivity (which we shall do), it should already be clear from what we have said to this point that the intersubjectivity of operations is strongly marked epistemologically, due to its being the indispensable condition for building objective *knowledge*.[27]

One could go even further and maintain that not only objective knowledge but any propositional knowledge always has praxis and operations at its roots. We do not wish to treat such a broad and engaging thesis here since it would involve us in discussions of psychology and of individual concept formation which are too far from our subject-matter. But the simple mention of this possibility should suffice to reveal how any position of mistrust towards the operational component of scientific knowledge could lead to undesired difficulties when one comes to solving certain problems in the philosophy of science.

Another point requires further investigation. How can operations function as tools for constructing objective knowledge? For example, we have said that it would be possible for me to ascertain whether a friend of mine has the same notion of red as I have by inviting her to select a red pencil out of a bundle of pencils; but how can I be sure that she has the same notion of selecting that I have? This objection is not very difficult to meet, since science (and knowledge in general) does not develop in a vacuum, and we can safely include language and gestures

(Footnote 26 continued)
doctrine of "language games" in which contexts are related with human behaviours and conducts, with acting rather than making. According to us, this strategy is still insufficient to give a foundation for intersubjectivity, and this is why we resort to the much stronger and effective notion of *operations* that are certainly a part of *use*, but are less vague and more suitable for breaking the circle of subjectivity, besides their decisive *referential function* which we are going to discuss in the sequel.

[27] This central thesis of our epistemology, which we have constantly and systematically developed since the publication of Agazzi (1969), and which will be duly articulated also in the coming chapters of this work, has been more recently advocated also in Hacking (1983).

among the most elementary operational tools that it has at its disposal.[28] But there is even more: when a certain notion is to be tested, as far as its objective status is concerned, a great variety of already tested objective notions is always at hand, and the different subjects do actually employ them. Nevertheless, this objection will be taken into consideration later due to another of its interesting consequences, namely because of the implicit reference it makes to a particular kind of relativity in every instance of scientific knowledge; but it is not our concern to anticipate such a discussion here.[29] There are, on the contrary, many other aspects of the operational side of scientific objectivity which at this point deserve a closer examination.

An important remark. The discussion we have devoted to the problem of *recognising* the place of *subjectivity* in knowledge, but at the same time of *overcoming* it in order to understand science as *public knowledge* must constitute from

---

[28] We should like to quote here an eloquent statement of Rom Harré which is in keeping with this view: "The power of certain elements of language and of certain gestures and the like to draw people's attention to things and the states of affairs obtaining among them is what binds language to the world, since it is thus that our attention is drawn to those states of the world which we are required to observe, that is to understand" (Harré 1970, p. 193; see, in general, the whole chapter 'Description and Truth' of this work). Concerning certain important consequences of the continuity between common knowledge and scientific knowledge as regards the issue of objectivity, see Buzzoni 1995 (e.g. pp. 108–109 and 120–131). Some readers may be surprised by the fact that, in the notes of this section, we have cited both realist and positivist scholars in support of our view, without saying whether or why we agree or disagree with each of them. This is not accidental, for we think that the convergence of differently-oriented scholars on the characterisation of objectivity that we are defending is a not-negligible symptom of its soundness. In the course of this work the reader will have abundant evidence concerning what our position is regarding positivism and realism, but we wanted to avoid that our present discussion be biased by such more engaging issues.

[29] An important philosophical question underlying this discourse is that concerning the possibility of disentangling knowledge from the whole of 'real life.' From an analytic point of view it is certainly possible and fruitful to *distinguish* cognition from other aspects of life, but this distinction cannot amount to a *separation.* As a matter of fact, our cognitive activity is a part of our 'life practices,' and humans (but not only humans) approach reality through action, and form in themselves representations of reality that are strictly bound to action. This deep embedding in real life, in *human existence,* has been particularly stressed by authors such as Heidegger, Husserl and the representatives of existentialist, phenomenological and hermeneutic philosophies. Unfortunately these philosophies have usually expressed a negative appreciation of science, often based on several misunderstandings, such that philosophy of science has considered them as useless or even misleading, and has adopted the mentality and methods of analytic philosophy. In this work he have decided to remain faithful to the style of analytic philosophy, since this has concretely provided the framework in which the majority of the problems we are going to study have been presented and debated. This, however, does not mean that we have accepted to remain prisoners of certain limitations of this approach; for example, our treatment of many questions will develop along the lines of a linguistic analysis, but we shall be ready to trespass the limits of this analysis when it proves insufficient for a full understanding of these same questions. For the same reasons one must recognise that certain basic intuitions of the above mentioned philosophies could and should be carefully taken into consideration when particular fundamental problems are envisaged. Therefore we shall make some reference to them in the course of this work, where such problems will be addressed.

now on a justification for the general methodological approach adopted in this work, namely, the leaving *out of consideration* the place of the subject. For example, when we speak of the intension or of the reference of a concept or of a term, we are certainly not ignoring that it is a certain subject who 'intends' something or 'refers' to an object; similarly, when speaking of a representation we are certainly aware that 'R represents X' for a certain subject S with a certain aim A. Concepts, terms, sentences, representations, and the like, however, have a circulation in science (and also in everyday language) only when they have attained a sufficient level of 'publicity' and, for this reason, can be used without any implicit reference to the individual subjects using them. For example, we can speak of the meaning of "triangle" not as the content of a single person's though, but as something that belongs to the concept (or term) considered in itself; similarly we can say that the referent of "Rome" is the city which is the capital of Italy not because someone is 'referring' to this city by using the term "Rome," but simply because this city is the standard referent of this term in English and French.

2.5 The Making of Scientific Objects: The Referential Side of Objectivity

In the preceding section our analysis of objectivity was performed from the general point of view of the theory of knowledge. It is true that, assuming this point of view, we have been able to indicate certain features of the notion of intersubjectivity which are of major interest in the specific domain of scientific methodology; but this has happened as a favourable and unforeseen *consequence* of results obtained within a much more general perspective.

Let us now consider, instead, the specific nature of *scientific* knowledge, and see what can be derived from the consideration of at least some of its inherent features. The task of identifying such features might, however, be rather complicated, since the differences between scientific knowledge and knowledge in general cannot be easily listed in a non-controversial way; and it might happen that, if we actually tried to compile such a list, we should need to include in it the features of testability and invariance which we have already treated in a different context. On the other hand, little help would be obtained through adopting the already-discussed assumption that science has dismissed the classical goal of providing a kind of knowledge which 'attempts the essence,' for this fact does not indicate any precise measure which ought to be implemented from an epistemological point of view in order to attain this goal.

Nevertheless, we are not left without any indication of the specific nature of scientific knowledge, for we can still consider one of the most remarkable features of science, one which correctly distinguishes it from common sense and everyday knowledge. This feature can be expressed as the fact that science does not make its statements generically but specifically; that is, no science has as its intended universe of discourse the whole of reality, or all possible worlds, but only a very restricted field of inquiry, a delimited *domain of objects*. However, every domain of objects, though being limited in scope, may contain a potentially infinite number of objects, as we shall see in the sequel. We have already stressed this fact earlier, but now we note that this feature is such an important one that it must be numbered among the few which prompted the birth of modern science at the hands of Galileo. In fact one of the basic points of view of the old 'natural philosophy' was that, in order to have reliable knowledge about some physical reality, one was obliged to rely on a general theory concerning the whole of nature. Galileo, on the other hand, stressed that such an enterprise must be considered intrinsically hopeless, whereas it is not hopeless to attempt to obtain certain sound results if we are content to study particular clearly circumscribed *aspects* of reality, without asking too much about what is before, behind or around them.[30]

The science of statics started from these premises and developed further into the more comprehensive science of mechanics, always preserving this character of limitation in scope. This remained paradoxically true even when mechanics became, at a certain moment, a kind of new philosophy of physical nature. For mechanics was not broadened to include every possible sort of object; rather, all the other aspects of natural reality were narrowed or reduced to mechanics.

But it is certainly not necessary to continue with examples, for the thesis that scientific research is always concerned with precisely limited domains of objects is generally admitted, and does not seem in need of any special defence. When a new science appears on the horizon, it is invariably because certain aspects of reality which were previously neglected, or were simply assembled with others in a wider domain of inquiry, suddenly become the specific objects of a specialised scientific

---

[30] We could summarise this feature by saying that modern science consciously presented itself as a programme of *piecemeal* knowledge, and this is again something which qualifies it as 'non-philosophical' knowledge. For philosophy has typically been a programme of *global* and general knowledge. This has to be understood not only in the above-mentioned sense, that an understanding of the general framework of nature was considered to be necessary for the understanding of particular entities or processes, but also in the sense that even one single entity or process was approached globally or 'as a whole' (since only in this way was its 'essence' supposed to be captured). The approach of modern science is the opposite, and this reflects itself in particular in the way science *explains* things and processes, such that they appear as *wholes* consisting of *parts*. According to the classical, philosophical view, the structure and behaviour of the parts was explained by considering the whole and its characteristic features (essence, form, internal and final causes); according to the new view, it is the structure and functioning of the whole which is explained as resulting from the behaviour and properties of its parts (in terms of efficient rather than final causes). This is the substance of the so-called *analytic method* that has become the backbone of all modern sciences (and which should be distinguished from the so-called analytic method of twentieth century Anglo-American philosophy). In the Chap. 10 of this book we shall be concerned with the question of whether or not global views or considerations can really be discarded from science, and shall see that they cannot. However, we shall arrive at that point only after having explored the features (and the merits) of the piecemeal approach on which scientific objectivity relies.

study.[31] This restriction of investigation to well-specified and clearly circumscribed domains of objects is so typical of the sciences that we might say without any arbitrariness that this is one correct way of qualifying scientific objectivity. In this case we could say that scientific statements are also objective in the sense that they concern only *particular* objects, and not reality 'in general.'

This admission does not in itself involve any particular ontological commitment, for it is capable of receiving a purely linguistic interpretation. For instance, one might express this condition by saying that scientific statements are always *relativised sentences*, meaning that they contain only a restricted list of technical terms, that their meanings are determined by the particular context in which they are embedded, that they obey certain established rules in order to be tested, and so on. However, working scientists would not feel fully satisfied with such a *purely* linguistic way of considering the statements of their science. They would certainly not contend that their statements are 'relativised,' and that this relativisation involves in particular several linguistic features of the kind mentioned. Rather, they would regard the existence of these features as a consequence of the fact that their statements *refer to* some specific *objects*, in an ontological sense of this word.

Of course, one could immediately say that the spontaneous and perhaps naive beliefs of working scientists by no means provide a justification or rational foundation for such an engaging *philosophical* thesis, and one might well recall that recent trends in the philosophy of language have not been particularly favourable with regard to a referential theory of meaning. It is usually considered naive and even completely unjustified to identify the meaning of a word with some (concrete or abstract) entity which the word is to designate. It is not our intention here, however, to embark upon a discussion of such a complicated issue in the philosophy of language. Let us simply say that the referent of a term cannot be *totally* excluded from the consideration of its meaning. If we do not admit this, we are led to the paradoxical conclusion that our language is being used to 'speak of nothing.' It follows that a minimal ontological basis must be preserved for every discourse; and scientists would certainly be particularly ready to admit this. This fact can be interpreted as an indication that every science is believed to have its own objects in *some* ontological sense (and here lies what may be termed the spontaneous realism of scientists). However, we shall say more on this issue in other parts of this work, particularly in Chap. 4.

Not as evident, on the other hand, is *how* such a referential basis can be provided for the individual sciences, a question we cannot avoid, since the very starting point of the present discussion has been that every science has its own specific objects. We are confronted, therefore, with the fundamental question: how can the objects of a science be given?

---

[31] The conviction that, in such a way—i.e., through a continuous extension of *scientific* criteria of objectification—*all* aspects of reality can be studied is the position of *scientism*, which we shall consider and criticise at the end of this work. At this point we are not entitled to exclude that criteria of objectivity different from those of science could be provided, though the question of giving examples of such criteria is too complex to be addressed here.

The answer to this question seems at first very easy, at least for the empirical sciences (which have the alleged luck not to be confronted with such intriguing questions as that of the nature of universals or abstract entities). In order to reveal the objects of a particular science, we simply have to single out a certain number of *things*, i.e. independent existents, and to declare the competence of this science to speak about them. So entomology is to speak about insects; zoology about animals in general; chemistry about elements, compounds, acids, and similar things, and so on.

How unsatisfactory this answer is becomes immediately apparent if we try to continue such a list. For instance, it would be very difficult to equip physics with its proper objects according to the above criterion since, in a way, every material body may be considered to belong to the subject-matter of physics, though not in all its respects, but only so far as some of its very general properties are concerned. However, if we were to follow this line it would turn out that not the material body, but some of its *properties*, are the objects of physics. But properties are not independent existents; in fact they are no less universal than the abstract entities of mathematics.[32]

But the inadequacy of the proposed criterion becomes still more apparent if we simply consider some 'thing' and ask what science is competent to deal with it. For instance, if we take a watch and ask what the area of its face is, we are considering it as an object of topology; if we ask what its mass is, or what the laws are that regulate the motion of its balance wheel, or what its influence would be on the magnetic field inside the room where it is located, we are considering it as an object of physics; if we ask what the composition of the alloy is out of which its case is made, or what the degree of purity is of the rubies that are inside it, we are considering it as an object of chemistry; if we ask its price relative to other watches and in relation to the present conditions of world watch production, we are considering it as an object of economics; if we ask whether wearing a watch of a certain kind might be an indication of its owner's having a certain sort of temperament, we are considering it as an object of psychology; or if our watch is rather old and we ask whether it once belonged to a certain prime minister whose biography we are writing, we are considering it as an historical object.

Here we shall introduce one technical notion and further clarify another, both of which are of great importance for the present work. As regards the preceding paragraph, we should say that each of the questions posed there is the expression of a particular *point of view* on one and the same independent existent, and that each such point of view makes of that existent a particular *object*.

---

[32] Paradoxically, this must be said even of those sensory properties that correspond to the 'secondary qualities' of things. For example, when I perceive (knowledge by acquaintance) the red colour of a cherry, I perceive *this* individual red, but when I say "this cherry is red" (propositional knowledge), I use an abstract notion of red of which the particular red of the cherry is an instantiation and, thanks to this fact, I can communicate with other people who do not share my knowledge by acquaintance.

Such points of view may be hierarchically ordered according to their degree of generality. The most general *categories* determine the domain of objects of different sciences (as in our example), but if we assume different *attributes* within the same category, we determine the objects of different theories within the same science. So, for example, considering reality from the point of view of matter, motion and force determines the objects of mechanics rather than those of biology (which considers reality from the point of view, let us say, of metabolism and reproduction). But then additional different viewpoints may be used in mechanics to study the objects of mechanics, or in biology to study the objects of biology, and this fact in turn implies the introduction of new, more specific concepts, at least some of which need to be operationalised. Let it suffice to say, then, that we have here illustrated that one and the same 'thing' can become the object of a new and different science every time a new specific point of view or viewpoint is taken of it.

Two results follow immediately from the above considerations. First, no object of a science is ever simply a thing in the everyday sense of this word; and second, every independent existent not only has the potential to constitute a variety of objects, but an *unlimited* variety of objects, since the number of objects can always be increased simply by taking new viewpoints on the 'thing' in question.

But now, what are the objects proper? Having excluded their being simply 'things,' it might seem, from our previous analysis of the watch example, that they may be *viewpoints*, but this is certainly not the case. For science, having earlier been characterised by its constant effort to attain objectivity, would now turn out to have as its subject-matter such subjectively flavoured entities as viewpoints. The situation need not be so peculiar, however, since what is meant by "viewpoint" here is not some sort of personal appreciation which obeys individual idiosyncrasies, but a particular 'way of conceiving of reality' and, if we had used such a stern locution in the above, it would not have aroused such an impression of subjectivism.

Still, the problem is to determine whether science can be identified as a form of investigation which has these points of view as the objects of its research, and one's answer is instinctively (and correctly) negative, since what any particular science restricts its interest to are certain *aspects* or *features* of reality (we shall call them *attributes*), that can be found (or not found) *in* individual '*things*.' It is in this *restricted interest* that a viewpoint consists; and the objects of a science are, therefore, made up of those attributes of reality that are of interest to the science. The problem, then, is to understand *how* an empirical science can single out in any 'thing' the attributes of reality that are of interest to it. In other words, how does a science practically determine the presence of its *intended* attributes (those of interest to it), i.e. *refer* to them in a way that is different from simple common-sense apprehension and from pseudo-science?

A first step may be taken towards answering this question by noting that each science can be characterised by its proposing and defending a certain system of statements. We do not maintain that science is *only* this, and we are open to the idea of considering science to be many other things as well, such as a social phenomenon involving many personal, social and historically determined commitments. Still, it is incontestable that one of the chief results of such an activity is that of producing a body of organised statements that are intended to make manifest the content of the *knowledge* that has been gained within that activity.[33] Moreover, we have already stressed that every science is characterised by its statements being *relativised*, so that each science becomes a *system* of relativised statements. Here we add that each science is actually intended not simply to be a system of statements, but more particularly to be a system of *propositions* (or of propositional functions) each of which *aims* at being true, either immediately, or after some suitable processing.[34]

On the basis of these remarks, our problem loses any flavour of a psychological nature (which seemed to be involved in the conception of viewpoints as determining scientific objects) and admits—at least at an initial stage—of an intrinsically linguistic treatment (i.e. a treatment concerning statements and the possible conditions of their truth). As a consequence, our first concern will be the following: how can we decide whether a certain statement belongs to a given science? The easiest way to handle this question is probably, again, that of considering an example.

Let us suppose that Mr. X is sitting in a room and says, "It is very warm here." We ask now whether Mr. X's statement does or does not belong to physics. From a certain point of view, one feels inclined to say that it does, for it refers to heat, and heat is one of the main objects of thermodynamics. From another point of view, however, one must deny that this statement belongs to physics, because physics provides us with no means for deciding whether it is true or false, and so the statement cannot be considered as expressing a proposition or a propositional function of physics. In fact, even if we correct the indeterminacy involved in the use of the free variable "here," by indicating instead the exact spatio-temporal location, the situation concerning truth or falsity would not change.

But why can we not say in physics whether this statement is true or false? Someone might be tempted to say that the reason is that science cannot accept as

---

[33] With this very general claim we do not automatically subscribe to what has been termed the *statement view* of scientific theories, nor to the thesis that science simply expresses knowledge. As we shall see later, these doctrines are partially correct, but are not adequate to cover all aspects even of the cognitive side of science. We also leave undetermined, at this point, the different kinds of statement which enter into a science (equations, hypotheses, laws and so on). It will be the task of the Chap. 10 of this book to pay due attention to those aspects of science that are not reducible to its providing a system of statements, and that even profoundly determine the way this system of statements comes about.

[34] This assertion does not enjoy general acceptance in present-day philosophy of science. However we feel entitled to make it here since we leave open, at this point, what ought to be intended by *true* scientific sentences. For our present purposes it is sufficient to recognise that no science exists where there is no *aim* to discriminate between those sentences which are admissible and those which are not. The intention of providing an organised system of *admissible* sentences is the same aim as that to which we are referring here, i.e. the aim of obtaining *true* sentences. A fuller discussion of the problem of scientific truth will be presented in what follows, especially in Chap. 8.

evidence individuals' (subjective) expressions of their own feelings or perceptions. But this answer is wrong. In fact a physician (i.e., a medical doctor) might take Mr. X's statement very seriously, as a real 'datum,' and try to infer from it some initial opinion about Mr. X's state of health. This means that such a statement is not devoid of *any* scientific value in itself, but that it is simply devoid of meaning for physics, while it has meaning for medicine.

The reason can now be easily given. If Mr. X had said, "The temperature here is 40 °C," his statement would have been accepted as physically meaningful, for physics admits of a certain number of criteria for stating the immediate truth or falsity of its statements, and among such criteria one finds the results obtained from the use of thermometers, but not the expressions of personal experiences of heat. The second formulation of Mr. X's statement is such as to be testable by using a measuring apparatus, whereas the first is not, and for that reason one of them belongs to physics and the other does not, although with regard to common sense they have nearly the same meaning.

This example has taken us very near to our point. The fact that a given statement can or cannot belong to a certain science depends on the criteria explicitly (or sometimes perhaps only implicitly) admitted by that science for testing the truth of its propositions.

This line of thought allows us to give a more exact explication of the rather vague concept of viewpoint which, as a kind of provisional notion, we adopted earlier when we said that every science is characterised by a certain viewpoint from which it considers reality. We can now state more exactly that every science is characterised by a certain set of specific *criteria* which are adopted in order to establish the immediate truth or falsity of its propositions (these criteria being dependent on the adopted viewpoint). This being the case does not imply that different sciences cannot in some contexts avail themselves of the same criteria, and it would not prevent particular sciences from possibly translating sentences of other sciences into those of their own. However, it is better, for the moment, not to take such very sensible exceptions into account and, instead, to consider the above-mentioned criteria as determining clear-cut distinctions between sciences so that, for example, a certain proposition reveals itself as belonging to medicine if it is formulated in a certain way, or to physics if it is formulated in another way, as in the example discussed above.

A closer scrutiny must now be made of the notion of immediate truth involved in our previous statements. It is intended to provide a more exact formulation of the intuitive idea of a *datum*, which is basic in every scientific epistemology. According to the view we are considering now, a science is regarded as containing a collection of propositions, while data (in the most usual sense of this concept that we shall adopt here, i.e., understood as sense-data) are not usually conceived of as propositions but as the contents of immediate knowledge. A rather obvious feature which characterises the notion of sense-data, however, is that *propositions* describing such data are immediately true, that is, true without need of any further justification, while other sentences in science, such as hypotheses, are not supposed to be immediately true, but to receive confirmation by a logical procedure connecting them to the data. Now, using a locution reminiscent of one that was rather common some decades ago, we may call those propositions which describe data "protocol propositions." As a consequence, we may call *criteria of protocollarity* those specific criteria which, within a certain science, permit the determination of which propositions are immediately true, that is, the determination of the science's protocol propositions. Hence, every science is in principle characterised by its own criteria of protocollarity.[35]

It appears that the shifting of the problem concerning the determination of the specific objects of a science in the direction of an analysis of the linguistic structure of science has in fact enabled us to remove the vagueness implicit in the idea of a viewpoint, since we can now say that every science selects its own criteria of protocollarity in order to fix its own immediately true propositions concerning reality. On the other hand, this transition towards an analysis of the linguistic structure of science is possible because we recognise that these criteria of protocollarity are suggested by the specific viewpoints of a given science. For example, if mechanics consists in a study of nature from the point of view of matter and motion only, and these intuitive notions are refined through the concepts of mass, length and duration, we are led to look for a criterion sufficient for establishing whether the proposition "the body A has a mass greater than that of the body B" is true or not. More than one criterion might come to mind: for instance, following a certain intuitive impression, we might believe that the right

---

[35] This claim is not affected by the 'theory-ladenness' of data, which we shall discuss in the sequel, since this condition would simply indicate the interconnection of this truth with the rest of the theory. Let us also remark that we are avoiding for the moment any technical standardisation of our language so that, for instance, we use "sentence," "statement" and "proposition" as synonymous, since they are often used as such in philosophical contexts, with the exception of the philosophy of language, where they receive a conventional technical diversification. The moment will come when we shall make use of such technical refinements, but we prefer to wait until we really need them. On that occasion we shall also call "state of affairs" (again in a technical sense) that which we here call a "datum."

We want also to stress that our use of the expression "protocol sentences" is only externally reminiscent of the same expression as used in the famous dispute over protocol sentences that took place in the Vienna Circle in the early nineteen-thirties. Indeed, protocol sentences were advocated by Carnap as the basis of scientific constructions, and were meant by him to be the report of individual mental phenomena. Because of this they were challenged especially by Neurath, who opposed to them "physicalistic sentences" (i.e. sentences formulated in the language of the physical sciences, referring to spatio-temporal features, and for that reason capable of overcoming the privacy of the subject). Protocol sentences were therefore essentially subjective (and indeed they reflected Carnap's initial 'methodological solipsism'), and are therefore very different from what we mean by the term here, where they are to provide the basis for *intersubjective* agreement. This they can do because they are not based on private perceptions, but on the performance of intersubjective operations. The criteria of protocollarity are in a way closer to the 'physicalistic' criteria, but even that is not true, since we also admit non-physical operations (as will be clear in the sequel). In other words, we are using the notion of protocol in a sense that is very close to that which scientists usually adopt when they simply mean a protocol to be a sentence strictly reporting the description of a datum, and we shall try to make this idea more precise and to derive from it some useful analytic features.

criterion of comparison be that of comparing the respective volumes of these bodies (such that, e.g., a cube of cork of 2 cm^3 would have a mass greater than that of a cube of iron of 1 cm^3). This, however, is not the choice made in mechanics; for (good) reasons that we shall not explore here, the criterion for comparing the mass of different bodies consists in putting them on the two plates of a balance and to attribute greater mass to the one whose plate sinks (through additional refinements this *operation* of comparison can be standardised such as to become an operation of *measurement* that will make of mass a *magnitude* capable of being assigned to a single body). This, as we shall see later in detail, happens because the protocollarity criteria are strictly bound to a privileged set of predicates (of which "mass" is an example) which enter the propositions expressing data.

However, before showing this, let us note how well the proposed solution works in the context of the problem of attributing single sentences to different sciences (under the 'idealised' assumption that these sciences have made their criteria of protocollarity suitably explicit, and that the context of the discourse avoids overlappings). If, in order to attribute to a sentence an immediate truth-value, we resort to using a balance, a chronometer and a metre stick, we can say that it belongs to classical mechanics; if we need to use a thermometer, it belongs (at least primarily) to the theory of heat; if we must use reagents, it belongs to chemistry; if we must consult documents in a general sense, it belongs to historiography; if we have to compare different kinds of texts, it may belong to philology; if we use some standard procedures known as psychological tests, it belongs to psychology, and so on.[36] It is clear, therefore, that thanks to the existence of these criteria of protocollarity, we can solve the problem of recognising which 'relativised sentences' belong to a certain science, for such criteria are at the same time criteria of relativisation.

One could note, however, that our solution is only partial, since it can only be used for sentences which are 'immediately testable' by means of some admitted criteria of protocollarity. What is to be done in the case of sentences which are correctly assigned to a science (such as physics) but which are not immediately testable by means of the testing methods applied in that science? A full answer to this question requires further preparation, which we shall provide in the sequel. But for the moment we can say that an expression belongs to a certain science as long as it is possible to accept or reject it, either directly or indirectly, on the basis of the protocollarity criteria which are admitted in that particular science. The two adverbs, "directly" and "indirectly," indicate the two possible conditions under

[36] These examples indicate that the notion of *datum* is actually broader than that of sense-datum we have referred to above for the sake of simplicity (indeed, it is common to speak of "historical data," "sociological data" and so on). In the case of the sciences, data are not constituted by simple perceptions, but, being the outcome of *operations*, presuppose a certain intellectual elaboration in order to relate these operations to the ascertaining of those specific attributes that are investigated by a given science, and must also conform to certain ontological presuppositions that underlie the conceptual framework of that science. Why this is the case will be explained later.

which a proposition can belong to an (empirical) science. Either it expresses a 'datum' (in which case it is immediately testable by means of the protocollarity criteria), or it contains, at least in part, some components which are not immediately testable. In this case, what is needed is that from this proposition some explicit links may be indicated which connect it (this connection being again typical of the science involved) with some immediately testable sentences. In this case we can say that the proposition in question has been indirectly tested on the basis of the protocollarity criteria.[37]

Because of the central role played by the protocollarity criteria, we can say that they 'make' the scientific object in the sense that *an object of a certain science is simply an aspect of reality capable of being described by propositions that can be directly or indirectly assigned a truth-value by means of the criteria of protocollarity of that science*. Because of this central role we can dismiss, from now on, the rather baroque expression "criteria of protocollarity," and substitute for it the clearer expression "criteria of objectivity."

Let us now reconsider the fact that every science admits of some standard criteria for obtaining its protocol propositions, that is, for obtaining the recordings of its data. This is obviously possible because these criteria are related to certain concepts which express properties, relations or functions in the broadest sense of these terms, and which we shall call *predicates* for brevity. These predicates are predicated of a certain 'thing,' and the role of the criteria discussed above is simply that of establishing whether this predication gives rise to a true or a false sentence. For instance, in classical mechanics we use predicates such as 'mass,' 'length' and 'duration'; and the use of a balance, a meter stick, and a chronometer are the standard procedures admitted for testing the truth of at least some sentences involving these predicates. The same kind of consideration can be repeated, with different degrees of effectiveness and explicitness, in the case of other sciences as well. It is because of this immediate and privileged link with the objectivity criteria that we must single out this kind of predicate and give them a special position. We shall call them the *basic predicates* of a certain science. They deserve this appellation because, as we have seen, all the sentences belonging to a given science must either be entirely constructed by means of them, or be explicitly bound to sentences which are so constructed.

We are now ready for the last step. Our previous point was that a scientific object is a 'thing' conceived from a particular point of view, the general nature of the object being determined by means of the criteria of objectivity of the science in question. Thus the adoption of a given set of such criteria '*clips out*' some particular object, while the adoption of a different set of criteria 'clips out' a different object, both from one and the same individual 'thing.' We can leave aside this metaphor of 'clipping out,' and express the matter in a linguistic form. Thus we

---

[37] This statement will be clarified in the sequel when we come to consider the positions of the various concepts in a theory, the distinction between operational and theoretical concepts, the function of models, and other related topics.

should say that an object of a given science contains only (and all) the aspects of a 'thing' which may be characterised by the basic predicates of that science. In this sense (i.e., from a *purely linguistic* point of view) *a scientific object is nothing other than a bunch of predicates.* Many philosophers of science sharing the 'linguistic turn' (that reduces any philosophical investigation on science to an analysis of the language of science) would probably feel happy with this conclusion. We shall see, however, that an object is by no means a purely linguistic entity, since the basic predicates which constitute it must be equipped with operations capable of providing the connection of the object with a *reference* (a notion that a linguistic analysis cannot ignore). This is why in our final formulation (in Sect. 2.7), where we go beyond the purely linguistic level of analysis, we shall come to speak of an object as a structured set of *attributes* rather than of *predicates*, predicates being only the linguistic tools for denoting the ontological attributes of reality.

2.6 The Operational Nature of the Basic Predicates

Let us now underscore a feature which may be the most decisive with regard to our criteria of objectivity. It is the fact that these criteria are necessarily *operational* in character. This is not surprising, after what we have said about it not being possible to establish intersubjectivity unless the circle of private sensations and perceptions is broken by means of operations. In addition, it is commonly admitted that nothing can be more intersubjective in a given science than its *data*. Now, let us ask how it is possible to regard as a datum, for example, that a certain board has a length of 2 m $\pm$ e (e being the margin of error). This means that *anyone* using a metre-stick of a prescribed kind, and placing it along the board in a standard way, must find that the said value is the length of the board. Similarly, if we say that a certain material body has a mass of 5 g $\pm$ e, we mean that anyone employing a balance of a specified type, must arrive at this value. As one sees, an operation (and especially all mensural operations that are typical of physics) always involves an instrument as well as precise instructions concerning its employment. Both must be given in order for the operation to be performed, and both must be understood in the same way by *everyone* wanting to know the datum. *At this stage* the instrument and the way of employing it must be taken as something *given*, as things of everyday experience, as non-analysed primitive entities. This implies, in particular, that the complexity of the instrument cannot be questioned at this stage and, hence, that even instruments much more complicated than a metre-stick must be accepted as being involved in the performance of *primitive* operations.

In this regard we may note, for instance, that modern astronomy, as a discipline distinct from ancient astronomy, is characterised by the fact that the images revealed through the telescope are accepted as data. This does not mean that the use of the telescope was (or is) in itself unquestionable, but as long as this use is actually questioned such an astronomy cannot begin. Indeed, modern astronomy could only begin with Galileo, when the employment of this instrument became accepted without question *within* this science. The same can be said with regard to the microscope. Microbiology is characterised, with respect to previously existent macrobiology, by its treating as data the results of observations made using this instrument. Although a microscope is a rather sophisticated instrument, it must be considered as something primitive *within this science*, leaving the possibility of discussing it at length, for instance, to optics.[38]

It is now obvious that the use of very sophisticated instruments in electromagnetics, nuclear physics, astrophysics and such does not present exceptions or difficulties from our point of view, according to which every science determines its data by resorting to operational criteria by means of which protocol statements can be established. This fact will later suggest many considerations concerning the historical determinateness and the collective nature of scientific research, but we must dwell on it somewhat longer at this point in order to clarify the idea of operation within a *scientific* context.

It might seem that we have made things too easy for ourselves in our presentation of the nature of objectivity in science, and some readers may not feel prepared to admit so promptly that the responses of complicated instruments must be accepted as data without question. It is indeed well known that certain scholars take the sophistication of modern scientific instruments as evidence of the impossibility of distinguishing between observational and theoretical concepts in

---

[38] The two examples mentioned here could be further expanded through an historical analysis. A good account of the conceptual, philosophical, and scientific difficulties which Galileo had to overcome to gain the acceptance of the telescope as a reliable instrument for observations may be found in Ronchi (1959). The fact of having consciously based the investigation of nature upon the use of *instruments* is therefore an additional capital mark which distinguishes modern science from philosophy, a mark which, once again, must be credited to Galileo, and which we did not mention earlier because its importance can be adequately estimated only after what we have said concerning the operational basis of science.

The extent to which scientific objectivity also depends on instruments in certain 'negative' respects may be clear from the second example. Indeed, historians of medicine and biology have sometimes been puzzled by the fact that several pictures, printed in eighteenth century books, showed certain bizarre details in the description of tissues and organs. This was not due to a lack of accuracy or some hidden dogmatic prejudice on the part of the scholars of that time, but simply to the fact that they were using microscopes with non-achromatic lenses, which made certain images or details of images 'apparent' which were later recognised to be aberrations, and which were removed with the invention of the achromatic microscope. This example can tell us many things. In the first place it shows the 'historical determinateness' of scientific objectivity (a feature to be discussed later); second, it shows that data are strictly 'instrument-dependent' and can actually change profoundly when different instruments are available; third, that in spite of this a 'correction' of the data is never possible by comparing them with the 'thing,' but only by resorting to new (instrumentally given) data; fourth, the 'unreliability' of certain data may hardly be discovered from 'within' the discipline in which they occur, but requires an external source of criticism. All this amounts to recognizing that data can be mistaken (which means that protocol statements can also be mistaken), as we have already noted. We leave other considerations aside and simply add that in order to do history of science correctly it is highly recommendable that the historian repeat observations and experiments using the instruments which were used during the epoch he is studying.

science while, according to our position, even the charge of an electron should be considered an observational (or better, according to our terminology, an *operational*) concept, if we had at our disposal an instrument designed to provide this value in a *direct* way, and if such an instrument were employed in the determination of the 'initial' operations.

This psychological difficulty notwithstanding, we claim that these are really data, and we can explain this rather easily. In the first place, the practice of scientific research confirms this interpretation. As science develops, its instruments become more complicated, but this does not prevent scientists from considering them as capable of providing *data* in a proper sense. What should they otherwise provide? The recording of an instrument is always a datum. Such a datum may be in need of a complicated interpretation (and the purpose of science is constantly that of interpreting data), but such an interpretation cannot help but accept these recordings as its starting points, as evidence which must be considered as *given*. This, of course, does not exclude that we can *question* the data if, for example, they are inconsistent with previous data or are in conflict with well-established theoretical assumptions. This, however, does not amount to a *rejection* of the 'questioned' data: the effort needed is that of understanding and explaining why they could occur. Sometimes we might discover that the operations were not correctly performed, that the instruments were imperfect, that certain unnoticed perturbing circumstances were at work, and so on. In such cases the 'aberrant' datum will be isolated and put aside. This issue will become much clearer when we come to see that, at least in the natural sciences, not individual data, but *regularities* are the matter investigated.

Furthermore, these are also data in the most intuitive and even naive sense of the word. They show themselves to the observer without asking of her any particular effort of mental processing, simply as the result of the correct execution of certain operational instructions. The only difference between the operations that one must employ in order to obtain an objective agreement about the use of the concept *red* (as in our previously discussed example), and the operations required in order to establish objectively the applicability of the notion *electric current*, is that in the first case she presupposes (as we have seen) that her interlocutor already agrees about the use of the notions *pencil* and *selecting from a bundle*, while in the second case her interlocutor must agree about the criteria for the applicability of the notion *ammeter*, as well as about the way of performing certain manipulations with this instrument. But this difference is only due to the fact that the first establishment of agreement has nothing to do with the science of electricity, while the second does; or, to put it differently, the first interlocutor is supposed to be just a person in the street, while the second is supposed to be someone trained in science. This means that, in order to be 'admitted into the discourse' of the science of electricity, one *must* know what an ammeter (or some equivalent instrument) is, just as one must know what, for example, a pencil (or some equivalent material thing) is, in order to 'enter into an (everyday) discussion' concerning the colour red.

It is patent from what we have said that the central thesis of operationalism, according to which scientific concepts are strictly bound to operations so that their meaning depends on them, is here accepted to a considerable extent. The large number of examples that Bridgman has presented in his book *The Logic of Modern Physics* as well as other papers constitutes a precise and well-known illustration of this position. By saying this, however, we do not intend to accept *all* the consequences that Bridgman and other operationalists have drawn from this central idea, and we shall have the opportunity to discuss certain of them later.[39] Yet it seems that there is a sense in which everyone involved in experimental science must accept being an operationalist. It is the sense according to which operations enter in a prominent way into the construction of the scientific object itself. We are going to see that the structure of the object is much more complicated than the simple bunch of determinations which can be operationally uncovered, precisely because it is a structure and not simply a bunch. But the basic truth must not be overlooked that, whenever we need to indicate the kind of object we are speaking about, or within which science or theory a certain statement is formulated, the most appropriate way we have for answering this question is to trace the concepts involved in this statement back to concepts which are *operationally defined*, that is, which are bound to particular criteria of objectivity that specify their *domain of immediate reference*.

Before going any further, it may be advisable to meet a couple of objections which are sometimes addressed to operationalism, and which might also concern what has been maintained here to the extent that it shares certain features of the operationalist epistemology.

The first objection is that the idea of tracing every scientific concept back to operations is patently contradicted by actual science, which is full of concepts whose definitions are almost completely theoretical and which, in any case, have no direct link to operations. Operationalists believe themselves to have escaped this objection by introducing the idea of 'pencil and paper' operations, claiming that every scientific concept is definable at least through operations with pencil and paper. This idea does not seem a very happy one, as we shall see later, and we believe that operationalists are mistaken in pretending that *every* scientific concept must be operationally defined, while only some of them (and perhaps only a few of them) are such as to require an operational definition. After this clarification, we can say that, when we too affirm that the concepts of a certain science must be, in a certain sense, traced back to their operational basis, we do not mean that this tracing back has something to do with reduction, or with a more or less liberalised

---

[39] In particular we shall see that the 'dependence' of meanings on operations must be understood in a careful 'intensional' way, which allows for the possibility that concepts having different meanings may be related to the same operations, while concepts related to different operations must necessarily also have different meanings. The reason for this is that operations are decisive and determinant with respect to the *reference* of concepts, and affect their *meaning* only as far as the reference has intensionally to do with the meaning, that is, only if it is considered to be a part of it.

idea of definition, but, actually, that there must be the possibility of a logical *analysis* by means of which one can explicitly see how non-operational concepts can be connected with operational ones. More details about this logical analysis will be advanced in the sequel, but let us state that it is just a question of an analysis, which does not mean the possibility of discovering these links by following, for example, the historical development of a discipline, or even its usual systematic exposition in a textbook. To give an idea of what we are referring to taken from a familiar field, we mention the example of the ordering properties of the points on a straight line, which were implicitly understood by traditional geometry already in Euclid's *Elements*, but were first singled out by Pasch only in the nineteenth century. These were discovered as the result of a logical analysis of what was really implied by the propositions of traditional geometry, although no historical recognition or careful study of the textbooks of this discipline would have revealed them.

The second objection has to be handled more delicately, for it charges operationalism with a methodological mistake, pointing out that operations can be envisaged as useful for testing or verifying a *statement*, while operationalists conceive of them as capable of determining the meaning of a *concept*. According to this criticism, a confusion between meaning and testability is hidden here. Meaning is something which is pertinent to a concept, while testability does not have to do with concepts, but with statements, and is posterior to the institution of meaning. One can admit that in Bridgman's and other operationalists' declarations such a confusion is sometimes to be found; but, on the other hand, one should always try to see whether such weak points are necessarily included in what people intend, or whether they are simply a consequence of the fact that an idea is not very carefully expressed. Undoubtedly, in our case the second alternative is true. In fact, if one thinks of the usual conception of the meaning of a scientific term, one can easily discover that it is taken to represent a certain set of qualities or properties, or (to express it more technically) it is taken to be an 'intension' in which a certain number of features are, so to speak, summarised.

Now, sometimes some of these features can be attached to operational procedures when testing propositions in which the concept in question occurs, while some other features cannot. The operationalists' proposal can safely be understood as a prescription not to take into consideration, when the statement is tested, those components of the intension of a scientific concept which cannot undergo operational manipulation. If one restates the operational point of view in this manner, no methodological incorrectness remains, since everything now appears properly considered on the level of meaning. Testability comes into consideration only in order to privilege certain of the components of the meaning. As we already noted, the disputable point is whether such a procedure is to be advocated for *every* scientific concept, and we have already said that this does not seem to be the case. Yet we can admit that, for those concepts which have to play a foundational role for the objects of a certain science, this prescription seems sound; and we should not be diffident even towards the expression "operational definition" which is frequently employed for them. If definition may be conceived of in general as a procedure for explicitly fixing the meaning of a term (and in such a way we accept an 'analogical' sense for it, not requiring it to be restricted to a linguistic procedure), speaking of operational definitions of operational terms may be accepted.

Let us stress here an interesting consequence of what has been said thus far (though it will be the object of a more detailed analysis later). The birth of a science (or of a subdomain of a science, or sometimes also of a new theory in a given science) appears as something 'contingent,' in the sense that there is no intrinsic necessity for it to occur. It is a *historical* event, that is, something that happens when a number of persons come to agree about the use of certain instruments with which they are or become sufficiently familiar, and which they employ in the same manner. Such a fact could even be conceived of as being conventional; and it is conventional, if only to a certain extent. But much more will be said concerning this aspect of the birth of sciences when we speak of the historical dimension of science.

The history of science clearly shows that this is really how things are. Modern astronomy and modern microbiology, as has been stressed, could only begin when a sufficient number of people had agreed to investigate nature using particular instruments; and the same could be said of scientific psychology, economics, and so on. This fact helps us to appreciate a statement which risks being misunderstood when it is received without preparation. It is that, in a science, data too are conventional.[40] Certain people find this declaration very puzzling, for data seem

---

[40] This flavour of conventionality concerning data may also be found, at least to some extent, in Popper's characterisation of his 'basic statements' (which play the role of data in his philosophy of science): "It is fairly easy to see that we arrive in this way at a procedure according to which we stop only at a kind of statement that is especially easy to test. For it means that we are stopping at statements about whose acceptance or rejection the various investigators are likely to reach agreement" (Popper 1959, p. 104). However, his *Logic of Scientific Discovery* (and in particular Sect. 5.4, 'The Relativity of Basic Statements') shows rather clearly that what is involved is not conventionalism in a strict sense, but rather a reference to that intersubjective agreement among specialists in the field which we too have advocated in the present section. What is missing in Popper is an indication of the elements which can make this agreement objective and reasonable, rather than dependent on individual judgements, which could lead to a real conventionalism. In general one must recognise that Popper lays great stress on the intersubjective nature of tests, and because of this he rejects the scientific relevance of the evidence provided by personal observation (see his criticism of "'our own' observational experience" in Popper 1963, p. 267).

On the other hand, one cannot deny that on other occasions he stresses certain affinities of his doctrine with conventionalism: "From a logical point of view, testing of a theory depends upon basic statements whose acceptance or rejection, in turn, depends upon our *decisions*. Thus it is *decisions* which settle the fate of theories. To this extent my answer to the question, 'how do we select a theory?' resembles that given by the conventionalist; and like him I say that this choice is in part determined by considerations of utility. But in spite of this, there is a vast difference between my views and his. For I hold that what characterises the empirical method is just this: that the convention or decision does not immediately determine our acceptance of *universal* statements, but that, on the contrary, it enters into our acceptance of the *singular* statements—that is the basic statements" (Popper 1972, pp. 108–109). Hence it is not altogether incorrect to say that Popper advocates a view of science according to which there is a certain 'conventional

necessarily not to be open to acceptance or rejection in science (*qua* data and not *qua* correct; as we have seen, their correctness can of course be questioned, but this does not 'eliminate' them), perhaps to be the only things in science which cannot be subject to conventions.

But now we can see at least one sense in which conventions are actually unavoidable here. To accept something as a datum depends on the criteria of objectivity which are admitted by a certain community of researchers—it depends on the kind of operations which have been selected for 'clipping out' the objects of a certain science. On the other hand, no arbitrariness of any kind is involved in this fact. Once the choice of the operational criteria of objectivity is made, what is found by applying these criteria *must* be considered as a datum. It is somewhat like the situation determined by one's preferring to look at a panorama from one window of a room rather than from another. This choice is surely conventional, as nobody could say that the panorama looks false if seen from any particular window. But this conventionality does not imply any *conventionalism* because, after having decided to look from a certain window, everything seen from it must be accepted as a datum, in the common-sense meaning of this term. The arbitrariness would occur if one denied the correctness of the other 'viewpoints'; but this is not the case in scientific practice.

Much more will be said on this point when we come to interpret this 'contingency,' or weak conventionality, as an historical determinateness. Let us also note that at this stage of our analysis we neglect the distinction between the existence of different sciences and the existence of different theories belonging to one given science, since the questions we are tackling here are still general enough to apply indifferently to both cases.

It is not without interest to compare the role and the task of operations as they appear in the treatment of intersubjectivity and as they appear in the determination of the specific objects of single sciences. In the first context, operations play the role of concrete procedures by means of which different subjects can reach an agreement about the applicability of certain notions. In the second context, they play the role of conditions to be followed in order to introduce basic predicates and, in such a way, to construct the specific objects of a given science. Clearly, these two functions differ. Yet they have a deep affinity. Indeed, we stressed that the criteria of objectivity are operational inasmuch as they are given through the indication of instruments and of *prescriptions* for using them. Moreover, the sense of these prescriptions, as we noted, was that *every* operator able to follow them correctly in certain conditions must obtain the same results, results which are the outcome of the operations constituting the definition of the predicate involved when *testing* the claim of any other operator concerning a given datum. This fact

(Footnote 40 continued)
component' in the data. (See also Popper 1963, pp. 278–279.) This conventionalist element depends directly on the fact that epistemological dualism is present in Popper and prevents him from appreciating the importance of the operational dimension in order to solve the problem of the empirical base (see, e.g., Buzzoni 1982, Chap. 2).

clearly indicates that the old conditions of universality, invariance and independence of the subject, which were the most typical marks of objectivity understood as intersubjectivity, are also included in the meaning of objectivity when understood as 'reference to specific scientific objects.'

This is only one aspect of the more general fact that, although the two different characterisations of objectivity follow independent lines, they turn out to be fully equivalent or interchangeable. This is so because the *same* operations, by means of which the objects of a given science are 'clipped out' of reality, are also those by means of which it is possible to reach that intersubjective agreement which is needed for the scientific treatment of these objects. Such operations offer, in this way, the foundation both of the epistemological and of the referential and 'ontological' side of scientific objectivity, since we can claim that *the conditions according to which the objects of a science are given are at the same time the conditions for knowing them objectively.*

The reader is likely to find a significant resemblance between the above statement and Kant's celebrated claim, "The conditions of the possibility of experience in general are at the same time conditions of the possibility of the objects of experience themselves, and thus possess objective validity in a synthetic judgement a priori."[41] Yet some differences with respect to Kant must also be stressed. First, the operations are, according to our conception, conditions which belong to the particular structure of scientific knowledge, and do not, as is the case with Kant's a priori, belong to the structure of our understanding in general. Second, in the case of Kant, the dualistic presupposition remains fully active and expresses itself through the famous "distinction of all objects in general into phenomena and noumena." For us, on the contrary, objects are *part of reality* (i.e., that part which has been 'objectified' through the operations), and are not something 'behind' which or 'under' which reality remains hidden, as in the case of Kant's noumena. What is not included within a certain objectification is by no means an unknowable, but simply something which has not been taken into consideration in that objectification, but which may enter some further objectification.

We shall develop this point later, when we treat more specifically the problem of the ontological status of scientific objects. For now, let us devote our attention to a deepening of the notion of the structure of a scientific object, of which we have thus far indicated only some initial features. Let us only mention, before entering upon this analysis, that the way of characterising scientific objects we have outlined here will provide us with a useful perspective when we come to interpret scientific change. We shall see, actually, that in several cases new disciplines, or new theories within the same discipline, may be interpreted as investigations of new objects, which depend on new 'viewpoints' on reality, and which are themselves equipped with suitable criteria of objectivity.

---

[41] Immanuel Kant, *Critique of Pure Reason*, A 158–159, B 197–198.

The clarifications given in this section provide a clear basis for a distinction which has too often been neglected in the philosophy of science inspired by logical empiricism, that is, the distinction between laws, hypotheses and theories, which is important especially in physics. We shall be specifically concerned with this distinction in Chap. 7, and shall offer some preliminary indications on this topic in the Sect. 2.7.[42]

2.7 The Role of Theory in the Making of Scientific Objects: The Object as a Structured Set of Attributes

2.7.1 The Scientific Object as an Intellectual Construction

What we have stated in the preceding section could easily be interpreted as the expression of an empirically-minded approach to the problem of objectivity in science. As a matter of fact, it represents the correct admission of the undeniably empirical aspects of that objectivity and, moreover, it would be strange if *empirical* science had little to do with *experience*. But also another side of science must now be investigated, which will show us how experience alone is insufficient for the construction of scientific objects (including experience extended in its operational dimension).

In order to open the way to this complementary discourse, more than one possibility is at hand. We select the one implicit in the thesis which has directed us from the beginning of our investigation, namely the fundamental identification of objectivity with intersubjectivity. We have already noted that what can be shared by a community of subjects is certainly not their 'experienced' knowledge of things, that is, their awareness of the various features that reality shows to each individual observer. It follows that, sensory qualities of things being private, they are not expected to constitute the content of intersubjective, or objective, knowledge.

Such a conclusion might sound rather strange, especially if applied to the 'empirical' sciences which seem completely immersed in the consideration of material things which reveal themselves through the testimony of the senses. But, notwithstanding this, we must admit that the actual situation is at variance with such an intuitive picture. Indeed, philosophers should be ready to find such a conclusion acceptable, and even familiar, for in the history of philosophy universality has always had to pay the specific price of not being related to sensations. The only possible way of avoiding this detachment from sensations is to declare the universality of concepts to be a pure fiction or some such thing. The reason for this impasse is that sense perceptions are inevitably private while intellectual concepts are normally considered universal.

[42] The distinction between *principles*, laws and theories constitutes the core of Dilworth (2007).

The consequence for our problem must then be the following: if the 'object' of a science is, by definition, something which must (in principle) be an object for all subjects, it cannot be but *an intellectually constituted structure*.[43] Here again we may point to a possible difficulty, namely that the idea of 'intellectually constructing' something seems to indicate a certain amount of possible arbitrariness in this construction; but we have already discussed the true sense of such an alleged arbitrariness, and need not repeat that discussion here.

Moreover, it is very easy to see how the operational character of the basic conditions of objectivity must result in the intellectual nature of the scientific object. When we perform an operation, what we can perceive by the senses are certain physical states of affairs such as the positions of indexes on the dials of instruments, or the change in colour of certain reagents and so on. But what we attribute to our object as a consequence of receiving these *sensory* impressions are *abstract* qualities, usually represented, moreover, by numbers or similar mathematical expressions.

Note that even the most vivid and direct sense perceptions, when they are translated into the language of physics, suddenly become so many abstract features. Think for instance of different colours which are perceived by our eyes in a beautiful variety of sense impressions, but which 'become' for physics just a series of electromagnetic waves of different frequencies. Should we affirm that such 'colours' of physics are not to be taken seriously? Quite the contrary, most people would be inclined to regard them as the 'true' colours, taking our sensibly perceived colours simply to be a result of our subjectivity. This position is mistaken, as it confuses two different levels: that of 'things' and that of 'objects,' as we have already noted. Colours as '*things*' (though not as independent existents) are perceived by our sense receptors (eyes), but they are for this very reason not objective and therefore fall outside of the domain of science; colours as electromagnetic frequencies are detected by instruments and thought by our minds, and as such they are not sensory, but can be *objective*. In any case, nobody could perceive colours *as* electromagnetic waves frequencies, but only *think* of them this way, the great advantage of this fact being that even a blind man, unable to perceive colours, can nevertheless know their objective representation as given by physics, if he learns optics and the optical theory of colours as electromagnetic frequencies.

This is certainly a decisive argument in favour of the view that scientific objects are really not bound to sensibility. This example tells us that all the resistance to

[43] Let us quote in this connection a particularly eloquent statement of Mathieu: "The publicity of the object decides upon the way in which it is constituted, as well as upon the nature of what it contains. The simple requirement of being ascertained by a subject would lead us to make the object a content of sensation, since the *sensibility* of the subject is indeed the means for revealing the object. For many subjects, on the contrary, the object is not related to sensibility, for sensibility does not belong to 'several' subjects, but to each subject individually. The requirement of being known to many—which is a simple step towards being valid for all—therefore confers a different value to objectivity, it makes the object as such something no longer related to sensibility, but to intellect" (Mathieu 1960, p. 25).

conceiving of objects as intellectual constructions comes from the confusion between the 'things' of everyday experience and the *objects* of scientific inquiry, and people who seriously take this difficulty as an objection against the intellectual character of scientific objectivity show by so doing that they are unaware of this fundamental distinction.

Of course all scientists, all researchers, when performing a set of operations accepted in their science, have sensory experiences. They see colours, read figures printed on tapes, record positions of indexes and so on. Yet as soon as they express the results of these operations in an objective way, all becomes colourless and abstract but, at the same time, sharable by a multitude of fellow researchers every one of whom, on the other hand, when performing the same kind of operations in order to test these results, will usually reconstruct a sensory picture of the world, will reincarnate those abstract entities in some sensorily well determined way. But this last state of affairs will again be subjective, as no one will be able to perceive the perceptions our new observer has when performing the relevant operations, just as no one was able to perceive the perceptions of the first individual to perform them.

After having examined one fundamental reason for the object's being an entity determined by means of 'abstraction', that is, the reason implied by the requirements of intersubjectivity, we can proceed now to a further one, which we already began to consider when we noted that the result of applying operational criteria is that of providing ourselves with abstract or mathematical representations. That this is the case is already a non-negligible indication of the kind of ontological status that might be attributed to the scientific object; but there is more to be said about it. In fact, every operational procedure reveals one *single* feature to be attributed to the object so that, after performing all the operations we need, we have a set of such features. But no object of any science is represented by a *pure* collection of features; it is always a *structured* collection, in the sense that all these features are mutually connected by certain mathematical and/or logical relations, which are not obtained directly from any instrument, but must be arrived at through the intellectual activity of the researcher. To this end, it seems clear that *besides* the capacity of forming sensory perceptions, we must be provided with a capacity for synthesising them. That these capacities cannot be identical may be seen from the fact that the same perceptual elements can be assembled in very different ways. This has been clarified even at the level of ordinary perceptions through the research of Gestalt psychology, but it is even clearer at the level of scientific concept formation, where much greater freedom in such an 'assembling of elements' occurs.

A rather delicate epistemological situation seems involved in what we have said, that is, that the different features which we can express operationally seem to require being considered as belonging to an entity which we need to denote as 'something' through the use of a name. Intuitively speaking, one might say that all this is obvious, since the objects are always obtained as the result of applying operational criteria (linked with certain specific viewpoints) to certain 'things'; and what is found in such a way must be conceived of as belonging to these 'things.'

This explanation is not psychologically unacceptable, and can also give rise, moreover, to important philosophical considerations regarding the nature of science, but it is of little use 'within' science itself. The alternative, which science has to offer, is therefore a system of logical connections providing a sufficient set of links between the different isolated features made evident by the empirical or operational inquiry in question. We call these links 'logical connections' in order to include under a rather general rubric not only mathematical relations, but also those non-mathematical kinds of relations which may be found in science.

Through this intervention of logical connections, *theory* makes its first appearance in science, in a sense which is more primitive and more basic than that which is most common in the literature. For when reference is made to theory in science, what is usually meant is a system of statements, some of which are accepted as hypotheses while the others are logically bound to these hypotheses and to one another so as to provide us with explanations, tests and so on. This conception is correct as far as it goes, but it seems more significant to consider as a distinguishing characteristic of theories that the *logos* is called into play, as something distinct from *experience*; therefore we must recognise that theoretical involvement in science begins much earlier than at the level of *theory construction* proper. Indeed, it appears already at the stage of *concept formation* and is a prerequisite for that effort of explanation that is considered as the most specific *task* of theories. This point will become clear in the sequel, when we specifically devote our attention to theories proper. For the moment let us simply see how, by reflecting upon this fact, we can acquire insight into the much debated question concerning *theoretical terms*, which we shall start to examine here and submit to further scrutiny in Sect. 3.2.[44]

It is due to the presence of the 'logical links' among operationally established features of reality that what are called theoretical terms make their first appearance in science, for at least two reasons. The first is that, when a set of predicates is

[44] What we are saying in this chapter does not provide a full account of our view, since it can give the impression that we intend that scientific investigation proceeds by *first* establishing loose operational features, *then* bringing them to a conceptual unity, and *finally* by reaching a theoretical picture of the domain of objects involved. This impression may result from the fact that we are obliged to expose the parts of our discourse in a succession. But we do not maintain that *unity* really comes *after* its elements. We think rather that a certain unity is there at the stage of 'clipping out' the objects, and that its elements become *explicit* by means of an appropriate introduction of operational and theoretical means. This is in some way expressed by the idea of *viewpoint* which we have frequently used, and which will be the object of a detailed presentation when we come to speak of the 'hermeneutic' nature of theories and of a particular meaning of the notion of *model* connected with this nature. Let us also note, incidentally, that our distinction between thing and object—in which the object is conceived as a ting considered from a certain point of view—is reminiscent of the distinction between "material object" (corresponding to our "thing") and "formal objectr" (corresponding to our "object") that was common in traditional epistemology. We are not resuming that old doctrine because its terminology (being strictly linked with the Aristotelian theory of the matter-form relation) could be hardly understood in its proper sense today, and in addition we have enriched it with several new elements that were absent in it (especially the operational constituent of the object).

structured so as to produce a unity and reveal their interdependence, it becomes practically inevitable, as we have just stressed, that we give this unity a *name*. While we have advocated the naming of such a unity for certain psychological reasons, we can now recognise that it is implied by the notion of objectivity itself. Since the object is the result of applying certain criteria of objectivity to reality, when those features of reality are objectively established and linked together, the object *exists* and, given that it exists, why should one not be entitled to *give it a name*? A mistake would occur if, by this name, one were to conceive of the object as something that *has* these properties, as some underlying hidden reality that shows certain features, while in fact the object simply *is* these properties or attributes (remember that in this work the term "attribute" is used in its technical ontological sense, that is, not as a particular element of language—as we use this term in in grammar—but as a feature of reality, such as a property, a function, or a relation which may be denoted by a predicate of the language) or, better, the structured set of them. But, on the other hand, one is also obliged to recognise that the object must necessarily be designated by a *theoretical term*, for individual operations can show that such-and-such attributes are present but, as we have already stressed, such individual operations cannot show that certain attributes belong together; nor can they show, more particularly, why they ought to be grouped in one way rather than another. For this reason not only is the object necessarily the result of a theoretical construction, but this fact already implies that theoretical terms are *necessary* in science, at least for naming its objects and for 'making sense' of the operationally obtained *empirical* results. Obviously, theoretical terms are even more indispensable at the *theoretical* level proper, that is, when 'theoretical entities' are postulated in a theory in order to provide explanations.[45]

Let the following examples clarify our point. If the sun, the earth, the moon (which are 'things' of common experience) are considered *only* as far as they have a mass, a position in space and time with regard to some frame of reference, a certain velocity, and are subjected to the force of gravity, while we disregard *all* their other possible features, including their having a volume, we are considering them as 'material points' and studying them as objects of classical particle mechanics. The term "material point" is a theoretical term in our sense because it

[45] We cannot express our view with the necessary completeness at this point, but in the sequel we shall introduce a more detailed semantical analysis, according to which it will be possible to see that a scientific object can be understood, in one sense, as an *abstract* object, univocally and exactly determined by the properties it 'encodes,' and in another sense as one of the many *concrete* objects which 'exemplify' the said abstract object. The language used in science (and not only there) must possess terms for denoting abstract objects, no less than terms for denoting concrete individuals, and this simply because terms denoting abstract objects are also needed for referring to individuals exemplifying such objects. For example, we not only need proper names such as *Rome*, *Napoleon* and *The Iliad* for denoting particular individuals of different kinds, but we also need general names such as *house*, *dog* and *electron*, in order to say, "this is *a* house," "I have seen *a* dog," "*an* electron has been emitted." This semantical analysis will be introduced in Sect. 3.4, and continued through Sect. 4.1.

is not the name of some operationally determinable attribute,[46] but of the result of the particular way such attributes are connected in a unity. If we consider a table, and in addition to the above attributes we also consider its dimensions in space, it is viewed as a system of material points. If in addition we consider the fact that the distances between two arbitrary points of this system will remain constant during its motion, the table is thought of as a rigid body, and this is again a theoretical concept.

Note that the question whether a 'thing' may become an object of a certain kind must be operationally tested and receive a positive or negative answer within the limits of approximation which are determined, from the one side, by the accuracy of the instruments and, from the other, by the specific problem in question. For example, the sun, the earth and the moon can be considered as material points not because they have no volume, nor because we are unable to measure them within certain limits, but because the consideration of their volume is irrelevant from the point of view of celestial mechanics. Similarly, a table may be considered as a rigid body in many contexts, while it might be considered as an elastic body in others. On the other hand, a liquid in a glass container could never be considered a rigid body since operational tests show that it does not satisfy with any accuracy the defined conditions for being one. Let us explicitly note that what we have said applies not only to objects which are more or less intuitively related to things, but also to *processes*. This means that not only are 'material point,' 'rigid body,' 'ideal gas,' 'electric current' and 'perfect fluid' examples of theoretical concepts, but so are 'elastic recoil,' 'adiabatic transformation,' 'uniformly accelerated motion' and so on.

Until now the contribution of theoreticity to the construction of scientific objects has been seen essentially as a consequence of the need of linking together, of bringing to unity, certain operational predicates. Many more objects, however, are usually admitted in the sciences by means of explicit definitions or contextual definitions, depending on the enrichment of the meaning of scientific concepts deriving from their occurring in laws and theories. This point will be fully clarified later, but we want to offer an initial appreciation of its importance through the critical examination of an interesting issue.

2.7.2 Scientific Objectivity and Idealisation

The conception of objectivity presented in this book fully captures the basic ideas of the 'idealisational' approach which has been especially developed by the Polish school of philosophy of science in recent decades, and of which the contributions of Wladislaw Krajewski and Leszek Nowak have been particularly significant. Our

---

[46] It is not important that in our specific example of the celestial bodies these attributes, such as mass and velocity, cannot actually be determined operationally, but can only be calculated.

position shares their thesis that mature science is characterised by the increasingly developed use of idealisations, for this means, in our terms, that it is characterised by an increasingly precise use of objectifications (since objectifications, as we have seen, result from isolating only certain features of reality included in a given point of view and disregarding all the rest, which amounts to considering only certain of reality's attributes and accepting only the respective predicates in the scientific language).

There is, however, an important point on which the idealisation approach differs from ours. Indeed, for that approach idealisations are *not* to be confused with theoretical concepts, since on that approach the former have no referents, while the latter do (e.g., there exist no material points—they say—while electrons do exist). Some misunderstandings must be clarified. First of all the meaning of "theoretical concept" accepted by these scholars coincides with that of the logical empiricist tradition, in spite of the fact that they want rather to oppose that philosophy of science; our meaning is different. But apart from that, a substantial and not a terminological issue is at the root of the difference, and has to do with a philosophical misunderstanding.

Indeed, if we take the claim of the philosophers emphasising idealisation seriously, we must say that only *individual* concepts (such as 'Socrates,' 'Rome,' 'the sun') can have a reference. In fact any general concept designated by a definition contains only a finite number of characteristics, and there exist no things which contain *only* those characteristics. However, the mistake here is the confusing of 'abstracting from' or simply 'disregarding,' and *denying*; abstraction is by no means negation. So, if we define man as a rational animal, we abstract from (or disregard) the fact that concrete men have eyes, legs and hands. But this would never lead us to say that the concept 'man' has no referents simply because 'real men' have eyes, legs and hands. Similarly, all idealisations must be seen as concepts in which *abstraction* is made from a number of features; but what characterises these concepts are their *positive* marks. Therefore, it suffices to see whether—within the context of our investigation—the said positive features are testable—with the accuracy required by that *context*—in order to find the referents of these concepts.[47] Hence material points exist no less than electrons in the sense that the criteria of referentiality needed for finding them are essentially not different from those needed for electrons. More will be said regarding this question when we come to discuss the issue of realism.[48]

We would like to say rather that even the difference between 'empirical laws' and 'idealisational laws,' proposed by the idealisation philosophers, should be handled with care. If we accept (as we do) that an empirical law is the expression

[47] A pertinent distinction between abstraction and idealisation is made by Dilworth, who is very sympathetic to the idealisational approach (see e.g. Dilworth 2007, pp. 123–127).

[48] To use a terminology adopted by Edward Zalta (see his 1988), we can say that scientific objects exist as *abstract* objects (i.e. as intellectual constructions) that *encode* certain properties, while not being *purely* abstract since they are *exemplified* (within certain margins of accuracy) by concretely existing objects or 'ordinary objects,' that is, by things of ordinary experience.

of some ascertained regularities in the *results* of certain operations, this law cannot help suggesting a certain relation between the attributes connected with these operations, and hence also between the predicates which denote these attributes, introducing in such a way a certain unification that (as we have seen) produces a theoretical concept, a concept that does not consist in a single predicate, but in the apprehension of a more or less complex intellectually conceived structure. At this moment the law already begins to be idealised, because it is understood that in every concrete application it will be satisfied only within a certain degree of approximation (and this happens, as we shall see later, because not all of the properties of the 'thing' are included in the concept of the object).[49]

The decisive step, however, occurs when we try to *explain why* the empirical law holds, and possibly why its application is less satisfactory under certain particular conditions. At this stage a *conceptual model* is introduced, which *causally explains the laws* governing the process in question; it is again a theoretical construction and its effective elaboration and clarification constitute the *theory* explaining the laws involved. In this new context *all* laws become idealised, since they must be *exactly* explained in the model (or theory); and the limited scope of their application is also explained as a limited applicability of the model (or theoretical concept, or idealisation) being used. To remove these limitations a new model must be produced. The perhaps simplest and best studied example is that of the Boyle's and van der Waals' laws for gases, concerning which we refer to the discussions in Krajewski (1977) and especially Dilworth (2008, pp. 101–107).

At this point it may be clear that, using our terminology, we should rather say that empirical laws have been transformed into *theoretical* laws (that is, laws justified within a *theory* which, in particular, entails that their expressions involve theoretical terms). Moreover, it may happen that the theory itself allows for the discovery (by means of a suitable deduction) of new laws which are therefore *theoretical*, and may turn out also either to be empirical (if they can be operationally tested *directly*), or to remain theoretical (if their testing is only indirect and coincides more or less with the admissibility of the theory as a whole).

This comparison with the idealisational school provides us with the opportunity to clarify how our sense of objectivity does not entail any incorrect form of essentialism. Indeed the representatives of this school (Nowak more so than Krajewski) explicitly advocate essentialism. There is an aspect under which this essentialism is unquestionable, and it corresponds to the fact that idealisations in general and idealisational laws in particular fully determine the 'essence' of the ideal model. This is unquestionable in the sense that they specify exactly 'what this model is.' However, beside this claim we find another one, that is, that science allows us to grasp the essence of things: "theoretical science penetrates through this surface into the essence of the world" (Krajewski 1977, p. 25). Here the 'dualistic' conception of the essence as something lying behind the surface of

[49] This line of thought can be found in the distinction between phenomenal and mensural experience proposed in Dilworth (2007), e.g. pp. 93–94.

appearances is patent, and is even made more explicit by a distinction of "main (primary)" and "side (secondary)" features of reality, while it is declared that "an idealisational law takes into account only the main factors."[50]

It is now clear why the representatives of the Polish school have maintained that Galileo was an essentialist. Krajewski says, for example, regarding Galileo's law of falling bodies, "Galileo's law grasps the essence of falling" (ibid., p. 26). They have done so because they have given an essentialistic meaning to the Galilean way of doing science, not only in the unproblematic sense of determining the essential features of certain 'affections' of physical reality, but in the more engaging (and untenable) sense of a dualistic conception of the essence, which might return their philosophy of science to a pre-Galilean epistemology.[51]

If readers feel some psychological difficulty in following the above arguments, we might suggest that they analyse them well in order to see whether this difficulty is not bound to the confusion between object and 'thing' and, perhaps, also to what we term the dualistic presupposition. As a matter of fact, in everyday language we are used to saying not that a thing *is* its properties, qualities, features and so on, but that it *has* them. The reason we do so, and the conditions under which we do so without giving rise to misunderstanding, have already been discussed in the section we devoted to the problem of substantialism and essentialism. However, no possibility of introducing a distinction between 'constitutive' characters and 'predicated attributes' is possible in the case of scientific objects in any ontological sense. (We shall see in the sequel another sense in which it is possible.) It follows that only an implicit presence of the dualistic presupposition might incline us to think that objects *have* attributes instead of *being* (the totality of) them. Objects would be conceived of as the hidden 'substrata' of their properties which is, as we have seen, an incorrect way of thinking even in the case of 'things,' and is *a fortiori* not tenable in the case of scientific objects that are endowed with no other attributes than those explicitly recognised as constitutive of them.

2.7.3 Operational and Theoretical Concepts

The preceding discussion helps us to appreciate a second reason for the presence of theoretical terms in science, namely that, in order to assemble or give structure to operational or observational properties, many steps are usually needed before we obtain a satisfactory intellectual picture, i.e. one capable of representing an object. Such steps are, in themselves, of the same conceptual nature as the one we have

[50] For this discussion see especially Krajewski (1977), pp. 25–26.

[51] As we have already seen in our discussion of essentialism, Locke's notion of essence is clearly dualistic, and can be considered in keeping with modern science. Therefore we are not surprised to find it also in many present philosophers of science. We have also shown, however, that this 'dualistic presupposition' is intrinsically unjustified and that it hinders a fully realist interpretation of science.

been illustrating, and follow an increasing order of complexity, giving rise to successive intellectual syntheses. They result at the very beginning from the making of certain logical or mathematical links between operational terms, and may later result from the making of logical links between operational and already established theoretical terms. At a later stage they may result from links between purely theoretical terms, and so on. Since such a procedure in actual science has usually taken much time to develop, one may well have, for example in textbooks, certain systematic expositions of a science in which only theoretical terms are put forth at the beginning as the 'primitive' ones. We have already explained however that this does not prevent us from analysing these concepts and tracing them back to their operational origins when our interest is in a foundational inquiry.

What we would like to avoid, however, is the impression that some kind of reduction or even elimination of theoretical terms might be implicit in our analysis. As a matter of fact, what we said about the relationship between *operational* and *theoretical* concepts in our discussion is reminiscent of another classical discussion, in which the reciprocal position of *observational* and theoretical *terms* has been given close scrutiny. We need not recount the details of that discussion, which originated with the claim (which one can find already in Russell) that it is only a question of patience and logical skill before one discovers a chain of definitions by means of which every theoretical term in an empirical discipline may be connected with purely observational terms. In such a way, definitions being by their very nature tools for dispensing with the concepts they define—once one has established the defining terms (i.e., those which occur in the '*definiens*')—the conclusion seemed obvious that it is possible, at least in principle, to 'eliminate' theoretical concepts and to restrict science to using only observational ones.

It is well known that the concrete execution of this program, started by Carnap and developed by him and several other scholars, eventually led to a negative result which was due not only to certain formal logical difficulties, but especially to the impossibility of drawing a sharp distinction between theoretical and non-theoretical terms (as Hempel and Quine, in particular, have shown).[52]

The trouble with this long story is, in our opinion, that it has somehow obscured the very nature of the problem, for it has given the impression that it is only as a matter of fact that we are unable to bring the reductionist programme to its intended end, and that this programme is in itself and in principle sound and philosophically tenable.[53] In order to see that this is by no means the case, let us ask what it would have meant had the programme been successfully carried out; or, more simply, let us consider those cases where a certain theoretical term actually happens to be immediately definable, or analysable, or decomposable, in

---

[52] See Agazzi (1981a) for a survey of this issue. It must be noted, however, that a deeper reason for this failure was the fact that these authors were unable to appreciate the difference between laws and theories. We shall explore this issue later.

[53] By saying this we do not intend to underestimate the numerous criticisms regarding the observational-theoretical distinction elaborated also within analytic philosophy. Let us mention, just as an example, Putnam's paper 'What Theories Are Not' (in Putnam 1975 I, pp. 215–227).

terms of certain operational constituents which cohere by virtue of some simple logical or mathematical connections.

Would this mean that the theoretical concept has been proven to be eliminable from science, or that its meaning is reducible to the operational concepts? By no means, because, as a matter of fact, we should simply have found the operational *constituents* of the theoretical concept, while their 'cohering' in that particular way could never be the simple result of their own natures, and hence we should miss this cohering factor after the 'reduction.' To express this in greater detail, we should say that the *logical network*[54] by means of which the different operational (or observational, or empirical) constituents of a scientific object are assembled into a unity, such that the object can be identified and receive a name, allows us to perform an *analysis* of the theoretical concept. This analysis, on the other hand, cannot be a *reduction* simply because, after the analysis, we are in the presence of *two* things, namely, the separate operational components *and* the structure (logical network) in which they were embedded. Therefore it is incorrect to say that the meaning of the theoretical concept might lie *completely* in its operational or empirical constituents, simply because the very same constituents would give rise to a different (theoretical) concept if they were given a different structure. Thus we conclude that if their operational components are *not sufficient* to distinguish the meanings of different concepts, then it is possible neither to reduce this meaning completely to these components, nor to eliminate its theoretical side.

The interesting fact is that the above discussion, which recognises the relevance of both operational and theoretical elements in science, does not only entail a critical revision of the extreme empiricist conception of science, but also affords a decisive criticism of the extreme 'idealistic' conception of science, according to which scientific objects (and especially objects of physics) are simply mathematical structures.[55] It is therefore worthwhile to consider briefly in which sense we can claim that scientific objects *have* a 'mathematical structure,' without being identical with such structures themselves.

[54] We use the expression *logical network* (as we used *logical links* earlier) in a loose sense, without specific reference to well-defined formal features. Moreover, at this stage of our presentation, several elements are still missing which could make the meaning of this "logical" more precise, and link it with the 'gestaltising' function of the intellect, rather than with the deductive patterns which are more commonly associated with the notion of logical.

[55] This view sometimes surfaces in Heisenberg's pages, for example when he says, "The elementary particles in Plato's *Timaeus* are finally not substance but mathematical forms.... In modern quantum theory there can be no doubt that the elementary particles will finally also be mathematical forms, but of a much more complicated nature" (Heisenberg 1958b, pp. 71–72). However, one should not overemphasise these claims; a few pages later he says, "When modern science states that the proton is a certain solution of a fundamental equation of matter, it means that we can from this solution deduce mathematically all possible properties of the proton and can check the correctness of the solution by experiments in every detail" (ibid., pp. 74–76). This mention of experimental check certainly mitigates the 'idealistic' flavour of the former statements.

It is certainly possible to give some credit to the Pythagorean conception of physical objects, according to which number is the essence of physical reality. As a matter of fact, we recognised that the objects not only of physics but more generally of every science must be conceived of as a network of relations, the basic ones being established among attributes which are operationally determined through the use of instruments, and which are often already expressed in the form of abstract quantities, and the subsequent ones being obtained by a more or less complicated mathematical processing of these initial data.[56] Now, if we consider mathematics as it is conceived of today, that is, as a 'science of relations' rather than as the 'science of quantity,' it follows that a scientific object must be considered, in a way, a mathematical construction.

This conclusion is valid provided that, first of all, one does not take it in an epistemologically 'dualistic' sense, which could be expressed, for example, by saying that the scientific object is *nothing but* a mathematical construction because only such a mathematical 'surface' can be studied, while the 'deeper reality' must be left untouched. Nothing of this kind, on the contrary, has been claimed here. What emerges from our considerations is that every scientific object, in being a network of relations, is particularly well suited to being *studied mathematically*, but not that each such object *is mathematics* and nothing else. We could say that it is *thoroughly* abstract without being *wholly* abstract. In other words, just as we have admitted above that empirical constituents represent *only a part* of scientific concepts, here we claim that the same is true with regard to their theoretical structure. And again, just as we have noted that the same operational components give rise to different objects when they are framed within different theoretical structures, now we must note that the same theoretical structure gives rise to different scientific objects when it is 'filled out' by different empirical elements.

2.7.4 The Nature and Structure of Scientific Objects

The question just hinted at is rather intriguing, for one cannot see at first glance how to distinguish a mathematical structure from a mathematically structured realm of, say, physical properties. The answer comes from a distinction between the *nature* and the *structure* of a scientific object.

We must remember that, according to our view, operations determine the *nature* of the scientific object—or its ontological status as we shall call it later—(as they 'clip it out' of reality, and determine the basic attributes that constitute it), while logical and mathematical constructions determine its *structure* (that is, the structure of the set of operational and non-operational attributes involved). When

[56] As we have said, this is especially the case in physics, where a special terminology is used, according to which what we have called *attributes* are called *parameters,* and *magnitudes* when they are measurable properties.

we have a 'mathematical model' of some aspect of reality, we have in a certain sense the structure, without as yet having a definite realm of objects to which this structure can be assigned. On the other hand, if we simply have a collection of data, obtained by employing certain accepted operational criteria, we have a material whose nature is already determined (in the sense that its pertinence to a certain science is already established), while its structure is still in need of further determination. The proof that the nature and the structure of a scientific object are really distinguishable may be offered by considering that one and the same mathematical model can often be successfully applied to very different fields of research—that is, to different kinds of scientific objects—while, on the other hand, the same set of data may frequently be structured according to more than one mathematical model.[57]

The importance of this remark should be clear. We can fully appreciate the similarity or even identity of structure (isomorphism) of different scientific objects, without misconceiving it as meaning that the objects themselves are identical. In order to recognise their difference, we simply have to check whether the *operations*, by means of which this common structure is *referred* to reality, are actually different in the different cases. We could also express this view by stressing that a mathematical structure simply indicates the *possibility* of a physical object, but its existence as a physical object must be ascertained operationally. A significant example in this sense is offered by quarks, the notion of which had been introduced on a purely theoretical basis for solving many difficulties in elementary particle physics. For a while practically 'everything' was known of these quarks (charge, mass, spin, magnetic momentum, and so on), so that they had the status of a satisfactory 'mathematical model.' This was not enough however for qualifying them as physical objects; and indeed there were physicists who believed that quarks existed as physical entities and 'searched' for them (i.e., they performed those operations which could allow physicists to 'observe' them), while other physicists believed that they 'existed' only in a mathematical model. Only the operational discovery of actual quarks could eventually prove their existence as physical objects; until then they 'existed' only in mathematical models.

The question now becomes clearer and, more particularly, we now find ourselves in a good position to grasp the possibility (and therefore also the

[57] For an example of the first kind we can consider Coulomb's law concerning the attraction of electrical charges, the analogous law for magnetic poles, and Newton's law of gravitation; they have the same mathematical form but apply, respectively, to electric charges, magnetic poles, and masses, which are very distinct physical attributes. For an example of the second kind consider any physical law which can be satisfactorily embedded in different mathematical models, such as the empirical laws of geometrical optics, which can be formulated in wave-like and particle-like mathematical formalisms. More generally, the mechanical models of the electromagnetic field constructed at the end of the nineteenth century showed (to the extent that they were successful) certain formal or structural features which they had in common with the electromagnetic field, while their respective natures remained distinct.

methodological necessity) of distinguishing a *mathematical model* from an *empirical (operational) structure*—two notions which can easily be confused.[58]

The possibility of conferring on a mathematical model the role of expressing, say, the structure of a physical object is given by there being actual operations among those accepted as providing protocol criteria for physics which give results in agreement with the mathematical model. If this is not the case, we are simply left with a mathematical model, but not with a model of a physical object (i.e., this model does not express any physical structure). But one could now ask: what is then a mathematical model, considered in itself? What kind of object does it constitute? Since, after all, we are able to grasp mathematical models in themselves, we know that they have a kind of autonomous life, which is often useful (and sometimes also dangerous), because such models are fully independent of the reality which they model. This seems to indicate that such models deserve to be considered as objects as well. Is this claim correct?

The answer is that it *is* correct and that one simply has to understand that a mathematical model is a *mathematical* object whose existence and structure may be investigated by means of mathematical criteria of objectivity. As we are concerned here with the empirical sciences, it is not our present task to explain how mathematical objectivity might be conceived. For the sake of simplicity, we could say briefly that mathematics must also have its own criteria of protocollarity, which must also be operational in character (we have already given some hints in this regard earlier). And we might accept that such operational criteria be identified (in a pictorial sense) with those pencil and paper operations suggested by Bridgman and other operationalists. But now the misunderstanding involved in the operationalists' claim that *every concept* is operationally defined in every science by simply resorting to such pencil and paper operations comes to light.[59] The mistake lies in the fact that while operations with pencil and paper are suitable for the definition of a *mathematical* object, in physics the problem is that of defining a *physical* object, and what makes an object physical are not *such* operations, but *other* ones. When the problem is that of defining the structure of a physical object, mathematics surely comes into play, but again not as an operational means of definition; it functions simply as a tool for the construction of the mathematical model.

We can also see that our identification of the scientific object with something which is mathematically structured can eliminate a certain difficulty that was put

---

[58] We ignore, for the moment, that besides mathematical models also *physical* (conceptual) models are used in science. Mathematical models can consist merely of equations; physical models require the representation of a physical reality. We shall address this question later, when speaking of the different meanings of "model" in science.

[59] This claim is to be found, for example, in the following statement of Bridgman: "Most of these non-physical operations are the operations of mathematics or logic; it is particularly obvious in the case of modern wave mechanics that many of the constructs are of this sort. 'Paper and pencil' operations is perhaps a suggestive name for many such operations. The variety of such possible 'paper and pencil' operations is doubtless greater than the variety of conventional operations of the laboratory.... Many of the 'paper and pencil' models constructed in this way are of great value" (Bridgman 1950, p. 15).

forward when we suggested invariance to be a typical mark of objectivity. There we mentioned one possible objection, which was that invariance is a property that may reasonably be attributed to the mathematical formulation of physical laws or to the mathematical description of physical events, but not to the events themselves. We can now see how easily this objection can be overcome once the physical object itself is recognised as having a mathematical structure: the invariance that one spontaneously admits for the objects can very naturally be conceived of as a *mathematical* invariance.

We shall conclude this section with a couple of remarks whose importance might not be fully appreciated now, but which will become apparent later. Scientific objects, as we have characterised them, can be said to be *abstract*, since they are fully and unambiguously determined as sets of *selected attributes* organized in an intellectually designed *structure*. The linguistic expressions we use in relation to them are therefore names of complex *concepts* that we can call *abstract* in a different sense, that is, in the sense of being thought-contents. We can say therefore that each abstract concept is *designated* by a certain complex predicate-term. However, such terms are not devoid of concrete *referents,*[60] since we are usually able to operationally find concrete individual entities (i.e., concrete things) which *satisfy* them (within the margin of mensural error required). We call the set of referents of these concepts their *extension*, and say that these concepts *refer to* their extension. This point will be expanded in the discussion of the ontological status of scientific objects.

2.8 The Independence of Scientific Objects with Respect to Visualisation

A still greater advantage in accepting the idea that a scientific object be constituted through a set of definite, precise, but *conceptualized* characteristics is represented by the freedom one acquires from always having to have *visual pictures* of scientific objects in order to be able to accept them as objects. This freedom not to remain prisoners of intuitive conditions has been practically accepted, by now, by almost everyone conversant with modern science. But this has often happened at the price of advocating an 'idealistic conception' of science according to which scientific objects and structures are *nothing but* mental constructions. We suggest that it is unnecessary to adhere to an idealistic conception of science. Rather, one

[60] We are obliged to compact our discourse here. Later we shall analyse in detail (particularly in Sect. 4.1) the relationship between a term, its meaning (the concepts it designates), and its referents. It will become apparent, then, why for us sense and reference are different things, and the notion of 'abstract object' will be clarified. For the moment let us simply stress that concepts constitute the *meaning* of certain terms, the referents of which are *intended* to exist in reality—even when it is known that they do not. Thus when I refer to the earth *as* a material point, I am referring to reality, not to my conception of reality, that is expressed by the concept of material point designated (in English) by the term "material point."

should be aware that while visual requirements are usually available for the representation of *things*—which we actually perceive in our sensory intuition—scientific objects are (as already pointed out) defined through abstract structures, though they receive empirical referents by means of operations (we shall see in the sequel, however, that operations are not the only means for securing referents).[61]

This point is important, since it stresses that what is *abstract* may well have (and usually actually has) *concrete referents*, and that these referents do not reduce, therefore, to a purely intellectual construction. Indeed a general condition is that, in order to study concrete things, we must make certain abstractions (i.e. envisage only certain partial aspects of such things, *abstracting* from innumerable other aspects). This allows us to construct *abstract models* of these things or, as we have said, to transform them into *objects* of a given kind. We then investigate the properties of these abstract models, and correctly say that these represent properties of the concrete things to which these abstract objects refer. For example, we do not study individual wolves in zoology, but wolves in general, that is, we study an abstract model of a wolf, and we may arrive at the discovery that, for example, a certain chemical substance produces cancer in wolves. Now the fact that this discovery has been made *within the model* (i.e., from a certain 'viewpoint' of which the model is a part and which we may qualify for brevity as the 'biological viewpoint') does not mean, of course, that the cancer may affect the *model*; but it obviously may affect the concrete individual wolves to which the model or the abstract object *refers*. Similarly, in physics we define the term "electron" through a structured set of mathematically formulated properties which together constitute a certain abstract object. But this does not entail that these are meant to be properties of the abstract object; they are meant to be properties of the single electrons which are the intended referents of the mathematical model we have constructed.

To use a traditional distinction, the concretely existing things, which are immediately present to us in an *intentio prima* (knowledge by acquaintance), cannot be investigated without the elaboration of a conceptual picture of them which can be intellectually scrutinised and is universal and abstract (*intentio secunda*). However, the results of our scrutiny do not concern the conceptual picture, but the concrete referents of the *intentio prima*. In the case of modern science, the *intentio prima* does not properly consist in *perceptual acts*, but in *operational procedures*, starting from which we elaborate a conceptual model which we then proceed to study (*intentio secunda*). As a result of our study we attribute to certain referents those properties which are compatible with the

---

[61] What we have discussed here amounts to rejecting the spontaneous idea that models in general (and scientific models in particular) are 'good' or 'better' to the extent that they have a close "resemblance" to the things they are intended to be models of. In Chap. 9 of Agazzi (1969) we have abundantly argued against this naive view, and shown that the most abstract mathematical models are by far superior to the intuitive models in the study of new fields of scientific research (the requirement of resemblance reducing to the minimal level of structural isomorphism). Also van Fraassen (2008) has stressed that resemblance is by no means a fundamental requirement for scientific representation.

operational procedures constituting the real tools of our *intentio prima*, and which do not necessarily meet the usual requirements of the *perceptual* (typically, visual) structure of this *intentio*.[62]

From what we have said it becomes clear that the term "abstract" actually has a plurality of meanings in ordinary language. It can mean either *general* (as opposed to *individual* or *particular*), or *non-sensory* (as opposed to *concrete* or *material*). In the case of ordinary experience, we do not hesitate to think that properties which we have recognised by reasoning *abstractly* in the first sense may refer to concrete things, because our criteria of reference are bound to sensory perception (i.e. our discourse is not completely abstract in the second sense). In the case of scientific discourse, however, many are hesitant to admit the existence of referents for the *abstract* constructions developed in research, because such constructions are not only abstract in the first sense, but also in the second; they cannot be perceived. However, no real reason is given for this perplexity, and the only way to understand it seems to be to see it as the consequence of a tacit presupposition, according to which only entities endowed with sensory qualities *exist* in a proper sense. But in this way the real reason for the perplexity appears to consist in an arbitrary collapsing of ontology to a few perceptual parameters.[63] We shall return to these considerations when we discuss the issue of scientific realism.

The distinction between 'things' and objects which we have discussed at length should suffice to clarify this issue. However, a few elementary historical considerations will confirm the substance of our explanations. Everyone knows that the difficulty, or even the impossibility, of 'visualising' the entities and processes involved in physical states of affairs was for a long while a serious obstacle to quantum mechanics. There were, however, historical reasons for this. We could view modern science as having been for two centuries a more rigorous way of considering the same realm of 'things' as we are presented with by everyday perceptions. In such a way, even if some simplifications or idealisations were made (such as those implied in the concept of a rigid body or an ideal gas, or in the principle of inertia), the scientific picture of reality did not conflict with the everyday picture, it being easily understood that the real world can only approximately match the idealised statements of science. Therefore, when the well-known difficulties of visualisation emerged, it was clear that they were grounded in the constant tendency to consider physical objects as 'things.' But, as has been

[62] The doctrine of the *intentio* was developed in a very profound and articulated way during the Middle Ages. Practically forgotten during the time of modern philosophy (the sixteenth through the nineteenth centuries), it was resuscitated by Franz Brentano, who explicitly recovered it from Scholastic philosophy in his theory of *intentionality* (cf. Brentano 1874). Since then, intentionality has become one of the central concepts of contemporary phenomenological philosophy, starting with the work of Edmund Husserl (see Husserl 1913) who had been Brentano's student. For a rich body of quotations regarding the history of this notion, see the article 'Intention' in the dictionary: Eisler (1927). We shall return to the topic of intentionality and intension in a much more detailed discussion later on in this work.

[63] This is, e.g., the position of Bas van Fraassen (1980), which we shall have the opportunity of discussing in detail later.

noted several times, a scientific object *must* be something different from a 'thing.' As a consequence, some properties belonging to a physical object, which are not thinkable as 'cohering' within a 'thing,' that is, which do not show themselves to be associated in the domain of everyday experience, are not prevented, for this reason, from being able to unite within a specific 'object' of a certain science. Such is the case, for example, with the corpuscular and undulatory character of physical particles. If we try to conceive of these two features as coexisting in one 'thing' by an effort of the imagination, we shall hardly succeed. But if we simply accept that what is confirmed by the admitted operational criteria of quantum mechanics has the right to be ascribed to quantum mechanical 'objects,' we are already out of the difficulty, without any need for the 'complementarity principle,' or anything of the sort, since this theory also provides the 'logical network' necessary to move from speaking of *attributes* to speaking of *objects*.[64]

Similarly, just as the existence of intuitively discrepant features does not prevent a scientific object from possessing them, so the impossibility of establishing the coexistence of certain intuitively plausible features by means of accepted operational criteria does not imply that they cannot belong to the object of a certain science, or that they are 'objectively non-existent.' This again may be relevant to quantum mechanics. If, in accordance with Heisenberg's principle of indeterminacy, conjugate magnitudes such as the position and the momentum of an electron cannot be determined at the same time with an accuracy greater than a certain value, one must admit that such magnitudes are, taken together, 'objectively undetermined', at the same time. within the 'domain of quantum objects.' As can be seen, this way of considering the issue condemns as misleading such questions as whether this uncertainty concerns the state of the physical world or simply our knowledge of that state. This question is misleading because, for physics, there is no such thing as a *real* world different from the *objective* world. (In the sequel we shall consider that difference between reality and objectivity which consists in objectivity's not exhausting reality. But that part of reality which is not included in e.g. physical objectivity is not, on the present view, investigated by physics.) What is not ascertainable by means of the accepted criteria of objectivity does not exist as an 'object' of a certain science, and if it is not ascertainable by means of the operational criteria of *any* existent science, it does not *yet* exist as a scientific object at all.[65]

Therefore, no conflict exists in something's being objective according to a certain science and not being objective according to some other science. The

---

[64] I have devoted some attention to this issue in my book on the philosophy of physics (Agazzi 1969, Chap. 8), and in other papers: see e.g. Agazzi (1988).

[65] One must be careful not to interpret this statement 'dualistically,' that is, as if we were saying that operations reveal reality 'as it appears,' and that we try to discover reality 'as it is' through theories (at least according to one realist interpretation of theories). What we mean is that by means of theories we aim at fully understanding and knowing the very same aspect of reality as has been envisaged within a certain viewpoint and made empirically accessible through certain operations. No dichotomy of 'reality' and 'appearance' is therefore introduced.

alleged conflict is manifest, again, because of the common-sense sound of the words, according to which it is impossible that something be objective in one sense and not in another. But we must remember that "objective" here means 'relative to the object(s).' Thus it must be regarded as very natural that, when we are dealing with different objects, some or all of the properties we are dealing with will also be different (as objects *are* their properties or, more precisely, a particular *synthesis* of their properties), and that what was objective from one point of view (i.e., what was a property of the objects of a certain science) may well no longer be objective from another point of view (i.e., may not be a property of the objects of another science, or of a different branch of the same science). This may be of relevance, for instance, to the fact that for macrophysics certain properties are objective while for microphysics they are not. This implies of course that the locution *physical object* is too general, and that one should speak instead of macrophysical and microphysical objects. This is moreover very sensible since the operational criteria for testing statements are different in the two domains (e.g. in microphysics one would never measure lengths by using a metre stick, or masses by using scales).

Another consequence of this fact is that one does not describe the situation properly when one says that classical mechanics was *falsified* by quantum mechanics, or some such thing. This description is mistaken at least because we are actually dealing with different disciplines and not different theories within one discipline, such that the two sorts of mechanics had to do with, and still have to do with, different objects; and it cannot be the case that they properly conflict, since in order to do so they would have to say opposing things about *the same* objects. Rather similar is the situation regarding the relations of classical mechanics and special relativity, that apparently concern the same "physical world." A closer scrutiny, however, shows that the notion of spatial distance is not linked with the same operations in these two theories because in special relativity distances are (ideally) measured by means of light signals, and not by (ideally) displacing a rigid rod, and it is known that precisely by analysing this way of estimating distances Einstein arrived at the most "surprising" consequences concerning, in particular, the elimination of absolute time. This issue, however, deserves a closer examination that will be offered later.

2.8.1 First Conclusions

The content of this chapter is a presentation of the general features of the conception of scientific objectivity proposed in the present work. In the Chaps. 3 and 4 several considerations of a more technical and analytic nature will be developed that intend to offer detailed arguments for founding the *realist* view of science that encompasses our conception of objectivity. These two chapters may be skipped by those readers who are not interested (at least immediately) in such technical deepening, because Chap. 5 (specifically devoted to the theme of scientific realism) will recapitulate and expand the results of such investigations in a much simpler discourse.

Chapter 3
First Corollaries in the Philosophy of Science

3.1 The Relativisation of Scientific Concepts

An objection may be raised to the solution we proposed in the preceding section to the problem of eliminating the clash between, say, classical and quantum mechanics. One might note that, after all, the *concepts* we use in both disciplines are often the *same*, and that the clash is due to the fact that these concepts *behave* differently in the two disciplines (i.e., that they do not conform, for instance, to the same mathematical conditions expressing physical laws). Therefore, it is truly difficult to see how such a clash might be considered different from the falsification of one theory by another.

In order to tackle this problem we must discuss the more general question whether the same concepts can or cannot retain the same meanings inside different theories. The question is by no means new, and in recent decades it has been the subject-matter of dozens of papers and books which have discussed the thesis of the 'meaning variance' and the 'theory-ladenness' of scientific concepts. We are going to see how this issue appears from the point of view of the theory of objectivity we are proposing in this book, and some particular aspects of the issue will be pointed out which are not to be found in usual treatments of it.[1] In anticipation, we can say that the question is usually formulated in an ambiguous way, since it does not say how one has to distinguish between a *concept* and its meaning while it is much more commonly understood that a *term* may have different meanings. Therefore it can be easily admitted, in general, that the same terms in different disciplines (and perhaps also in different theories within a given

[1] The considerations that will be developed in the present section and in the two following sections have been outlined by the author in diverse papers, and were given a synthetic presentation in Agazzi (1985).

E. Agazzi, *Scientific Objectivity and Its Contexts*, DOI: 10.1007/978-3-319-04660-0_3,
© Springer International Publishing Switzerland 2014

discipline) do *not* have the same meanings, owing to the 'holistic' nature of meaning.[2] However this does not amount to the fact that the same *concepts* are designated by these terms in the different contexts, simply because concepts are meanings and, as a consequence, the same terms do not designate the same concepts when the terms are endowed with different meanings. It follows that, this meaning always being context-dependent (as will be seen), it must be different in the different sciences (or theories), and it is therefore natural, for instance, that magnitudes such as position and momentum can be 'exactly' measured in classical mechanics, while being subject only to a certain 'undetermined' measurement in quantum mechanics. This happens because *it is not the same position or the same momentum* that are *meant* in the two theories, owing to the theoretical contextualisation they receive, and quite independently of the practical difficulties in performing the 'exact' measurement.

This differentiation has two sources or reasons (which are a direct projection of the two elements which we have seen enter into the determination of every scientific object), the presence of basic predicates endowed with an operational character, and the presence of a logical network which connects different predicates, conferring to their totality a certain structure. As has already been stressed, basic predicates deserve to be singled out as those which really 'make the object' inasmuch as, being directly bound to the operations, they concretely manifest the viewpoint which 'clips out' the objects of a given science from reality. Moreover (but we are not going to deepen this question here) basic predicates have a naturally privileged position due to their direct link with operations, which endows them with a *referential privilege* that other concepts may receive only indirectly.

As a consequence, one must say that if the basic predicates occurring in two theories are at least partially bound to different operations, though they may be expressed by the same terms, they are operationally defined in different ways, and their *meanings* therefore cannot actually be the same. Therefore, they do not designate the same concepts; they do not denote the same attributes; and this condition is already sufficient for saying that *any* further concept that may appear in the theory must *at least to a certain extent* (and, let us add, usually *only to a certain extent*) be affected by this very difference in the context in which it is embedded. The relation of classical and quantum mechanics seems to offer an example of this kind. The above point will become much clearer if we come to the second possible source of differences in the meanings of scientific concepts designated by the same terms in different theories/disciplines. In order to better appreciate the relevance of the

---

[2] It may be noted that in the textbooks of traditional logic, the theory of the *terms* was developed considering them as constituents of propositions, and it was usual to distinguish three kinds of term: mental, oral and written. Mental terms were concepts. Already for Aristotle, however, oral and written terms were considered *signs* of the mental terms, and this justifies the modern practise of considering a *term* only as a linguistic expression, that *designates* the concept. This is the convention we shall adopt in the present work. Other authors might prefer, e.g., "express" or some other equivalent terminology.

logical network (of a theory/discipline), let us first present an additional reason for introducing theoretical terms into our discussion. It is all too natural and obvious that an empirical science obtains its objects by means of empirical operations; but it is neither natural nor obvious that it is compelled to *speak* about these objects *only* by means of empirical predicates. Assuming an empirical science *not* to be so compelled, it follows that *theoretical* (i.e., non-empirical) predicates are needed not only in order to unify operational predicates and denominate that unity, but also in order for one to be able *to go on speaking* about the unity, and to predicate of it something which, though being *related* through a logical network to the operational predicates, is not *expressed* by them.

In order to avoid any infiltration of epistemological dualism into this discourse, and to avoid giving the impression that theoretical terms here play a kind of metatheoretical role with respect to operational terms (since we have said that they are also used for 'speaking about' the objects constituted by the operational attributes), let us present an example from everyday life. Suppose that, in order to identify a certain man in a given group of people, I need three or four empirical properties, such as the colour of his eyes and hair, his height, the shape of his nose or mouth, and so on. Once I have identified him in this way, I am not obliged to speak about him only by adopting these few predicates. I can clearly attribute to him (i.e., predicate of him) several additional features, some of which may be empirical (such as the colour of his skin, or the fact that he is laughing) and others of which may be non-empirical (such as his being a doctor, or a religious man, or his being of some particular age).

Clearly, the initial 'basic predicates' were empirical, and we needed them to *fix the reference* of our discourse. This means that, for example, if we say that Mr. X is a doctor, we can claim that we are predicating this property of the man we *intend* only if this Mr. X is also the person in our group who is identifiable through our basic predicates. But many kinds of properties can be attributed to Mr. X, both empirical and non-empirical, and we can use any of them in speaking of our intended man if there is a network of sentences which, if needed, can show how to relate what we are saying to the basic predicates we selected for identifying him. In this sense we can also say that we are broadening the description of our 'object,' because all the new predicates we use actually improve its determination by bringing new elements into the logical network that expresses the structure of the theory, and thereby restricting the range of entities capable of being this particular sort of object.[3]

[3] The empirical attributes which are being considered here as tools for fixing the reference of the discourse should *not* be regarded as properties giving rise to a *definite description* of a certain Mr. Smith, but as *ostensive criteria* which we select in order to intersubjectively point to the referent. Their role is therefore that of offering a complex of conditions that functions as a *rigid designator* in Kripke's sense. What is interesting, however, is that these predicates express at the same time some property of Mr. Smith (which his name alone would not do), so that they play a double role, that of ostensively providing the reference, and that of providing a description at the same time. The combination of these two functions, which is not really transparent in the case of this everyday example, will be apparent in the case of the operational predicates of science.

Let us now return to the question at issue. We can maintain that the basic predicates 'determine' the scientific object, not because they have an intrinsically privileged nature, but simply because they were explicitly assigned the task of *identifying* it. But then many other predicates may enter the discourse, provided we can, if needed, show that they are related to those which are basic. In every theory proposed in an empirical science a logical pathway must enable one to go from any sentence containing only predicates different from the basic ones (e.g., from sentences composed entirely of theoretical terms) to at least one sentence which contains only basic predicates. This fact assures us that we are still speaking about our *intended objects*. Furthermore, this is why *testability*, as already stressed, is such a fundamental requirement for all empirical theories. As a matter of fact, without testability we would never be sure that the theoretical propositions of the theory concern its objects.[4]

In the above discourse a presupposition seems to have been too lightly made, that is, that a clear distinction can be drawn between *operational* and *theoretical* (or non-operational) predicates, since one of the most respectable claims in recent philosophy of science is that such a distinction does not exist between *observational* and *theoretical* terms (and it seems that our *operational* terms may be equated with the observational terms referred to here).[5] This question will be discussed later. For the moment let us assume that it makes sense to distinguish (sharply or otherwise) between operational concepts (which we shall call O-concepts for the sake of brevity) and theoretical concepts (which we shall call T-concepts). We now would like to discuss the case of a term "P," that is, of a

---

[4] To use our Kripkean analogy again, theoretical statements may be valid in several 'possible worlds' and, by means of their links with the operational predicates, they are shown to refer to the particular world in which our objects reside, being rigidly referred to it by the operational predicates, which are equipped with an ostensive role.

[5] The already-mentioned thesis of the theory-ladenness of all scientific terms obviously amounts to the negation of the said clear-cut distinction, but the question has a much longer history even in the empiricist philosophy of science. Indeed, this philosophy has its roots in the doctrine of the Vienna Circle, that made 'empirical verification' the condition of the meaningfulness of sentences, which in turn entailed that only observational terms have meaning. Since it soon became apparent that many scientific statements were doomed to be meaningless if this criterion were to apply, a progressive liberalisation began which tried to remain faithful to the verifiability principle while attributing increasing importance to the formal and syntactical features of 'the language of science,' which was to constitute a means for the circulation as well as the true creation of meaning. In this way, while it was originally supposed that meanings could percolate up from their genuine source (the observational terms) to the other terms (the theoretical ones), the development of the investigation led to conceiving of meaning as something global, pertaining to the theory as a whole, rather than to single terms. In this way the thesis of the theory-ladenness of all scientific concepts had already been explicitly elaborated within the logical empiricist epistemology, especially in the late 1950s and early 1960s, and was ready to be used just a few years later by those who made of it a cornerstone of the incommensurability thesis. The steps in this development may be clearly perceived in Carnap (1936, 1952, 1956), Hempel (1952, 1958), and Sellars (1961). A very good survey of this story is provided in the chapter 'Empiricist Criteria of Cognitive Significance: Problems and Changes' of Hempel (1965) (which is almost a reprint of two earlier papers).

generic predicate, which is connected with other predicates within a certain theory, and compare it with the analogous situation involving another term "P′".

We observe, first of all, that the meaning of "P" *is*, at least in part, the logical network which relates it to all the other terms in the theory, so that this meaning is simultaneously determined by the *presence* of certain O-terms and T-terms in the network, and by the *particular connections* among these terms (or, if one prefers, by the *structure* of the network). Let us now consider the case of "P" and "P′" being connected with the *same* O-terms and T-terms by means of *different connections*, and the case of "P" and "P′" being connected by means of the *same* logical network with at least partially *different* O-terms or T-terms. It will be clear that in these cases the meaning of "P" and "P′" cannot be the same. This sort of situation is schematised in the following figures.

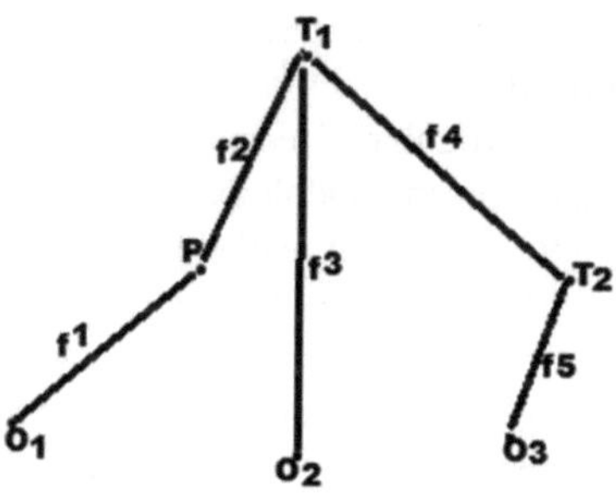

Fig. 1

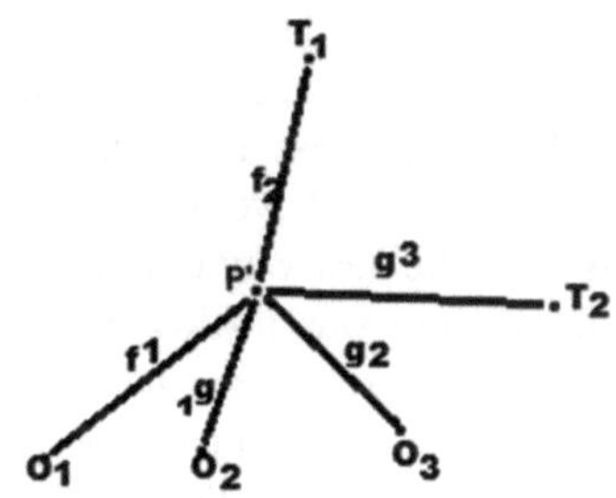

Fig. 2

Case A (Figs. 1 and 2). We assume the concepts to remain the same (O_1, O_2, O_3, T_1, T_2), but the logical relations to change (f_3, f_4, f_5 disappear, and new functions g_1, g_2, g_3 connect P′ directly to O_2, O_3, T_2).

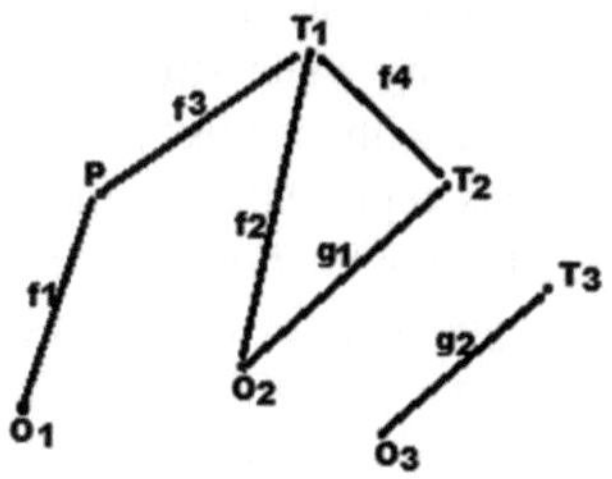

Fig. 3

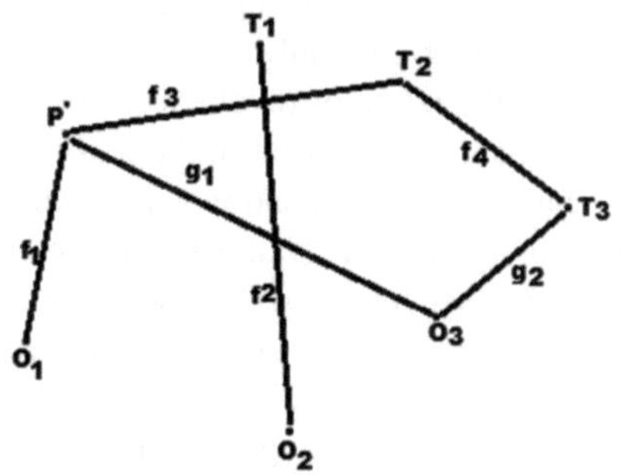

Fig. 4

Case B (Figs. 3 and 4). We assume the concepts and the functions to remain the same, but to be connected in a partially different configuration.

In both cases the meaning of P and P′ cannot be the same. Other similar cases can easily be constructed.

In the above diagrams the lines between the points indicate certain logical relations between the respective predicates (which we can think of as consisting in equations or functions in which these predicates occur), and we see that, even if the predicates could be said to remain the same, the change in the logical correlations which occurs in going from Figs. 1 to 2 or from Figs. 3 to 4 will alter the meaning of P and make it different from that of P′ (but the other meanings would also be modified, despite the fact that we have left the other terms unchanged). On the other hand, it is obvious that if the structure is preserved, but at least one of the predicates is changed, this will alter the meaning of P′ (and of the other concepts as well).

The fact that the meaning of a concept is context-dependent may therefore be expressed by saying that it depends on the *meaning* of other concepts and on the logical links it has with them, which also implies that it in turn influences the meanings of all the concepts with which it is linked.

The awareness of such a dependence is not particularly recent; it may be considered to have officially entered contemporary methodology of science through the new way of conceiving of the axiomatic method in mathematics when this method was no longer regarded simply as a tool for introducing deductive order in a discipline, but as something which was able to create, at least to some extent, the very objects of the discipline. The difference between these two positions is rather patent. If one considers axiomatisation as a way of deductively ordering a discipline, one regards at least a certain number of the terms occurring in the axioms as *names* for the entities which the discipline is supposed to describe, and the ‘meaning’ of these terms may be regarded as their *reference* to these objects. But if one considers axiomatisation as something that must ‘create’ a certain discipline, no objects are presupposed as existing, and the axioms must in a way be able to have meaning even without having, properly speaking, a reference.

If this is the case, the meaning must necessarily arise from the reciprocal links that the different concepts have with each other; and, if a question of reference is advanced, it can only concern the *possibility* of discovering some structure of objects the relations of which can be put into correspondence with the links between the concepts expressed in the axioms, so as to be faithfully represented by them.

Such an axiomatisation was explicitly proposed by Hilbert in his *Foundations of Geometry* of 1899. That book differed from the traditional geometry textbooks of the time, not because it advocated a new geometry (in fact its content is still comparable with that of ordinary geometry), but for its conceiving of the axiomatic method in a new manner. “Point,” “straight line,” “plane,” and so on were no longer presented as names for specific geometrical entities, but as terms the meanings of which were ‘contextually’ defined by all the axioms, and which were thus capable of having as referents *whatever* objects as could satisfy those same axioms.[6]

[6] This is the most usual way of presenting this issue. Historical accuracy however leads one to recognise that this new way of conceiving of the axiomatic method had already been prepared by

3.2 The Operational–Theoretical Distinction

The claim that *every* scientific concept is context-dependent may also be expressed by saying that every scientific concept is necessarily *theory-laden*. This later claim has become widely accepted in recent philosophy of science, marking the extreme reached by the pendulum motion which had seen its initial opposite position in the empiricist claim that all scientific concepts must themselves be either observational concepts or be 'reducible' to observational concepts. In particular, the above

---

(Footnote 6 continued)
Pasch (see Pasch 1882), and had been completely developed by Peano and his school between 1889 and 1899 (i.e., in the decade preceding the first publication of Hilbert's *Foundations of Geometry*). In particular, Mario Pieri (a disciple of Peano) explicitly defended the idea that the meaning of the primitive concepts is 'defined' through the postulates (see Pieri 1899, 1901). While Pieri called this a "definition through postulates," it became customary later to call it an "implicit definition." However, this latter expression was not considered a very satisfactory way of characterising this contextual interdependence of meanings, and nowadays it is used in a much more restricted and technical sense in mathematical logic. The reason for this dissatisfaction was that it is not really clear how the meaning of the *single* primitive concepts could be fixed on the basis of the simultaneous presence of the postulates.

In several papers devoted to Hilbert's *Foundations of Geometry* which appeared in the *Jahresbericht der deutschen Math.-Vereinigug* in the years 1903 and 1906, as well as in a couple of letters to Hilbert (see Frege 1969, vol. 2), Frege correctly pointed out that the totality of the postulates may at most define a 'second order' concept (of which the primitive concepts occurring in the postulates are so to speak the ingredients), but cannot establish the meaning of these concepts themselves. Frege's criticism remained uninfluential (owing to the growing favour of the formalist trend in mathematics), and at most led later to a 'readjustment' of the issue. As was suggested by Bernays in a review of the then newly discovered correspondence between Frege and Hilbert (published in the *Journal of Symbolic Logic* 7, 1942: 92–93), the postulates of elementary geometry, for example, represent an *explicit* definition not of the single concepts occurring in them, but of the concept of Euclidean three-dimensional space.

Even so, Frege's correct criticism has not been met; and if we give it the attention it deserves, we must at least refrain from saying that the axiomatic context (or *any* linguistic context) *entirely* determines the meaning of the concepts. We are certainly entitled to say (as we have said) that this meaning depends also on the context on an intensional level; but this dependence cannot mean the dissolution of the meaning in the context, otherwise no meaning at all could emerge. This is why the thesis of total 'meaning variance' is already untenable for *semantic* reasons.

We are not particularly interested in discussing the semantic 'stability' which must exist to some extent even if context is given its fullest role. On the other hand, we shall later present specific arguments in favour of the existence of a 'stable core' in the meanings of the operational concepts, based on *referential* reasons. What we may say here is that every concept enters into a scientific theory equipped with a meaning whose structure is well articulated and depends on many factors. It is therefore wrong to say that terms receive their meanings totally and only through the theoretical context. This is actually an authentic formalistic fallacy which goes back to Carnap's proposal to consider physical theories as interpreted logical formal calculi (see Carnap 1934). In fact empirical theories do not begin to exist as formal systems, but may at most be 'formalised' after they have attained a certain stage of development. At this stage it may also be possible to detect the 'variance' of meaning which occurs as a consequence of a term's being located in a different context, but this 'variance' is always *partial*. Therefore, when a term is used in a particular sentence, it is normally used only according to a part of its meaning, and it may well happen that the part concerned is *not* affected by 'meaning variance'.

statement has led to a couple of logical consequences which appear rather disastrous for any empiricist philosophy of science (and not only for such). They are, first, the thesis that no reliable distinction may be introduced between observational and theoretical concepts, simply because there are no purely observational concepts in a strict sense. And, second, the thesis that, all concepts being in a way theoretical, their meanings are relativised to the theory in which they are embedded to such an extent that they do not allow of any comparison between propositions belonging to different theories nor, consequently, between the theories themselves (which is one sense of the thesis of the incommensurability of scientific theories[7]). We shall discuss the first of these two issues in this section.

Our first point is that we do not speak here of *observational* predicates, but of *operational* predicates, this fact being justified by our previous discussion concerning the role of operations in science, which can by no means be put on the same footing as observations. To see this, consider the fact (already stressed in Chap. 2) that observations are necessarily referred to the privacy of the observer, and so cannot meet the requirement of intersubjectivity. (Another reason for the discarding of observability in favour of operationality will appear soon.)

We can now formulate our problem in the following way: we are confronted with the fact that all concepts in a science must be conceived of as context-dependent, while at the same time the intimate conviction of every scientist, as well as of the uncommitted philosopher of science, is that operational concepts are not context-dependent, as they are related to something which lies 'outside the theory,' and which is even a prerequisite for the existence of the theory, in the way that operations are.

The way to reconcile these two opposing claims can be found if we resort to the *intensional* conception of meaning. This conception does not maintain that the meaning of a concept *reduces* to its intension, or that the notion of extension is vacuous or useless; it simply says that, despite the good services that the extensional viewpoint has offered in mathematical logic, intension has a no less important role to play in the methodology at least of the empirical sciences (but actually not only of these).

By the *intension* of a concept we mean what one *intends to express* through the concept when, for example, one predicates this concept of a certain thing in a judgement. To put it differently, intension is the complex of attributes (such as qualities, properties, and relations) which are 'meant' by that concept and included in its *meaning*. Such attributes, of course, are aspects of 'reality,' but they are

[7] This is actually a particular way of conceiving of incommensurability that is strictly bound to a linguistic view of scientific theories. Other ways of conceiving of incommensurability, however, relate it to a 'gestaltic' switch and are presented, e.g., in Dilworth (2008). They are also more in keeping with the general conception of science proposed in the present work, as will be seen later.

universal only as 'abstracted' by the mind. The intension is therefore (at least in a certain sense) a set of 'abstract' entities, while the extension is a set of individuals which are often concrete and which are those entities to which the intension can correctly be applied.

A large literature has been devoted to this subject, thus we are not going to enter into more details here, remaining content with the distinction as we have presented it (more details will be provided later, when we consider intensions for other reasons). The usefulness of this way of conceiving of intension lies in the fact that we can distinguish several *components* in it. For example, in the intension of the concept of man we can find as components the concepts of the properties of being an animal, of being endowed with reason, of being able to speak, of being two-legged, and so on, or the relations of being the user of language, of being the inventor of numbers and the alphabet, of being omnivorous, and so on.

A difficulty with the concepts of everyday language is that their intension is sometimes too far-reaching and is always rather indefinite; the advantage with scientific concepts is that, at least in principle, their intension can be fairly well determined and limited to a restricted number of components (which are usually identifiable depending on the presence of other concepts with which the given concept is linked, and by the logical network through which it is bound to them).

If we now consider an operational concept, we see that among the components which are constitutive of its intension is included the property of being linked to certain specific operations. For instance, if we assume that the concept of mass is 'operationally defined' in classical mechanics with reference to a balance (in the sense already discussed when speaking of operationalism in general), we can say that this part of its intension remains well distinguished from other of its components, such as the theoretical links which bind mass to space, time and force via the fundamental laws of mechanics. But now we can go a step further and note that this operational component remains *unchanged* even when we modify some other parts of the theoretical context. We are therefore entitled to say that every operational concept is endowed with a 'stable core' of intension, which we might also call its *ground intension*, while the other parts of its intension are 'mutable,' in the sense that they may change according to different theoretical arrangements.[8]

We have in this way the solution to our problem: operational concepts are certainly context-dependent and theory-laden as far as the *whole* of their intension (*including* the mutable part) is concerned. But they are context-independent, and thus not theory-laden, as far as their ground intension is concerned, because this is solely related to the operations which enter *directly* into the constitution of the concept. (Incidentally, our view relieves us of having to admit such hybrid features

[8] This has nothing to do with essentialism, as we do not claim that the ground intension is more important than other parts of the meaning, as will be better explained in the sequel.

as 'correspondence rules' or similar devices, whose nature is obscure since they should be linguistic entities—i.e., practically propositions—endowed with the magic ability of allowing us to jump outside language. Our operational concepts, whose intension is partly related to referents of a clearly empirical nature, and partly to the rest of the theoretical structure proper, appear to perform the task much better.) This discourse, of course, does not entail that our general view of theories reduces to this presentation of their 'linguistic aspect.' But this will be clarified in due course.

In this way we have what we need in order to draw a distinction between operational and theoretical concepts, because we can safely admit that no concept in science is totally operational; but this does not prevent us from being able to recognise those concepts which have an operational ground intension—which we shall legitimately call *operational concepts*—and to distinguish them from those which are only indirectly related to operations (i.e., by means of the logical network), which we shall call *theoretical concepts.*

We can schematise this distinction by means of the following diagram:
The meaning of the operational concepts in two different theories[9]
T *and* **T′**

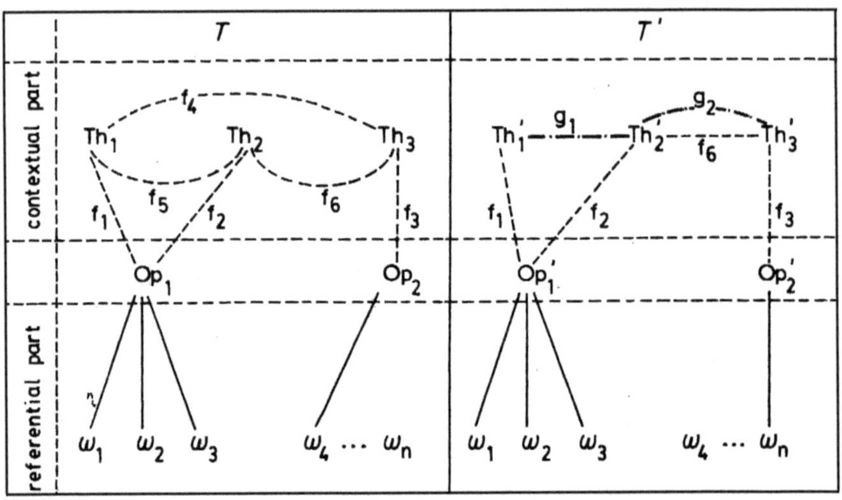

Explanations

(a) Th_1, Th_2, Th_3, and Op_1, Op_2; and Th'_1, Th'_2, Th'_3, and Op'_1, Op'_2 are the theoretical and the operational concepts, in **T** and **T′** respectively.
(b) The lines - - - - and -.-.-. indicate formal (i.e., mathematical or logical) relations existing between the different concepts.

[9] This diagram will be taken up again literally in Sect. 7.2.8 in a wider discussion concerning theory comparison.

(c) The continuous lines ________indicate the non-formal (i.e., referential) relations existing between the operational concepts and their 'defining' concrete operations. These operations are distinguished by the different ϖ with which they are related.

(d) It is supposed that all concepts are indicated in **T** and **T**′ by the same *name* (or *term*). Still their *meanings* are different, at least because of the different formal contexts of the two theories.

As is clear from the diagram, the stable core, or ground intension, of an O-concept is 'stable' because it expresses a relation of the concept to something which is *external to the theory*. This makes it rather trivial that, as a consequence, this component of the intension is not theory-laden. On the other hand, the fact in itself is not at all trivial, as it reminds us that operations belong to *praxis*, even if it is a 'noetically oriented' praxis, that is, a praxis intended to secure knowledge and not (per se) other advantages. This has much to do, by the way, with the fact that it is through these operations that operational concepts are endowed with a reference; as a matter of fact, reference (conceived of as something related to meaning, but not coinciding with meaning) must in a way lie outside the context in which meaning is elaborated, though introducing some information into this context by virtue of referential links (i.e. by means of the ground intension). This is why we are entitled to call this 'stable core' or 'ground intension' of the O-concepts the *referential part* of their (intensionally conceived) meaning, as we have done in the diagram.

A legitimate suspicion which may surface at this point is that theoretical concepts should be devoid of reference. We are far from claiming this; what we have shown here is that operational concepts are endowed with a *direct* reference, but this does not exclude other concepts' possibly being endowed with an indirect reference. Indeed, we shall see (after a number of further considerations) that the aim of science is also that of establishing an indirect guarantee of reference for its theoretical concepts as well. However, there would be no point in anticipating these considerations here.[10]

---

[10] Certain authors, such as Dilworth, lay stress on the difference between *reference* and *referent*, and maintain that reference is not, properly speaking, a property of terms or concepts, but rather an attitude of the speaker, who uses the terms with the intention of referring his listener to certain objects. We shall briefly discuss this issue in Sect. 4.1 and shall indicate why we prefer to stick to the more traditional view according to which reference is a property of terms and concepts. We should like, however, to lay stress on the intrinsic *pragmatic* side of science, which is the root of its unavoidable operational dimension. We have very schematically underscored this dimension by attributing a specific role and position to *operational concepts*, but we are aware of having left unexplored the complex nature of operations themselves and, in particular, their fundamental difference from *observations*. Nor are we going to go deeper into this issue in the remaining parts of this work. An accurate exploration in this direction, however, can be found in Chap. 2, of Stepin (2005), especially pp. 68–89, where a presentation of different 'layers' is made necessary for relating theoretical schemes with experience, via 'instrumental situations' and 'empirical schemes'.

A very important feature of the distinction sketched here between operational and theoretical concepts is its clearly being a *relativised* one. No concept is operational or theoretical in itself; its being the one or the other depends on the theory in which it occurs. If in this theory the concept in question happens to be introduced by an operational definition, it is then operational, and is endowed with a referential intension which is not theory-laden, and to which a further component of contextually determined (or theory-laden) intension will be added. If this is not the case, the concept is simply theoretical. Therefore, one and the same concept (or rather term, as we have already noted) may be operational in one theory and theoretical in another.[11]

At this stage, it is possible to see why the traditional distinction between 'observational' and 'theoretical' could not do the job of securing to science a 'neutral' basis for comparing theories. In fact, observations as such do not provide us with any recognisable intensional feature to be attributed to concepts, so no 'stable core' or ground intension could be given by using them. In addition, observations are observations and nothing else, and this obliges us to conceive of the observational-theoretical distinction as an *absolute* one, leading to well-known impasses—impasses which can easily be avoided by relativising this distinction through referring to explicit, clearly describable and well-delimited operations. This is why we shall not use the notion '*observational*,' considering its positive aspects to be equally well provided by the notion '*operational*.'[12]

---

[11] The notion of the position of a single particle is an operational concept, for example, in the classical mechanics of material points, while it is a theoretical concept in the kinetic theory of gases.

[12] We have said that we do not accept the 'traditional' partition of observational and theoretical terms. However, a much more complex use of the notion of observation is very much in keeping with our way of characterising the operational concepts in which complex instruments enable us to 'observe' entities that are unobservable in the everyday meaning of this term, which strictly relates it to perception. This is in keeping with the well known claim that our instruments can be seen as 'amplified human senses,' so that it is correct to say that we can 'observe' thanks to them much more than we could observe without them.

This extension of the notion of observability (that, e.g., plays a significant role in Harré 1986) becomes even more important if we consider that the ability to 'observe'—in this much richer and more interesting sense—increases with the development not only of sophisticated technology, but of scientific knowledge as such. An excellent presentation of this enlarged sense of observation is to be found in. Shapere (1982) and, in a sense directly related to the operational approach offered in the present work, in Buzzoni (1987). However, since in the great majority of the literature the dichotomy 'observational-theoretical' is still understood in its old empiricist form, we shall avoid using "observational" and use "operational" instead, except for some special and explicitly declared purposes. Let us note, however, that we are not doing this in order to stay faithful to an alleged 'genuine' sense of the notion of observation, simply because we do not share the 'radical empiricist' tenet that endorses such a view. This tenet, for example, permeates van Fraassen (2008) and is expressed through such sharp declarations as "in the sense in which I use that term: observation is perception, and perception is something possible for us, if at all, without instruments" (p. 93); and in the subtitle "Observation by instruments': our bewitching metaphors" (p. 96). Modern natural sciences have been characterised by being *empirical* and not purely speculative, precisely because they adopted instrumental observation

The position advocated here is the inverse of the position of J. D. Sneed, who maintains the relativisation of what he calls *theoretical* concepts. According to his view, theoretical terms are simply T-theoretical (i.e., theoretical relative to a particular theory *T*, according to some criteria which we need not mention here), while nothing is said regarding what is *empirical* about *non*-theoretical terms. According to our approach, operational terms are meaningful relative to the particular theory in which they occur, and provide a foundation for the empirical claims of the theory, while the theoretical terms are simply those which are non-operational (of course, relative to the said theory), so that the idea of T-theoreticity is fully accounted for in our perspective as well (an additional 'positive' characterisation of theoretical terms will be presented later, when we shall discuss the proper aim of theories in science). This inversion seems to us to be justified by the fact that the requirement of being empirical must play a fundamental role in every investigation of the nature of *empirical* theories; i.e., theories intended to apply to empirical reality, conceived of as consisting of attributes made evident through concrete operations. Owing to the 'analogical' conception of science we are advocating, we have no objection to the fact that this 'empirical' component be constituted by 'secondary qualities' rather than by 'primary qualities' though, in the paradigmatic case of physics, we have to do with operations of measurement, referring to primary qualities and determining magnitudes. The fact that the traditional empiricist philosophy of science overemphasised this requirement does not justify its nearly total rejection, which has become rather fashionable in philosophy of science. What really matters is to recognise the precise limits of the empirical requirement, as well as its indispensable role which, in particular, offers a reasonable clarification of the question of scientific data (be they e.g. the data of a physicist or an historian).

A significant symptom of the need to give an appropriate place to the empirical and referential components of theories seems to emerge from the price Sneed has to pay for not having given a 'positive' characterisation of empirical concepts. In fact he is led to include the referents of a theory within the theory itself, since on his conception a theory is an ordered n-tuple consisting of a set-theoretical predicate, certain sets of its possible models, and finally the set of those empirical states of affairs which are 'intended' to be a model of the predicate (see Sneed 1971). But this view unfortunately obscures the distinction between a discourse and the referents of the discourse (a discourse which also includes possible abstract

(Footnote 12 continued)
(with the decisive advantages of intersubjectivity and precision that everyone recognises); therefore it sounds surprising that such a fundamental fact is declassed to the status of a metaphor within an approach that intends to offer a good interpretation of the nature of modern science. This, however, might be precisely a significant symptom of the intrinsic fragility of the radical empiricist tenet itself.

models of the referents). Therefore, Sneed (and his follower Stegmüller) are obliged to adopt *ad hoc* strategies for applying their 'structuralist' epistemology to actual science, such as turning to the beliefs of scientists for singling out the intended domain of a theory, or to vague 'family resemblances' with the intended model, or to the pragmatically inspired notion of 'availing oneself of a theory.' The difficulty resides not so much in the fact that such notions may sometimes be unclear, but in the fact that they transcend the 'structuralist' perspective, and are in a way alien to it. On the other hand, however, these notions find an appropriate place in the view advocated in this work.

Indeed a direct corollary of the preceding is the solution of a much debated problem: do *data*, i.e. immediately true sentences (in the sense already discussed), exist in science? The fashion at present is to deny that such data exist. Data, it is said, are always theory-laden, and in such a way are not essentially different from hypothetical statements. We can remark that such an answer does not take into account that the notion of a datum must also be *relativised*. Of course no sentence can be claimed to express a datum *in itself*, but within an *empirical* theory there *must* be data according to the criteria of objectification admitted for that theory. This does not prevent these data from being obtained by means of very sophisticated instruments, nor from their presupposing, therefore, at least the theory of those instruments. But this is not our problem; as we shall see later, this has to do with the historical determinateness of scientific knowledge, which in any scientific context implies the presence of pre-existing 'available knowledge'; this knowledge certainly includes many scientific theories (as well as other elements, such as ontological and metaphysical principles).[13] What is at issue here is not whether a datum depends on *some* theory, but whether it depends on *the theory in which it is considered as a datum*. And in science it does not so depend and ought not so depend; even though in practical science there is always feedback between the instruments and operations which 'make' the objects on the one hand, and the developing theory on the other. *At least a certain number* of sentences must be recognisable as data that are independent of such feedback, in order for science to provide basic criteria for *testing* sentences and theories. More will be said on this point, however, when we return to the problem of the historical determinateness of scientific objectivity.[14]

[13] Having distinguished between theories and laws, we may even say that the 'available knowledge' is represented by the accumulation of the expressions of known laws rather than of laws and theories (the latter concerning the understanding and explanation of laws). However, this issue is not particularly important here, and we can safely admit that even theories belong to the said 'available knowledge,' according to the broad sense of knowledge we have already accepted, and according to which understanding and explanation are constituents of knowledge despite their being hypothetical.

[14] We have made much use of intensional ways of speaking, but this should not be seen as peculiar, since all discussions related to 'theory-ladenness' are in fact of an intensional nature, since the context-dependence of meanings cannot help but be primarily related to their intensions. Therefore, we do not criticise the structuralist view of theories for having resorted to intention in order to give a sense to the notion of "intended applications" or of "target system." We simply note that the structuralist approach is essentially constituted by a sophisticated use of set theoretical formalisms

3.3 Comparing Theories

The above considerations concerning the possibility of recognising, within a *particular* theory, its operational concepts, and therefore of affording a stable core of meaning for them which in turn justifies the possibility of recognising data in that theory, has immediate consequences regarding the problem of the possibility of comparing different theories, with respect, for instance, to their relative superiority.

Before entering into the details of this discussion, let us make a general remark of a purely philosophical nature. What we expressed in the preceding section is simply the application to scientific discourses of a requirement which ought to be satisfied in the case of any discourse, and which we might refer to as 'the stability of the semantic logos.' By this stability we mean that terms must not be allowed to change their meanings simply as a result of the changing configurations of the discourse. In other words, a meaning (or some basic portion of meaning) must be attachable to a term in such a way as to remain with it independently of the contexts in which the term is used (a condition we believe to have satisfied through our notion of intension, which in the case of basic predicates rigidly designates particular referents).

How can we justify such a claim? A first justification might already come from considering that such a stability is actually a prerequisite for everyday discourse, and this is *a fact of life*. In other words, if such a stability were not available, no interpersonal communication would be possible (because, otherwise, the fact that a certain term is used by two parties to a dialogue would simply constitute an *homonymous* use of this term, that could not help one express what one *means* and intends to communicate to the other party); and since we have *evidence* of this communication, it follows that at least some stability of meaning does obtain.

To this reason, which is based on factual evidence, we add another, based on a logical argument. If the meaning of a term were *always and totally* context-dependent, contradictory statements would be admissible, and the principle of noncontradiction would be deprived of any function in our discourse. For example, take two contradictory sentence-forms, such as "A = B" and "A ≠ B." If the meaning of "A" and "B" were not established independently of these two sentences (each of which represents a kind of 'microcontext' for them), we should say

(Footnote 14 continued)
– instead of the traditional formal-logical tools—in the metatheoretical analysis of empirical theories; and there is no way of characterising intentionality by means of such instruments. Therefore, in the last analysis, the fact that a certain model M represents or applies to a certain target system T only depends on the "intention" of some scientist to consider it to be able to do this. This obviously entails subjectivity, and this does not disappear even if we concede that such an intention is that of a certain scientific community, because what still fails is the indication of *how* the scientist or the scientific community can evaluate *whether* the model M represents the target system T or not. The operational criteria we have insisted upon play precisely this decisive role. For a more developed presentation of the structuralist view one can consider, besides Sneed (1971), also the classical works Stegmüller (1979) and Balzer-Moulines-Sneed (1987) and the survey by Diez-Lorenzano (2002).

that they do not express a contradiction, because the A that we claim to be equal to B in the first microcontext is not the same A that we claim *not* to be equal to B in the second. Why is it not the same? Simply because it is posited to be equal to B in the first case and different from B in the second. One sees therefore that, unless we are ready to claim that contradictions are altogether impossible, we must accept that any given concept is provided with some independent and stable meaning.

How this stability may be attained can be understood in terms of our 'intensional' theory of meaning. Indeed, one has to admit that the intension of a term (i.e., the ensemble of attributes that the term is 'intended' to express and which constitute its meaning), globally understood, necessarily changes with a change of context, and even with the growth of knowledge in which some particular concept is involved. For instance, if we compare the intension of the term "man" today with its intension at the beginning of the nineteenth century, we should certainly recognise that we 'intend' a lot of attributes by this term which were not even thinkable before: what the theories of evolution, psychoanalysis, and neurophysiology have contributed to our present conception of what man is. This means, therefore, that the intension (i.e. meaning) of "man" has changed (if only through being enriched), and this clearly seems to speak against any claim of stability for the semantic logos. However, we can see that this very example implies a certain stability of meaning. The reason is simply that we do not say that we have *substituted* the concept *man* with another concept, or that some unspecified concept has been enlarged. We say that it is the concept of *man* which has been modified, improved, enlarged and so on, which means that the concept preserves a certain permanence through its variations.

Are we led back to essentialism or substantialism by virtue of these arguments? Not necessarily. It seems that we can overcome the difficulty, provided we take into account the relationship between meaning and reference. The natural solution, on this way of thinking, could be the following. We accept that an enlargement, or even a modification, of the intension of a concept does not undermine the stability of the concept's meaning, provided that the intended referents (and, consequently, the extension) of the concept remain the same. This means, in our example, that we are entitled to say that we shall be dealing with the concept *man*, even after our increased knowledge has enriched its intension, because we still intend the same individuals to be referents of this concept as were intended to be referents of it before. Therefore, we recognise through this example that it is the *referential part* of the intension that is constant, not all or other parts of the intension, meaning or concept.

This remark recognises the importance of the 'descriptive' concepts in every empirical discourse. They are usually neglected in the philosophy of science, and it is also the case that the 'descriptive sciences' are less highly estimated than the sciences which are able to provide explanatory theories. However it is undeniable that the descriptive concepts play the fundamental role of securing the connection of a given context with its referents, and in such a way deserve full respect, even though they may appear to be of lower status from other points of view. For example, defining man as 'a featherless two-legged animal' might sound a little ridiculous in comparison with other definitions which much better capture his 'essence' (such as, for instance, 'a rational animal'). Nevertheless, the first characterisation (jointly with other similar ones) may help us fix the reference of the concept of man rather satisfactorily, while the second does not. This example may clarify the sense in which the role attributed to operational predicates here does not imply essentialism, and how the stability of the reference may be secured through certain 'humble' descriptive predicates in spite of the variability of 'high-ranging' theoretical predicates.

Our remarks clearly show that even in everyday discourse, concepts (or at least a good number of them) possess a *ground intension* or stable core of meaning which does not change, being that part of the meaning which is directly related to reference. In the case of everyday language, however, it may as a matter of fact be very difficult to pick out these concepts, and, even more, to identify their 'stable core.' We know that such ground intensions actually exist simply because we use concepts rather unproblematically in our verbal communication, but it might prove a hopeless enterprise to try to make them explicit (the concept of man perhaps constitutes a good example of how difficult this task may be). And this may not be the last reason that has pushed many scholars (e.g. Kripke) to return to the doctrine of essentialism in recent years.

But in the case of science we are luckier, for (at least according to the analysis advocated in the present work) we have specific (or specifiable) criteria for establishing reference and, therefore, for fixing the stable core of the operational concepts.[15]

On the basis of the foregoing considerations—which were needed in order to explain the general sense of our position—we may now turn to the problem of comparing *theories*. In order for this comparison to be possible, the theories in question must be concerned with the same 'domain of objects,' and this fact is not easy to clarify in the usual literature because this notion is assumed in a very vague sense, and thus cannot play a role in the actual discussion. It is therefore understandable that a trend developed in twentieth-century philosophy of science when (in keeping with the 'linguistic turn') the Vienna Circle proposed philosophy of

[15] The above considerations explain how we can satisfy a very reasonable requirement concerning the relative stability of reference expressed by Harré: "Our theory of reference must not make the achievement of a referential relation between a person and thing so fragile a link that every change in the meaning of the vocabulary with which we describe the things we believe to exist requires us to revise our ontology. Nor must we make that link, once achieved, so robust, that we are obliged to hold on to it no matter how much the meanings of our descriptive vocabulary has changed" (Harré 1986, p. 99). We have seen that the indispensable 'stability of the semantic logos' is granted by the permanence of the referential core of the intension of concepts that is compatible with significant changes in the linguistic-contextual part of this intension. That this stability also entails a stability of ontology will become clear in the sequel, when the decisive ontological role of reference is discussed. A more detailed discussion of this issue will be presented in Sect. 5.3.5.

science to be a metalinguistic-methodological study of science. Within this view (which has remained typical of analytical philosophy and of the logical-empiricist tradition, and which we shall provisionally 'assume' here without discussion), the problem of theory comparison receives a linguistic formulation. This may be schematised—with several simplifications not affecting the substance of our argument—in the following way: if theory T' is able to explain the *empirical* statement E, which theory T is unable to explain, we can say that T' is to that extent better than T. As we have said, we shall omit from our considerations here the possibility of having more than one empirical statement involved, as well as '*ceteris paribus* conditions' or other methodological requirements, which have been amply discussed in the pertinent literature.

Assuming also the conception, typical of the logical-empiricist approach, according to which to explain an empirical fact E using a theory T is to provide a formal deduction of E from T in conjunction with suitable statements of conditions, we can express the above 'comparison' by saying that T' is better than T inasmuch as E is formally deducible from T' but not from T.

But now the objection arises that, in order for this requirement to be of any significance, it has to be understood that the *meaning* of the terms involved in the formal deduction is the *same* in both cases. This was precisely what was taken for granted by the empiricist tradition, and this is what became more and more controversial later.[16]

For a while, people accepted that the theoretical concepts which enter theories T and T' are different in meaning even when they are expressed by the same words, because it was supposed that 'observational' concepts, which are the only non-logical components of E, are in any case endowed with the same meaning in both theories. But doubts about the possibility of clearly distinguishing observational from theoretical concepts first—and eventually the claim that all concepts (including so-called observational or empirical ones) are theory-laden—inevitably eroded the original confidence felt with regard to the possibility of comparing theories. Theory comparison presupposes inter-theoretic stability of meaning (which was supposed to be provided by observational tools); once this disappears, the meanings of all concepts are strictly dependent on the theory in which they occur, and theory comparison is in such a way impossible. This is the basic argument for the alleged 'incommensurability' of theories, which is one of the favourite theses of the 'new' philosophy of science.[17]

---

[16] Feyerabend in particular has insisted on this point, already in his (1963), pp. 16ff.

[17] To be fair, one should recognise that neither Kuhn nor Feyerabend suggest that incommensurability should imply incomparability. Kuhn allows that theory comparison may take place, but on grounds *other* than those suggested by the logical-deductive scheme accepted by empiricists and Popperians. We are not interested here in examining these other grounds, such as accuracy, scope, simplicity, fruitfulness, and the like, most of which Dilworth has incorporated into the Perspectivist conception of science (see Dilworth 2008, Ch. 9, pp. 66–88), but simply want to show how the incommensurability thesis is in itself untenable, and therefore how it makes *some* sense also to compare theories according to the logical-deductive scheme. In parts of this book to come, on the

According to the perspective maintained in this book, however, the assumptions of the above argument cannot be taken as justified since, if we introduce operations in the proposed way, we indeed have the tools we need to be able to say:

(a) there are at least some predicates that can be recognised as empirical (viz. the operational ones);
(b) these predicates have a ground intension or a stable core of meaning which is not theory-dependent;
(c) nothing prevents the same operational predicates from occurring in two different theories T and T' with their ground intensions unchanged;
(d) it is possible to formulate empirical sentences E in which only such O-predicates occur, and which involve only their ground intensions (to the extent that, in order to test E, *only* the operations implied in the operational definitions of the O-predicates are involved).

Under these conditions—the stability of the meaning that is *actually involved* being secured for E—the two theories *are comparable* in the above specified sense. Let us also note that, owing to the function we have attributed to the O-predicates, this happens because the two theories concern the *same objects*, thanks to the *referential* components of the intension of these predicates.

(Footnote 17 continued)
other hand, we shall also clearly indicate the *limits* of this scheme, and propose a more comprehensive approach. Let us therefore quote a passage of Kuhn in which all the above considerations are present:

> The point-by-point comparison of two successive theories demands a language into which at least the empirical consequences of both can be translated without loss or change. That such a language lies ready to hand has been widely assumed since at least the seventeenth century when philosophers took the neutrality of pure sensation-reports for granted and sought a 'universal character' which would display all languages for expressing them as one. Ideally the primitive vocabulary of such a language would consist of pure sense-datum terms plus syntactic connectives. Philosophers have now abandoned hope of achieving any such ideal, but many of them continue to assume that theories can be compared by recourse to a basic vocabulary consisting entirely of words which are attached to nature in ways that are unproblematic and, to the extent necessary, independent of theory. That is the vocabulary in which Sir Karl's basic statements are framed. He requires it in order to compare the verisimilitude of alternate theories or to show that one is 'roomier' than (or includes) its predecessor. Feyerabend and I have argued at length that no such vocabulary is available. In the transition from one theory to the next words change their meanings or conditions of applicability in subtle ways. Though most of the same signs are used before and after a revolution—e.g. force, mass, element, compound, cell—the ways in which some of them attach to nature has somehow changed. Successive theories are thus, we say, incommensurable.
>
> Our choice of the term "incommensurable" has bothered a number of readers. Though it does not mean 'incomparable' in the field from which it was borrowed, critics have regularly insisted that we cannot mean it literally since men who hold different theories do communicate and sometimes change each others' views. More important, critics often slide from the observed existence of such communication, which I have underscored myself, to the conclusion that it can present no essential problems (Kuhn 1970, pp. 266–267).

Another way of expressing the same idea might be the denial of the possibility of comparing theories based on the assumption that there are no independent facts or data to provide us with a criterion for comparison, since facts and data are always such 'relative to a given theory.'[18] The interesting fact is that we too have been claiming the 'relativity' of facts and data, in the sense that facts and data depend on the particular operational criteria employed in a discipline and, as a consequence, also by any theory proposed in that discipline, but we did not take the further step which consists in saying that data and facts are relative to *every single* theory; on the contrary, they remain constant for *all* theories belonging to a given discipline. This is fully compatible with the admission that the meaning of a concept or a statement be 'in general' relativised to theories, since this does not prevent two (or more) theories' having the *same* tools of relativisation *with respect to that concept or statement.* According to our perspective, this may actually be the case with respect to a restricted class of concepts and statements, that is, for the *operational* concepts and for the statements which contain *only* these concepts. This is the case when the two theories are grounded on the same *basic predicates* which are related to the *ground intensions* by means of the same *operations*, and differ only because of the different logical networks they apply (and therefore, also because of the different theoretical concepts they use).

What we are maintaining here is not that theories are *always* comparable, but only that they *may be* comparable as a matter of principle, and *sometimes* actually are comparable as a matter of fact. Therefore, we shall not deny (as pointed out earlier) that concepts labelled with the same name in classical and quantum mechanics respectively actually possess different meanings, such that one is entitled to say that it is not the same energy, position, velocity, and such, which is being considered in each theory.

We take this position for two reasons. The first is that, the two theoretical contexts being different in the case of classical and quantum mechanics, they induce differences in the intensions of their respective theoretical and operational concepts. From this point of view, the situation is not much different from that in the case of Euclidean and non-Euclidean geometries, where we should always bear in mind that it is not the same space which we say admits exactly one, more than one, or no parallel line to pass through a given point, because the axiomatic contexts defining space are different in the three cases. It is because of this, by the way, that no violation of the principle of noncontradiction or of the excluded middle is to be found here (i.e. there is no theory conflict), as both principles

[18] This claim has often been made, especially by Feyerabend, who has consequently denied that two different theories may "*refer* to the same objective situation." See, e.g., Feyerabend (1978), p. 70. This stance, however, is incompatible with his acceptance in his (1975) of gestalt-switch phenomena as constituting instances of incommensurability (as has been pointed out in Dilworth 2008).

presuppose a stability of meaning. In addition to this, we may say that help in making a comparison cannot be derived from operational concepts in the case of classical and quantum mechanics either, since the mensural operations involved in quantum mechanics are not the same as those in classical mechanics. One can therefore say that the two disciplines refer to different 'objects,' and are thus not comparable, as regards their relative superiority, since they do not have the same intended domain. The fact that they share certain terms is a consequence of the fact that several intensional components are preserved more or less unchanged in the concepts expressed by these terms; but these components are related to each other in a different way, and are also bound to different components in the two theories.[19] Therefore, we should not say that quantum mechanics has to be accepted *over* classical mechanics, but *besides* classical mechanics.

We might here consider another example which seems to offer an instance of comparable theories. Let us take the wave and corpuscular theories of light (T and T') as they were conceived in the first decades of the nineteenth century, at which time the corpuscular theory was discarded from physics. In this case we cannot help but admit that, though the theoretical frameworks were really at variance with one another, they were based on the same operational criteria for making light objective. Actually, both theories enabled the deduction of testable sentences concerning actual beams of light travelling through holes, being reflected, refracted and diffracted by means of suitable devices, passing with different velocities through media of different densities, and so on. Due to this *common* stock of empirical facts an empirical statement E was eventually found that was formally deducible from T while its negation was formally deducible from T', and this[20] eventually led to the rejection of one of the theories and the acceptance of the other.

---

[19] Just to give a brief example, the notion of velocity retains its most intuitive intensional features in quantum mechanics, being understood as the rate of change of the position of a particle in its trajectory with respect to time. However it is precisely because the assimilability of a particle to a material point localised in space and time, or the notion of trajectory, are problematic, that the concept of velocity also undergoes modifications. Heisenberg's uncertainty relations may be seen as the new 'contextual' situation which modifies the composition of the traditional intensional pattern of concepts imported from classical to quantum mechanics.

[20] The tools of relativisation for the operationally definable concepts of optics in the first decades of the nineteenth century were the same both in the context of the corpuscular and of the wave theories of light, and it was thanks to them that empirical laws and experiments could be accepted with the same (operationally determinable) *meaning* and with the same *reference*, in spite of the fact that these laws were differently interpreted and explained by the two theories. However, it was precisely because of this 'common relativisation' of the operational concepts that the theories could be *compared*, and that one superseded the other *at that moment*. (At least in this case we believe that the result of the comparison was more decisively determined by this 'deductive-empirical' procedure than by anything else).

Of course, one must not be naive and believe that it was the 'corpuscular' feature which was defeated in the corpuscular theory. This was believed to be the case simply because, owing to the already mentioned visualising attitude of physics at that time, the intuitively most distinct points of disagreement between the two theories were thought to be the really decisive ones. Today we realise that at least some part of the corpuscular conception of light has to be retained, along with certain of light's wave-like qualities.

The conclusion is, therefore, the following: when two comparable theories *A* and *B* are in fact compared, if the results of the global performance of *A* are more satisfactory than are those of *B*, it is *A* that must be considered the superior theory. If we want to use our 'intensional' way of speaking in regard to this issue, we might say that, when crucial experiments seem to have condemned one particular concept as being inadequate with respect to the object under consideration, the theory as a whole should be condemned, while the weak point ought not be localised in the intension of some particular incriminated concept. Indeed, the intensional part of the theoretical concept which was thought to be responsible for the failure may well be innocent, and may be rescued by further developments of that science.

The discourse concerning the comparison of theories is, however, much more complicated than it may appear from what has been said in this section, as it is deeply rooted in the much more highly elaborated discourse on *theory change*. In any case, it represents a rather clearly identifiable sub-problem of that more general problem, and is characterised by some logically and epistemologically crucial questions which we have tried to identify and discuss here. We shall speak about the more comprehensive sense that this problem has in its relation to other questions when we come to consider theory change more specifically. At that time we shall revisit the issue of comparability in order to make explicit certain non-essential presuppositions of the present discussion. In particular, some points of the present discussion seem to depend on the acceptance of the statement view of theories, and on the deductive model of theory comparison, but this is not really the case. The really important thesis is that theory comparison is based on *reference* rather than on *meaning*, and this is why the operational criteria are so important in this respect.[21]

[21] Hence we could say that the positions of Kuhn and Feyerabend represented a progressive step in that they revealed the need to transcend the syntactical narrowness of the Deductive Model, and open the door to semantical considerations (see especially Kuhn 1974, p. 504). However, they failed to take the other step, that of proceeding to praxis, which would have shown them that theory comparison is made on the basis of 'practically' (we say *operationally*) determined referents.

3.4 The Notion of the 'Universe of Discourse'

The different aspects of the complex structure of scientific objectivity we have explored thus far have provided us with sufficient analytic tools to tackle the problem of the ontological status of scientific objects or, if one prefers, of the ontological commitment of science. Before directly studying this question, however, let us give a more exact idea of the general position maintained in this book by discussing a notion which is often used in the philosophy of science, but which has a special meaning in our perspective. This is the notion of the *universe of discourse* of a science, which in the literature is usually made equivalent to the notion of the *domain of individuals* of that science. Such an equivalence is not a simple accidental coexistence of linguistic expressions. Indeed, it covers at least two tacit presuppositions.

The first presupposition is that by "universe of discourse" we must understand a set of entities, equipped with properties and relations *about* which a certain discipline or theory is intended to speak. As we have already observed, this way of thinking corresponds to conceiving of these individuals as 'things,' if one adheres to the intuitive picture of science. If one considers instead certain more sophisticated approaches, such as those represented by model-theoretic treatments of both the formal and empirical sciences, one can see that properties and relations are considered there extensionally as sets of individuals, sets of ordered n-tuples of individuals, and so on. Both in the intuitive and in the sophisticated conception, a kind of hidden Platonism is involved, in the sense that the individuals and their attributes (be they conceived of intensionally or extensionally) are supposed to *exist* in themselves and to *be given* independently of the science which attempts to 'speak about' them as faithfully as possible.

The second presupposition, on which the first is actually based, is the identification of meaning with reference. In fact, when one speaks of a 'domain of discourse,' one employs an expression which, in itself, is of a linguistic character, and as such simply stands for something like "the framework inside which the discourse is intended to be meaningful." Only if one identifies meaning with reference can "domain of discourse" be considered to be synonymous with "the set of designata to which the discourse is intended to refer."

As should be clear from the preceding parts of this book, we do not think that either of these two presuppositions is correct or acceptable, and so we propose an alternative interpretation for the notion of the universe of discourse.

The most intuitive notion we might call into play in order to establish a first affinity is perhaps that of a *conceptual space* which is characteristic of every single science and, within a science, of its different theories. Indeed, as we have already explained at length when we refuted the naive conception according to which every science is characterised by selecting a certain domain of 'things' as its proper field of inquiry, what is typical of a science is rather its 'way of looking' at things or, better, the restricted 'thematic field' to which it limits its inquiry. This means that each science has to do with a specific 'domain of concepts' rather than with a specific 'domain of things,' and it is by means of these *concepts* that it formulates its questions, its problems, its conjectures, its predictions, and its testable sentences. This is what we have expressed earlier in saying that, when a science is considered as an organised set of statements, its specificity is expressed by the particular set of *predicates* it adopts, predicates being the names of concepts in a given language.

We say that the expression "conceptual space" only approximates that of a scientific domain of discourse because we think that the two are related by a genetic link and by a process of technical refinement which constrains us to maintain a certain distinction between them. More on this point will be said when we come to speak of the historical determinateness of science; but we can already express at this stage the main lines of this relation.

When suitable historical conditions are 'ripe' (these conditions being a result of internal and sometimes external factors with respect to science), certain new vistas become current in the scientific community, or new ideas begin to take shape in the mind of a single scientist, ideas which lead to a (more or less) new way of looking at reality. Such new perspectives tend to organise themselves around a restricted number of fundamental concepts regarding entities, properties, relations and processes, and come to constitute a unity which we could compare to a new *Gestalt* in which several already known details are organised in a different shape, or are suddenly shown to be relevant to one another in a way not realised before.

Transitions of this sort occurred, for example, when the Copernican revolution took place, when the mechanistic worldview became widespread in the seventeenth century, when the first steps in the scientific interpretation of fossils were taken in the eighteenth century, when the idea of biological evolution was proposed by Lamark and Darwin, when 'scientific psychology' was begun by several scholars at almost the same time in the nineteenth century, when the idea of the quantum of action occurred to Planck as regards the nature of radiation. In other words, when a new scientific discipline is founded, or when a new theory is about to be proposed within an already existing discipline, this event is prepared for by a process of 'gestaltisation' which we propose to call the construction of the "conceptual space" of the new discipline or theory.[22]

But the constitution of this conceptual space is not in itself a *sufficient* condition for producing a science. In order for this to happen the concepts included in the conceptual space must undergo a process of purification, simplification and

[22] This notion of 'conceptual space' has some affinity with the notion of 'paradigm,' but differs from it inasmuch as it is prescientific. For the same reason it also differs from the 'logical network' of which we have spoken in previous sections, since it is not yet articulated into explicitly defined concepts and explicitly formulated sentences. When this happens, we have a transition to the construction of a *theory* proper, which may be considered a linguistic presentation of the *Gestalt* (and as such is always only partially successful); and one of the most typical features of a theory is indeed the establishment of the 'logical network' just mentioned.

explicitation, and must be reduced to a small and manageable group, in which at least some of them appear to play a strategic central role, while others have to play the indispensable role of providing the testability of the whole *Gestalt*. Concepts of the first sort are almost without exception what will become the *theoretical* concepts in the new discipline or theory, while those of the second sort should give rise to its *operational* concepts. In any case, such a transition is not automatically guaranteed, and it may take years for it to be properly accomplished.[23]

Only if and when this transition has been effected and we have a structure of concepts fairly well related through a logical network—some of them also being endowed with a recognised operational procedure for testing the sentences in which they occur—can we say that we possess the *domain of discourse* of the new discipline proper. Without this distinction we should either confuse any worldview or metaphysical interpretation with science itself (by saying that every 'conceptual space' is already a 'domain of discourse' in a proper sense), or leave unanswered the question of how the domain of discourse of a science is determined (as is the case in most of the current views in the philosophy of science, which either neglect the problem of the genetic starting point of theories, or interpret it, paradoxically, in a non-genetic way, that is, according to an alleged discontinuity in theory change).

We could summarise the above by claiming that giving the domain of discourse of a science is a *semantical* problem which amounts to describing the structure of the *meanings* involved in that science. This, on the other hand, cannot be identified with the problem of describing the set of referents of the science, since this is rather a *pragmatic* question (in a sense of "pragmatic" which is not the usual semiotic one, but is related rather to the idea of operating or doing something, and is in this sense faithful to the original conception of pragmatism introduced by Peirce, which had an explicit operational connotation).[24]

If someone should object that we are dogmatically anti-referentialist, and that nothing actually prevents us from equating the meanings of scientific concepts with their referents (or with their extensions) we could simply invite our would-be

---

[23] For example, the 'conceptual space' of classical mechanics was taking shape when Galileo first proposed characterising nature in terms of its quantitative features (primary qualities), which should provide knowledge of the nature of the motion of material bodies in space through the discovery of laws. To this Newton explicitly added the notion of force, that is, a very particular form of cause (or, if one prefers, a particular manifestation of efficient causality) which was not meant to 'produce' things, but only to modify motion by acting upon material things from the outside. This general framework or *Gestalt* had to be refined and analysed into a set of concepts which were really of use (e.g., some of the Galilean primary qualities, such as 'shape' were not retained), such as those of position, duration, mass, velocity, acceleration and force; and these concepts had to be equipped with certain operational procedures of measurement. Some work in this sense was already done by Galileo, and the rest was done by Newton with whom the actual discipline of particle mechanics was inaugurated, by means of the introduction of such theoretical concepts as that of material point, absolute space, absolute time, and so on, and the explicit formulation of theoretical laws (such as the laws of force and of gravitation).

[24] We shall return to this question, and analyse it with the necessary detail in Sect. 4.3.

opponent *actually* to show us, for example, the 'domain of discourse' of physics conceived of extensionally. As we have already remarked when criticising the 'thing-object' confusion, no one would really be able to indicate to us the *individuals* which are *specifically* the objects of physics. This means that, even with the best of wills and the most tolerant of attitudes in semantic matters, we could not accept that the domain of discourse of a science is given extensionally or referentially, simply because such a domain *does not exist* at all independently of the science itself. The conclusion is therefore that when we give the domain of discourse of a science, we simply *give* (at least in principle) a list of concepts and some *criteria* of referentiality for at least a few of them.

The mention of criteria of referentiality clearly indicates that we are far from being insensitive to the problem of reference (the discussion of the preceding sections should have made this sufficiently clear in any event, and we shall be specifically concerned with this problem again in Sect. 4.3). As a matter of fact, every science, aside from having its domain of discourse, also has its *domain of referents*; but this domain cannot be identified with the domain of discourse which is intended in current philosophy of science. The reason is not merely that we cannot conceive of these referents as being simply 'things,' i.e. individual existents. More complex, and perhaps more interesting, features are to be found if we further explore the structure and the conditions for the making of this 'domain of referents.'

The first remark is that, contrary to the common view, the domain of referents is not *given* for a certain science, but is rather *constructed* step by step, and is a function of the *predicates* which enter the logical apparatus of that science. This fact is simply another way of expressing what was said in describing how predicates (and more precisely, basic predicates) 'clip out' objects from things. One has only to add the obvious remark that, although in principle every 'thing' may become an 'object' of a given science (and thus *enter* the domain of its referents), this is not to say that it actually or in fact does so.

In order to see this with some clarity, let us suppose that the 'domain of discourse' of a particular discipline has been fixed. This means that a *given* list of operational basic predicates $O_1, \ldots, O_n$ has been advanced, together with some theoretical predicates $T_1, \ldots, T_p$. The basic predicates are also provided with their respective 'operational definitions,' which amounts to saying that, for every such predicate, an instrumental device is indicated, with a list of instructions stating how it is to be used and which results must obtain in order for the predicate to be said to apply. As we have already mentioned on more than one occasion, an object is 'clipped out' of a 'thing' as a result of an application of *all* the basic predicates to that thing.

Let us now suppose (going back to an already used example) that we have some 'thing' as concrete as a toothache (only those who never experienced a toothache could say that it is not a real 'thing,' because it cannot be seen or touched), and that we want to know whether it can be an object of mechanics. Assuming that the basic predicates of mechanics are mass (to be measured by a balance), length (to be measured by a metre stick) and duration (to be measured by a chronometer), we see that only one of these basic predicates of mechanics can be applied in our example (i.e. duration). As for the other two predicates (mass and length), a toothache simply cannot be submitted to the operational procedures devised for them, and we must say, therefore, that it has neither a mass nor a length in the sense of mechanics. Therefore it does not belong to the referents of mechanics or, equivalently, it is not an object of mechanics.

It is therefore clear in which sense the referents of a science are not 'given,' but are simply 'constructed' by applying to 'things' the operational criteria of the basic predicates. This implies, in particular, that the domain of referents itself is obviously not given, but that it is rather under continuous construction, in the sense that it is an *open* and potentially infinite set. This corresponds well to the actual situation in science where we have to do with open domains of referents, and never with the alleged infinite domains of individuals of which current analytical methodology often speaks with more fantasy than realism.

More interesting features connected with this fact will be seen when we consider the problem of the semantics of formalised empirical theories. It will then be clear that practically no tool of the usual model-theoretic semantics used in mathematical logic is applicable, and that the alternative semantics we are going to propose has certain commendable features, such as those of decidability and non-ambiguity.[25]

With this clarification, we can now accept an expression which we have refrained from using hitherto in this section, that is, "domain of objects." The reason we preferred not to use it is that it is commonly understood as being synonymous with "domain of discourse" inasmuch as the latter is understood in the extensional and referential sense just discussed. However, if we adopt for the notion '*domain of discourse*' the intensional interpretation proposed here, we could use the expression "*domain of objects*" to designate not the domain of referents of which we spoke above, but the domain of *abstract objects* which constitute (as we have already explained at the end of Sect. 2.7) the *denotations* of the predicates admitted in the domain of discourse.[26] Indeed, when we speak of 'an object,' according to our point of view, we know how many complex conceptual features it involves, and we therefore cannot confuse it with an independent anonymous 'referent.'

---

[25] See the appendix: 'The Semantics of Empirical Theories'.

[26] No uniformity of use exists in the literature regarding the term "denotation," though it is most frequently employed to indicate a word-world relation, and is considered synonymous with "reference." For reasons which have already been explained, at least in part, we speak of denotation also to indicate the relation between a linguistic expression and an abstract object, or *intensional object* (which does not belong to 'world' in the everyday meaning of this expression), while we prefer to speak of reference when the relation is established with an object for which we have 'referential procedures' at our disposal. This terminological convention will be further clarified in the next section.

This is why we maintain that, once the domain of discourse of a science is given, its domain of objects will soon emerge, since, as we have seen, the concepts constituting this domain of discourse are used by the scientists' intellectual creativity to construct those *Gestalten*, those 'models,' which are *intentional entities* (or *entia rationis* according to the classical terminology), and which are those *abstract objects* which we have already seen at length to be the *scientific objects* in a proper sense.[27]

However, we cannot rest content with only this notion of object. Indeed, when we began our discussion of scientific objectivity in general, we noted that it is very common and natural to say that every science investigates its own objects, and we have also repeatedly said (even in the course of this very section) that the criteria of referentiality clip objects out of 'things,' or that a 'thing' may or may not become an object of a given science. Now it is patent that in all these expressions the object is conceived of in a referential sense, as something which is *related* to the abstract object and at the same time *distinct* from it. Should we try to eliminate this second meaning?

It is neither necessary nor advisable to do so. It is much better to recognise that the notion of scientific object has a *bipolar* or two-faced meaning. On the one hand, it denotes the (unique) *abstract* object and, on the other, it denotes a *referential* object. The first is an *ens rationis*, an *intentio secunda*, a *noema* (in Husserl's sense), a *Gestalt* or a model (according to our previous terminology), or the object encoding certain properties (according to Zalta's terminology). The second is (in the case of the empirical sciences, to which we limit our attention here) a *concrete individual* which falls under the abstract concept (as Frege would say), or is a referent endowed with the properties constituting the abstract object (according to our previous terminology), or is an ordinary object exemplifying those properties (in Zalta's terms), or is part of the intended domain (in Dilworth's sense). In other words, we have already seen that the referents of a science are only those things which satisfy its predicates; thus we shall call them, more precisely, *objectified referents* (i.e. referents *inasmuch as* they satisfy the conditions of the given objectification). We can however equally well call them *referential objects*, and in such a way we shall have done justice to the sound belief that every science studies its specific objects not only in a conceptual, but also in a referential sense.

This is by no means a peculiar feature of sophisticated modern science. If I read an article about dogs, I must say that its domain of discourse is constituted by concepts that can be applied to dogs, and that its domain of objects is constituted by dogs, but this in two senses: in one sense the article speaks of dogs only in general, and therefore what it says applies to an 'abstract dog' (or to a model, a

[27] There is a difference between *intention* and *intension* (and between the related adjectives), but at the same time these two notions are both historically and conceptually related. We shall examine this issue in the next section.

concept, a *Gestalt*, a *noema* of dog and so on). But in the other sense it is supposed to be pertinent to concrete dogs inasmuch as these individual entities instantiate the features expressed in the abstract concept. In short, the article *refers* to actual concrete dogs, *through* the abstract concept of dog, so that we can legitimately say that the object of the article (or its subject-matter, that of which it treats) is in one sense the properties constituting the abstract concept of dog, while in another sense, and at the same time, it is the dogs themselves.

We have seen that the domain of the *abstract* objects of a science is open (since new objects may be included in it in the course of the investigation); much more open is the domain of *referential* objects, since membership in this domain is *contingently* decided on the basis of the results of testing procedures, which concern single individuals, so that it seems clear that this domain is potentially infinite and constantly expanding. However, what may seem surprising, but is indeed a confirmation of the futility of the extentionalist point of view, is that we have absolutely no need to know the composition of this domain of referents in order to pursue a scientific investigation.

In order to see why this is so, let us introduce a distinction useful in philosophical discussions, and which we shall make use of later in this work. It is the difference between the *whole* and the *totality of individuals*. When we speak of the whole, we do not speak of a content proper, but rather of the horizon in which all possible contents of a certain line of thought, discourse or investigation are to be included. In this sense, when we speak, for example, of the whole of physics, we do not mean by that the totality of physical objects or of the referents of physics, but the horizon within which every possible physical object or referent must be included. In this sense, while it would be impossible to give or even assume the enumerability of the totality of such individual referents, it is by no means impossible to determine the whole of physics (at a given stage of its evolution). To this end we only need to make explicit the predicates which constitute the domain of discourse of physics. We may correctly say that these predicates determine the whole of physics simply because nothing could conceivably belong to physics as one of its objects unless it is describable by means of these predicates, and everything satisfying this condition must be numbered among the objects of physics.[28]

The totality of things which belong, as referents, to any discourse is on the contrary unmanageable, if one excludes the trivial cases of finite totalities which are of little interest to most sciences. We can therefore say that, while it would be pretentious and even ridiculous to claim that physics is the science in which one knows or tries to know *the totality of physical objects*, it makes sense to say that physics tries to determine and study *the whole of the domain* which they constitute.

[28] A synonym of "whole" in the sense we are using it here could be the more usual term "scope." We have preferred the more exotic terminology, however, in order to underline the global and 'transcendental' purport of those conditions that actually envisage 'the whole of reality' from a 'partial' point of view.

While the importance of this distinction may not be apparent at this point, let us simply mention an issue with respect to which it is illuminating. One of the most fashionable topics in contemporary criticism with regard to science involves calling the category of totality into play. In other words, science is claimed to be an inadequate or even deceptive form of knowledge because it ignores the 'totality.' Unfortunately, this category is often misunderstood, as it is sometimes taken in the sense of what we have here called the *whole*, and sometimes in the also-mentioned sense of the *totality of things*. As a consequence several misunderstandings arise, some of them particularly serious. For instance, one of the main breakthroughs of modern science has been, as we have already seen, that of disconnecting the problem of knowing and understanding some particular features of reality from the task of determining their interrelations with the totality of things. Strangely enough, some modern scholars seem to suggest returning to such a prescientific way of conceiving of a totality, being unaware that most scientific progress has been attained through a transition from the investigation of totalities to the investigation of wholes.

Plurals are used here on purpose, for they indicate (as does the analogous use of the expressions *horizon*, *conceptual space*, *thematic field*, and *domain of discourse*) that, despite a certain paradoxical flavour, the wholes are always *partial*. Because of this partiality, they can complement one another, be compatible, and even, so to speak, be mutually embeddable. The whole of physics, for example, may be thought of as being composed of the union of certain subdomains or *wholes* which are otherwise separated from one another, such as the whole of mechanics, the whole of electrodynamics, the whole of atomic physics, and so on. Note that the whole's being a complex of constitutive conditions, and not a collection of entities, enables one and the same entity, or 'thing,' to belong to different wholes, according to the possibility of its being envisaged by means of the predicates determining the structure of this or that particular whole.

In summary, 'things' are approached and described in different sciences by means of different objectifications which express the viewpoints or the 'wholes' of single sciences. This takes place through the application of operational criteria of objectification to 'things.' But this is not the logically primitive fact, since these criteria are devised only within a particular *Gestalt*, in which several concepts are organised into a unity. Sometimes this *Gestalt* is of a low level and is almost entirely constituted by empirical and sensory features; but in almost all the sciences much more complex *Gestalten* are introduced by virtue of an intellectual synthesis. In the case of these more complex constructions, some of the features entering the *Gestalt* must be equipped with operational conditions for testing. Only if these requirements are satisfied can a general perspective on things, a conceptual space, be promoted to the level of being the domain of discourse of a particular science, and make it practically possible for 'things' to enter the domain of objects of that science, and actually be investigated by it.

What we hope to have sufficiently clarified thus far is therefore the concept of a scientific object as we see it. Still, some understandable dissatisfaction could remain as far as the notion of 'thing' is concerned. Although it does not play a positive role in our discourse, it nevertheless constitutes a kind of prerequisite for the notion of object itself, and thus merits further investigation.

[illegible]

Chapter 4
The Ontological Commitment of Science

4.1 A Semiotic Framework[1]

In the course of our presentation we have often used expressions such as "meaning," "reference," "denotation," "intension," "extension," and so on. All these (and related) terms are far from having unique, standard meanings in the philosophical literature, and so we must clarify the way we are using them in this work. Moreover, we have presented several theses which presuppose, or imply, certain ways of conceiving of the relations between language, thought and reality that diverge at points from certain widespread ways of envisaging them (e.g., our thesis that an 'intensional' rather than 'extensional' semantics is appropriate for the treatment of scientific theories from a linguistic point of view). Without entering into details (especially concerning the critical evaluation of the most controversial points), we shall simply try to make explicit the general lines of the semantic framework in which the discourse of this book is situated, and at the same time fix the terminology we shall adopt for the treatment of certain special topics.

4.1.1 Sense and Reference

In the field of methodology and philosophy of science, semantics is most usually conceived of as consisting in the task of assigning an 'interpretation' to 'meaningless' symbols of a given language. This interpretation in turn is seen as an association of certain appropriate *referents* or objects (individuals, sets of individuals, and so on) to the different kinds of symbols, which in such a way are supposed to receive a *meaning* and to *become* meaningful. This approach has become standard for the semantics of formal systems, and constitutes the basic perspective on which model theory is grounded in mathematical logic, while also

[1] Some parts of this section have been published in Agazzi (2012).

E. Agazzi, *Scientific Objectivity and Its Contexts*, DOI: 10.1007/978-3-319-04660-0_4,
© Springer International Publishing Switzerland 2014

being the approach presupposed by almost every current work on the semantics of empirical theories.[2]

Against this perspective, we maintain that the task of semantics is certainly that of providing the *meaning* of linguistic expressions, but this task is different from the problem of providing the *referents* of these constructions, since it requires as a precondition that of providing for them a *sense*. These two problems are analytically distinct, even though they are strictly *related*. There is a difference between simply *referring to* a thing and *saying something about* this thing, this difference being especially evident in those cases in which the referent can be spoken about in different true statements (while remaining the same referent), or in those cases

[2] It would take us too far afield to reconstruct the history of this approach. However, at least certain points deserve mention since they throw light on issues relevant to our discussion. As we have noted on earlier occasions, twentieth century positivist philosophy of science was deeply influenced not only by the 'linguistic turn' of contemporary philosophy, but especially by the creation of mathematical logic and the 'formalistic' trends of Hilbert's philosophy of mathematics. The adhesion to the linguistic turn produced the conviction that a full understanding of science could be provided by an analysis of the *language* of science. The fascination of mathematical logic led to the creation of what has been called *the mythology of deductivism* (see Harré 1970, Chap. 1), which in particular has led to what has been termed *the statement view of theories* and *the hypothetico-deductive model of scientific explanation* (Hanson 1958, p. 70 ff.), and in general to an abhorrence of the idea that "vehicles for thought are not wholly propositional but 'pictorial' as well" (Harré 1970, p. 2). In short, the empire of formal deduction, which was seen to cover all mathematical disciplines, was extended to encompass all of the exact sciences, including the empirical ones. But the last step was even more drastic. In his *Logical Construction of the World*, Carnap explicitly says that "science is concerned only with the structural properties of objects" (see his 1928, p. 12 ff.), and makes explicit reference to the spirit of mathematics and to Hilbert's doctrine of "axiomatic definitions" (which we have already mentioned, particularly in Sect. 2.8 of this work). But this doctrine had been created in order to dispense with *reference* in mathematics, and how could it be applied to the empirical sciences? We know that a possibility in this sense is not excluded, and that already Poincaré, e.g., had maintained that science is able to discover objective relations in nature (therefore considering a 'content' overstepping the simple syntactic domain). This is the central idea of that 'realism of structures' that has known interesting developments recently but was not the idea that Carnap was advocating in his work. Carnap says—and logical empiricists have repeated after him—that a scientific theory is a formal calculus to which 'interpretations' may afterwards be assigned (by means of 'correspondence rules' or otherwise). But how can these interpretations actually be selected and applied? Mathematical logic (and in particular its branch known as model theory) has a reply to this question, since the interpretation of formal systems is conceived of extensionally and in purely set-theoretical terms, so that the *nature* of the elements of the set is immaterial, and their properties and relations may be (extensionally) defined in an arbitrary way in order to construct the interpretation. But this cannot be satisfactory for the interpretation of *empirical* theories, in which the referents are *intended* and endowed with attributes which the theory is meant to express and describe, and in which no formal tools are able to single out this 'intended' model among the other infinite models of the theory. For more details regarding the efforts and the difficulties which emerge in the realisation of this programme, see Przelecki (1969); for a more general survey see Agazzi (1981). For a discussion of the inadequacy of the model-theoretic approach in order to single out the 'intended model' see Agazzi (1976), reproduced also as Appendix in the present work, whose arguments are similar to those presented later by Putnam in Chap. 2 and the Appendix of Putnam (1981).

in which the same sense can be legitimately predicated of different referents in true statements where only the denotation of the referent changes. This distinction has been clearly elaborated by the Scholastic logic in the distinction between *intentio* and *suppositio* and has been recovered in the Fregean distinction between *Sinn* and *Bedeutung*, of which we are going to speak below. As a consequence, it is very far from obvious that, when we give an 'interpretation' of a formal system by associating its expressions with certain referents, we provide these expressions with a 'full' *meaning* (i.e., including a *sense*). Of course, we *can* provide them also with a sense, but this requires associating them with other such senses, and not with referents, as we shall soon see.

This resistance to conflating meanings with referents has a long tradition in the history of philosophy. It is implicit, for example, in all criticisms of the so-called ontological argument for the *existence* of God; and it is at the root of Kant's claim that some "synthetic" (i.e., empirical) condition must be present in order for a statement to be credited with providing knowledge, since this is tantamount to saying that even a perfectly 'meaningful' thought or sentence is not a warranty for the existence of the corresponding referent. But it is also to be found at the origin of the contemporary treatment of semantics, a treatment which does *not*, however, begin with the explicit introduction, or codification, of the *term* "semantics" in the technical vocabulary of contemporary philosophy that is usually traced back to Charles Morris' subdivision of semiotics (the general theory of signs) into syntax, semantics and pragmatics, where "semantics" had an explicit referential connotation.[3] Indeed, if we consider semantics to be the study of meaning in general, we certainly cannot maintain that it was only in the 1930s that this problem was approached in a 'modern' way. Indeed, according to a widely accepted historical

[3] Actually Morris does not define semantics as the study of the *meaning* of signs, but says that "semantics deals with the relation of signs to their *designata* and so to the objects which they may or do denote" (Morris 1938, p. 35 of the 1971 reprint). Therefore, semantics is introduced with a tacit referential connotation. But this is not surprising, if we consider the general behaviouristic background of Morris's thought, in which meaning could not really play any role. It is worth noting, however, that Morris was considering *meaningful* languages, that is, languages for which the problem is not that of *attributing* a meaning (be it understood in a referential sense or otherwise), but that of explaining in what meaning consists. On the other hand, the use of the term "semantics" in the field of logic, methodology and philosophy of science, introduced by Tarski in the 1930s, was then, in a certain sense, codified in Carnap's *Introduction to Semantics* (1942). Carnap makes direct reference to Morris's work in the first pages of his book, where he introduces his discourse on semantics. But he is also particularly sensitive to the different trend in the philosophy of mathematics and formal logic characterised by the *formalistic* outlook, according to which formal systems are meaningless constructions to which a meaning may be artificially assigned through interpretations, as we have indicated in the preceding note. To a certain extent this was also Tarski's attitude when he first introduced semantic considerations into the methodology of the exact sciences—though, on his view, this expressed the need to overcome the purely syntactic approach that still characterised Carnap's work at that time. In particular, a merit of the Polish School is that of having established and vindicated semantics as a particular part of the methodology of science and as a specific discipline of logic (see Tarski 1933 and 1936, and Kokoszynska 1936).

reconstruction, the 'modern' consideration of the problem of meaning begins with Frege who, for every linguistic expression or *sign* (*Zeichen*), proposes to distinguish a *sense* (*Sinn*) and a *reference* (*Bedeutung*).[4] Sense and reference are both included in Frege's semantics. Moreover, if one considers that his chief

[4] Actually the German word *Bedeutung* is normally translated as "meaning," (in the sense of 'significance') which would make it synonymous with *Sinn*, or "sense," like the corresponding English terms. However, Frege, in his famous paper *Über Sinn und Bedeutung* (1892), wanted to avoid all 'identities' being tautologies on his view, and used these terms to make a distinction which he though would allow him to do this. As a consequence, it has become customary in recent literature to reflect this technical distinction by translating the Fregean "*Bedeutung*" as "reference" (even in German it is now replaced by the term "*Referenz*"), leaving "meaning" to be synonymous with "sense," as it is in everyday language. We can note, however, Feigl's translation of that paper, where he, following Carnap, uses the term "*denotatum*," and we may also remember that early authors used to translate Frege's terminology as the difference between "meaning" and "denotation" (see, e.g., Russell in 'On Denoting').The issue is not trivial because "reference" can be considered a poor translation of *Bedeutung* if one means by reference the 'act of referring,' since Frege's *Bedeutung* means the object, not the act, of referring. A few authors advocate this view, as a consequence of their having taken into special consideration the concrete activity of speaking. By "term," therefore, they mean something written or spoken which can be used (by speakers) in referring (listeners) to referents or in expressing senses. Referents (more generally, *objects*) are what listeners are referred to, using terms, by speakers. Therefore "reference" is one linguistic expression of the more general category of *intention*; it consists in the directing of a listener's attention (referring the listener) to a referent by a speaker. In conclusion, refer*ence* is an act (involved in e.g., stating, judging and describing), while a refer*ent* is an object. This view, proposed by Strawson and shared, e.g., by Austin and Searle, is certainly of interest as far as it concerns the use of statements, but its limits consist precisely in making of reference just a property of such a use, strictly depending, in particular, on the concrete circumstances of this use. It seems to us, however, that if one wants to draw attention on this concrete use, one might better speak of *referring* (that has all the normal features of an act), leaving *reference* to indicate in general the domain of objects denoted by a term, a domain to which one or more referents can belong. (By the way, the title of the paper where Strawson proposed his thesis (Strawson 1950) has the title "On referring" and not "On reference.") Owing to this situation, one must admit that "there is ambiguity about the term 'reference,' an ambiguity which threatens the validity of many observations usually made on the nature and conditions of reference. But we need not worry too much about it if our question is what it is, viz. whether definite descriptions would ever allow singular reference. If the question is understood as a question regarding the reference of the definite description, it is a purely semantical question,... If, on the other hand, the question is understood as a question regarding the speaker's reference, it is also a question about the speaker's ability to refer." (Sen 1991, pp. 25–26). After having recognised this ambiguity, however, we prefer to eliminate it in our work by deciding to intend reference according to its semantical sense. This choice is recommendable for the same reasons that induce us to speak of the sense or the meaning of a term without caring about the particular speaker or listener that is exchanging these private mental contents. Therefore we shall conform to the more widespread practice in this work for the sake of simplicity, and use "reference" almost as a synonymous of "referents" or, more precisely, as indicating the domain to which the referents of a term (if any) belong.

Coming now to the second expression of Frege's paper (*Sinn*), our preference would be to use "meaning" in the most general sense, so that one could include in the meaning of a linguistic expression (as different 'aspects' of the meaning) both the sense and the reference. We shall refrain from adopting this more complicated convention here, and shall only make an occasional mention of it when we speak of intension and extension.

philosophical intention was that of studying *thought* in the sense of 'thought-contents' (*Gedanken*), and that he was firmly convinced that such a study was possible not at a psychological level but through the study of *language*, we can conclude that the major weight of his semantics was on such (objectively conceived) thought-contents, that is, on *sense*.[5] This is so much the case that he maintained that referents may be obtained only through sense, and for that reason also attributed a sense to proper names, which are the typical linguistic signs having individuals as referents. But this three-level semantics (where the three levels are those of sign, sense and reference) lost its intermediate level already with Russell, and the meaning of linguistic signs was reduced to their referents or *denotata*, in spite of Russell's remaining Fregean in certain respects. This tendency was reinforced in the *extensionalist* semantics for formal languages introduced by Tarski, and developed in model theory in mathematical logic. As will be clear in the sequel, we too are going to defend the thesis of a 'three-level semantics' essentially in the same spirit, and for this reason we shall propose to consider *meaning* as a composite entity consisting of *sense* and *reference*; therefore, we shall accept the common consideration of semantics (in a general sense) as the theory of meaning of linguistic expressions, but pointing out, at the same time, that the problem of reference oversteps the scope of linguistic analysis, and calls into play an extralinguistic *practical* dimension (that of *operations*).

However, this is not the entire story. If we consider the tradition attached to Brentano, Husserl and Meinong (a tradition that was contemporary with Frege, and which explicitly influenced and was influenced by his doctrine), we see that interest is focused on the cognitive acts in their *intentional* aspect (intention and intension are not the same thing; however they are significantly related, as we shall see). This also leads (e.g., in the case of Husserl) to a three-level semantics in which the content of the cognitive acts is a world of meanings very similar to the Fregean *sense* (the world of *noemata*, or of intentional objects *qua* intentional), while the *referents* (which Husserl calls *objects*) remain outside the interest of phenomenological research (in fact Husserl never provided a theory covering referents).[6]

---

[5] Already before his 1892 article, 'On Sense and Reference,' Frege had very clearly expressed his views on the objective nature of concepts and on the primacy of understanding concepts over the task of indicating referents. See for example this passage: "The concept is something objective that we do not construct and which does not construct itself in us, but rather that we try to comprehend and, hopefully, to really comprehend, if we do not erroneously seek something which is not there" (Frege 1891, p. 158).

[6] Valuable accounts of the historical and conceptual links between semantic conceptions which originated and were developed in the field of phenomenology on the one hand, and in the logico-analytic tradition on the other, are presented in two articles by Guido Küng (1972, 1973). In these papers one may also find a useful explanation of certain systematic and terminological distinctions which we have hinted at only briefly in some passages of this work. It is perhaps not superfluous to note that the term *noema*, that we relate especially to Husserl because of his important analysis of the nature and relevance of this notion, is already present, with nearly the same meaning, in Aristotle.

Looking at our own time, there is, on the one hand, a rich body of research based on the semantics of *sense* and *intensions* (intensional semantics, the semantics of non-existent objects, and so on),[7] as well as a no less influential trend involving the development of a *referentially* based semantics (that of the so-called anti-Fregeans), which attempts even to dispense with the world of sense. This latter trend takes the only interesting problem to be that of clarifying how, without relying upon mental representations, linguistic expressions can have *referents* thanks to a system of socially determined contexts of communication.[8] Still more recently, however, in parallel with the so-called "cognitivistic turn" prompted by functionalism, Chomskian innatism, and Fodor's theory of mental representations, the internal (although not necessarily psychologistic) aspects of meanings have been emphasised again; and since these internal aspects need not be incompatible with the external aspects studied by Donnellan, Kripke, Putnam, Dretske, etc., some have proposed dual-aspect theories of meaning.[9] From the short account just given it appears that the contemporary concept of semantics is truly ambivalent, and actually covers two different approaches or intellectual interests, the one having to do with the study of sense, the other with the problem of reference.

At this point one feels entitled to propose an even more radical historical reference, and go back to the Aristotelian distinction (which was also preserved in the Western philosophical tradition) between *semantic discourse* and *apophantic discourse*, the first being related only to *meaning*, i.e., the *understanding* of linguistic expressions, and the second having to do with the *reference* of these expressions, which is implicit in the fact that a certain kind of expression (namely, statements) may be affirmed or denied, giving rise to the problem of their truth or falsity.[10] In this way we can say that the primary semantic problem is that of meaning understood as *sense*, while the problem of reference is rather secondary and indirect because (as we have already explained and shall develop more extensively in the sequel) 'capturing the referent' is an operational or pragmatic enterprise, which is certainly related to sense, but also largely independent of the *full statement* of sense. On this point we agree with much of what the anti-Fregeans say about reference, without sharing, on the other hand, their opposition to the Fregean *sense*, which has a substantial role to play outside the particular

[7] A documentation and discussion of this trend may be found in Zalta (1988).

[8] A recent presentation of this latter trend is provided in Wettstein (1991). This work not only offers a reconstruction of the central views and aims of the Fregean and Russellian approach, but also presents an illuminating interpretation of the philosophical insights which are only partially made explicit in the referentialist approach of people such as Kripke, Donnellan, Kaplan, Perry and Putnam who, developing Mill's conception that proper names have only reference without any meaning (contrary to Frege's claim) have elaborated a 'new semantics' in which mental contents, intensions, and thoughts tend to be dispensed with (we are not interested here in analysing the possible exceptions to this rule).

[9] For some information regarding these more recent trends let us simply refer to a few works such as McGinn (1982), Davidson (1986), Block (1986), Lepore and Loewer (1987).

[10] We shall consider this Aristotelian doctrine again in Sect. 4.4, where we shall also give a few textual references.

problem of determining referents.[11] Let us note that the 'primacy' we are attributing here to sense over reference in *semantics* depends on the fact that semantics is an investigation regarding *language* and, for this reason, must follow the inverse path with respect to the process that has led to the formation of language. Indeed, from a *genetic* point of view, it is obvious that sense and reference have a common root in our primordial acquaintance with the 'external world.'[12]

4.1.2 Intension and Extension

A distinction which does not coincide with that between sense and reference, but has clear affinities with it, is that between intension and extension. This distinction was introduced in modern philosophical literature by Carnap (see Carnap 1947), and was in many respects a revival of the traditional distinction between the *comprehension* and the *extension* of a concept. Intension is the sense, the content of thought, the set of properties expressed through a concept, while extension is the class of individuals to which the concept applies.[13]

---

[11] Wettstein, for example, maintains that the central role attributed by Frege to sense is a consequence of the fact that Frege and Russell (and in his view even several 'conservative' anti-Fregeans who are unable to renounce a 'cognitive fix') are still prisoners of that 'representationalism' which he (correctly) traces back essentially to Descartes. However, it is certainly mistaken to attribute such a conception to Frege. Indeed, not only is there no plausible evidence for this in his major works, but in other writings he strongly criticises the thesis that representations are the objects of our knowledge (see, for example, "representations cannot be seen or touched, neither smelled, nor tasted, nor heard. I take a walk with a friend. I see a green meadow; I have in such a way the visual impression of the green. I *have* it, but I do not *see* it." Frege 1918, p. 67; my italics).

But it is after all not that important to decide this question regarding Frege, since the admission—and indeed the elaborate and rigorous introduction—of the ontological world of meanings was performed by a scholar, namely Husserl, who certainly counts among the most decided opponents of representationalism and epistemological dualism in contemporary philosophy. Precisely because he realises that that which we intend in a perceptual act is the referent, but that in a certain different sense we also intend the meaning of such a referent, Husserl introduces the notion of the intentional object which, starting with his *Ideen* of 1913, becomes the *noema*. This remains very distinct from the referent, since the referent belongs, so to speak, to the external world, while the *noema* belongs to the world of meaning. A clear presentation of these passages is offered in the already cited article, Küng (1973).

[12] As a conclusion of our discourse we can say: if one considers 'meaning' as a general notion, including as its 'aspects' both sense and reference, one can consistently maintain that semantics studies meaning, and that in so doing it has a legitimate (and necessary) part that is concerned with sense, as well as a legitimate (and necessary) part concerned with reference. Let us also recall that in presenting certain of our views on 'intensional semantics' we have explicitly stressed that the 'relation to the referents' is part of the intensional meaning of concepts (we have qualified it the 'referential part,' as distinct from the 'contextual part,' of this meaning).

[13] This is not only conceptually, but also historically true. In fact William Hamilton, starting his lectures on logic in 1837–1838, introduced *intension* in his discussion of the "quantity of concepts," explicitly equating it with the notion of comprehension, which was common in the tradition. Let us consider a few remarks from his *Lectures on Logic*, published after his death:

Carnap himself mentions Frege's distinction between sense and reference in this context, and seems to accept the whole of the Fregean approach. In particular, he strongly affirms that the investigation of intensions is no less important, and need be no less rigorous, than the study of extensions. However, in his consideration of intensions he was concerned with them as indispensable for treating logical difficulties that affect the truth-conditions of sentences belonging to modal or epistemic contexts, and his ideas developed in this direction in the context of systems of what are called *intensional semantics* and *intensional logics*.

Neither Carnap nor his followers regarded the consideration of intensions as necessary, or even useful, in the semantics of usual declarative contexts (such as

---

(Footnote 13 continued)

> This quantity is of two kinds; as it is either ... Intensive or Extensive .... The former (the Intensive Quantity) is called ... by the Latin logical writers the *comprehension* (*comprehensio, quantitas comprehensionis, complexus*, or *quantitas complexus*). The latter (the Extensive Quantity) is called ... by the logical writers of the Western or Latin world, the *extension* or *circuit* (*extensio, quantitas extensionis, ambitus, quantitas ambitus*) and likewise the *domain* or *sphere* of a notion (*regio, sphaera*) .... The Internal Quantity of a notion—its Intension or Comprehension, is made up of those different attributes of which the concept is the conceived sum, that is, the various characters connected by the concept itself into a single whole in thought. The External Quantity of a notion or its Extension is, on the other hand, made up of the number of objects which are thought mediately through a concept. (See Hamilton 1860. The quotation is made from the 2nd revised edition, 1866, vol. 3, pp. 141–142.)

The Scholastic tradition had actually known the use both of "*intentio*" and "*intensio*." The second term had been used by certain authors in order to eliminate the ambiguity of the meaning of "*intentio*" which, after having for centuries meant 'purpose' or 'goal,' had started to be used to indicate the representational content of a concept in general. For this second meaning, they proposed to use "*intensio*." Therefore Hamilton's terminology was in fact resuming a certain tradition, and his influence was such that it practically replaced the use of "comprehension" in the English-speaking logical community. This explains why Carnap, when promoting the circulation of this notion in the literature of contemporary logic, did not even mention "comprehension," and made use of the term "*intension*," which was then adopted, as is clear from what he says:

> Now we shall introduce the terms "*extension*" and "*intension*" with respect to predicators.... The use of "intension" varies still more than that of "extension." It seems in agreement with at least one of the existing usages to speak of the same intension in the case of L-equivalence. Thus we lay down the following *conventions*.... The *extension of a predicator* (of degree one) is the corresponding class.... The *intension of a predicator* (of degree one) is the corresponding property [pp. 18–19].... The *extension of a sentence* is its truth value.... The *intension of a sentence* is the proposition expressed by it. (Carnap 1947; from the 2nd enlarged edition of 1956, pp. 26–27).

In fact, it suffices to recall, for example, that Tarski speaks of the 'extension' and 'intension' of the concept of truth in his 1944 paper (Tarski 1944, p. 342), without feeling the need to explain the sense of these words, which means that they were in common use.

those of scientific theories), where the extensional methods continued to be dominant. The reason for this reluctance appears to be the same as that which had led to the reduction of meaning to reference, that is, diffidence towards admitting the legitimacy of something (i.e., meaning) not strictly identifiable with a subjective mental state, which at the same time was to be the manifestation of an abstract, speculative, and perhaps 'metaphysical' form of existence.[14] Extensions, in other words, can be empirically given in the same way as referents, which are individual objects that constitute extensions by being grouped in sets or set-theoretically manageable structures. Intensions, on the other hand, have a much more elusive status. This philosophical reason is more fundamental than the (non-negligible) factual reason that, while set theory had provided a powerful technical tool for expressing extensional semantics, it was unable to provide an exact treatment of properties and relations.

We may note that an advantage with using the terminology of extension and intension is that both may be considered as constituting two aspects of *meaning*. In this case, of course, "meaning" would be taken very generally, and would no longer be synonymous with "sense" (the term "sense" actually not being used in this context), as we have hinted in footnote 13.

From what we have said it appears that, in order for the distinction of intension and extension to be really significant, one must attribute to intension a kind of existence different from that of a pure and simple mental state. This was of course admitted by those scholars whom we have already mentioned, such as Frege (who considered *Sinn* to be the objective content of thought, which remains the same independently of the different psychological acts of thinking with which people apprehend it), and Husserl (for whom every mental state has a 'content'—the *noema*—that is different from the 'object' to which the state is directed, and which might not even exist). In particular, the admission of such entities is advocated in order to understand the common experience of such normal mental states as thinking, believing, hoping, desiring and seeking, which are 'directed' towards objects or states of affairs that very often do not exist. There must be 'something' towards which these mental states are directed, and this something, in spite of not belonging to the world of concrete things, is not obtained through self-reflection. Therefore it is not part of the mind but, in a way, part of the world. When we think of Pegasus—Brentano would say—we do not think of our idea of Pegasus, exactly as we do not think of our idea of the moon when we think of the moon—as Frege would say.

In order to fix its position, we can call this entity an *abstract object*. This terminology indicates, on the one hand, that it is not part of the mind (it is an 'object' towards which the 'subjective' mental act is directed) and, on the other, that it is not part of the 'concrete' world. Let us note, however, that we do not

[14] In this respect, one can note that, while Carnap relates extensions directly to *linguistic expressions*, and not to the properties or relations they express, Frege, on the contrary, associated the extension with *concepts*.

consider "abstract" as synonymous with "immaterial." For example, institutions, laws and several cultural entities are not material, but are 'concrete' and—as we shall better see later—they can be referents of 'abstract' entities such as concepts. The same holds for literary or mythological individuals that have a 'concrete' though non-material existence in novels or legends, and are 'intended' by the respective conceptual representations. All this will become clear later when we explicitly enter into an ontological discourse. It seems that equating this abstract object with the *intension* of a term would be a reasonable solution. The so-called 'non-existent objects' are therefore pure intensions to which no concrete referent corresponds, while, of course, concretely existing objects are referents and are also thought of through an intension.

The justification of this proposal comes from the consideration of the most accepted examples of intensions. They are typically concepts of properties and relations, so that in a semantics that truly wants to take intensions seriously, we must say that basic predicative expressions *refer to* the contents of their intensions (i.e., properties and relations). But now we can easily see that a compound (or complex) predicative expression formulates a certain connection of properties and relations so that it is simply a matter of consistency to say that the abstract object which corresponds to this combination is the intension of the predicative expression. The process through which the concepts of properties and relations are combined so as to lead to an abstract object is a construction concerning which we have indicated several details in earlier sections, and which—being an intellectual activity—for this reason results in a *noema*. To use a terminology we have already adopted, we say that this noema *encodes* the properties which enter into its construction.[15]

We must recognise that what has been said thus far, while explaining how it is possible to have a world of meanings objectively structured independently of the concrete world, is not sufficient to let us reach this concrete world. Here the reference vindicates its role, which may be expressed in several ways. Thinking of something does not imply the existence of this something; not all conceivable properties of an object need to be satisfied by the object; an abstract object 'encodes' a complex of properties that may fail to be instantiated in any concrete (material or non-material) object; and so on. In such a way we recover the standard terminology (i.e., we have room for reference no less than for sense or intension, and conceive of reference in the most usual way). However, it is clear at the same time that the referent is not an object which is determined through *predication*. Predication will serve to identify which properties the referent should exemplify, but whether it instantiates them or not (and even more radically, whether there is an object instantiating them) is not a question that can in practice be answered by a purely conceptual assessment. As we have already stressed several times, the contact with the referent has to do with an operational intervention, and this is the

[15] For example, the complex predicative expressions, "the red coloured sky at sunset" and "the golden mountain" are both abstract objects or *noemata*, resulting from the composition of concepts that, individually taken, have their reference, but while the first complex *noema* has obviously a referent, the second does not.

novelty which enables us to say that the object so identified is *concrete*. We shall not insist on this point now since we shall return to it in the Sect. 4.1.3.

We have found here again reasons for advocating our three-level semantics (language-sense-reference). While scholars who maintained the reducibility of meaning to reference (i.e., the advocates of an 'extensional semantics') eliminated the 'intermediate' level of sense and intension, those who eliminated the autonomous level of reference (i.e., the advocates of the 'contextual' nature of meaning) lost the means for linking language and thought to the concrete world. Moreover (as we shall see in Sects. 4.4 and 4.5) there is no possibility of speaking properly of truth if reference is not taken into account. All this is valid in general, but it is also directly relevant to our investigation regarding science. Models, hypotheses, theories are all, at least in many respects, the contents of thinking acts, and therefore are given in 'intentional states.' As such, they are *abstract objects*, encoding certain properties, and are subject to the 'intensional logic' of admitting, believing, proposing, and so on. However, they are not simply investigated as abstract or intentional objects, but are *intended* to relate to the concrete world, and are actually evaluated according to their ability to cope with this intention. This implies the transition to referential procedures, and to the operational semantics of concrete exemplification.[16]

---

[16] Radical empiricists are not ready to admit the 'intermediate level' of sense and intensional objects, but this can bring them into difficulties when they elaborate their most sophisticated doctrines. A case in point is van Fraassen's *empirical structuralism* whose clear understanding should obviously rely upon a satisfactory definition of the notion of structure itself. However no such definition is provided by the author, who simply uses the undefined word "structure," and expressions such as "abstract structure" and "mathematical structure." Let us consider, however, a significant passage (van Fraassen 2008, p. 238):

> Essential to an *empiricist structuralism* is the following core construal of the slogan that *all we know is structure*:
>
> I. Science represents the empirical phenomena as embeddable in certain *abstract structures* (theoretical models).
> II. Those abstract structures are describable only up to structural isomorphism.

This way of speaking would incline one to conceive of these 'abstract structures' as intensional objects, as noemata, to use our preceding terminology, but a strict empiricist—as the author is—does not want to admit this ontological qualification. Thus one does not know what 'kind of reality' these abstract structures have since, after all, they are different from nothing and can even be embedded into one another. Moreover, why are they called "abstract" if there is no mention in that book of "concrete structures"? We are told that phenomena are "embedded" into abstract structures, but this might have something like an analogical sense if phenomena also had a structure, but this is explicitly excluded a few lines below this quotation. On the contrary, we would say that models and mathematical structures are *abstract objects* that can be *exemplified* by phenomena that become their *referents* thanks to certain operations.

Van Fraassen, however, neatly rejects such an idea. For instance, discussing the example of an assertion stating that a table top is square, he says, "That is true, but simply because this table top is square—*c'est tout!* It is true because the top's sides are of equal length and the

4.1.3 Intensionality and Intentionality

Let us now clarify a terminological issue. In the philosophical literature (but also in our present work) the terms "intension" and "intention" are used in ways that sometimes seem to imply a difference in meaning, and sometimes seem to imply almost a synonymy between them. How can one avoid such ambiguities?

We have already provided, in notes to this and to foregoing sections, certain historical elements of clarification, recalling in particular that already in classical and medieval philosophy *intentio* originally meant *proposal* or *goal* (which is also the meaning that survives in today's use of the word "intention"). However, later in the medieval philosophical tradition, *intentio* received a more general meaning, being identified with the representational content of mental acts, a content which is no longer bound only to acts of volition, but which also concerns cognitive or epistemic acts of different sorts. This second sense was explicitly recovered by Brentano and his followers, so that in contemporary philosophical language both meanings of "intention" are present. Therefore, *intention* (even when it is not understood in the everyday sense of proposal or goal) is a specific characteristic of psychic *states*, and indicates their being intrinsically 'directed' towards something. This something need not exist concretely, and as such has simply an *intentional reality*.[17]

On the other hand, *intension* is understood in logic and the philosophy of language as that part or aspect of the meaning of a term which is different from its *extension* (and we have mentioned in a note that this term also had a late Latin

(Footnote 16 continued)
angles between them are right angles. It could be paraphrased as 'the table top instantiates the Euclidean square form,' but the cash value of the assertion carries no metaphysical commitment: it is just that the table top is square" (2008, p. 249). One could object that the operations of measuring the sides and the angles of the table can allow me to say that the table top is square *because* they lead to exemplifications of the properties *encoded* in the concept of square, and not, for instance, of circle. We think that an empiricist could come to accept this position, provided that some (for him) palatable explanations were offered regarding how such abstract concepts are arrived at, but in such a way the discussion would be limited to the domain of epistemology, and would not concern ontology. What is transparent in van Fraassen's position, therefore, is an ontological stand, in which one could recognise the features of *nominalism*; and if one also notes his allergy to "metaphysical commitment" (which is declared also in other passages of his work) one can interpret his position as a development of the neo-positivist tradition. This is by no means a negative appreciation; on the contrary, the intelligent, original and creative elaborations that this philosopher has been able to produce during many years attest to the internal wealth of that tradition, a wealth that can be gladly recognised also by those who criticise the tradition, point out its limits, and advance different proposals.

[17] We note that recognising the 'intentional' nature of cognitive acts and their products (i.e. representations) can be limited to recognising the 'directionality' of such acts and products, and to these products even the qualification of 'intensional' can be applied without the further step of admitting an 'intensional reality' as such. This is typical of radical empiricists and is well demonstrated in van Fraassen (2008) where intention and intension are very parsimoniously mentioned in this limited sense (pp. 21–22), whereas an 'ontological status' for such intensional entities as representations and models is not provided, and this is a critical point in his doctrine that we have considered in the preceding note.

ancestor, "*intensio*"). Therefore, intention (which is a characteristic of psychic states), and intension (which is a semantic property of terms or concepts) are different things. However, despite their differences, the intension of a term is clearly something which can be grasped through the *intentional* act of thinking (indeed, it is the content of this act, though understood in an 'objective' and not just in a psychologically subjective way). Therefore *intentional objects*, that is, objects towards which any intentional act or state is directed, cannot avoid being a combination of intensions, and in this sense they are also *intensional* objects, which we must distinguish from concrete objects. This is why we also say that they are *abstract* objects.

As a consequence of this link, sentences that describe intentional states (e.g., those of believing, wishing, and so on) are intensional. However, this declaration may be taken in two senses. The more usual sense consists in calling them intensional simply because they violate certain principles (such as substitutivity or existential generalisation) of the standard logical calculi based on an extensional theory of truth. In order to circumvent these difficulties, special formal tools have been proposed, and the logical calculi using them have been qualified as *intensional logics* for this external reason. But an intensional logic can also, and more interestingly, be developed as a 'logic of intensions,' based on an appropriate 'intensional semantics.' This is what we have being proposing for many years regarding the semantics of empirical theories, and this is what has been more recently performed, on a much more general scale, by Edward Zalta. What is interesting with regard to this kind of logic is that it permits the formulation of theories about the objects involved in intentional states.[18]

4.1.4 Sentences and Propositions

What has been said thus far shows that an intensional semantics truly deserving of the name explicitly recognises the ontological legitimacy of *intensional* objects, which are at the same time *intentional* objects, rather than trying to circumvent them or explain them away, as has been done for decades. Concretely speaking, this amounts to giving a kind of primacy to *attributes* (i.e., properties and relations) rather than to individuals, as we have been maintaining for many years,[19] as well as in this book. And, not surprisingly, this is also what characterises the approach of several contemporary intensional semanticists.[20]

[18] In the first chapter of Zalta (1988) one finds a discussion of some additional reasons which support the merging of 'intentionality' and 'intensionality' in contemporary logical research.

[19] See, for example, Agazzi (1969, 1976).

[20] The most significant example is Zalta (1988). His book provides for the first time an extensive and rigorous formal treatment of relations, stating in particular the conditions under which there are relations, and the conditions for their identity. In this way, the traditional set-theoretic treatment (for which relations coincide with their exemplification-extension) is no longer the only rigorous tool available for the study of relations, not to mention the fact that the exclusive use of

Intensional semantics also shows its features in the way it treats the meaning and the truth of sentences. Unfortunately, a rather large spectrum of proposals exists in the specialised literature as to the distinctions that must be drawn between the notions 'sentence,' 'proposition' and 'statement.' Since we shall reconsider this issue in Sect. 4.4, here we shall simply relate the core of our position. We mean by "sentence" a *linguistic* expression through which a content of thought is formulated (a sentence is constructed by predicating certain properties or relations of certain objects, possibly in combination with the use of logical operators). By "proposition" we mean the *intentional* content of the sentence, which as such is an abstract object in the already explained sense; and we say that a sentence *expresses* a proposition. A question we meet now is whether a proposition can also have a referent. Our answer is yes, and we identify this referent with the *state of affairs*, or the *fact*, which is described in the proposition (as will be clear later, this state of affairs is not necessarily a fact of the material world, but it depends on the 'regional ontology' referred to).

We are aware that in holding this position we do not conform to the ideas of many scholars—not only of those who simply do not distinguish between sentences and propositions (they are usually the 'extensionalists'), but also of certain 'intensionalists,' who accept the Russellian identification of proposition and state of affairs.[21] We advocate this difference as a result of the consistent application of our idea of a three-level semantics. Besides language (sentences) and thought (propositions), we also want to consider the world (states of affairs). Let us also note that if we did not introduce this distinct level we would fail to distinguish between encoding and exemplifying in the case of sentences. In fact we can say that a proposition is an abstract object 'encoding' what a sentence says, but this does not imply that there is a state of affairs which 'exemplifies' the fact expressed through the proposition. It is obvious that for a semantics of *scientific* theories it is important to distinguish between simple propositions (i.e., abstract intellectual constructions) and concrete states of affairs to which they may or may not refer.

The notion of *statement* is (for us) simply that of a proposition which is asserted. We could also say that a statement is a *declarative sentence*. The addition is not futile, since one and the same proposition may occur within the context of different 'attitudes'; it may be the object of questions, beliefs, thoughts, hopes, and so on. If it is the object of an assertion, then it becomes something which may be true or false. This is why statements (and only statements) must be considered when the problem of truth is analysed. However, since in scientific theories (at least ideally) only declarative sentences are used, it is legitimate, in this context, to speak of truth and falsity as properties of propositions. We shall be devoting special attention to this topic later.

(Footnote 20 continued)
this tool had reinforced the wrong idea that relations are 'strongly extensional,' while it is intuitively clear (and Zalta's theory accounts for this) that even logically equivalent properties or relations are *not identical* from an intensional point of view, that is, from the point of view of their sense.

[21] Also Zalta, for example, makes this identification.

4.1.5 A Few Summarising Schemes

It should be clear from our presentation that while on the one hand we maintain a well-defined position in semantic matters, on the other we recognise certain good reasons for holding contrary views. The consequence of this is that we may sensibly introduce a few general schemes that include almost all the notions we have discussed in this section, and at the same time show where each of them is particularly significant, and where each is used only in a limiting sense. Moreover, these schemes will give us the opportunity of distinguishing certain levels of analysis which may have appeared to overlap in some of the foregoing discussions.

We shall begin with a strictly *semantic analysis*. Following the approach inaugurated by Frege (and never really rejected after him), we distinguish three types of linguistic expressions, namely, proper names, predicates and sentences. For each of them we shall define "sense" and "reference":

Sign	Sense	Reference
Proper name	Individual concept	Individual entity *i*
Predicate	General concept	Attribute (property or relation)
Sentence	Proposition	State of affairs

Besides the distinction between sense and reference, another may be introduced which expresses an *epistemological analysis*,[22] that is, the cognitive position of the notions listed above, as well as of others:

	Through the senses	Through thinking
Subjective knowledge	Perceptions	Mental representations
Intersubjective knowledge	Material things	Senses (concepts, propositions), truth-values, numbers, abstract objects

Finally, we consider an *ontological analysis*:

Mental reality	Representations, intentional states
External-world reality	Material things, signs, states of affairs, attributes
Objective intentional reality	The 'contents' of intentional states: senses, truth-values, abstract objects, *noemata*, numbers, etc.

[22] Let us note that we are here going to use the concept 'knowledge' in its broadest sense, i.e., including both 'knowledge by acquaintance' and 'propositional knowledge,' therefore admitting the existence also of a subjective knowledge, as well as the fact that knowledge does not necessarily entail the subsumption of the particular under a universal. This 'tolerant' attitude is adopted because we do not need, here, to enter into more detailed issues that might impose a more refined analysis.

This schematisation accepts several of Frege's basic positions: the difference between the subjective status of 'representations' and the objective status of thoughts, as well as the fact that this objective status is different from that of material objects since it is not apprehended through sensory perception. This led Frege to introduce in his last writings the 'third realm' of immaterial objects (as we have done here). However, let us also note that the three schemes are not in a one-to-one correspondence. For example, it is true that 'senses' are the most interesting inhabitants of this third world, but they are not the only ones, as we have indicated. Also, referents need not belong to the external world. Many of them do, but others belong to the world of abstract objects, while signs are parts of the external world.[23]

We can also accommodate in our schemes intensions and extensions by revising in a more Carnapian spirit the above scheme for semantic analysis:

Sign	Intension	Extension
Proper name	Individual concept	Singleton $\{i\}$
Predicate	General concept, *noema*	Set of individuals (n-tuples of individuals) exemplifying the concept
Sentence	Proposition	Truth-value

Let us now note where these schemes function naturally, and where they are more or less forced. In the case of proper names it is rather clear that to attribute them a sense is rather problematic if we want to distinguish this sense from the accidental mental picture which any subject associates with the name. In other words, it seems rather problematic to say which 'objective' meaning should characterise the senses of "Napoleon," "Rome," and so on. In this case we could either say that such a sense is constituted by an infinite collection of properties, so that only the individual in question possesses all of them simultaneously, or that no real sense is bound to a proper name. Both theses have had their advocates, and this shows that attributing a sense to a proper name is just using in a limiting case something which is justified only by the fact that we 'understand' the name. The real semantic feature eminently related with a proper name is reference. Even extension is used rather vacuously in the case of proper names, since the difference between i and $\{i\}$ is introduced in a formalistic vein. All of this explains why those semanticists whose theories have been developed essentially as solutions to the problem of the reference of proper names have been led to disregard sense.

[23] An analysis of the Fregean 'tripartition' is presented in Thiel (1965), where the author sees a one-to-one correspondence between the distinction 'sign, sense, reference,' and the distinction 'subjective-real, objective-non-real, objective-real,' and charges Frege with having introduced a "completely unacceptable contamination" by his "allowance of a participation of ontology in the doctrine of sense and reference" (pp. 151–152 of the English edition). A persuasive argument that Frege did not perform such a contamination, accompanied by a vindication of the legitimacy of considering the ontological counterparts of a semantic analysis, is contained in a critical appraisal of Thiel's thesis in Carl (1982), pp. 61–65.

Therefore, they have offered a less satisfactory treatment of the semantics of other linguistic expressions (practically, the only additional questions they are able to treat satisfactorily are those concerning indexical expressions. i.e., linguistic entities intended to speak just of individuals). Let us note that, on the contrary, in the case of definite descriptions the presence of a sense is unproblematic, but this is so because a definite description is, after all, a predicative expression.

As to predicates, what we have presented in the first scheme only partially reflects the way of speaking that we used in the foregoing chapter, where we maintained that a predicate "denotes" an attribute, and that its "referents" are the objects which exemplify it. What we are proposing in the first scheme tries to be faithful to this idea, without its being able, however, to use the difference between denotation and reference, owing to the simple fact that denotation does not appear in our schemes. Therefore we are now going to spell out a difference between denotation and reference that we are considering only here in our restricted context (recognising that the two notions are usually considered as synonymous). Saying that predicates 'refer' to attributes would have the advantage of recognising the 'concreteness' of many properties, without necessarily becoming involved in too narrow a position concerning the ontological status of properties and relations. After all, properties and relations may also be exemplified; this certainly cannot occur without an individual *in which* they are exemplified, but this still remains a matter of the attribute or property being exemplified.[24] For example, the general concept of red is exemplified by *this* actual red, which is the colour of the pen lying on my table, and which is not only *distinct* from the red of the cherry I have just seen on the tree, but also something *different* not only from *this* other attribute of the pen consisting in the relation of lying now on my desk, but also from many other attributes, some of which might

[24] This does not overlook what Frege stressed concerning the 'unsaturated' character of concepts, as contrasted with the saturated character of objects, but at the same time accounts for an ontological feature which was somewhat obscured by Frege's purely linguistic analysis, that is, that properties often have a rather unproblematic *intrinsic* determinability. (Frege recognised, on the other hand, that concepts may also become objects saturating a second-order concept.) As to the ontological analysis, we think that the traditional distinction of *esse per se* and *esse in alio* still provides a sensible tool for the ontological analysis of properties, which does not endow only individuals with ontological relevance. Let us point out, by the way, that our semantic analysis is partially in keeping with Frege's theory, at least in the sense that he claims that a predicate, being an unsaturated expression, cannot have as referent an object, and therefore must denote a *concept*. But since a concept is an entity whose kind of existence consists in being true or false of some object, the result is that for Frege a concept is a property or a function (in our vocabulary, an attribute). As a result, both Frege's semantics and ours state that the referent of a predicate is an attribute. However, our agreement with Frege is only very partial, for concepts are for him the referents, and not the senses, of predicates (contrary to what we maintain). Frege is obliged to maintain this position in order to be consistent with his fundamental distinction between concept and object. However, unfortunately, he never formulated a clear view on the ontological status of concepts, remaining content with the fact that their existence is justified by the possibility of quantifying over them (which is a purely linguistic justification). On the contrary, we do have such a concern for ontology, and give to concepts the ontological status of *intensional objects* (on the level of sense); and at the same time we say that they refer (or may refer) to concretely existing attributes.

not even apply to a pen. In a sufficiently large ontology, such as the one we are advocating, attributes may not only be *denoted* through *noemata* or abstract entities, but also *referred* to concretely via their exemplifications.

On the other hand, we are aware that speaking of the 'reference' of a predicate is rather problematic, since the fundamental conception of a referent is that it be an *object* in the sense of an individual. We find here something like a counterpart of the difficulty of speaking of the sense of a proper name: the reference of a predicate is spoken of in a 'limiting sense' and this is why we prefer to use the expression "denotation" (which—as we noted—is actually synonymous with "reference") when we speak of the relation of a predicate (which is a 'sign,' a linguistic entity) with an *attribute* (which belongs to the 'world'), leaving the use of "reference" for indicating the relation of a predicate with the individual *objects* satisfying it (which too belong to the 'world'). This move is justified by the fact that it is pertinent to speak of *extension* in the case of predicates, and the extension is the class constituted by the concrete individuals which have the attributes denoted by those predicates and which, for this reason, can be said to exemplify these attributes. Therefore, we should have preferred to call these individuals the 'referents' of the predicates in the more extended discourse in which we had denotation and reference at our disposal (denotation regarding attributes and reference regarding the individual objects). However we shall not make use of this conventional difference in the sequel, though recognising that reference and extension are related notions, but are not identical, so that a Fregean and a Carnapian semantics are not really equivalent (as we have already noted).[25]

After the specifications given here, we feel authorised to defend our terminology, which is somewhat personal, but not arbitrary. Having explicitly introduced *abstract* objects, which are *intensional* and belong to the sphere of *objective sense* or to the *noematic world*, we say that linguistic expressions *designate* such objects. When, in addition, the abstract objects are also exemplified by objects of the concrete world, or *concrete* objects, we call these the *referents* of the corresponding *noematic* entities (and, by extension, of their linguistic expressions). In such a way we say that an expression *designates* its sense and *refers* to its

[25] This explains why it would be problematic (contrary to one's first impression) to say that the reference of a predicate is its extension. In fact, for Frege, while the sense of a predicate is "its way of being given," and its reference is the concept (equated with a property or relation), the extension (which he calls *Umfang*) has no semantic relation to the predicate. Carnap, on the other hand, does not use the distinction of sense and reference. The classical tradition was actually more sophisticated than contemporary approaches: the 'comprehension' of a concept (that is, its 'content,' its 'intention,' to use our vocabulary) was the class of its characteristics, that is, the class of those upper-concepts which occur in its definition (let us say, e.g., concepts such as 'animal' or 'rational' in the case of the intension of 'man'). The 'extension' was instead the class of its lower-concepts (such as, e.g., 'European,' 'musician'), among which appear, on the bottom-level, the *individual* concepts. Therefore, the classical notions of extension and intension only denoted relations between concepts and did not involve such an 'ontological jump' as is implicit in considering the extension of a concept as constituted by concrete individuals, instead of concepts.

referent(s). From what has been said, denoting is clearly analogous to referring in the semantics of *encoding*, while referring is more demanding outside this semantics, since it requires exemplification as well (and this accounts for the fact that "denoting" and "referring" are usually considered synonymous—which they are, if no difference between encoding and exemplifying is introduced).

Finally, coming to sentences, the novelty we have introduced is that of considering states of affairs as the referents of propositions. This has some similarity to the doctrine which considers sentences as the 'names' of propositions. But under closer scrutiny this similarity disappears, for our view expresses rather the idea that, in general, a referent is what exemplifies an abstract object. Having accepted identifying the extension of a proposition with its truth-value conforms to a tradition which goes back to Frege and was accepted by Carnap, both of whom tried through argument to dissolve the paradoxical impression it creates. However, we think that the paradox is really dissolved only if one is able to secure a proposition's referent 'in the world,' as distinct from something abstract such as a truth-value. Therefore, provided that we (again) distinguish reference from extension, we can say that the extension of a proposition is a truth-value *because* its reference is a state of affairs. Our justification of this is, in brief, the following: being true or false is a *property* of a proposition; therefore, as in the case of any property, being true or false has an intension and an extension. The intension is that the sentence is exemplified (or not exemplified) by the state of affairs, and the extension is just the corresponding truth-value. This enables us to put all true sentences in one class, and all false sentences in another. It is clear that this is very different from saying, as Frege does, that truth-values are the referents of propositions (and it is also different from the reasons for which Carnap maintains that they are their extensions). We gladly admit that a certain degree of conventionality is present in our proposal (essentially because we have generally defined extensions as classes of referents). But this peculiar meaning of extension in the case of propositions is only an example of the 'limiting case' use of certain semiotic notions that we have already noted in other cases, and that corresponds to our aim of keeping distinct, while still related, the three levels of language, thought and world.

What we should still clarify is the notion of *external world*, which we have included in our scheme of an ontological analysis without any previous discussion. We shall simply say, at this point, that the external world is, on our approach, the world of *referents* and, in a more elaborate sense, the world of those entities which can be reached by using the operational *criteria of referentiality* of which we have often spoken in this work (that we have also called criteria of protocollarity/objectification). As we have already noted, this does not imply that this world consists entirely of material entities. Since the criteria of referentiality are of very different kinds, a corresponding variety in the ontology of the furniture of this external world logically follows. What makes this world different from the world of intensional abstract objects is that abstract objects simply *encode* properties, while objects of the external world *exemplify* them; and the most efficient criteria for ascertaining whether or not this exemplification occurs are (especially in the sciences) the criteria of referentiality. This is why we shall consider reference again in Sect. 4.3. But, before doing that, we find it opportune to examine a concept which has not so far been analysed, the concept of *thing*.

4.1.6 What are Things?

This question might well be the title of a treatise on general ontology, but we are certainly not going to tackle it in such a broad sense. We can avoid such a discussion because, while in ontology one would consider this question as an absolute one, we are going to consider it only *relatively*, that is, as relative to the distinction between objects and 'things' which we have already explained at length. The straightforward consequence of this approach is that the concept of thing itself will be relativised, since one and the same entity may be a thing in one context (i.e., from one point of view) and an object in another. What we are saying is tantamount to claiming that "thing" expresses the notion of a *'functional' role* rather than that of an intrinsic feature of certain entities.

In order to make this role explicit we can say that an entity plays the role of a thing in a scientific discipline when it can be *identified* and thus be *referred to* unproblematically, independently of the theory being considered, and also independently of the form the theory eventually takes. It was in this sense that we said earlier, for instance, that a telescope, a microscope, and an ammeter are to be taken as 'things' in order for us to use them as instruments for constructing, respectively, modern astronomy, modern biology, and modern electrical science. They play the role (they have the *function*) of things simply because they are to be identifiable, recognisable, and capable of being manipulated without recourse to astronomical, biological, or electrical notions respectively. This does not imply, of course, that one can use or even identify these things without any knowledge whatever. In order to distinguish an ammeter from a clock, which might have a rather similar external appearance, one must have a sufficient notion of what an ammeter is and be able to use it and read its results. This means, therefore, that the ammeter (as far as it is a thing) can be submitted to study and become the object of another *theory* (the 'theory of the instrument').

But there is more to this concept than what we have expressed thus far. Indeed, the fact that we can take a telescope or an ammeter and let it play the role of a thing is a consequence of the fact that these entities are already the result of a preceding *construction*, in which they were either the objects of a proper disciplinary investigation or its 'applied offspring.' They are therefore *data* from one point of view (from that of the theory which uses them as *starting points*, that is, as its own operational tools) and *constructs* from another point of view (i.e., from that of the genetic process which led to their actually being produced and becoming so familiar and trusted as to be taken for granted).[26]

[26] We are aware of the double sense in which we are using the term *data*. In the preceding sections we have qualified as data the results given by the immediate application of operational

One should also be careful to explicitly recognise the feedback between the instruments used in a theory and the theory itself, so that the instruments can often be improved, redesigned or even designed from scratch as a consequence of the progress reached in the discipline in which they are applied. This is true, but it does not affect our position in any essential way. For, if the *new instruments* (which for the very fact of being 'new' have been obtained as 'objects' of at least certain parts of the theory) are such as to provide us with new operational predicates, this simply means that they have practically given rise to a new domain of objects, with respect to which they play the role of unproblematic things. (This must not imply that such a new part of a discipline, or even a new discipline created in this way, should be recognised officially as such—this is why we often have the impression that the instrument modifies itself *within* the theory). On the other hand, if the new instrument happens to provide us with a new method for testing, *within* the theory, already existing operational predicates, this will not harm our perspective, as we have shown elsewhere in a deeper discussion of operational concepts from a more technical point of view.[27]

But we can go a step further and say not only that physical instruments, such as telescopes and ammeters, can play the role of *things* with respect to a discipline, but that abstract entities such as differential equations or potential can as well. The reason we are entitled to make such a claim is that such abstract entities are well individuated and can be recognised and manipulated, and their results can be ascertained no less than in the case of an ammeter or some other physical instrument.

On the other hand, abstract entities too, while playing the role of identifiable 'data' in this specific situation, can be problematised and submitted to inquiry in another discipline, such as mathematics, where they become the *objects* of inquiry; and, in the case of applied mathematics, they may even take the form of 'constructs' or 'artefacts,' of technical offspring specifically designed for application to particular physically relevant states of affairs.

What has been said here about the particular 'things' which are used in order to construct theories and to determine domains of objects may now be repeated in an even less problematic sense for those 'things' which actually become the objects of different theories. Here again we stress that, in order for an entity to play the role of a 'thing,' it must be something *identifiable*, and be such as can be *referred to* when the discourse of that discipline applies. This means that a pipe, a piece of chalk, a book, the moon, the beard of Mohammed, and so on, can be considered as things as far as they are identifiable within the community of people who speak about them, and may serve as referents of the discourse which is developed within

(Footnote 26 continued)
devices; here we include among data also the instruments and all non-problematised things. We are doing this on purpose, in order to underscore the common character of 'givenness' of all these different entities. This will not be done in the sequel, however, when we return to using "data" in its more technical sense of operationally established states of affairs.

27 See Agazzi (1969).

that community. But here too it is not only concrete material entities that are entitled to play the role of 'things.' A toothache, a feeling of love, a Beethoven symphony are also 'things,' for they are identifiable by a very large community of people, and can become referents of a variety of discourses, some of which are also of a scientific nature.

What we have said thus far shows how improper it would be to identify 'things' with so-called material bodies (one can do this, of course, as a matter of convention, but one would then be left with a nearly useless and inapplicable concept), but also because it shows us how the domain of 'things' is in itself changeable and subject to historical evolution. This is not only trivially true since, for example. that thing which is Leonardo's *Gioconda* did not exist before being painted by him, or those things which are cars did not exist two centuries ago, when people were not yet able to construct them. This is also true for a great many intellectual constructions that concern empirical reality. This point is not immediately clear, and deserves closer scrutiny. If one is not a confirmed Platonist in the philosophy of mathematics, one would not find it difficult to accept that, for instance, differential equations did not exist before a certain historical time, when Newton, Leibniz and other mathematicians created them. From that time on, they began to be 'things' in the (intellectual) furniture of the world. But leave mathematics aside and ask, for example, what we can say about electric current or electric fields or atoms. Today, no one belonging at least to the Western world would find it difficult to say that an electric current is an existing 'thing,' no less than are his shoes or his car. Yet electricity was not a 'thing' 300 years ago, when even scientists had not yet found an acceptable theoretical image of it.

As for the other two notions, 'electric field' and 'atom,' it is clear that only scientists take electric fields for granted and are so familiar with them that they consider them as 'things,' while for most people electric fields still appear to be problematic intellectual fictions. On the other hand, the scientifically much more complex notion of an atom is currently, for purely cultural reasons, much more familiar, and atoms are easily considered as 'things' also by non-specialists.

Two lessons must be drawn from the above considerations. The first is that whatever is afforded the status of 'thing' or object is historically determined in the sense that it depends on the cultural and historical situation in which the notions appear to be sufficiently well identified as to admit of a referential discourse pertaining to them. The second is that we are free of any ontological presupposition at this stage. Indeed, if someone should say: of course, the electric current was *discovered* by science not so very long ago, but it has always *existed* in nature, we should reply that, from the point of view of the distinction between 'thing' and object that we are discussing, this current could not 'play the role of a thing' before the historical time at which it was clearly exhibited and knowledge regarding it became sound.

This in particular shows that today's scientific objects may easily become the 'things' of everyday life tomorrow. This is also a reason why it would be useless to try to find the difference between objects and things on the basis of a discrimination between the abstract and the concrete. This distinction does not work because what was abstract yesterday may be concrete tomorrow. (Think for example of the many particles in physics which were hypothetical constructions for a long while before being 'observed,' after which time they began to be regarded in the scientific community—and gradually also in the more general human community—as being things.) This observation is important for another reason as well: it confirms that the character of being a thing is not really dependent on any material, sensory or concrete feature one might attribute to an entity, but only on the entity's being believed to have an autonomous existence, i.e., to exist independently of any particular theory. For example, almost all the elementary particles of physics, when they are conceived of as existing as things and are taken as such in many contexts, are not actually endowed with more or other attributes than those they are taken to have in the physical sciences, so that the same predicates which we use to describe them as objects are those which we use to describe them as things. Only their role has changed.

The present discourse should show that science not only studies already admitted 'things' under many viewpoints, making them the objects of its inquiry, but also that it is able to provide an enlargement of the world of things themselves, by making us acquainted with the existence of things of which we were before totally ignorant. This is already a hint regarding science's ontological commitment to scientific objects.

This discussion concerning things and objects returns us from a general consideration of our 'semantic framework' to the more specific topic of our investigation, with which we shall now continue.

4.2 Are Scientific Objects Real? The Ontological Side of Objectivity[28]

A question that remained open at the end of the preceding section was whether or not 'things' can be entirely accounted for by a 'functional' interpretation. More precisely, the question is whether a 'thing' is simply a role (namely, the role of 'givenness' and of 'referentiality'), or whether it is an entity 'endowed with a role.' In our discussion we have always used, implicitly or explicitly, the second interpretation. But one must say that we were practically obliged to do so for the sake of the discussion, since it is in keeping with common parlance to employ the notion of thing in an ontologically committed sense. It was already fairly difficult to claim that the 'role' of things is simply relative; and it was not advisable to make the situation more complicated by raising additional questions about the 'status' of things. But what could be overlooked for the sake of a first simpler presentation cannot escape an appropriate critical investigation which might well lead to the conclusion that the ontological status of things is not that sound.

[28] Certain parts of the present and following sections have been included in Agazzi (1997b).

However, we shall not tackle the problem of the ontological status of things directly, and this for a couple of reasons. The first is that, owing to the already indicated relativisation of the concept of *thing*, it will not make any difference if we begin our ontological inquiry with the concept of *object* instead. The consequence of this choice will be that, if we should come to endow the object with a certain ontological relevance, our problem could be practically solved, at least implicitly, for things as well. The second reason is that we are, after all, interested in objectivity, and it therefore makes sense to explore the ontological side of this notion. If we were to succeed in discovering some acceptable perspectives on it, we would be satisfied even if this should not enable us to illuminate the whole of the analogous problem concerning things. If this should be the case, we could leave this problem to the specialised care of ontologists.

Let us now explicitly pose the question: are scientific objects real? The correct answer to this question presupposes a proper understanding of the meaning of its concepts. As for the concept of a scientific object, our task is completed in this regard; but what about the concept *real*? There is, notoriously, a wide spectrum of meanings for this concept, and it is clear that, according to some of them, one should deny the reality of scientific objects. Let us only mention the everyday use of "real" according to which it means 'concretely perceivable,' especially by means of sight and touch (a use where epistemological criteria are endowed with an ontological power). If one retains this meaning, it is obvious that many scientific objects (e.g., those studied by microphysics) cannot be said to be real. However, that we cannot adhere to such a restricted meaning is already clear if we consider the vagueness and ambiguities to which it can give rise in the context of everyday language itself. For instance, acoustic perceptions are already considered with some mistrust within this context (actually there is a certain inclination to deny reliability to *pure* acoustic evidence, and to admit that 'it might easily be wrong'). But, on the other hand, people are also easily persuaded of the existence of viruses, of bacteria, and of atoms and electrons, most of which have not only never seen, but which in many cases are by their very natures excluded from being seen.

In order not to be confronted with similar difficulties we shall not try to specify the notion of reality by means of other *particular* requirements. Rather, we shall resort instead to a very basic characterisation which we have already hinted at in an earlier section (Sect. 2.3). We shall say that *real* is what is *different from nothing*, the only requirement we need for reality being therefore that of existing, and not that of existing *as* a reality of a particular kind. It follows that, whenever we are in the position of stating that there would be a difference in the world depending on whether a particular entity exists or not, that entity, if it exists, deserves to be called real. As we remarked in that section, dreams and hallucinations are thus real, since there is a difference between having and not having them. All this amounts to saying that the 'kind of reality' one can attribute to an entity depends on the 'point of view' from which reality is considered.

What we are saying here is reminiscent of the traditional thesis of the 'analogical' meaning of being, which goes as far back as to Aristotle. This doctrine has a powerful analytic purport, and the contemporary discussion concerning realism would benefit greatly from giving it more consideration. But even in contemporary philosophy the thesis of the 'analogical' meaning of being has revealed great vitality. In particular, it has surfaced in connection with Brentano's rediscovery of the *intentio* which we have already mentioned. It was indeed Brentano who stressed that one of the most significant features of *intentionality* (considered to be the specific characteristic of mental phenomena) was the possibility of intentional states being directed towards objects which—as he said—do not necessarily exist. The task of clarifying the puzzle of the nature of intentional objects was pursued along different lines by Husserl and Meinong, the first through a deeper scrutiny of the nature of intentionality (and in particular through the introduction of the notion of the *noema* as something different from the object), the second through the delineation of an articulated ontology of 'objects,' some of which may not exist (*Gegenstandstheorie*).[29]

However, we cannot completely subscribe to Meinong's distinction between reality and existence, simply because, on the one hand, there appears to be no sound reason for claiming that something real does not exist and, on the other hand, it might lead to serious misunderstandings to say that something that exists may not be real even from a particular 'point of view,' though this could be in keeping with certain ways of speaking (e.g., when we say that hallucinations are not real from the point of view of physics). In such cases it is more appropriate to say that such entities do not belong to the objects of physics. Thus Meinong's restriction of existence to concrete material things is not to be recommended if it is to be taken as a convention; and it seems incorrect if it is taken as a substantial claim, as we shall soon see. It seems much more reasonable to claim that there are different *kinds of reality*, and that to each kind corresponds its *kind of existence* (as well as its kind of meaningfulness, truth, and so on). So, for example, in the kind of reality consisting of material things, the Minotaur and numbers do not exist, while they exist, respectively, in Greek

[29] See in particular Husserl (1913) and Meinong (1904). The solutions adopted by Husserl and Meinong are often said to be divergent. But they are not really. Rather, they correspond to the differences in approach we have just mentioned, and it is possible to reconcile them if a more positive interpretation of Meinong's claims is made by resorting to certain elements of his ontology which are not really clear in his writings, but were fully specified by his student Ernst Mally (see Mally 1912). This is shown in a short but valuable presentation of the debate about intentionality, which took place at the turn of the century, provided in Chap. 6 of Zalta (1988). Zalta himself (as we have already said) develops his own theory of intentionality by using both Meinong's distinction between being and existing (so that 'abstract objects' do not exist), and Mally's distinction between determining and satisfying, which he reproduces in his notions of encoding and exemplifying. Thanks to this elaboration Zalta can give a theory of 'abstract objects' (which do not exist, in spite of encoding precise properties), and 'ordinary objects' (which exist and exemplify certain properties), as well as a fully-fledged theory of relations, which enables him to eliminate (in the most direct and convincing way we know of) all the usual difficulties connected with the principles of existential generalisation and substitutivity in intensional contexts. As we have already occasionally noted, there are many points of similarity between our theory of scientific objects and Zalta's theory of intentionality. However, there are also differences, which will be discussed soon, and which in particular are also differences between our conception and Meinong's, in spite (again) of several points of similarity.

mythology and in the realm of mathematics (and in these domains it makes sense and is even true to say that the Minotaur lived on Crete, and that 2 plus 2 equals 4). Following this line, one could even reverse Meinong's classification and say that we are usually prepared to call real beings which we are ready to accept as existing in at least one kind of reality (i.e., from one 'point of view'). It is often said that numbers are not real *because* they do not exist, and that they are 'nothing but' mental constructions. In this view, which is also expressed by Harré,[30] the different kinds of reality are determined according to the criteria we accept for *admitting* something as existing. Our position is still different: we strictly connect reality and existence, and say that the criteria for *demarcating* a certain kind of reality are at the same time those which enable us to ascertain which individuals, properties, processes *exist* or occur in *that kind* of reality.

More details on this issue will be given in the section devoted to the question of realism. However, it is useful to briefly discuss this doctrine of 'non-existent objects' because it can be shown (in our view) that the advantages that have been obtained in the recent developments of this theory can be preserved without separating existence from reality, as has in fact been done there. We shall proceed to this discussion through a comparison of our view with Zalta's theory, which has the special merits of clarity and rigour. A first point is that Zalta does not analyse the notion of existence (while he very carefully analyses several other notions), but simply defines it incidentally in an almost parenthetical way: "By '*exists*,' we mean 'has a location in space'" (p. 21),[31] or: "one will prefer to distinguish *being* (that is, logical or metaphysical existence) from *existence* (that is, physical existence)" (p. 103).

In this second quotation a kind of self-punishment is at work. In fact, in the very effort of distinguishing being from existence, it is said that being is a *particular kind* of existence (i.e., logical or metaphysical existence). But why is this distinction to be recommended? Apparently because it permits us to qualify abstract objects as 'non-existing,' so that one can assign a content to intentional states even when they cannot be referred to physically existing objects: "If we define abstractness so that it implies non-existence, then abstract objects that encode properties prove useful for understanding directedness towards non-existents" (p. 18); "The properties abstract objects encode characterise them, and so encoding is a kind of predication.... By encoding properties of whatever kind, abstract objects have content, and can serve to characterise the content of representations and images" (p. 18).

However, it is far from obvious that these advantages cannot be obtained without the gratuitous limitation of existence to physical existence or existence in space and time. It seems that the same result could be equally well obtained by saying, for instance, that intentional states may happen to be directed towards abstract objects encoding certain properties, even when there are no physical objects instantiating these properties. In such a way we should attribute a particular kind of existence (let us call it, e.g., *intentional existence*) to the abstract objects,

---

[30] See Harré (1964), p. 48 ff.

[31] The page numbers quoted are from Zalta (1988).

without equating it with the *physical existence* of other objects. But let us immediately add that there are many other kinds of existence which are different both from the physical and the intentional, such as existence in a story, existence in a dream, existence in a musical composition, existence in a project, and so on, which must be differentiated and characterised, while they risk disappearing if only physical existence is qualified as existence. It is certainly not accidental that Zalta finds himself committed to explaining the notion of "true in a story" through a rather cumbersome device, while if on the other hand one admitted the idea of a *fictional reality* one could introduce criteria of referentiality relativised to a given story, and in such a way easily explain such elementary claims as "Hector is a Trojan warrior in *The Iliad,*" or "Figaro does not exist in Shakespeare's *Hamlet.*" These examples show that it makes perfect sense to say that a certain individual (Hector) exists in *The Iliad,* and that he instantiates certain properties in that story, despite his not existing in space and time, while another individual (Figaro) does not exist in *Hamlet*, not because he does not exist in space and time, but because he is not one of the characters in the play.

Last but not least, one should avoid having to claim that there are non-existent objects. As a matter of fact, Zalta states that the possibility of making this claim without turning it into a logical falsehood is an advantage, but one can be very doubtful about this, while no one would be uncomfortable saying, "there are non-physical objects," "there are non-intentional objects," "there are non-historical objects," and so on, where "there are" retains its familiar sense of "there exist." We gladly acknowledge that the doctrine of non-existing objects was pushed by the desire to give a satisfactory account of the linguistic use of such expressions as "round square," "the golden mountain," "the present king of France" and "the fountain of youth," expressions which certainly do not denote physically existent objects. We also acknowledge that it was a real achievement to establish that these expressions nevertheless denote *abstract objects* of an intensional nature. But now, why should we call these objects *real* and at the same time *non-existent*? We do not find such a move acceptable; but, on the other hand, we shall indicate a way of recovering the correct intuition behind it, which wants to preserve a difference between, and a distance from, *purely intentional* existence and 'autonomous' existence. Our proposal consists in giving full value to the difference—but at the same time to the intimate relationship—between intentional objects, or *noemata*, and referents.[32]

In short, an abstract object exists *as* abstract object, as *noema*, and has a kind of intentional reality. But this kind of reality is such that it *points*, so to speak, towards

[32] We should not like to give the impression of being particularly critical of Zalta. In fact his position not only has its clear and acknowledged historical roots in the works of Meinong and Mally, but is amply consonant with a rich production regarding 'non-existent objects' which is current in modern literature, where an emblematic title could be, for example, that of Parsons' (very valuable) book *Non-existent Objects* (Parsons 1980). The reason we have discussed Zalta at length is that his approach constitutes, in our view, perhaps the best treatment of this kind of problem, and at the same time the one in which it is possible to see how this strange denigration of existence to certain things which 'there are' can actually be dispensed with.

another kind of reality, where a *referent* should exist satisfying it (or where an object of that particular kind of reality should instantiate the properties the *noema* encodes). Now, by using the specific criteria of referentiality or protocollarity characteristic for that kind of reality (e.g., by using certain scientific instruments, by consulting archives, by carefully reading a story or a play, by administrating psychological tests, by resorting to sociological inquiries, by making a mathematical calculation, and so on) we can sometimes discover that such an intended referent actually exists, and sometimes that it does not. It should be clear, therefore, that certain physical objects may not exist (such as, e.g., the golden mountain) because it happens that there are no referents in the realm of physical reality for those abstract *noemata* which characterise them. On the other hand, non-physical objects may exist, having that kind of non-physical reality which corresponds to their being represented through a *noema* whose intended referent should belong to a certain kind of non-physical reality (e.g., to a fictional reality), and they actually satisfy that *noema* in the realm of *that* reality (such as Hector in *The Iliad*).

Because of this, we must be careful in explaining what we mean when we say that dreams and hallucinations, though *real* as dreams and hallucinations, do not actually refer to any *real* state of affairs. The most spontaneous way of understanding this statement is to see it as expressing the distinction between mere *images* of reality and reality itself, or between the images and their *referents.* Although the two ways of speaking seem equivalent, they may actually cover two very different conceptions, the one being epistemological dualism, and the other being the doctrine of the *intentional* nature of knowledge.

The difference between these two positions may perhaps be sketched in a synthetic but sufficiently effective way by saying that both share the view that our cognitive activity has an intentional nature in the sense that it is 'oriented towards something' as a goal or, if one prefers, that it is conceived of as being completed by a *terminus*, with which we try to come in contact. Both doctrines also agree that this intentional effort leads to the production of certain 'images,' and that for this reason we can say that such images constitute the *intensions* of our cognitive performances. However, at this point the two doctrines diverge, for the one maintains that this very 'image' constitutes the endpoint of the cognitive act itself, which simply means that the 'image' is *what is known*. The other doctrine maintains that the endpoint is not the image, but some entity which we are able to approach *through* the image; this entity may be called the *referent.*[33]

[33] As an example of the second doctrine we could mention the Scholastic doctrine of the *phantasmata* (which we could translate as "sensory images"). These are not *id quod cognoscitur* (that *which* is known), but *id quo cognoscitur* (that *through which* one knows), and this in spite of our intellect not being able to know things unless it passes through these images (*nisi convertendo se ad phantasmata*). The other doctrine is explicit in Descartes, who claims that we know our 'ideas' and sees them as something whose *origin* must be *causally* explained. From the logical analysis of these ideas he believes it possible to infer the existence of things (e.g., of God and the external world) from their causes. Only the thinking subject escapes this fate, since it is present to itself in the act of *cogito*. This approach, as we have already said, was retained by the majority of modern philosophers up to Kant, and constitutes the philosophical doctrine that we have called

It is important to note that, according to the non-dualistic doctrine, the intensional image, the *noema*, always preserves its function of being the means *through which* we point to the referent, and never *replaces* the referent, not even in the case of non-existent referents, nor in 'epistemic' contexts (such as believing). For example, if someone believes that centaurs are wandering in the woods surrounding a certain city, he does not believe or think that *noemata* are wandering in the woods, but that certain concrete (though strange) individuals wander there.

As can be seen, one possible way of distinguishing the above positions is to say that epistemological dualism identifies intension and reference, while the other doctrine does not.[34] The feature to be stressed in this analysis of epistemological dualism (and it is indeed because of this feature that it is a dualism) is that it does not deny that the cognitive act has a referential orientation—that it aims at 'reaching' an ontologically independent entity. Only it believes that this basic aim remains frustrated and that, therefore, the referent must be looked for within a domain which is closer to the knowing mind, that is, within the realm of its own 'images of reality' while the second entity (the real object) cannot be known. In such a way we are not really confronted with the elimination of the referent, or of the referential side of knowledge, but with a reduplication of it, one referent (actually the 'proper' referent) remaining epistemologically inaccessible, and the other (the 'replacement' referent) being epistemologically accessible. This is why we said that it is typical of the dualistic position to conceive of images of reality as being separated from reality proper. The locution *image of reality* actually includes implicit mention of the inscrutable referent which we cannot hope to know. In our doctrine, on the other hand, we prefer simply to speak of an 'image,' because its being an 'image of' something can only be stated if we explicitly point to the referent; and in such a way the image preserves its proper role of being a tool rather than the terminus of knowledge.

Of course, the term "tool" may also produce some misunderstanding; so it is much more appropriate to say that the different images are *the ways* the referent has of *being present to the knowing subject*. So, for example, a tree's way of being present with respect to a house is that of being at a certain distance from it and exerting on it a gravitational attraction (which is neutralised by other forces). If the tree is in the presence of a camera with a photographic emulsion, its presence manifests itself also as the production of a chemical reaction on the emulsion

---

(Footnote 33 continued)
epistemological dualism or representationalism. As we already said in a preceding note, these themes have been revisited in the contemporary theory of intentionality, but they are also to be found, for instance, in the Fregean distinction of *Sinn* (sense) and *Bedeutung* (reference), in several works of Russell, and so on (sometimes in a 'representationalistic' sense as in Russell, sometimes not). In addition to the literature already cited, let us mention here Searle (1983) and Dreyfus (1982).

[34] Let us note, however, that this identification is very different from that which characterises the referentialist and extensionalist semantics we have discussed in the preceding section. In that case, intension was actually eliminated, and it was maintained that meaning coincides with reference. In the case of epistemological dualism, instead, intension is present, and is indeed so dominant that it precludes access to the referent, and itself *becomes* the referent.

which, if suitably handled, may lead to a two-dimensional *physical picture* of the tree. If the tree is in the presence of an animal equipped with organs of vision, its presence manifests itself to the animal also as a *visual image*. And as the physical picture may be stored even when the tree is no longer there, and may be submitted to different uses, so can the visual image. Generally, images are the different forms of *intentional presence* according to which a referent may be present to a knowing subject; and in this sense they are the referent itself *inasmuch as* it is present, for example, to sight, touch, hearing, memory or thinking.[35]

Being different from nothing, images belong to reality as well, that is, they are real, and as such they may also become the *referents* of a cognitive act. But in such a case they are not the 'replacements' of another, inaccessible, proper referent. Rather, they are themselves the proper referents of the *new* act of knowledge, which is 'intended' towards the image and not towards its referent (an act of knowledge which primarily consists in a *self-reflection*). In other words, if I look at a table, the image of the table which is being formed in my mind has the table as its referent, and I know the table *through* it. But nothing prevents me from being interested in considering not the table, but its image, and asking questions about *this image*; for example, how it comes to be, or how it can be preserved and called to mind even in the absence of the table, and so on. In all such cases, the image is the referent of my cognitive efforts, and I am trying to acquire knowledge about it (e.g., by means of a psychological inquiry).[36]

The whole meaning of this discussion is that epistemological dualism, besides being intrinsically untenable, is also perfectly useless, because any positive function it might be believed to have is conceivable without it. Indeed, such a positive function could be seen as consisting in the possibility of considering images as referents. As we have just seen, however, this possibility is safely preserved in the alternative conception which has, in addition, the great advantage of admitting other kinds of referents besides images. These considerations belong to basic epistemology and not specifically to philosophy of science. We shall see later, however, that they are relevant to the discussion of the issue of scientific realism.

As we have had the opportunity to present epistemological dualism here in the guise of a 'doubled referentiality,' it might be useful to restate for the last time the

---

[35] Due to this fact, *noemata* may be at the same time 'objective,' that is, independent of the subjective 'act of thinking,' and from mental or psychological images of different kinds. But they cannot help being at the same time reality's 'way of being present to thinking' in its various aspects. This was expressed by Frege when he tried to explain in what sense not only concrete referents, but also thoughts, are endowed with objective existence: "I understand by objectivity an independence from our perceptions, intuitions, and ideas, from the establishment of internal images, from the remembering of earlier perceptions, but not an independence from reason: for to say that things are independent of reason is to judge them without judging, which is like trying to wash the fur without getting it wet." (Frege 1884, p. 36). In this Fregean statement we can find a short but significant formulation of the spirit of a non-naive realism, i.e. of a realism that does not reduce reality to simple ideas but at the same time does not overlook that we can state existence only of known realities.

[36] What we are explaining here is a brief presentation of the traditional distinction between *intentio prima* and *intentio secunda*, which we have already mentioned in a preceding section.

reasons for the untenability of this doctrine by taking this point of view into special consideration.

Reduced to its actual core, the notion of a referent is simply that of an entity, of something *existing*, to which a certain intension applies. It follows that, whenever one calls a referent into play, one commits oneself to a claim of existence. Now, only two ways are at our disposal in order to support the *cognitive* claim that something exists, either by evidence (of some kind), or by argument.[37] In the case of the hidden, inaccessible referent, which is supposed to lie behind the 'images of reality' according to the dualistic tenet, neither evidence nor argument are produced in order to support this claim, which is, therefore, a purely *dogmatic* one. But the interesting fact is that if one were able to show that this claim is not dogmatic, one would *ipso facto* prove that it is contradictory. Indeed, if some justification could be provided in favour of the claim that the 'unknown' referent actually exists, this would amount to saying that we had some means for *ascertaining* its existence. But at that moment we would have contradictorily admitted that the *unknown* referent is actually *known*, for knowledge cannot be anything different (or anything more) than the ascertaining of an existence, be it the existence of an individual, a property, a relation, a state of affairs, or what have you.

Let us stress that this criticism also applies to the Kantian doctrine of the *noumenon*, despite the fact that Kant declares this noumenon to be thinkable but not knowable. This distinction between knowing and thinking seems prima facie to solve the problem, as the hidden noumenon is presented here as the referent of an act of thinking and not of an act of knowledge. Yet this solution is ineffective, because Kant maintains that the noumenon *exists*, and in such a way he cannot help admitting that we at least *know* this much about it. In order to avoid this conclusion, Kant should maintain that we *think* that it *may* exist. But this is not what he says. According to him (at least in one part of his doctrine) the noumenon exists, and is not merely possible.[38] This, however, is a well known problem regarding Kant's philosophy.

---

[37] We recognise, therefore, that existence statements may also be advanced, and are actually often advanced, on the ground of faith or belief, that is, on grounds that we here call *noncognitive*.

[38] See the discussion on the Kantian theory of the noumenon sketched in Sect. 1.7. It is interesting to note, in this context, that Husserl, precisely because he was anti-dualist, could not admit that if an external world exists it may be constituted by things-in-themselves that are in principle unknowable. This is possibly one of the reasons why, in the last stage of his philosophy, he became an idealist. A possible explanation of this fact (which, however, is not admitted by certain interpreters of Husserl) is that he thought that realism was untenable because the external world can be known by human beings only *partially*, in the sense that such knowledge could never be complete. Our comment is that this conclusion would not be logically correct. In fact, the incompleteness of our knowledge of the external world does not provably depend on the fact that there are, so to speak, parts of this world that are epistemologically inaccessible, but rather on the fact that the objectifications we can make of this world are potentially infinite, and that, as a consequence, our knowledge is *inexhaustible* even though it is always *realistically referred* to this world. We shall return to this issue in our discussion of realism (and, in any case, we do not maintain that this is a precise criticism of Husserl's doctrine).

This discussion is important in that it allows us to conclude that knowledge can by no means and under no circumstances be ‘knowledge of nothing,’ because if we can say that we have some kind of knowledge, it is because we are at least in the presence of certain ‘images.’ Now, at least a few images in our minds are there because they constitute the manner in which some referent has been *presented* to us, and in this sense they are witness to our knowledge being knowledge of something. Anyway, this is true in general, but it does not guarantee that *every* image I have has a referent. For example, if I am dreaming of an apple, the image I have now was stored in my mind on the occasion of some referential situation in which the apple was actually there, but this does not imply that there is a referent of that image *now*. The same can be repeated if I think of a dead person whose face I remember well, or of a past experience. The situation is similar when one imagines some entity or some pattern of action as a pure invention of her mind, such as when one writes a novel, or a book of science fiction. In such cases, the ‘ingredients’ of the story are constituted by ‘partial’ images, which have had a referent in the past experience of the person who is inventing the story, but which are put together through a more or less free association, giving rise to new complex images which are not meant to have a referent.[39]

The obvious question is therefore: how can we be sure that in similar situations our knowledge is not ‘knowledge of nothing’? Our answer is the following: if we distinguish between *intension* and *reference*, we can say that our knowledge is never knowledge of nothing because it is at least knowledge of an intension or, if one prefers, knowledge of a ‘world of meanings,’ of some *noemata*, of ‘abstract objects.’ This is in particular a point on which certain modern discussions concerning so-called ‘non-existent objects’ have contributed very useful insights, as we have seen. This inten*s*ion, in turn, owing to the inten*t*ional nature of knowledge, is so to speak projected towards a possible ‘world of referents,’ and this means, as we have explained in the preceding section, that the intentional act ‘seeks’ objects which *exemplify* the properties, relations, propositions, *encoded* by the abstract objects constituting the realm of intensions. But other requirements must come into play in order for this possibility to be realised.

We shall discuss these referential requirements later, but let us now return, from the general analysis sketched here, to our specific problem of determining whether scientific objects are *real*. After the foregoing analysis we can say that there is a sense in which they are (rather uncontroversially) real, and this is the sense in which scientific objects, being structured sets of concepts which encode certain properties and relations, are at least intensions or *noemata*; hence, they are different from nothing and, accordingly, are real (real as *abstract objects*). We can therefore concentrate the strength and the challenge of this question on the following further question: are scientific objects also referents?

[39] In order not to complicate our discourse, we do not mention here those particular images which may be obtained by abstraction, and which could lead us to apply the present reasoning to abstract entities such as those of mathematics. Therefore we are not adhering to a basic empiricist view, as might have been suggested by the elementary examples used here.

Our way of formulating the issue has the advantage of dispensing with the difficult and perhaps not completely sound problem of determining what 'metaphysical reality' is, in order to see whether a *noema* does or does not correspond to such a reality. In other words, we need not make use of the distinction between ontology and metaphysics that was presented, for instance, by Roman Ingarden,[40] and is often evocated in several books on ontology. A distinction which would be implicit even in an idealistic position which maintains that nothing exists behind the *noemata*. Indeed the claim that nothing exists behind the *noemata* is of a metaphysical character, since it pronounces on what does or does not exist in an *absolute* sense. Our position, instead, makes (at least in the case of scientific objects) the existence of the referent different from and independent of the existence of the *noema* on the basis of 'criteria of referentiality,' which are of many sorts and therefore inscribe the referent in different sorts of ontology, so that its existence is *relative* to a particular ontology, an ontology that depends at the same time on the particular 'point of view' adopted, and on the criteria of referentiality linked with this point of view. However, owing to the *analogical* nature of existence which we have already mentioned, this is authentic existence as well. (But this issue will be our specific concern in the section devoted to the problem of realism.)

The complete answer to our question of whether scientific objects are also referents will come at the end of a detailed inquiry, but we can already note that we have an initially sound basis for answering the question positively. As one may remember, the general feature of our conception is that objects are clipped out of things, and that a thing is essentially that which in a given situation is entitled to play the role of a referent (i.e., of something which does not have a purely intentional reality). It follows that the process of objectification takes place in a referential situation, and is carried out under strictly referential conditions (for not only 'things,' but also instruments and operations are to be 'given' in a referential sense, as we have already explained). Hence we can say that it would already be absurd to claim that something which is not only known, but even *must* be known to every subject, could be non-existent. This purely a priori consideration (which amounts to recognising the autonomous existence of 'abstract objects' in the sense of intersubjectivity) is reinforced by the remark that the actual conditions by means of which the object becomes known to a subject are of a *referential* nature.

All the same, we said that we have an *initially* sound basis for endowing objects with a referential nature as well. In saying this we were implicitly pointing to the fact that several scientific objects correspond to theoretical constructs, and in such

---

40 See, for example, Ingarden (1964/1965), vol. 1: *Existentialontologie*, p. 33.

a case it is not clear how referentiality could be preserved. This is why we now need to devote a specific analysis to the problem of referentiality itself.[41]

4.3 Some Additional Remarks About Reference

We would now like to deepen the analysis of the nature of referents, and in particular to provide an explanation and justification of the claim we expressed in a paradoxical vein in Sect. 4.1, where we said that reference does not constitute a semantic problem proper, but rather a pragmatic one. A major difference between the notions of sense and reference becomes manifest through the different tools we need for finding them. Let us consider the simple (but central) case of a predicate. In order to determine, to catch, or to discover the sense of this linguistic expression—not in the subjective sense of what *we* mean, imagine, or feel when using it, but in the sense of understanding what it *objectively* means (so that we might even be ready to change our way of conceiving of it accordingly)—what we primarily (perhaps not exclusively) need is a *conceptual analysis*, that is, an intellectual activity which aims at *understanding* what content of thought is expressed through the concept designated by the predicate. In view of this, we explore the linguistic context of this concept, that is, we see how its sense emerges out of the relations it entertains with other concepts in the language, to the extent that we know this.[42] At

[41] Certain authors have insisted that realism (be it common-sense or scientific realism) is a 'metaphysical' position and, as such, must not be confused with an epistemological or a semantic view, nor made dependent upon a particular theory of truth (such as a correspondence theory). This, in particular, is the position advocated in Devitt (1984). We can agree, to a certain extent, with this view, especially because in the examination of the literature on realism and anti-realism it often appears that anti-realist positions are taken as a consequence of the holding of particular semantic or epistemological tenets. This fact, however, cannot prevent one from recognising that a simple 'metaphysical' defence of realism, such as the one advocated by Devitt in the form of a very general 'naturalistic defence,' falls short of providing cogent *arguments* in favour of realism (though he is often rather convincing in his criticisms of particular forms of anti-realism). This is why we are convinced that a good deal of semantic analysis (and in particular a clarification of the issue of reference), as well as a close scrutiny of the notion of truth, are needed for a correct understanding of the issues at stake in the realism debate, and for the evaluation of the 'arguments' produced by the opposed parties in this debate.

[42] This 'holistic' conception of meaning has been particularly stressed, as we have seen, by those recent scholars who have made of it the fulcrum for advocating the theory-ladenness of every scientific term, the incommensurability of theories due to meaning variance, and so on. It is fair to recognise, however, that this doctrine is not that new. In fact it was already contained in the conception Frege himself considered to be the major novelty of his logic with respect to his predecessors, and even the reason why his logic was (to a certain extent) in opposition to that of Aristotle and the tradition.

This conception maintained the primacy of the proposition over the concept. The primitive unit of meaning is the proposition, in which a judgment (i.e., a content of thought) is expressed, and only by analysing this primitive meaning can we determine which concepts occur in the proposition, and establish their meanings. Traditionally, things were seen the other way round: by

a second stage, we take into consideration the speaking community that uses the language; and finally we consider the referents to which this concept is commonly applied. In order to determine or discover a referent, instead, we fundamentally need a *practical interaction*. We must meet the referent, we must 'have to do' with it, we must feel its difference from ourselves and our mental products, and its resistance with respect to us (all this without denying that we may have been 'guided' towards the referent by the consideration of a sense). Only at a second stage, as we shall see, might we be entitled to claim the existence of certain referents on the basis of arguments (but still not on the basis of arguments *alone*).

This non-mental (or non-intellectual, non-linguistic) but rather 'practical' character of referentiality is clearly implicit both in the common-sense way of thinking and in several philosophical doctrines (such as, e.g., Peirce's pragmatism). For example, in order to explain why a dreamed apple is not a real one (i.e., one to which reference can be made), one is spontaneously tempted to say, "because you cannot eat it," pointing in such a way to some kind of practical interaction with the referent. The same can be repeated when we say that this table is a real table and not simply a creation of our mind, as idealists would allege. We might possibly support our statement by adding that we can touch the table, move it, sit on it, considering all these features not so much as ways of *perceiving*, but as ways of acting, or of *meeting a resistance*. The same way of thinking is to be found behind the vague and equivocal expression *external world*, or behind the statement that the referent is external to the subject. It is clear that this way of being external has no spatial connotation proper, but hints at a kind of confrontation or difference which the referent shows with respect to the subject.

As for the history of philosophy, the most interesting confirmations of this non-intellectual nature of the relation between a person and a referent come not from the empiricist tradition, but from the idealist tradition. Let us only mention in this connection Fichte's conception of the *Non-I* as being a perpetual 'obstacle' which the transcendental *I* has to overcome. This led him to give primacy to action rather than to thought, thus qualifying his idealism as an ethical one.[43]

(Footnote 42 continued)
abstraction we first obtain concepts, and then we combine concepts to form judgements. The rejection of this conception certainly begins, *philosophically*, with the primacy of judgment explicitly stressed by Kant; but Kant was unable to translate this view into his way of conceiving of *logic* (his logic remains structured according to the traditional patterns, and so does his doctrine of judgment, which he still sees as consisting in the attribution of a predicate to a subject). It was only with Frege that the Kantian novelty found its recognition in logic (and in fact the relation subject-predicate is replaced in Frege by the relation function-argument). Concepts are regarded as unsaturated logical entities, and as such cannot be the primary bearers of meaning. Of course, this does not prevent, in the subsequent steps of the construction of a language, compound concepts' being obtained by combining the senses of already available concepts and, similarly, determining the sense of a proposition on the basis of the senses of its constitutive parts, including concepts.

[43] As Schopenhauer pointed out, the German word for reality, i.e. "*Wirklichkeit*," involves the idea of action (*Wirkung*). The same is true of the English "actual."

If all the preceding is true, it follows that the referents must be met in our practical way of acting, and this is why we could say above that the only proper characteristic of a thing is to be an identifiable referent (where identifying is meant to be a concrete act of 'meeting,' and not to be part of a process of describing. In other words, in the present discussion "to identify" means "to single out"). This means that whatever has sufficient unity to make of it an identifiable entity with which we can have some practical interchange is by this very fact a *thing*, an entity which we can consider, which we can ideally 'point to,' *about* which we can start thinking and speaking and which, therefore, can become a *referent* of our discourse. If it happens that a certain number of such things are at our disposal, so that we can establish an interaction with them which becomes so to speak *standardised* within a particular community, then these things may become the conditions and the starting points for introducing operations by means of which we determine scientific objects in the way already explained. This therefore justifies the claim that *operations are the basic conditions of referentiality* for scientific objects: operations are themselves referents, which are applied to referents.[44]

To see this point more clearly, let us reconsider the relationship between 'things' and 'objects' under the new light introduced in this section which enables us, in particular, to consider it also as an expression of the difference between sense and reference. In order to consider something as an object of a certain science, we must first of all identify it; and this happens, in most cases, because we are able to perform such an identification in space and time by means of our sense

---

[44] This conception of the pragmatic and operational nature of reference, which we have presented and defended for many years, has several affinities with doctrines that have been elaborated more recently. For example, it is not accidental that the 'new semantics,' or anti-Fregean semantics, of which we have spoken in Sect. 4.1—and which is typically a semantics of reference as opposed to a semantics of meaning—explicitly maintains that reference has ultimately to do with ostension, since all that matters as regards a referent is to identify it. This identification occurs in the context of a social communicative practice whose task, however, is not that of improving our understanding of a meaning: "the circumstances of utterance do help to provide us with an identification of a referent, but not by providing some descriptive characterisation of it" (Wettstein 1991, p. 26). The context of utterance, moreover, is only partially linguistic, not only because natural languages are conceived of as social institutions governed by a complex system of rules and conventions, but also because in concrete situations the identification of the referent is primarily bound to material gestures: "the pointing gestures not only provide cues as to the reference but actually determine the reference" (Wettstein, op. cit., p. 78; in general this work offers a well-developed account of these concepts).

But also Rom Harré has expressed a similar view of reference: "Referring is a human deictic practice, by which, with any means at hand, one person tries to draw the attention of another person to a being in their common public space" (Harré 1986, p. 97; the title of Chap. 4 of this work has the significant title: 'Referring as a material praxis'). With respect to these and similar positions, we note that our theory is more elaborate and has the essential feature of relating reference to the more precise notion of operation, which is much more than simple gestures or a general 'picking out' as, for instance, in Harré's example. This is why, in particular, we believe that our position is better suited for treating the problem of reference in the sciences, and better equipped for granting the stability of the referent—in spite of 'meaning variance'—which we have already discussed.

perceptions, and to help other people make the same identification by resorting at least to gestures, and more often to other more complex concretely available tools. Once a thing has been identified, it is possible to *refer* to it, and to speak about it using different languages, that is, by resorting to different sets of predicates. Everyday language employs predicates which have a non-specialised meaning, while different sciences introduce, as we know, 'relativised' sets of predicates, by means of which it is possible to 'clip out' a bunch of objects from one single thing.

We shall now consider the relationship existing between these two 'worlds of predicates' (the non-specialised and the specialised), which is usually understood in an incorrect sense. Actually, the most widespread view is that the predicates of everyday language are imprecise or, at best, primitive and naive, while the reason for introducing scientific predicates is to provide precision and explanation. In other words, a very general view is that we certainly cannot help starting from everyday experience and ordinary-language predicates, such as those expressing colours, sounds, and so on, but then scientific discourse (e.g., physics) *explains* these features in terms of its *exact* predicates. The consequence many people draw from this is that, owing to this fact, everyday predicates become, at least in principle, dispensable, and we could safely do without them (this consequence clearly has a *reductionist* character).

Yet, despite its prima facie soundness, such a reductionist claim is totally groundless. Not only because it ignores the relativity of the different discourses (e.g., it is true that we can dispense with using secondary qualities in physics, but this does not entail that they are eliminable in an absolute sense). As a matter of fact, it is easily seen that no explanation in any proper sense occurs when we make the transition from ordinary-language predicates to the specialised predicates of an exact science. Let us consider this in the context of an example, and suppose that our primitive and 'naive' predicates are the names of certain perceived colours such as red, yellow, or green, while the corresponding 'exact' predicates are the expressions of these colours in the form of electromagnetic waves of different lengths. We shall also assume that the physical theory of light as an electromagnetic phenomenon is accepted and known. What would now be, say, an explanation of red in terms of frequencies of electromagnetic radiation? As we know, to explain means to give an answer to the question "*why?*"; and in our case we could imagine our question to be, for example, the following, "Why is this pencil red?" (Or even, if we want to be very strict, "Why do I experience the colour red when looking at this pencil in a white light?")

The alleged explanation might go as follows: "This pencil is (appears) red because it absorbs electromagnetic waves of all lengths, except those with a length of approximately 650 nm (or 10^{-4} cm), which are reflected by its surface (reaching the retina of your eye, the impulse travelling to your brain and resulting in your experiencing the colour red)." One should really be very simple-minded and naive to accept this as an answer to the question, because the answer gives no reason whatever (even in terms of the electromagnetic theory of light and of the physiology of visual perception) *why* this particular frequency should give rise to the experience of red, rather than to that of yellow or green. Someone might hope to overcome the difficulty by replying: "Of course, we cannot pretend that the electromagnetic theory of light tells us *why* red corresponds to a wavelength of approximately 650 nm, yellow to a wave-length of approximately 580 nm, extreme violet to a wave-length of approximately 400 nm, and so on; the theory simply shows us *that* red, yellow, green, and so on, do actually correspond to such-and-such wavelengths."

We have nothing against this reply, provided that one is aware that it implies abandoning the hope of an explanation. For to state *that* something is the case is by no means to explain *why* it is the case. If we investigate more carefully what actually happened in our alleged explanation, we can easily see that something, which has been identified (i.e., singled out) in a primitive and independent way as being red, happens to be *expressible* or predicable (not *explicable!*) within a different context (the *point of view* of modern science vs. the *point of view* of everyday experience) as something which reflects electromagnetic waves of a certain length. But this amounts simply to a change of vocabulary (that also implies a certain categorial shift from sensory to intellectual predicates), and not to a deeper insight (unless one is a reductionist); and the whole enterprise sounds very much like a *translation* rather than like an *explanation*. By this, of course, we are not denying that, thanks to such translations, many phenomena that we describe in ordinary language can receive an explanation once they have been duly 'objectified' in a certain science.

This mention of translation is significant at this point in our discussion because it introduces the proper context in which the problem of *meaning* takes shape (for translation is essentially the operation of establishing identity of meaning across different linguistic frameworks). Moreover, our example already suggests that the different sciences might constitute different 'contexts of meaning' in which reality is considered. Indeed, this conclusion does not come too unexpectedly, as we have already spoken of a 'conceptual space' determined by every discipline, which is itself not very far from the notion of a context of meaning. Also, certain common ways of speaking already point towards this solution, such as when we say that, in physics, red is 'seen,' 'interpreted,' 'conceived of' *as* being an electromagnetic vibration of a certain frequency. These ways of speaking are typical of the discourse of translation, or of meaning interpretation, rather than of explanation proper.

But now we shall apply to our problem something which, according to us, is a common feature of every translation between two languages, namely, that one of the languages must provide the reference (besides the sense) of at least a good many words, that is, of those words which are supposed to denote 'things,' while the other language may be thought of as being (in a loose sense) a renaming of these referents which 'parallels' the first language (besides being a 'reconstruction' of the senses of its words). For example, if we are simply told that the Italian word "*cane*" is translated by the English word "dog," we have received almost no information, unless we know in at least one of these languages what the reference of the word is. In this case, we shall be able to say that the same referent is also referred to by the word in the other language. But note how obvious it is that no one would say that "dog" is an *explanation* of "cane" in English; it is simply a linguistic *expression* for denoting the same referent as that of the Italian word.

A rather interesting remark is that these referents of everyday language are 'natural kinds,' rather than individual entities denoted by proper names (we shall not develop this remark; entering into the discussion of the nature of natural kinds would lead us too far afield). In fact, individual entities are often denoted in all languages by practically the same name (the differences being mostly due to phonetic reasons), such as Paris, Rome, Caesar, Jesus, while natural kinds are normally denoted by terms which may be very different in the different languages.

Now, if in order to understand the *meaning* of a general term (such as a term denoting a natural kind in a given real language) we *also* need to relate it to certain referents, this means that for general terms as well the meaning cannot be *entirely* provided by the linguistic context. This has a few consequences. First, the contextual theory of meaning is mistaken, if it is taken as maintaining that the *linguistic* context can *fully* determine the meaning, since this is not true not only (trivially) in the case of proper names, but also in the case of most general names; a complete determination of meaning through linguistic context is only possible for those concepts which denote purely 'abstract objects,' as we have seen, that is, only in *special cases*.[45] Second, *meaning* cannot be equated with *sense*, because sense (as distinct from reference) does not allow us to really 'understand the meaning' of a word, unless we are *also* provided with some referential information (as the example of the two languages has shown). Third, *meaning* does not coincide with *reference* either, not only because what we actually mean by a concept largely oversteps what we can 'point to' in a referential act, but also because much of this meaning comes from the linguistic context. All these points support our thesis (that we hinted at in Sect. 4.1) according to which meaning embraces both sense and reference.

Let us now consider the case of different sciences being applied to the study of one and the same 'thing' (or of a given collection of things). Each science may be conceived of as using a specific language which receives its reference from a pre-existing language.[46] This is in the simplest cases an ordinary language, which contains predicates related to sense experience, as well as the names of tools, concrete entities, operations, and so on. When our 'thing' is identified by means of this ordinary language, it becomes possible to perform several kinds of

[45] This explains why in the *Foundations of Arithmetic* (1884), where Frege was concerned only with the meaning of mathematical concepts, that is, of abstract objects, could he defend a purely contextualistic theory of meaning, while he opened a space for the consideration of the referent when he enlarged his considerations, for example in the paper 'On Sense and Reference' (1892).

[46] Let us recall that, according to the view proposed in this work, every scientific discipline is characterised by the specific 'point of view' from which it considers reality, which entails the adoption of certain specific predicates. They make the 'specificity' of the language of this discipline, in which the presence of several elements of an everyday language is simply a tool for communication and referentiality. Therefore, the 'discourses' formulated in a discipline certainly contain many expressions of common language, but are disciplinary only to the extent that they contain the specific predicates of the given discipline.

'translations.' Some may transpose predicates of the ordinary language into 'disciplinary' predicates of a given science, some others may consist in translating disciplinary predicates from one discipline to another. In both cases, the possibility of *connecting* the different discourses is provided by the existence of the *common reference*, so that we are entitled to say that we are in the presence of a diversity of *meanings* that are each referred to this unique referent.

A slightly different situation arises when we are not speaking of a 'thing' within different disciplines, but when we are directly confronted with a translation between two disciplinary languages. In this case too it is necessary that one of these languages play the role of a 'referential' language, in the sense that its criteria of protocollarity or referentiality be recognised (at least partially) as constituting admitted testing procedures for the other discipline and its language as well. If this is the case, it may be possible to perform the translation and to 'interpret' the facts of one theory in terms of the concepts of the other (this is actually the case with certain physical theories). If this is not the case, the only way is to go back a sufficient number of steps in order to find a third language (disciplinary or common) which can serve as a referential language for both disciplines.

All this can be made clearer through an example. The introduction of food into an organism, its transformation and assimilation, constitutes a process which is describable using ordinary language; and such a description 'identifies' a certain *referent* (namely the process of digestion on the part of a particular organism). Then, the *meaning* of this process—resulting from the meaning of the terms used to describe it—can be expressed by 'talking about' it in different ways (i.e., in different disciplinary languages) according to the different 'viewpoints' assumed in considering it within the different sciences. So, the meaning of the process can be *expressed* either in terms of physiological predicates, or in terms of chemical reactions, or in terms of thermodynamic transformations, and so on (each form of expression being part of a different language). No one such expression or interpretation of the metabolic process can really *explain* the others, though each of them is of relevance to the others.[47]

We can now pass from the intuitive discourse used till now to a more technical way of speaking that employs certain notions we have already introduced in other sections. A thing or a referent can be presented as an entity which *does not encode* any property, but may *exemplify* many properties. This assertion sounds puzzling at first, since it is obvious that concrete things not only possess properties, but are normally described by mentioning a certain list of properties which 'characterise' them. In spite of this, it is clear that a single, individual thing cannot be characterised by any finite list of properties, because any such list may in principle be satisfied by other things as well.

[47] It is obviously possible that a certain 'point of view' takes shape within another point of view, so that the two languages are more significantly interconnected than just through a community of reference; but we are not interested in discussing this point here.

As a matter of fact, we have said in the earliest sections of this work that a thing is a 'potentially infinite bunch of objects' (while at the same time being an independent existent), meaning by this that a thing may be considered under a potentially infinite number of points of view, and in such a way also be considered as being endowed with a potentially infinite number of properties. But precisely for this reason it would be arbitrary to say that a thing is *totally characterised* by any particular set of properties (which is the proper meaning of *encoding*). On the other hand, this does not conflict with the fact that there may be a set of properties (also potentially infinite) which a given thing does *not* possess. For example, a toothache does not posses mass, as we have already noted. This shows that possessing (i.e., exemplifying) a potentially infinite set of properties does not mean that a thing is not determinate, so that anything could be said of it. On the contrary, a thing has the determinateness of its *individual concreteness*, and for exactly this reason it cannot be captured through a finite list of properties, which necessarily leaves too much undetermined. On the other hand, nothing prevents a thing from *exemplifying* a few properties, or even just one property, if it happens to possess this property among the many it exemplifies.

This is an interesting feature which already distinguishes referents (or things) from abstract objects. Indeed, an abstract object, as we have seen, is totally and univocally characterised by the properties it *encodes*; but this is exactly the reason why it does not exemplify these properties. For example, the abstract object consisting in the concept of dog encodes the properties of having four legs and barking, but the concept itself does not bark, nor have four legs, that is, it does not exemplify these properties. They are exemplified, on the other hand, by an indefinite number of concrete individual dogs, each of which exemplifies several other properties (e.g., having a certain colour, a certain weight, and so on) as well.

When we defined "scientific object," we said that a referent of this term is a structured set of attributes; and at the same time we stressed that, as such, it is an 'abstract' object. This would be tantamount to saying that a scientific object encodes certain properties, and is completely determined by them. But now we have been led to recognise that no scientific object (conceived as referent) can exemplify the properties it encodes. Is this a paradox? Absolutely not. On the contrary, this fact helps us understand the correct meaning of certain common claims, such as that "the rigid body, or the perfect gas, *does not exist.*" Obviously these claims implicitly presuppose that existence means exemplification, and they stress that no concrete object exemplifies a rigid body or a perfect gas. This is true, but not at all strange, since requiring of a concrete object that it exemplify exactly and exclusively the properties of a rigid body or a perfect gas would really mean that such a concrete object *encode* these properties, and we have seen that things, concrete objects, referents, do not encode properties. A concrete object always *exemplifies* many kinds of properties whose simultaneous coexistence usually leads to the impossibility of *exactly* exemplifying all of them.

This reflection shows that the relation between encoding and exemplification is not straightforward. It is not a trivial mirroring or point-to-point correspondence. Except (perhaps) for the most elementary empirical properties (such as secondary qualities), all complex encoded properties are exemplified only to certain degrees, due to the simultaneous presence of other exemplified properties in the concrete objects. This explains why, in the sciences, many properties may be seen to be exemplified only in some very sophisticated experimental set-up in which the concomitance of other 'disturbing' factors can be eliminated. This confirms that the transition from the encoded properties to their exemplification occurs, in the sciences, through the mediation of operational procedures, each having its *margin of approximation and accuracy* (to be discussed later). However, we must be aware that what is of interest with respect to abstract concepts is that they encompass many possible referents, each characterised by its possessing different properties (most of them purely contingent), as well as that it is possible to reason in terms of logical necessity and sufficiency only with respect to abstract objects.

On the other hand, there is nothing really disturbing in this situation since, while it is impossible to have a referent univocally encoding the properties of an abstract object, it is always possible to envisage an abstract object exactly encoding certain properties exemplified in a concrete referent. For example, we can explain the behaviour of a concrete physical system which deviates from that which is described in an abstract model of the system by taking into consideration a great deal of actually ascertainable perturbations, and in such a way constructing a new abstract model 'tailored' to the explanation of the behaviour of this particular concrete system. The founder of modern science, Galileo, was fully aware of all this.[48]

From what we have said it follows that the referent is not only characterised through the 'negative' but interesting feature of *not encoding* properties, but also through the 'positive' feature of *exemplifying* properties. This circumstance is also

---

[48] In the Second Day of the *Dialogue Concerning the Two Chief World Systems*, Sagredo (who expresses Gaileo's position) celebrates the use of mathematics in the study of natural phenomena, and the Aristotelian Simplicio objects that to use mathematics is wrong, since in the realm of material things no geometrical properties really hold, due to the imperfections of matter. For instance, in the physical world it is never the case that a sphere touches a plane at a single point, as geometry would demand. This objection is rejected by Salviati, who remarks that this happens because in the material world no perfect spheres or planes exist, but this does not invalidate the geometrical reasoning, for "even in the abstract, an immaterial sphere which is not a perfect sphere can touch an immaterial plane, which is not perfectly flat, not in one point, but over a part of its surface, so that what happens in the concrete, up to this point, happens the same way in the abstract" (Galileo 1632, *Opere* VII, p. 233; English translation, p. 207).

Here we see that, while it is recognised that no concrete object could encode abstract properties, but only exemplify them within a certain limit, it is also possible to mirror the properties exemplified by a concrete object in an abstract object where they may be studied in full generality and in terms of logical necessity. The discussion in Galileo's work continues with its being said that "the mathematical scientist (*filosofo geometra*), when he wants to recognise in the concrete the effects which he has proved in the abstract, must deduct (*diffalcare*) the material hindrances, and if he is able to do so, I assure you that things are no less in agreement than in the arithmetical computations." (op. cit., p. 234; English translation, p. 207). Here we find the indication of the necessity of non-trivial efforts to find the exemplification of encoded 'abstract properties' in the realm of 'concrete objects.'

interesting as it vindicates, after all, the importance of the *noema*, of the sense, in the determination of referents, and in such a way at least partially justifies what Frege (and many others) claimed, that the sense determines the referent. However, we would like to stress immediately that this justification is only partial, and we can explain this by saying that the sense is necessary in order to *recognise* the referent, but it does not *provide* us with the referent. The sense is indispensable for evaluating whether what we have met or found is actually the referent we were looking for, but the sense does not produce such an encounter. The sense certainly suggests what we should *do* in order to meet the referent, but this *doing* is not part of the sense.

This necessity and this insufficiency are again readily understood if we distinguish encoding from exemplifying. Exemplifying is certainly related to encoding, since a referent necessarily exemplifies properties which are encoded by some abstract object (hence, there is no referent which is not also the exemplification of some sense). But on the other hand encoding, as we have already noted several times, is a form of *predication*, and it is not the case that whatever is predicated is also exemplified (this is why sense alone does not secure reference). As a result of our analysis we arrive again at the conclusion we reached through our intuitive reflections. The semantics of exemplification is not a semantics of predication; and what remains is that it is a semantics of doing, of operating. If we want to test whether a property is exemplified, we must introduce some *testing procedures* for it. We cannot limit ourselves to conceptual or linguistic elaborations, since these would not bring us beyond the level of predication.

Our last remarks could give rise to the impression that operational testing procedures can overstep the purely linguistic and conceptual level because they are 'material,' and therefore that only material properties are exemplifiable. This is not so. Without repeating what we have already said several times regarding the extremely various spectrum of operational procedures (which are to a certain extent bound to immediate sensory experience, but may be rather far from this experience in their most characteristic features), we would like to approach this question here by briefly indicating that material properties may (sometimes) be non-materially exemplifiable, and that non-material properties may be exemplifiable. To use a famous example, Pegasus certainly encodes the property of being a (winged) horse, and therefore of being a physical body; but Pegasus does not exemplify this (material) property in the (material) world of common experience. (Pegasus does not exist in the realm of physical bodies, because it is not intended to be detectable through our physical criteria of referentiality.)

However, Pegasus *exemplifies in mythology* the property of being a physical body. We cannot ascertain this physical property of Pegasus by means of any material testing procedure; we can test Pegasus's exemplifying this property only through a non-material kind of operation, such as reading mythological stories where Pegasus is described. This example, by the way, shows that *noemata* may well be referents, and this confirms from another vantage point our remark that objects and things are only relative concepts. In fact, Pegasus is certainly and, so to speak, primarily a *noema*, an abstract object. However, it can be referred to up to the point that it may exemplify many material properties, and this is possible since it has received a particular status in a story, a status which allows testable statements to be made concerning it. But we are not saying that *noemata* may become referents only in this way. For example, as we have already noted, the *noemata* of our intentional acts may be referred to through self-reflection, and this is a kind of *mental operation*, which is different from a *predication*. Mathematical objects are also usually referred to operationally, so that we can say, for example, that a certain numerical system not only fails to exemplify a particular material property, but also a certain particular mathematical property (for instance, it might not be closed with respect to subtraction or division.)

The considerations of this section have provided us with a first answer to the intriguing question of how we can know whether a *noema* is just a *noema*, or whether it also corresponds to a referent. In order to decide this question we must see whether we can simply *refer to it* through self-reflection or in purely intentional states (such as thinking of, believing, wishing). If we can, we must say that what we have to do with is (at least for the moment) a simple *noema*. But if we can devise certain not purely intensional operations by means of which we ascertain that this *noema* is exemplified, these very operations entitle us to maintain that there is a referent to which this *noema* 'corresponds.' This correspondence, however, is nothing more than the exemplification of the *noema* by the referent. Moreover, the ontological status of the referent—of which we may even say, now, that it belongs to the 'external world' in the sense of not being purely an internal representation of the mind—strictly depends on the criteria of reference. The referent may not only be a physical object, but also a mythological being, a character in a novel, an historical person, a mathematical object, a legal prescription, a moral imperative, and so on.[49]

[49] The considerations presented in this section enable us to propose a critical appraisal of a famous distinction advocated by Wilfrid Sellars in his paper 'Philosophy and the Scientific Image of Man' (1963, pp. 1–40). According to Sellars, the self-consciousness through which man-in-the-world realises the identification of his proper nature depends on a complex idealised view in which perceptual elements, conceptualisations, classifications, categorisations and theories of various kinds are deeply interrelated to form a global 'image,' an image which has gradually evolved from pre-historical times up to the present, and which we take for granted and consider obvious. This is the 'manifest image' of the world. With the creation of modern science, however, a new image of the world has rapidly emerged, the 'scientific image,' which is not in keeping with, but at variance with, the manifest image shared by common sense. According to Sellars, however, the manifest image (though being unavoidably the accepted frame of reference of our daily life) is *wrong*, and should be replaced by the scientific image, which gives us the true representation of reality.

Several criticisms have been levelled against this doctrine. They include that the manifest image is the ground on which the scientific image itself is founded; the different sciences offer different images of the world, so it is hardly sensible to speak of *the* scientific image; and science is in a continuous state of revision such that the scientific image capable of replacing the manifest image is at best an ideal and rather utopian end-state whose features and time of realisation we cannot even imagine.

We are not interested in discussing these (and similar) criticisms, since we see certain deeper

One may have received the impression that we have proposed enough criteria for finding referents (perhaps too many), but the story is not over. Even though we admit that the referent is characterised by means of operational criteria of referentiality, which specify certain (not all, as we have said) of the properties it exemplifies, we are not saying that having recourse to the operational criteria of protocollarity is the *only* means for ascertaining the existence of referents. This is tantamount to asking whether operationality is a necessary or only a sufficient condition for referentiality. Empiricist tradition believes that it is also a necessary condition; but in order to obtain a satisfactory answer to this question we must first explore (still in a very approximate way in order not to become involved in too lengthy a discussion) another crucial issue, that concerning truth. Before entering upon this discussion, which will occupy the Sect. 4.4, we would like explicitly to note that the conception of truth which we shall advocate strongly underlines the

---

(Footnote 49 continued)
reasons for disagreement with Sellars. The first is that his separation fails to consider the distinction between things and objects and, in particular, ignores that the sciences investigate only a limited number of specific attributes of things (e.g., *physical* attributes). But also common sense cannot do otherwise: the manifest image of the world is that which is elaborated starting from a rather large, but still limited, number of attributes, and contains those conceptualisations and theories that are considered adequate for understanding and explaining the world as characterised through such attributes. Moreover, both science and common sense do not limit themselves to producing intellectual constructions, but try to secure their access to reality (and truth) by means of their criteria of referentiality. In the case of the sciences they are standardised and few in number, but common sense too relies upon a variety of commonly shared operations and ways of doing things that enable people to 'meet' things and refer to them in the concrete.

This common situation becomes even clearer if we consider that the operational criteria of referentiality are of a *practical* nature, not only in the elementary sense of constituting a concrete way of acting, but also in the sense of belonging to *praxis*, i.e., of being determined by the needs and the *ends* of a particular human activity. Therefore their adequacy must be measured according to their ability to satisfy these needs and fulfil these ends. Once this is clear, it is also clear that the manifest image of the world is cognitively *right* and adequate for the conduct of man in the multiple activities of his ordinary life (and it also easily absorbs several contents of the scientific image when this is needed). These considerations, by the way, capture the reasons for which social/pragmatic relativism is partially right. In other words, the *pluralistic ontology* we have advocated as a consequence of considering the different sets of attributes of reality and their different conditions of referentiality justify a view according to which the manifest and the scientific images of the world are complementary rather than in opposition. The deeper reason for Sellars' position is that it is a combination of metaphysical realism and partial epistemological dualism. It is obvious that he is a 'scientific realist' (in the strongest sense of maintaining that science gives us a true representation of reality as it is). But at the same time he is also an epistemological dualist, since he maintains that the manifest image of the world that humans have in their most common way of knowing is not a true portrayal of reality. This amounts to saying that reality exists independently of our knowledge of it (metaphysical realism), and that knowledge cannot be had of reality by ordinary means. However, this reality can be faithfully represented (though only in an idealised limit-situation) by science.

It is not accidental that Sellars himself recognises the affinity of his position with that of Kant. He agrees with Kant that the world of common sense is a 'phenomenal' world, but he submits that 'scientific objects,' rather than the metaphysical unknowables, constitute the true things-in-themselves, which science is able to know (see Sellars 1968, p. 143).

referential nature of truth, and will do so in what follows with such insistence that one might be tempted to believe that, in our view, truth is a defining condition of referentiality. This temptation must be resisted. Truth and referentiality are *fundamentally related* but remain distinct, more or less as sense and reference are distinct, though being so related. Indeed, we have proposed an *operational* or *pragmatic* theory of reference in which no mention of truth has been made. But we are not alone in this way of treating the issue. Let us only mention that Kripke and Putnam, for example, have developed a well known *causal theory of reference*, which they have applied not only to proper names but also to 'natural kinds,' and in which no use of the notion of truth is made.

4.4 Some Considerations Regarding Truth

The distinction between meaning and reference is useful in that it throws light on a controversial issue, namely the problem of obtaining an acceptable conception of truth. The mere mention of such an issue might produce a sense of discomfort in our reader for, if our treatment of scientific objectivity were supposed to settle this general question before coming to some conclusion, we might abandon the hope of ever seeing an end. However, we are not pretending here to engage in the most controversial parts of the dispute over the nature of truth. We would simply like to point to certain general aspects of the question which are probably less controversial, and which could help us solve some of our problems.

4.4.1 The Adjectival and the Substantival Connotation of Truth

The notion of truth is referred to in two basic ways in ordinary language. One is by means of the substantive "truth," and the other via the adjective "true." Both of these uses, however, are susceptible of different applications. "Truth" is often used to denote something like a *substance*, that is, something endowed with a kind of existence *per se*. According to this sense, for example, we often say that we want to know 'the truth' about President Kennedy's murder, or we even speak, in the plural, of 'the truths' of physics, of mathematics, of Christian faith. However, this 'substantialistic' conception of truth—which might come to be bound to certain rather deep ontological presuppositions—is not the only possibility left open for the use of "truth"; indeed, this term is often used simply to denote a *property*, that is, the property of 'being true.'

In such a way, we find a link between the substantive and the adjective, and we may believe that, giving primacy to the adjectival use (and considering the substantival use simply as a way of denoting the corresponding abstract property), we could avoid any ontological commitment. But this is not necessarily the case; it depends on what we consider to be the entity of which it is correct to predicate that it may or may not be true. In ordinary language we find expressions such as "he is a true friend," "the true way to happiness is such and such," "a true Christian should love every human being," and so on, and we also find "relativity theory is true," "this sentence is true," and similar expressions. The first examples are reminiscent of a 'substantialist' view to the extent that being 'true' is predicated of several entities which are expected to conform to some ideal example or paradigm which operates like an ontological essence these entities should realise.

The last examples, instead, use "true" as a property which can be attributed to linguistic expressions, such as a sentence, or a theory, that is, something that (at least according to a certain view) is thinkable as a set of sentences. We are not interested here in discussing the precise features and the possible justifications of these different uses.[50] We can limit ourselves to saying that, our interest being in scientific knowledge, and this knowledge being at least to a large extent expressed in systems of sentences, it suffices that we restrict our attention to the problem of the truth of sentences or propositions. This implies that we shall consider the adjectival meaning of truth (restricted to sentences) as primary, and the substantival meaning, understood as denoting a property, namely, the property a sentence possesses if and only if it is true, as secondary. All other uses will be considered as being 'improper,' not in themselves, but in the context of our considerations. The reason for considering them improper is that it not only seems difficult to give them a clear meaning outside our context, but inside it they certainly do not possess much sense, and may actually lead to certain misunderstandings, as we shall try to show in the sequel.

4.4.2 *Statements, Sentences, Propositions and States of Affairs*

Our troubles are not completely over once we have decided to adopt the use of "true" and "truth" only with reference to sentences or propositions since, at this stage, we have intended "sentence" (according to everyday language) to be synonymous with "proposition," "statement," "assertion," and similar expressions. However, in linguistics, logic, and philosophy of language, these expressions are not taken as synonymous, and their differences are exploited for introducing rather subtle, but sensible and important, distinctions. As a consequence, a difficulty emerges. It has not been sufficiently agreed upon in the literature which meaning is to be attached to each of these terms, and this in particular leads to the problem of knowing whether "true" is to be predicated of a sentence, a proposition, a

[50] A rather detailed analysis of this issue may be found in Agazzi (1988b). We shall return to this discussion in a later section.

statement, or some other entity. It would be meaningless to propose our choice among the different options after having discussed the strong and weak points of the various positions on this issue. We shall simply introduce a convention which, apart from performing the necessary task of fixing our terminology, has the merit—we believe—of capturing the strongest points of the most common theories, by complementing them in a more complete framework. In particular, our convention will result in an application of the three-level semantics we have already presented in Sect. 4.1, and which we shall take up again in a discourse already outlined (in that context) in Sect. 4.1.4. By so doing we are going to indulge in a few repetitions, but this will make the present treatment self-contained and allow for certain new expansions.

A very common distinction, which we accept, is the following. A sentence is a linguistic construct, while a proposition is the conceptual or intellectual objective content expressed through this construct. However, here certain problems appear: must this content be identified with the meaning of the sentence or not? There are certain indications that are against identifying a proposition with the meaning of a sentence. For example, a proposition is something of which we say it may be true or false, while the meaning is something of which we say it may be only clear or unclear. Or, two sentences such as "The teacher of Alexander the Great was a philosopher," and "The most famous disciple of Plato was a philosopher" are clearly distinct, and their meanings are also different, because the property of being a philosopher is predicated once of someone qualified as the teacher of Alexander, and once of someone qualified as a disciple of Plato. Yet, we can say that both sentences express the same *proposition* in the sense that they express the same informative content, thanks to the fact that, the individual Aristotle being both the teacher of Alexander and the disciple of Plato, what the two sentences say is one and the same thing, that Aristotle was a philosopher.

In spite of these (and similar) objections, we propose to identify a proposition with the *sense* of a sentence, explicitly relying upon our distinction between meaning and sense, according to which the sense is the *objective* thought-content which represents only a part of the meaning (the other part being the reference). In this way, we are not disturbed by the first objection. It is true that a sense may be clear or unclear, rather than true or false (we consider here the common tendency to speak of the "true meaning" of linguistic expressions as being 'improper'). But we say that if (and only if) the sense of a linguistic expression—in our case of a sentence—is clearly *understood*, we can fix the proposition, and therefore we may start asking whether the sentence is true or false. This corresponds to the fact that, in our three-level semantics (the three levels being language, thought, and the world), the link between language and thought is secured by understanding a sense, and once this sense is understood, it constitutes the second of this structure's three poles.

As to the second objection, it expresses a rather widespread conception, namely that which identifies a proposition with a 'state of affairs.' According to our three-level semantics, this is tantamount to saying that a proposition belongs to the third pole, that is, to the world. We believe that this conception is unnecessarily counterintuitive, and therefore prefer to consider propositions as belonging to the level of thought. In this way one easily sees that our difference with respect to the second objection is only terminological. This objection says that two different 'meanings' (in thought) may correspond to one single 'proposition' (in the world). This we maintain, saying that two propositions (i.e., two 'meaning contents,' or senses, or thought-contents) may correspond to one and the same state of affairs (in the world). It is also easy to recognise that our choice is in keeping with Frege's identification of a proposition with the 'sense' of a sentence, but also with Husserl, and in general with those philosophers who recognise a specific dimension for the world of sense as distinct from the 'external world' (or the world of reference, as we have proposed to call it).

But now let us consider the third level. Here we have introduced 'states of affairs' as *referents* of sentences, and in so doing we not only differ from Frege (for whom the referents of propositions were their truth-values), but also from Carnap (who retained the same doctrine, but used the terminology of 'extension'). Our position, instead, is practically that of Husserl, who was actually the first to propose a 'modern' systematic consideration of states of affairs in his *Logical Investigations* (1901) which preceded not only the more celebrated (and also more elaborated) treatments of this topic by Russell and Moore, but also the less famous, but not less significant or elaborate, treatments provided by Meinong and Marty.[51] For this reason we consider it advisable to leave the word to Husserl, who explained the affinities and the differences between this way of considering states of affairs as referents of propositions, with respect to the similar case of object names:

> If we consider, e.g., statements [*Aussagesätze*] of the form "S is P" we generally regard the subject of the statement as the object 'about' which the statement is made. Another view is, however, possible, which treats the *whole* state of affairs which corresponds to the statement as an analogue of the object a name names, and distinguishes this from the meaning of the statement. If this is done one can quote as examples pairs of sentences such as "*a* is bigger than *b*"—"*b* is smaller than *a*," which plainly say different things. They are not merely grammatically, but also 'cogitatively' [*gedanklich*] different, i.e., different in meaning-content. But they express the same state of affairs; the same 'matter' [*Sache*] is predicatively apprehended and asserted in two different ways. Whether we define talk of the 'object' of a statement in one sense or the other—each has its own claims—statements are in either case possible that differ in meaning while relating to the same 'object' (*Logical Investigations*, I #12).[52]

---

[51] See Meinong (1902) and Marty (1908).

[52] See Husserl (1901), pp. 288–289 of the English translation. An interesting study of this view of Husserl on states of affairs is contained in Mulligan (1989). It is not sufficient to admit autonomy to the world of meaning in order to preserve the difference between sentence, proposition, and state of affairs. Indeed, states of affairs and propositions may be identified not only when one neglects the autonomy of the intermediate level of sense, but also if one overlooks the specificity of the world of referents. This was already done by Frege. A more recent example is to be found in Zalta (1988). We can say that such an identification of proposition and state of affairs results in a kind of trivialisation. Instead of designation, of reference, an *ontological*

We now come to a final distinction, that between sentence and statement, which we believe should play a role in a treatment of the problem of truth. A sentence, we have seen, has a sense (a proposition) and a referent (a state of affairs). If we recall our discussion of what are termed 'non-existent objects' (Sect. 4.1), we know that a proposition may represent an 'abstract object' *denoted* by a sentence, even if there is no concrete state of affairs that constitutes the *referent* of the sentence and/or of the proposition, that is, even if there is nothing in the 'concrete world' which exemplifies the thought-content of the proposition. (This is why we can say with equal right that such a putative entity can be the referent of the sentence and/or of the proposition.) The most spontaneous way of picturing this situation is probably to say that, in this case, the sentence (or the proposition) is false, while it is true if the state of affairs referred to actually obtains or exists. We believe that this way of presenting the situation is not very accurate, and that we should say, instead, that in the first case the proposition "has no referent," and that in the second it "has a referent." In our opinion, the proposition's being true or false is an *additional* feature which only makes sense when the proposition is *stated* or asserted or declared. This is why we say that truth and falsity concern particularly *statements*, meaning by "statements" sentences, or propositions, which are 'stated,' that is, asserted or declared. This is no pure Byzantinism. Indeed, if a proposition is simply contemplated with respect to what it means, it is neither affirmed nor denied, and it may simply be questioned, doubted, made the object of a wish or an imperative, and so on. In all these cases it would be impossible to say that the proposition is true or false, but it would preserve both its sense and its reference (if it has a reference).

This doctrine is very old, and was expressed in the assertion that truth belongs to *judgment*, judgment being not just the composition of two concepts, but the statement that such a composition 'holds,' or obtains. In modern times, the most convinced advocate of this theory was notoriously Frege (and this is not accidental, if we remember what we have said in a preceding note regarding the absolute primacy he gave to judgment over concepts). Frege, already at the beginning of his *Begriffschrift* of 1879, even introduces a special sign to distinguish affirming a statement from merely expressing its thought-content, and correctly says that truth only concerns asserted judgements. Several authors have found this distinction superfluous, but we shall not examine their arguments. We shall instead present a few additional reasons for this distinction in the present section when we come to discuss the difference between *semantic* and *apophantic* discourse. Let us point out, however, that, in the case of a (completely formulated) scientific discourse, which is the only one which is of interest for us in this work, it is implicitly understood that we have to do with *declarative sentences*, that is, with *statements* (even hypotheses are declarative sentences, since they are tentatively

(Footnote 52 continued)
identity is introduced, instead of an identity which is only *intentional*. Husserl was aware of this difference (which was already clear to Scholastic philosophy), and did not make this mistake.

'proposed as true'). This is why, in this work, we shall speak indifferently of the truth of a proposition, sentence, or statement, it being understood not only that we consider only sentences that are supposed to be stated or asserted, but also that every sentence must, in any case, be 'equipped' with its proposition (i.e., this very sentence must be taken with its sense). On the other hand, no proposition can be investigated if it is not clearly expressed linguistically in a sentence. It will also be tacitly understood that, when we speak of a sentence, we refer to a *closed* sentence, that is, to one containing *no free variables*, since, as is well known, only such sentences express propositions.[53] A question to be addressed later is whether 'linguistic' candidates other than sentences or statements may be taken into consideration as being concerned with truth, and in particular whether theories may be such candidates (as distinct from the simple expressions of laws). We shall see that the question is to be handled with care, since theories are not, in a proper sense, identifiable with sets of sentences or statements, though they also have, in a broader sense, this kind of feature.[54]

4.4.3 The Truth of a Sentence

Let us now consider a famous example of a sentence: "Snow is white." In it, a certain *property* is predicated of snow, that of being white. Now, if we say that this sentence is true, we predicate of the sentence itself a particular property, namely

---

[53] It is perhaps not superfluous to compare our approach with Carnap's, which is still the most commonly followed in standard epistemology of the empirical sciences. In Carnap (1942), an appendix is devoted to terminological remarks, where the terminological variations concerning the meanings of "sentence" and "proposition"—to be found in different authors and dictionaries—are reviewed. Carnap particularly recommends using "(declarative) sentence" as a linguistic expression, and "proposition" as the *designatum* of the sentence, that is, "that which is expressed (signified, formulated, represented, designated) by a (declarative) sentence" (p. 235), which is explicitly equated with Wittgenstein's notion of state of affairs. Here we find on the one hand a certain recognition that only statements (i.e., declarative sentences) matter in the epistemology of the sciences, and on the other hand an identification of propositions with states of affairs, which reflects the elimination of the 'second level' already discussed in Sect. 4.1. An account of the complexity of the topic sketched here may be profitably read, for instance, in Cohen (1962).

[54] We shall see that theories are closer to the semantic than to the apophantic logos to the extent that they have a 'representational' rather than explicitly 'declarative' character. In this sense a theory—like a concept, which typically belongs to the semantic and not to the apophantic logos—is always incomplete, and involves a certain degree of implicitness. It is incomplete in the sense that it always intends to express a certain 'way of seeing' a given field of reality, leaving many other aspects out of consideration. It is largely implicit because it consists, as we shall see later, in the proposal of a certain global *Gestalt*, which can only partially be translated into a finite set of explicit declarative sentences, though it must undergo such a translation if it is to be submitted to tests or, in general, to a critical examination. These considerations will become clearer in Sects. 5.4 and 5.5, where more will be said about the nature of theories.

that of being true. Still, there is a non-negligible difference between these two kinds of 'properties' which becomes clear if one considers what is actually being claimed when uttering the two sentences. When we say "Snow is white," we are clearly speaking about snow, and are making one of its constituent features or attributes explicit. What do we really claim when we say, "'Snow is white' is true"? Are we speaking about the sentence or about snow? Both possibilities are included in this way of speaking, and an indication of its ambiguity may be seen in the fact that one would hardly use it in actual conversation. As a matter of fact, one would probably either say, "It is true *that* snow is white," or "'Snow is white' is a true sentence," where it is clear that in the first case we are speaking about snow, and in the second about the sentence.

Yet we must note that the first way of making the initial statement unambiguous actually introduces a use of "true" which is not the one we have assumed in our present discussion, for it is applied here not to a sentence or to a proposition, but rather to a state of affairs (the 'fact' that snow is white), so that the first interpretation of the initial statement should have better received, for example, the following form: "Snow is *actually* white," or "It is *a fact that* snow is white," rather than the form "It is *true that* snow is white" which contains an 'improper' use of "true."

That this use is really improper may be seen from the following simple remark: in the said sentence "true" is not used in a strictly *adjectival* way since we must say "true *that*," and this is linguistically different from "true" because we have to do with a 'substantivised adjective.' In fact, to say "it is true that," like saying "it is good that," is more or less the same as saying "it is a true thing that" or "it is a good thing that," where these adjectives are clearly predicated of an undetermined entity which we may identify with a fact or a situation but not with a sentence. Surprisingly, many authors, who have advocated what has been termed the "redundancy theory" of truth (the theory which says that "*p* is true" has the same meaning as "*p*") have overlooked, or at least underestimated, the difference between saying "true," and "true that."

Let us now ask what it *means* to say: "'Snow is white' is a true sentence," or, equivalently now, because we have critically rejected the first colloquial version of this expression, to say:

"Snow is white" is true.

Is this really to assert something about the sentence "Snow is white," that is, to claim that there exists a particular attribute which can be predicated of it by simply inspecting it, such as its being an English sentence, or consisting of three words? Certainly not. To predicate of a sentence that it is true means to *relate* it to something different from itself which, however, is not its sense. Indeed, we must even say that we cannot speak of the truth of a sentence if we have not assumed that truth or falsity cannot be attributed in a proper sense to a sentence but only to the proposition it expresses, to the proposition which constitutes its sense, and only indirectly and, so to speak, by extension, to the sentence itself.[55] But even so, no one should have any difficulty in admitting that understanding the sense of a sentence, that is, understanding a proposition, is in general insufficient for saying whether the proposition is true or false (this should only be possible in the case of analytic propositions). In order to say whether a proposition is true or false, we have to look outside it, that is, as we have said, we have to look to its *referent*.

This claim is not a dogmatic one, but is based on a simple analysis of what actually happens, at least in most cases, when we accept a sentence as true. If I say, for instance, "The pen I have in my pocket is black," I can very well understand the meaning of this sentence, and therefore the proposition it expresses, without knowing whether it is true or not. In order to know this, I do not need a conceptual, linguistic or logical analysis of the sentence; I need only to remove the pen from my pocket and look at it. Moreover, the sense of the sentence remains unaffected, independently of the fact that, as a consequence of this inspection, it is found to be true or false.

This is what we mean by saying that to claim that a sentence/proposition is true means to predicate something of the sentence/proposition, but in an incomplete sense, as when we predicate fatherhood of someone. Being a father is, in a certain sense, a *property* of an individual, but *only as far as* this individual is in a certain *relation* with other individuals (his children). In exactly the same sense, being true is a property of a sentence, but *only as far as* this sentence is considered as being in a certain *relation* with something else. Moreover, just as the *meaning* of the concept of being a father cannot be expressed unless an allusion to the relation to his children is made, in the same way the *meaning* of the property of being true consists, for a proposition, in the explicit mention of its relation to this 'something else' which, as we have seen, must be conceived of as extralinguistic, as belonging to 'the world,' as we have said in our three-level semantics. We gladly admit that the present considerations are of a very general character, and that they do not seem to have much to do with modern science, but this relevance will appear later.

The nature of this 'something else' is specified in several different ways according to the various conceptions of truth. For example, it is rather usual to call it a "state of affairs" (or sometimes a "fact"), as we too have done, but it can also be a 'contextual dependence,' as some other theories of truth would prefer. In calling it a *referent* we are taking, for the moment, a neutral position that is

---

[55] This fact should be obvious: "London is a city," and "*Londres est une ville*" are different sentences (they even belong to different languages), but have the same sense, that is, they express the same proposition. It is because this common proposition is true that we can say of these sentences (in an extended but acceptable sense) that they are true, as are all other sentences, in all possible languages, that express the same proposition. This seems to be a small detail, but it removes the difficulty of defining truth only *relative to a given language* which is, as we shall see, a shortcoming of the usual definitions of truth, which define truth directly for sentences, rather than indirectly, through the proposition which is expressed or denoted by the sentence. This is also the case as regards the Tarskian definition of truth, and Tarski is fully aware of this fact (see Tarski 1944, p. 342). The necessity of in some way obtaining a proposition when using only a sentential semantics is often alluded to in the philosophical literature in terms of the notion of *semantic ascent*, which will not be discussed here.

probably compatible with all the various doctrines about the nature of truth. For the supporters of the correspondence theory of truth, this referent is something belonging to the structure of the world; for the supporters of the coherence theory, it might be the logical connection a sentence has with other accepted sentences; for others it may be its being related in a particular 'rational way' to a body of pre-existing knowledge, or to its fitting well in a certain linguistic game played by a particular community of speakers, and so on.

What is important for us is to stress that all conceptions presuppose a confrontation of the sentence with something *given*; and, as 'givenness' is, according to our conception, the mark of referentiality, we are entitled to say that it is intrinsic to the structure of truth to envisage the relation of a sentence (or, better, of a proposition) to some kind of *referent*.[56]

[56] This is tantamount to saying that sentences (and, we say, propositions) are *truth-bearers*, but that there must be something *in virtue of which* they are true. This something may be called the *truth-maker* of the sentence. As to the determination of these truth-makers, different proposals have been put forward. On our view—which distinguishes a purely linguistic level (sentence), a *noematic* level (proposition), and a referential level (state of affairs)—what makes a sentence true is not its related proposition (which is only its sense), but the state of affairs to which the sentence refers. Indeed, we maintain that the state of affairs makes the proposition true (or false) and, indirectly and automatically, makes true or false all sentences which express this proposition in any possible language.

However, in those conceptions in which the three levels are reduced to two, this option may be impossible to adopt. For example, Zalta gives full space to the world of *noemata*, but fails to seriously provide a space for the world of referents, so that he accepts the identification of propositions with states of affairs, and claims that they are both 'abstract,' since both are constituted by properties and relations, which are abstract objects. As a consequence, he is led to say that propositions possess in themselves a "metaphysical truth or falsity [which] is basic. If a proposition is true, there is nothing else that 'makes it true.' Its being true is just the way things are (arranged)" (Zalta 1988, p. 56). According to this view, the basic 'metaphysical' truth of propositions is what assures the derivative 'semantic' truth of sentences. Needless to say, we have here to do with a rather peculiar and unclear notion of 'metaphysical truth,' which creates certain difficulties even with other aspects of Zalta's theory. But the reason for adopting this position seems to reside in the fact that referentiality has been considered impossible in the case of properties and relations.

This does not happen, for example, in the case of those authors who have developed a theory of 'truth-makers' in which attributes (in our terminology; or *moments* in the authors' terminology) function as truth-makers (see, e.g., Simons 1982, Mulligan et al. 1984). This theory may be considered a significant refinement of a position which Russell expressed in a well known passage of *The Philosophy of Logical Atomism*, where he said: "When I speak of a fact… I mean the kind of thing that makes a proposition true or false" (Russell 1918, p. 36), a position which is interesting to the extent that it shows that also for Russell (at least at a certain stage in his intellectual development) propositions were truth-bearers and not truth-makers. More about this problem will be said in the Sect. 4.4.5.

4.4.4 The Tarskian Definition of Truth

What we have maintained seems also to be well in keeping with the most rigorous, and famous, definition of truth offered in contemporary philosophy, that is, with that of Alfred Tarski.[57] We are not going to present and comment on this definition, but rather adopt its well-known basic features as a kind of guideline for developing our considerations.

The concept "definition" receives a variety of meanings in contemporary logic, methodology, and philosophy of language, where different 'kinds' and procedures of definition are treated. Yet the most basic function that we attribute to a definition is that it explains the *meaning* of a concept; and the most standard and elementary way of performing this function is expressed in the form of the so-called "explicit definition." Such a definition is expected to provide necessary and sufficient conditions for recognising this meaning. If we are lucky, the fulfilment of this task can be expressed *linguistically* through a sentence having the form of a biconditional in which the *definiens* is declared equivalent to the *definiendum*. This occurs, in particular, also in the Tarskian definition of truth.

The sentence expressing the definition must obviously belong to a language in which both sides of the biconditional can be formulated, and this is why, if we intend to define truth for a sentence, we must do so in a metalanguage in which, on the one hand, the sentence of the first language (the 'object-language') is identifiable and denoted by a name and, on the other hand, the condition formulated through the *definiens* is also expressible. In our case, using for example English as metalanguage and Italian as object-language, we could say:

"*La neve è bianca*" is true if and only if snow is white.

We do not need quotation marks for "snow is white" because in the metalanguage this sentence is intended to be about the world and not about the object-language.

The more usual procedure, as is well known, is to use only one language, and to employ quotation marks to indicate the metalinguistic mention of a sentence, that is, in our case:

[57] Two fundamental works must be taken into consideration in order to understand the Tarskian theory: Tarski (1933, 1944). The first contains the complete and technical development of Tarski's definition of truth for formalised languages; the second is devoted to the discussion of some relevant philosophical aspects of the notion of truth. For our purposes it will be sufficient to consider the 1944 paper.

(T) "Snow is white" is true if and only if snow is white.[58]

What should be clear, in any case, independently of the visual devices employed, is that in the metalanguage we have the mention of a sentence on the left side of the biconditional, and the mention of a referent on the right side. This means, in other words, that the metalanguage is here playing the role of a *referential language*. This point needs to be well understood. One might be inclined to say, prima facie, that the purpose of definitions being that of connecting *meanings*, what the biconditional does is simply to make the meaning of its left side identical with that of its right. This is correct, in a way, but we must not forget that our task here is not to explain the meaning of "Snow is white," but of "'Snow is white' is true." The difference between the two situations is undeniable, though it might require a rather detailed discussion to explain it adequately; in any case, it is apparent if we resort to the two languages explicitly. In this case we should say:

"*La neve è bianca*" means 'Snow is white'

where it is clear that this statement is about sentences, since in its second part we again mention a sentence (actually a sentence of the metalanguage itself taken into consideration metalinguistically). Hence we are left only with the issue (which is in perfect agreement with what we have said about the fact that the requirement of truth postulates something *in addition* to the making explicit of meaning) that the *meaning* of "'Snow is white' is true" is identified in the metalanguage not with the meaning of the object-language sentence "Snow is white," but with the referential performance of this sentence (that is, with the state of affairs it refers to).

4.4.5 Semantic and Apophantic Discourse

We are aware that some points are still obscure as regards this matter, and we shall try to clarify them by resorting to a traditional distinction, one which has unfortunately been rather neglected in recent times, namely that between the *semantic logos* and the *apophantic logos*, a distinction we have already evoked in Sect. 4.1.1. The level of the semantic logos is that on which one tries to establish, analyse, or clarify meanings; the level of the apophantic or declarative logos is that

[58] We are here following Tarski (1944), with the following slight modification. Tarski introduces (T) as a sentence 'schema':

(T) *X is true if, and only if, p*,

where *p* is a sentence and "*X*" is the name of this sentence. From (T) actual sentences may be obtained by substituting for "*X*" and "*p*" in the way just indicated. Our example involves such a substitution and (as Tarski himself says) may be considered for that reason to be a "partial definition of truth," that is, a definition of truth for the particular sentence considered (here, "Snow is white").

on which one makes assertions, affirming or denying something.[59] It is completely clear from what we have said thus far that the 'problem of truth' (understood as the problem of establishing *whether a given sentence is true*) is of an apophantic nature since it is the problem of connecting a sense with a referent (with an occurring fact), and not that of analysing a sense. But now we can ask the subtle question, in the *definition* of truth itself, have we to do with a semantic or with an apophantic issue? At first this question seems unnecessary, since everyone knows that definitions have the role of providing meaning, or sense, which seems to be the semantic problem *par excellence*. However, things are not so simple, since this is the proper task of *nominal* definitions. But the philosophical tradition has also considered *real* definitions, whose task is rather that of explicitly expressing what an entity 'really is.' For this reason, the task of real definitions was not simply linguistic (and as such also open to a certain amount of conventionality), but it was at the same time bound to 'objective' linguistic conditions. In other words, a real definition is something between the semantic and the apophantic logos, since in a way it must be true (or, better, *adequate*).

We do not wish to purse this line of investigation since it would lead us too far afield. We have simply mentioned this perspective in order to call attention to two things. The first is that, even without pretending that a 'real definition' expresses the essence of a thing, we are entitled to require that it at least *correctly express* or *make explicit* the meaning of a concept (while a nominal definition simply *establishes* or *stipulates* this meaning, more or less conventionally). This meaning may be thought of, for the sake of simplicity, as being couched in the way the concept is commonly understood, as emerges from its use. The second thing is that this general requirement also applies to the definition of truth. Let us note that, when offering his definition of truth for formalised languages, Tarski explicitly said that he wanted to make explicit and rigorous the common way of understanding the concept of truth and, moreover, that he wanted his definition to be not only "formally correct," but also "materially adequate," that is, to be such as to entail that all intuitively true sentences are also true according to his definition.[60]

---

[59] We have said that this distinction has been neglected, but we do not say that it has been totally ignored in contemporary philosophy. Indeed, we have already mentioned Frege's insistence on the difference between *declarative sentence* (*Behauptungssatz*), simple *thought* (*Gedanke*), and even *judgment* (*Urteil*). A declarative sentence is a kind of 'notification' or 'announcement' (*Kundgebung*) of a judgment (see Frege 1918, p. 62). And while declarative sentences must always be either true or false, this is not the case with thoughts. Another, even more explicit, recognition of the difference we are speaking about is to be found in Husserl, who uses the term "apophantics" to indicate that part of formal logic which is concerned with the predicative judgment, following in this way the Aristotelian and, in general, traditional terminology (see Husserl 1929, p. 63). One can also say that recognition of the difference between a statement and a sentence—which we have explicitly accepted, and which has a certain circulation in the philosophy of language—indicates an awarcness of this distinction. However, we still have the impression that this difference has not really been thoroughly exploited, as we shall try to do in what follows of this section. For more details on this distinction, see Agazzi (1989).

[60] Tarski (1944), p. 341.

With these premises in mind, let us reconsider the definition of truth (in the sense 'is true') for a single sentence:

(T) "Snow is white" is true if and only if snow is white;

and let us seriously consider that the definition must make explicit a meaning, namely, the meaning of its left side. The commonly admitted task of *explicit* definitions is to make the *definiendum* and the *definiens* completely interchangeable, so that the definition itself could be considered as a 'cancellation rule,' allowing us to eliminate from any context one of the two expressions (in particular the *definiendum*) by replacing it with the other. This was also Tarski's aim. As he explains in his 1944 paper, the ordinary notion of truth is in part ambiguous, such that its application may lead to contradictions (such as the antinomy of the liar). His aim was therefore to construct an explicit definition of truth such that the term "true" could be dispensed with, and replaced by conditions in which only certain totally unproblematic notions and features are used.[61] However, it seems obvious that (T) does not provide such an explicit definition of "is true," that is, a definition to be used as a cancellation rule. Rather, it directly expresses our intuitive notion of truth. The right side of (T) does not sound like a definitional replacement of its left.

The effective construction of an explicit and formal definition of truth is provided elsewhere (in particular starting with Tarski 1933), and is very briefly hinted at in the 1944 paper. It consists in a technically elaborate procedure thanks to which *truth* is actually 'defined' starting from the notion of *satisfaction* using rather complex instruments of the logic of classes. One may wonder whether these tools are really "completely clear and unequivocal," as Tarski says, but this requirement cannot be evaluated on the basis of intuitive familiarity. As a matter of fact, these tools are all explicitly and univocally determined in the technical parts of mathematical logic where they are used, and therefore they serve the purpose perfectly. The reason for which it has proved necessary to pass through the notion of satisfaction instead of trying a direct recursive definition of truth is

[61] After having explained that the definition of truth must be given in the metalanguage, Tarski says:

> It is desirable for the meta-language not to contain any undefined terms except such as are involved explicitly or implicitly in the remarks above, i.e., terms of the object-language; terms referring to the form of the expressions of the object-language, and used in building names for these expressions; and terms of logic. In particular, we desire *semantic terms* (referring to the object-language) *to be introduced into the meta-language only by definition*. For, if this postulate is satisfied, the definition of truth, or of any other semantic concept, will fulfill what we intuitively expect from every definition, that is, it will explain the meaning of the term being defined in terms whose meaning appears to be completely clear and unequivocal. And, moreover, we have then a kind of guarantee that the use of semantic concepts will not involve us in any contradictions (Tarski 1944, pp. 350–351).

briefly mentioned by Tarski, and is simply related to the technical impossibility of doing otherwise.[62]

If things are so, statement (T) simply says that any sentence can be 'asserted as true' in a given language if and only if it is 'asserted' in this language that things are so. But (T) does not say anything about the criteria of assertability, as one might have expected from a definition of truth. As a confirmation of this, let us simply quote what Tarski says in discussing a criticism of his definition levelled by Gonseth:

> In fact, the semantic definition of truth implies nothing regarding the conditions under which a sentence like (1):
> (1) *snow is white*
> can be asserted. It implies only that, whenever we assert or reject this sentence, we must be ready to assert or reject the correlated sentence (2):
> (2) *the sentence "snow is white" is true.*
> Thus we may accept the semantic conception of truth without giving up any epistemological attitude we may have had: we may remain naive realists, critical realists or idealists, empiricists or metaphysicians—whatever we were before. The semantic conception is completely neutral towards all these issues.[63]

This unequivocal statement is very significant, since it makes clear that even the complex and technically elaborate procedures used for recursively defining truth from satisfaction do not provide a *criterion* for asserting true sentences. Indeed, an examination of this recursive definition would end up (in the case of our example) with the following two statements:

1. A sentence is true if it is satisfied by all objects, and false otherwise;
2. object *a* satisfies the sentential function "*x* is white" if and only if *a* is white.

Clearly, the concept of satisfaction involves the same reference to the 'external' circumstance that was involved in the right side of equivalence (T), and does not contribute any kind of analysis or dissolution of it.[64] The definition of satisfaction does not provide *criteria* for satisfaction. Thus in conclusion we may say that Tarski's work provides an explicit definition of truth which permits the elimination of the notion of truth in favour of that of satisfaction. But this does not amount to the elimination of the reference to the 'external circumstance' appearing in the right side of the biconditional, nor does it provide a criterion for saying *whether* a sentence is true or not.

These conclusions may sound deceptive at first, but under a closer scrutiny they show us that we cannot hope to capture the meaning of truth simply at the level of

---

[62] "It may seem strange that we have chosen a roundabout way of defining the truth of a sentence, instead of trying to apply, for instance, a direct recursive procedure. The reason is that compound sentences are constructed from simpler sentential functions, but not always from simpler sentences; hence no general recursive method is known which applies specifically to sentences." (Tarski 1944, p. 353).

[63] Tarski (1944), pp. 361–362.

[64] A good analysis of this point is provided in Keuth (1978), pp. 64–72.

linguistic analysis, particularly at the level of an analysis which tries to restrict itself to sentences and ignore propositions and senses.[65] In order to advance, we must do something which Tarski himself has hinted at when, in Sect. 2 of his paper, he expressed his proposal to make precise (but not to discard or reject) the intuition implicit in certain common ways of expressing the notion of truth, such as by saying "*A sentence is true if it designates an existing state of affairs.*"[66] A reasonable idea for proceeding in this direction seems to be the following. Let us try to see what considering (T) as expressing an equivalence between two *apophantic sentences* would imply. The rationale of this strategy is clear: we would like to investigate under what conditions we are entitled to *assert* that *X* is a true sentence, expecting that the equivalence should suggest under what conditions we would also be entitled to *assert* the right side of (T).

This way of considering definitions is not unusual. On the contrary, it applies to all those definitions which do not allow the pure and simple substitution of one expression by another (typically by a term with a defining condition)—and which may thus be considered as cancellation rules—but which only allow the interchangeability of *whole sentences*. In such cases the substitution is admissible because the *truth-value* of the two sentences is the same (*apophantic logos*); and this in turn is possible to the extent that some *relation* is put between the logical subject of the *definiendum* and some other logical subject which occurs in the *definiens*. For instance, when we define "*x* is soluble" by stating, "*x* is soluble if *x* dissolves when put in a liquid," we mention a liquid, near *x*, in the *definiens*, and moreover, the *definiens* is clearly an apophantic statement.

---

[65] That Tarki's semantics is of this kind may be easily seen from the few lines where he explains why his definition of truth is to be called "semantic":

> I should like to propose the name "*the semantic conception of truth*" for the conception of truth which has just been discussed.
>
> *Semantics* is a discipline which, speaking loosely, *deals with certain relations between expressions of a language and the objects* (or "states of affairs") *referred to by those expressions*. As typical examples of semantic concepts we may mention the concepts of *designation, satisfaction*, and *definition*. (Tarski 1944, p. 345.)

It is evident that we here have to do with a 'two-level' semantics, considering only the language-world relation, and ignoring the mediation of the senses (and the existence of speakers and listeners). This was already clear at the beginning of the paper, where Tarski declines attempting to define truth for propositions, since,

> as regards the term "proposition," its meaning is notoriously a subject of lengthy disputations by various philosophers and logicians, and it seems never to have been made quite clear and unambiguous. For several reasons it appears most convenient to *apply the term "true" to sentences*, and we shall follow this course. (ibid., p. 342).

It is therefore embarrassing (but not surprising, as we know) that such a semantics remains unable to justify a notion of truth in which reference to the world is actually made, as we shall see in the sequel.

[66] Tarski (1944), p. 343.

Analogously, in the definition of truth we have in the *definiens*, which is an apophantic statement of the metalanguage, a description of the *referent* of the logical subject of the *definiendum* (in our case the *sentence* of which we are predicating truth). In brief, we can say that in the Tarskian definition of truth the biconditional serves to connect the sentence named on its left side with the state of affairs referred to on its right side. By noting this, we are also recognising that, in this definition of truth, the metalanguage is also playing the role of being a 'referential language,' as we have already discussed.

For this reason we must say that in Tarski's study two preoccupations are at work which are reflected in two different meanings of the connective "if and only if." The first preoccupation corresponds to the requirement of formal correctness, and leads to the interpretation of the biconditional which is found in explicit definitions and serves the function of a cancellation rule. In order to avoid ambiguities, it has become customary to use, in such cases, the notation "$=_{\text{df}}$" instead of "if and only if." The second preoccupation corresponds to the intention of making explicit the notion of truth (or at least one relevant sense of this notion) by stating necessary and sufficient conditions for characterising it; and this reflects the requirement of "material adequacy."

To clarify the difference, let us consider two examples. If we say "a polygon is equilateral if and only if all its sides are equal in length," we have an instance of the first use—and this is actually an explicit definition of the notion of an equilateral polygon. But when we say that "a polygon is equilateral if and only if it is equiangular," we are not giving a *definition*, but expressing a geometrical *intrinsic property* of polygons, which may be proved as a *theorem* in elementary geometry. Now, in the case of (T), we must say that the "if and only if" has the second meaning, since it is intended to express the *intrinsic property* a sentence must have if it deserves to be called true. And this simply because Tarski's intention was not that of *eliminating the notion of truth* (which he considered indispensable in ordinary language no less than in science), but that of offering a *suitable characterisation of the notion* which could be used in mathematics, science and rigorous portions of everyday discourse, while being at the same time free from contradictions.

To be sure (as we have already recalled in a note), Tarski introduces (T) under the general form:

(T) *X* is true if, and only if, *p*,

that is, as a sentence schema, which gives rise to a sentence whenever we replace *p* by a concrete sentence of a language, and *X* by a metalinguistic name of this sentence. Moreover, the task of (T) is that of expressing the *adequacy condition* for any possible definition of truth, rather than itself being such a definition; and in fact Tarski speaks of "convention (T)" in the 1933 monograph, while in the 1944 paper he usually speaks of "equivalence (T)." Therefore one expects that—since (T) is *not* a definition—a concrete explicit (i.e., 'eliminative') definition *D* be proposed for truth, and that it pass the test of (T), a test which in Tarski's view is very severe, since it consists in the fact that from D all the equivalencies of the form (T) must be derivable. The recursive construction of such an eliminative definition *D* based on 'satisfaction' actually passes the test and is therefore accepted.

However, we can ask why (T) has the privilege of being adopted as a criterion for adequacy, and here we actually have *two* answers. One is that (T) expresses the intuitive, pre-theoretic notion of truth, according to which a sentence is true if what it *asserts* corresponds to the facts. The other instead expresses the way we use "true" in normal language, and simply says that whenever we *assert* a sentence *p*, we are also bound to *assert* that *p* is true (independently of the reasons we might have for doing so, and even if we are wrong). Let us note, by the way, that this completely justifies our suggestion that (T) may be considered correctly only by taking it as an equivalence of *apophantic* expressions.

But if these remarks are correct, it turns out that, after all, (T) specifies what we *actually mean* by "true," and therefore it provides a *definition* of truth. Tarski does not discard this view, and indeed, while saying that (T) cannot itself be a definition, because it is only a sentence *schema*, he accepts that all substitution instances of (T) obtained as said above are *partial* definitions of truth for the single sentences p_1, p_2, and so on. He even says that "the general definition of truth has to be in a certain sense a logical conjunction of all these partial definitions."[67] This statement may sound strange, but a little reflection shows that it corresponds to the *purely extensional* way of defining a concept, which consists in constructing the set of its referents by enumeration. According to this strategy, one must say that a sentence in a given language is true if and only if it belongs to the set of those sentences which are *asserted* in that language, and this for no other reason than that it would not be in conformity with our usual way of speaking to *assert* a sentence and at the same time say that it is not true.

We can ask why Tarski introduced such a poor and even trivial 'extensional definition' of truth. The answer seems again to be double: primarily because in such a way there would be no risk of including in the set of true sentences paradoxical expressions such as the antinomy of the liar; but also because (as we have seen) Tarski was very diffident towards meanings and intensions in general, and with the meaning of the concept of truth in particular.[68] As a conclusion, Tarski actually provides two 'eliminative' definitions of truth: the rather trivial 'extensional' definition just mentioned, and the less trivial 'recursive' definition based on satisfaction. Both of them (apart from avoiding contradictions) pass the test of equivalence (T): the extensional definition simply automatically and trivially; the recursive definition somewhat indirectly (i.e., because satisfaction, as we have seen, is bound to the 'right-hand clause' of (T) as well).

Finally, the crucial problem surfaces. Is the sense of (T) simply that of respecting the common use of "true" in the language, or is it that of capturing the

---

[67] Tarski (1944), p. 344.

[68] Tarski says: "Much more serious difficulties are connected with the problem of the meaning (or the intension) of the concept of truth" (Tarski 1944, p. 342).

idea according to which a sentence is true only if it says how things really are? In Tarski's paper both requirements are present. We have already seen that, in responding to Gonseth's criticism, Tarski lays stress on the first aspect. But it is no less true that in many other passages he also points to the second aspect, especially since he wants to present his theory as a modern way of expressing the traditional conception of truth—sometimes called the "correspondence theory"—which goes back to Aristotle. This is why, in particular, Tarski and Aristotle have often been presented as supporters of the correspondence theory of truth.[69] The discussion of this issue is not simply of historical interest (in which case we would feel free from having to deal with it), but has direct impact upon our systematic investigation. Therefore we shall analyse it to some extent.[70]

---

[69] Among the most famous writers who have interpreted Tarski as a supporter of the correspondence theory let us mention Popper (especially Popper 1972) and Davidson (see Davidson 1969). Among those who have denied this qualification, let us mention Black (1949), Field (1972), Harré (1986) and Keuth (1978). At any rate, what is certainly undeniable is that Tarski himself explicitly intended his theory to be a specification of what he indicates as the classical, Aristotelian correspondence theory of truth, as is clearly stated not only in the more 'philosophical' paper of 1944, but already in the introduction of his extensive and technically complex 1933 monograph:

> I would only mention that throughout this work I shall be concerned exclusively with grasping the intuitions which are contained in the so-called *classical* conception of truth ('true—corresponding with reality') in contrast, for example, with the *utilitarian* conception ('true—in a certain respect useful'). (Tarski 1933, p. 153 of the English translation.)
>
> I mean a definition which we can express in the following words: *a true sentence is one which says that the state of affairs is so and so, and the state of affairs indeed is so and so* (ibid., p. 155).

All that is to be done therefore is to see whether the result he obtained actually corresponded to his intentions. What we have seen (and what his critics after all are able to show) is that his definition does not *necessarily* imply a correspondence theory of truth, that it is also *compatible* with other doctrines. But it would be unjustified to say that it does *not* provide a suitable framework for an acceptable formulation of that theory, and that it is naturally oriented in such a direction.

[70] Before passing on to this analysis, however, let us briefly discuss the question whether Tarski's truth-predicate corresponds to an analysis of the semantic conception of truth that has become involved in what is called "formal semantics." This does not seem to be the case, in spite of the fact that the technical tools introduced by Tarski in providing his 'recursive' truth-predicate have paved the way for such further developments, which on the other hand required the introduction of an *undefined* notion of truth. (For a valuable discussion of this and related issues see Etchemendy 1988.)

The reason for this—as we have already hinted—is that Tarski wanted to avoid recourse to senses and propositions, while the most natural way of understanding the whole issue (a way which is also accepted by modern formal semantics) is that the sentence "Snow is white" makes a claim which depends on the colour of snow, while the sentence "'Snow is white' is true" makes a claim that depends both on the colour of snow and on the *meaning* of "Snow is white." This implies that the left and right sides of (T) are not on an equal footing. However, it is for this reason that the biconditional expressed by (T) is informative, that is, that it gives us information

It is perhaps useful to remind the reader that the undoubtedly complex semantic considerations we are presenting here are not digressions from the central topic of this work, since they provide the foundation for fundamental claims we are going to advocate later regarding scientific realism and the capability of science to deal with truth.

4.4.6 The Aristotelian Conception of Truth

Before leaving our topic, it will be of some interest (though not strictly necessary for the economy of the present work, and this is why we shall try to be concise) to see the extent to which what we have been saying here about truth (including our treatment of Tarski) is related to the Aristotelian notion of truth (to which Tarski declares himself to remain close).[71] For the sake of brevity we shall limit ourselves to a couple of quotations.

We note, first, that Aristotle too conceives of truth as being a property of sentences. (Aristotle does not actually consider the distinction between sentences and propositions which appeared in the history of philosophy only with Stoic logic, so that it is fair to assume that his sentences express propositions just as ours do in the present discussion.) Moreover, he recognises the distinction between semantic and apophantic logos (and is indeed the father of it). This distinction may be found in its most synthetic form, perhaps, in the following quotation from the *Categories*: "Every affirmation or negation seems to be either true or false, while of things enunciated without any connexion, none is either true or false: as 'man,' 'white,' 'runs,' 'conquers.'"[72]

As for the question of making precise what it means for a sentence to be true or false, we can quote a most significant passage of the *Metaphysics*, where he says:

(Footnote 70 continued)
about the meaning of the sentence "Snow is white." The fact that Tarski, in spite of elaborating an 'eliminative' truth concept, introduces his equivalence (T) as a touchstone for evaluating the material adequacy of his definition of truth, indicates that he was actually working to some extent with "a hybrid of different pretheoretic conceptions of truth" (Etchemendy 1988, p. 62). In particular, the concept of truth that Tarski tries to analyse retains certain elements of that which is called the *propositional* conception of truth, which sees truth as a property of extralinguistic entities, independent of the linguistic or semantic features of any particular language (on this point, see again Etchemendy 1988, pp. 62–63).

[71] See Tarski (1944), p. 342–343.

[72] *Categ.*, 4.2a, 8–10. Other passages in which he speaks of the *apóphansis* are, for example: *De Int.*, 4, 17ª2; 5, 17ª22; 6, 17ª25; *An. Pr.*, I. 1 24ª16. According to Aristotle, the *apóphansis* is in general the declarative sentence, which may be either an affirmation (*katáphasis*), or a negation (*apóphasis*: notice the difference in the orthography). Latin authors translated the Aristotelian term differently, with *enunciatio, sententia, propositio*, which are the obvious etymological antecedents of our "statement," "sentence," and "proposition"; but in general, in the traditional textbooks of logic, one speaks of *judgment*, and in more recent textbooks one finds "apophantic judgment" in the sense of "predicative judgment."

"To say of that which is that it is not or of that which is not that it is, is false; to say of that which is that it is, or of that which is not that it is not, is true."[73]

The interesting features which emerge from these lines are essentially two. The first is the repetition that truth and falsity concern language, that is, sentences or propositions at large (*to légein*). The second is that the condition for truth is expressed by a relation to a *referent* and nothing more than a referent. This point is worth noting because it shows how the Aristotelian definition of truth (like the Tarskian definition, which explicitly claimed to be simply an adequate explanation and formalisation of Aristotle's) is rather uncommitted as far as further ontological conceptions of truth may be concerned. In particular, it is not said that this definition *must* be interpreted as expressing a correspondence theory of truth, as several scholars maintain (or at least a particular 'strong'—or perhaps 'rough'—form of the correspondence theory, of which we shall speak in the Sect. 4.5). The reason Aristotle does not seem to advocate such a view is that he maintains not only that it is true to say of *that which is*, that it is, but also that it is true to say of *that which is not* (*me on*) that it is not. Now, a correspondence theory (in the rough sense mentioned above) could somehow be appended to the first statement—which may be interpreted as indicating the reference to an existing concrete object—but certainly not to the second, because there can be no correspondence in the sense of a point-to-point relation between a sentence on the one side and something 'non-existing' on the other. This means that we cannot conceive of the Aristotelian *on* as meaning something like an individual object, but rather as meaning 'what is the case,' for in the first case we should have much trouble clarifying what we mean by its being true to say of a 'non-object' that it is not. On the contrary, it makes full sense to claim that it is true to say of 'what is not the case' that it is not (the case).

In other words, the Aristotelian definition cleverly leaves open the use of the verb "to be" both as a verb (*einai*) and as a substantive participle (*to on*) according to its *two* main uses, that is, as expressing *existence*, and as expressing *predication*. In such a way, we can say that the Aristotelian proposal covers *both* cases of *positive* instantiation: the one in which we claim that it is true to say of that which *exists* that it exists, and the other in which we claim that it is true to say of that which *is such-and-such* that it is such-and-such. Clearly, on the other hand, it can also be unproblematically applied to the *negative* instantiation only in *one* sense, that is, in that in which we claim that it is true to say of that which *is not such-and-such* that it is not such-and-such. It would be, instead, controversial to claim that it is true to say of that which *does not exist*, that it does not exist, because we might

---

[73] Here is the complete passage:

> This is clear, in the first place, if we define what the true and the false are. To say of what is that it is not, or of what is not that it is, is false, while to say of what is that it is, and of what is not that it is not, is true; so that he who says of anything that it is, or that it is not, will say either what is true or what is false; but neither what is nor what is not is said to be or not to be. (*Metaph.* 1011^{b}26–29; transl. by D. Ross.)

This is, by the way, the Aristotelian text which Tarski explicitly quotes in his article of 1944.

well face difficulties specifying 'about what' we are formulating this statement. All these reflections encourage us to say that the Aristotelian doctrine seems already to refer true and false discourse to 'states of affairs.'

This may be interpreted as a possible form of correspondence theory of truth, but not (as we shall see) of the one which is usually meant when this theory is criticised. Such a (doubtful) correspondence theory essentially conflates the predicative and the existential use of "to be," but there is no reason to attribute it to Aristotle. From the considerations expressed here, and also from others encompassing the whole of his ontology, one can clearly exclude his being a (perhaps unconscious) supporter of a correspondence theory of this kind.[74] To summarise what we consider to be the genuine Aristotelian conception of truth, we could say that for Aristotle truth is conceived of as a relational property of a discourse (sentence or proposition) definable in this way: "true is the discourse which declares that which is; false is the discourse which declares that which is not." Should this sentence sound linguistically awkward in English (in other languages it does not, owing to the relative 'independence' of the verb "to be"), we might also say: "true is the discourse which declares what is the case, false is the discourse which declares what is not the case."

4.5 The Referential Commitment of Truth

4.5.1 The Correspondence Theory of Truth

The correspondence theory of truth has no officially recognised formulation, and is presented, instead, in several approximate and different versions.[75] However, it

[74] Let us simply provide additional confirmation by quoting a few lines from the *De Interpretatione*, where the definition of truth is given through a clear reference to a state of affairs, or to something being the case, rather than to existence. The text is quite compact (but clear), and the translation should not be too literal in order to make it easily understandable in English: "If it is true to say that a thing is white, it must necessarily be white. Again, if it is white, the proposition stating that it is white was true; if it is not white, the proposition to the opposite effect was true. And if it is not white, the man who states that it is, is making a false statement; and if the man who states that it is white is making a false statement, it follows that it is not white. It may therefore be argued that it is necessary that affirmations or denials must be true or false." (*Hermen.* 9, $18^{a}40$–$^{b}4$; translated by E. M. Edgehill, W. D. Ross editor). A similar declaration is formulated more synthetically in *Metaph.*, IX, 10, $1051^{b}5$.

[75] To be more precise, the term "correspondence theory of truth" is rather recent, and was introduced by Russell when, against the doctrine of absolute idealism which claims that "truth consists in *coherence*," he wanted to maintain that "truth consists in some form of correspondence between belief and fact." In this sense, the correspondence theory of truth is a recent doctrine, which was probably defended in its explicit form by a few authors such as Russell, Moore, and the early Wittgenstein. However, it has predecessors in the whole history of Western philosophy, starting with Plato. But if we extend its scope to such a broad spectrum, it becomes almost impossible to characterise it univocally (e.g., it is hard to say whether the

seems correct to identify what is most often presented as its central feature as the claim that there is an absolute and fixed structure of reality which is mirrored by language, in the sense that the structure of *true* sentences or propositions is the same as the structure of that of which they are true, and that this is what makes them true. In such a way language builds up a kind of image of reality itself. When this image (expressed in a given sentence) *corresponds* isomorphically to reality, the sentence is true, otherwise it is false. Certain scholars see a more problematic aspect of this doctrine to be that it seems to imply the ontological subsistence of properties and relations. Other scholars, rather, are not happy with the fact that this theory should afford an ontological status to 'negative properties.' These and other issues are widely discussed in the literature, and we do not want to dwell on them here, partly because we intend to propose a conception in which many of such issues will appear under a different light.

After the clarifications provided in the foregoing sections, it should be clear that our 'referential' conception of truth should not be confused with the most usual way of portraying the correspondence theory of truth sketched above. The main reason for disagreement with such a version of the correspondence theory is that it conceives of reality (considered at a given moment) as being absolute and structured *in itself*, independent and alien to discourse and thinking, and that this claim is made in the very moment in which reality is being necessarily considered as being *in relation* to discourse and thinking. One could say, therefore, that this correspondence theory of truth is one of the many 'modulations' under which epistemological dualism offers itself, and that the contradictory nature of this dualism appears particularly clear in this case. For it is intrinsically impossible, as we have already discussed, to think in general of reality as something 'external' to thinking, because in order to do that we must actually think of it. In this particular case, we should say that reality allegedly remains the unaffected term of a *relation* (in which truth consists), while truth can be established *only* if reality is being considered not, so to speak, 'as such' nor 'in itself,' but exactly *as far as* it enters this *relation*.[76]

In order to clarify the issue, let us again use our earlier example and assume that we want to know whether X is or is not a mother. We have already stressed that, in order to ascertain this, we must find out whether there is at least one Y such that X has given birth to Y. But how could we recognise such a Y if we were to pretend to single it out by considering persons 'as such' or 'in themselves'? In other words, no Y could be recognised as being such as to justify that X is his or her mother,

(Footnote 75 continued)
Scholastic definition of truth as *adequatio intellectus et rei* should be qualified as a correspondence theory). For a condensed but excellent account, see Prior (1967).

[76] It is not without interest to note that Hegel's criticism of Kant's doctrine of the unknowable thing-in-itself was expressed, at least on one occasion, by stressing how this theory would make self-contradictory the very notion of truth. We would accept "the contradiction of a truth, which must be at the same time non-truth, a knowing of *what is*, which at the same time does not know the thing in itself" (Hegel 1812, 1969 ed., vol. 2, p. 500).

unless he or she is considered as being in this specific *relation* to X, and this relation actually holds between X and Y. To pretend to ascertain whether X is the mother of Y by considering Y alone would be as absurd as to pretend to ascertain this fact by considering X alone. Only if we consider both X and Y *within the context* of this specific relation (and in this order), may we hope to receive an answer to our question.

Coming now to the case of truth, we have already recognised that truth is only superficially a *property* of a sentence, since it is actually a *relation* between a sentence and, let us say for the moment, reality. But now we must immediately say that it is not reality 'in itself' which may be conceived of as being confronted with the sentence, but precisely reality *as far as it is put in a particular relation* with the sentence. Just as we could not say, for instance, that X is the mother of Y because, say, Y is rich, is tall, speaks German, but only if we could establish a *specific relation* (i.e., the relation of generation) between X and Y, so we are not entitled to say that a sentence S is true of reality R simply because R is 'made in such-and-such a way.' We must explicitly mention the *special relation* between S and R which *can* justify the claim that S is true of R, and this is the relation of 'referring to.' It follows that one cannot even ask the question whether S is true of R without automatically considering R as the *referent* of S, that is, without considering reality as being embedded in the referential relation.[77]

[77] This conception was already included in a thesis of Scholastic philosophy, which it is fashionable to discredit as being naively realist and committed to uncontrolled metaphysical enthusiasm. This thesis concerns the identification of the famous 'transcendentals,' that is, the identification of those features of reality which were thought to be endowed with such a degree of universality as to be 'commutable' with being itself. These transcendentals, as is well known, are *unum*, *verum* and *bonum*. Let us only consider *verum* (truth), which is of direct relevance to our discourse. Why was it claimed to have the same 'latitude' as being itself (*ens et verum convertuntur*)? Simply because *verum* was not conceived of as something existing besides *ens* (note that this would be impossible, because everything existing is *ipso facto* included in the domain of being), but as being *ens* itself *as far as* it bears a relation to thought. This doctrine was sometimes summarised by saying that the sentence *ens et verum convertuntur* expresses the 'intelligibility' of being.

This may be accepted, but it is perhaps more perspicuous to a modern reader if we say that it expresses three fundamental facts: (a) that there is no possible 'exteriority' of being with respect to thought (not in the sense that being coincides with thought, but in the sense that one cannot even think of a being as 'external' to thought, without including it in thought by this very act); (b) that it is not possible that thought be exterior to being (thought must necessarily be thought of something, otherwise it would be thought of nothing and therefore 'no thought' at all—this amounts to a radical rejection of epistemological dualism); (c) that this non-exteriority of thought and being does not imply their ontological identity, but only an intrinsic and necessary *relation* between them. This was sometimes indicated as 'the intentional identity of thought and being,' and we prefer to indicate it with the notion of the relation of referentiality. We can therefore conclude that this third fact may be seen as making explicit that being is the *referent* of our thinking activity and of the relational nature of truth.

Let us point out a curious fact. We are spontaneously accustomed to saying that truth is the property of a sentence, of a discourse, of thought (to the extent that it is expressed in a discourse). The ancient way of thinking analysed above seems instead to conceive of truth as a universal property of reality, and this can be somewhat puzzling. However, there is little reason to feel

4.5.2 The Referential Theory of Truth

We shall try now to draw some conclusions from the preceding points in order to better characterise the referential theory of truth being advocated here with respect to that form of the correspondence theory of truth we are claiming to be inadequate. The seemingly unshakeable foundation of the correspondence theory may perhaps be concentrated in the following statement: "Reality is what it is." Can one attack this statement? Certainly not, but, on the other hand, one is entitled to say that in order to *know* 'what reality is,' it is not sufficient that reality simply 'be'; it must enter into a certain *relation* with the knowing subject. If this knowledge takes, for example, the forms of perceiving, of thinking, of speaking about, and so on, we must say that, in these different relations, 'reality is what is perceived,' 'reality is what is thought of,' 'reality is what is being spoken about.' The naive way of conceiving of these situations (and the reason for which some persons are afraid of them) would be to imagine that they, so to speak, modify or deform reality. This view is as naive as that which would claim that M is being deformed by being considered in the relation of motherhood to Y, or by being considered in the relation of 'being seated on' with respect to a chair. On the contrary, the undramatic situation is that, within the relation of perception, a certain reality cannot help being a set of perceptual features; within a relation of thinking, it cannot help being a set of thoughts; within a relation of speaking, it cannot help being a set of predicates, and so on.

Where then is the difficulty? The difficulty may be seen by some to rest in the fact that we should have said, for example, in the three different situations envisaged above, that reality *has* perceptual features, *is the object* of certain thoughts, *receives* certain predicates. Unfortunately, this difficulty is nothing but another expression of epistemological dualism, and is gratuitous and untenable exactly as epistemological dualism is. To use our example again, the analogous situation would be to prevent us from saying that X *is* the mother of Y, or that Y *is* sitting on a chair, and pretend instead that X *has* the property of motherhood with respect to Y, and so on. As a consequence, we must say that a certain reality *is* a set of perceptions (i.e., *is* red, *is* round, *is* smooth, and so on) *within a perceptual context,* just as X *is* a mother, a brother, rich, young, etc., according to the different relations he or she enters with persons, things, etc.

As we are specifically concerned here with the relation of truth, we may well say that, this relation being expressible as that of a reference between sentences and reality, reality *is*, as far as this relation is concerned, the totality of the

(Footnote 77 continued)
puzzled. Truth is neither a property of discourse nor of reality proper, for it is a relation between them—so we might be more inclined to lay stress on one of the poles of the relation rather than the other; but both are necessary. It is a little like preferring to say that X is the father of Y rather than that Y is the child of X. Both sentences express the same relation from two different points of view, and this is why, though having different meanings, they express the same state of affairs, and are true or false under identical circumstances.

referents of all the linguistic features which constitute all possible sentences in general. One sees that, in such a way, the situation we have delineated elsewhere with regard to scientific objects is very general. What is known is always known as a bundle of attributes. The difference in the case of scientific objects is essentially that they are a bundle of *selected* attributes, related to some standardised operational procedures which endow them with the special advantages we are already familiar with.

Let us explicitly note that the entire discourse developed here concerns the definition of truth, or the structure of truth, but it does not engage itself—for the moment—in providing *criteria* for truth. This is why our analysis is compatible in principle with several different doctrines on this point (i.e., with a doctrine concerning how to ascertain if or how a sentence meets its intended referent). The coherence theory of truth may place these criteria in some logical relationships with established sentences; the acceptance theory of truth may relate them to a socio-cultural agreement, the pragmatist theory may identify them with usefulness, and so on. As for us, we have already indicated *one* basic criterion of referentiality, the one represented by operations, and now we have prepared the necessary elements for also exploring a second criterion, which is embedded in the structure of truth itself. We shall proceed now to an analysis of these criteria.[78]

4.5.3 Referentiality and Operationality

Some useful elements are already contained in the preceding analysis. The first, which is immediately implicit in the acknowledgement of the relational nature of truth, is that it simply makes no sense to say of a certain sentence that it is true (or false) without adding 'about what' it is claimed to be such. In other words, a sentence is always true or false 'of something.'

Although this seems an obvious requirement, in everyday speech we never follow this rule, but are instead spontaneously inclined to think that a sentence is true or false in itself. Why do we think this? Because common sense is easily inclined towards a spontaneous realism, in the sense that it seems persuaded that there exists an absolute, stable, and well-defined reality standing before us, which is the implicit (but inevitable and necessary) referent of *every* sentence, independently of the particular conditions and contexts in which the sentence itself occurs. All sentences are therefore meant to be true or false 'about' that reality, so it does no harm to omit mentioning it explicitly. This happens, however, because we are unaware that such a 'general' reality is indeed the *partial* domain of referents accessible through the criteria embedded in our present ordinary life context.

---

[78] We shall return to the distinction between the definition and criteria of truth in Sect. 4.6.

We have already seen how untenable such a tacit persuasion is, and we shall not repeat the arguments we gave earlier. We shall therefore presuppose that every discourse which aims at being qualified as true must, at least to a reasonable degree, make precise what we have already qualified as its 'domain of discourse,' and which we can also qualify, due to what we have said since, as its 'domain of reference.'

An easily drawn corollary to this is the 'relativity' of truth, to which we shall return soon in this section. But for the moment we would like to underscore that, if we take this relativisation (which we might perhaps better call a *limitation* or *restriction*) of the truth concept seriously, we can recover in a critical and correct way that basic intuition which Tarski declared himself to be willing to follow when he proposed his definition of truth, and which (as we have documented through some quoted passages from both his 1933 monograph and 1944 article) he expressed in terms of "correspondence to reality" in the sense that one says "that the state of affairs is so and so, and the state of affairs indeed is so and so."[79]

---

[79] It is certainly not accidental that in the passages just mentioned (Tarski 1933, p. 153 and 155) the author explicitly wants to *contrast* his conception with what he calls the *utilitarian* conception of truth, which we would now call the *instrumentalist* conception. Moreover, in Kokoszynska (1936) the author, who was a pupil of Tarski, very carefully analyses the differences between the Tarskian approach and the logical positivists' theory of truth, as it was expressed at that time, especially by Carnap in his *Logical Syntax of Language* (Carnap 1934). Kokoszynska qualifies the Tarskian view as "*The absolute truth concept*" (p. 149), and this expression is intended to oppose what she presents as the *coherence theory* of truth, which was actually implied in the logical positivist effort to eliminate the truth concept:

> The efforts to eliminate the truth concept from science proceed mainly in a determinate direction. One tries namely to replace this concept through a syntactic concept. Every coherence theory of truth, i.e., every theory according to which the truth of a sentence consists in its correspondence [*Übereinstimmung*] with other sentences may—fundamentally—be considered as such an effort. To this it is usually opposed, as a correspondence theory of truth, the theory that truth consist in a correspondence [*Übereinstimmung*] with reality (op. cit., p. 149).

It is not our aim here to discuss Kokszynska's paper, which clearly explains why the Tarskian conception of truth can "by no means be reduced to a concept related to the syntax of any language" (p. 153), and at the same time explains why it is not intended to provide *criteria* for truth (pp. 156–157). We would simply like to call attention to the fact that in his 1944 article Tarski himself makes reference three times (with approval) to this paper of his former pupil. This constitutes an important confirmation of what is already sufficiently clearly expressed in his direct statements, where the acceptance of the 'classical (i.e. Aristotelian) correspondence theory of truth' is indicated as the notion he accepts and wants to clarify.

Therefore it is simply fair not to put Tarski against himself, or to charge him with inconsistency or lack of insight, by saying, for example, as Max Black does:

> I cannot accept Tarski's claim that his definition favors the "classical Aristotelian conception of truth." I regard his view as neutral to this and other *philosophical* theories of truth (Black 1949, p. 105).

(Footnote 79 continued)

This interpretation of Tarski's theory has become rather common. Putnam, for example, also maintains that "Tarski's work is philosophically neutral; that is, it does not vindicate the correspondence theory of truth" (Putnam 1983, p. 83).

The fair way of interpreting Tarski, as we have already discussed in the preceding section, seems therefore to be the following:

(a) the intuitive 'correspondence' conception is pre-theoretically accepted;
(b) the effort is made to provide a definition for the *use* of the concept of truth which will protect it from introducing inconsistencies in the language;
(c) the definition is given in two ways: the first simply by 'codifying' the fact that a correct speaker cannot assert a sentence and at the same time say that it is not true; the second by constructing the concept of truth from the concept of satisfaction.

The result is therefore that we can rely upon this conception because it will not introduce inconsistencies if properly applied. Now the problem is whether the intuitive conception receives some kind of additional support or indication of plausibility. This support is not implied by the first 'extensional' definition, since the 'reason why' a speaker asserts "*p*" (or the 'criteria' for such an assertion) are irrelevant to the fact that this speaker cannot refuse to call "*p*" true, if he or she asserts it. The second, 'recursive,' definition, on the other hand, is closer to the intuitive conception, since the concept of satisfaction, as presented by Tarski himself, clearly expresses a *language–world* relation, in which, in the last analysis, words are put into *correspondence* with non-linguistic entities, with *objects*. This is also demanded by the meaning of the concepts one uses in doing this, such as that of 'denoting' in the case of a name or an individual variable, of 'applying' in the case of a predicate, or of being 'fulfilled' in the case of a sentential function or a sentence. This situation has been explicitly recognised, for example, by Davidson, where he admits that:

> The semantic concept of truth as developed by Tarski deserves to be called a correspondence theory because of the part played by the concept of satisfaction; for clearly what has been done is that the property of being true has been explained, and not trivially, in terms of a relation between language and something else (Davidson 1969, p. 758).

For this reason we cannot agree with those who consider Tarski's efforts as "philosophically irrelevant," because, to repeat with Black:

> The philosophical disputants are concerned about what *in general* entitles us to say "It is snowing" or "London is a city" *and so on*. In other words, they are searching for a general property of the designata of true object-sentences (Black 1949, p. 105).

In fact, it is already *philosophically relevant* to know that we can use the intuitive notion, call it "correspondence theory" or what you will (which was held responsible for contradictions such as the antinomy of the liar) as a legitimate content or sense of our concept of truth, and that we are 'guided' by this notion in selecting the definitional starting point for the construction of a formal recursive definition of truth.

However, we must also recognise that it was not within the reach of Tarski's conceptual approach to give a more satisfactory characterisation of the intuitive or 'classical' conception. This is simply because the classical conception was equipped with a *theory of knowledge* in which intentionality played a central role, while Tarski, in spite of having broken the constraints of a purely syntactic perspective, was still imbued with the empiricist atmosphere of his philosophical environment, and could not make more explicit what it means for a sentence to 'correspond to a state of affairs.' His semantics remains, as we have called it, an *extensional semantics*; and it is not accidental that its precious technical tools could be used for the

The crucial problem is clearly the following: we must be able to *indicate* the referents in a non-linguistic way, but at the same time without cutting the link which should enable us to *relate* the language to these referents. We are convinced that our operational criteria of referentiality fulfil this double task. For one thing, as we have explained at length, they are extralinguistic: they consist in *doing something*, and not in *speaking* (even though in most cases it is normal to use language in order to describe how these criteria must be concretely applied). For another, we have also stressed that these criteria are not given at random, but are devised as means for ascertaining whether or not a given predicate *applies* to a thing in the way that is expressed in a given sentence. This fundamental feature manifests the *intensional* character of our semantics. Referentiality criteria are elaborated in connection with *meaning*, they are introduced with the view of making this meaning precise under certain special circumstances, and of enabling us to relate the *sense* contained in the meaning with that basic component (the *referent)* which promotes the sense to the role of being a guide in the search for truth, as Frege had already correctly stated.[80] In this way, we no longer say (at least in sufficiently precise languages, such as those of science) that a sentence refers to 'reality' as such or at large, but that it specifically refers to those referents which are singled out by its criteria of referentiality (under the 'special circumstances' implied here), and *at the same time* these criteria of referentiality are *criteria of truth*, since they do not apply to isolated predicates but only to predicates in a propositional context.

(Footnote 79 continued)
elaboration of modern 'formal semantics' only when they were supplemented with the consideration of *meanings*, as we have already mentioned in a note in the preceding section.

Something similar also applies to the other great contribution due to Tarski, that is, his characterisation of the notion of *logical consequence* in terms which are not purely syntactic. Here again, successive 'model theoretic' tools, which are usually considered to constitute a coherent development of Tarski's original view, are rather at variance with this view. They could develop out of Tarski's techniques only because they were tacitly equipped with meanings they did not actually have in Tarski's original approach. In our terminology, we should say that they were to be interpreted in the spirit of an *intensional* semantics (see Agazzi 1976). In a more recent work, John Etchemendy has advanced similar remarks, and supplemented them with a thorough analysis, vindicating the role of a *representational* semantics which has essentially the same meaning (see Etchemendy 1990). By considering all these elements, one may adopt a more objective attitude towards Tarski than that which is based *only* on a strictly logical analysis of his papers. For example, most (or perhaps all) of the critical remarks advanced in Field (1972) are correct in themselves, but not equally pertinent as an interpretation of Tarski's programme, since it must be evaluated *historically*, that is, by considering what Tarski could express within the framework of the concepts available to him in the 1930s, and not on the basis of clarifications which have come much later.

[80] "The drive towards truth" said Frege, "is that which pushes us forward from sense to reference" (Frege 1882, p. 33). We do not maintain that meaning *reduces* to this (in fact we have already said that, for us, meaning includes both sense and reference). But we can agree to a large extent, for example, with Dummett, when he states that a "grasp of the meaning of a sentence of a language is to be taken as consisting in a knowledge of the conditions for it to be true" (Dummett 1973, p. 460).

This point deserves a deeper scrutiny and obliges us to overstep an analogy which may prove helpful if taken loosely (as we too have done at times) but may obstruct if taken strictly, namely the analogy between the denotation or denomination which relates a name to an object, and the reference which relates a sentence to a state of affairs. This analogy was already suggested (in a rather cautious way) in the passage of Husserl's *Logical Investigations* already quoted in the preceding section. However, already Russell, but also others, had claimed that sentences cannot be names of facts, since facts can be affirmed, denied, put in doubt, but not denominated.[81] We think that this objection is not particularly strong; indeed, it is not said that denomination is the only possible way to *refer* (it may be so in a strict Fregean perspective, but this is indeed one of the weakest, perhaps the weakest, point of Frege's semantics). For example, one could maintain that sentences *describe* facts, and that this is their specific manner of referring to them. In this way a fact could be conceived of with that kind of unity and identity which we usually attribute to individual objects, without needing to be 'denominated' by a sentence.

This view too has been challenged by those philosophers who have noted that facts do not possess that univocal relation to linguistic expressions which characterises names. In other words, a true sentence, which allegedly should be true in virtue of its corresponding to a well-defined fact or state of affairs, is logically equivalent to an infinity of true sentences. Each such sentence should correspond to a new fact. Thus we would open the way to the existence of an uncontrolled and undetermined infinity of facts. For example, if the following sentence is true "there are seven days in the week," the sentences "there are less than eight days in the week," "there are less than nine days in the week," and so on, are also obviously true; but then there should be a fact corresponding to each of these true sentences.[82] Neither does this objection seem very strong, and to see this we can exploit the analogy with names denoting objects. Also a single individual—say Socrates—may be denoted through a large set of equivalent (and even non-equivalent) definite descriptions, but no one objects here to the fact that definite descriptions denote individuals without 'multiplying' them *ad infinitum*. Rather, we find it normal that an individual be characterised by a potential infinity of properties, some of them significant, some of them simply tautological; and we are aware that each such characterisation remains *partial*. Now, we are entitled to demand the same (and nothing more) also for facts or states of affairs with respect to sentences. It is true that the same fact may be *described* in a great variety of ways (even those that are potentially infinite), but this does not prevent it from remaining the same state of affairs, in a way similar to the claim that Socrates remains the same Socrates when we describe him as the husband of Xantippes or as the master of Plato (the main difference may consist in the fact that we 'encompass' in a single state of affairs all its logical consequences).

---

[81] See the discussion in Prior (1967), p. 228.

[82] See Davidson (1969), pp. 752–753, and Keuth (1978), pp. 76–84, for this kind of objection.

In discussing this point we have come closer to the peculiarity (and the advantages) of our position. In fact, the above-sketched objection might have a certain weight if one does not distinguish things and objects, as we do, because in this case one becomes truly involved in the undifferentiated complexity of the infinity of aspects of unanalysed reality. It is therefore not accidental that Davidson, after having discussed this issue, concludes that a definition of truth based on what he calls "the strategy of facts" helplessly collapses when faced with the situation that "true sentences cannot be told apart in point of what they correspond to (the facts, The Great Fact) or are satisfied by (all functions, sequences)."[83] In other words, such a 'strategy,' if no better refined, simply amounts to the 'rough' picture of the correspondence theory of truth that we outlined at the beginning of our discussion, and which imagines the referent of *each* sentence to be in a way the whole of reality (The Great Fact, as Davidson says). We have a straightforward confirmation of this in a passage by Davidson on the same page:

> Seen in retrospect, the failure of correspondence theories of truth based on the notion of fact traces back to a common source: the desire to include in the entity to which a true sentence corresponds not only the objects the sentence is "about" (another idea full of trouble) but also whatever it is the sentence says about them.[84]

Now, our conception avoids precisely these two difficulties. First, because the partial, and circumscribed, criteria of operational referentiality enable us to eliminate the 'trouble' of making precise what it is that a sentence is *about*. The sentence is about whatever 'thing' may be handled with the admitted criteria of referentiality or protocollarity. Second, this very procedure ensures us that our sentences are not intended to say about a thing 'whatever' may be said, but only what can be said within the 'domain of discourse' permitted by the predicates of the given discourse.

What is still unsatisfactory in our way of speaking is that we mention 'things,' while we have earlier said that sentences refer to 'states of affairs.' This is true, but we have reverted for a moment to our previous way of speaking only in order to recall certain significant analogies. We shall now abandon this colloquial style also for an additional reason: Davidson, for example, precisely because he was impressed by this (alleged) consequence of the uncontrolled proliferation of facts, decided to reject the 'strategy of facts' and to remain closer to Tarski's strategy of satisfaction which, in his view, permits a sparser ontology in which only objects, or pairs of objects, of the common world are involved. However, this is exactly what does not deserve special appreciation. If a sentence must have a reference (and it must), this reference has to be something *different* from an individual entity, be it a 'thing' or even an 'object' in our technically refined sense. A sentence may refer to a *fact* just in the same provisional and vague sense in which a term refers to a *thing*. But, in the same way as in a rigorous discourse terms refer to *objects* which are obtained from things in a special way, so sentences refer to *states of*

---

[83] Davidson (1969), p. 759.

[84] Ibid.

affairs, which are what may be said of facts under specialised and rigorous conditions.

To show how this happens, let us first give an intuitive and rough picture. In our metalanguage we can speak, in a way, of anything, provided we are aware of what we are doing. We can speak of a sentence such as "Snow is white" and predicate of it a property (we say that it is true). When we take this metalinguistic assertion at the 'semantic logos' level and want to explain what it *means*, we say that the object denoted by "snow" possesses the property denoted by "white," and that this constitutes the *state of affairs* denoted not by the one or the other of the two terms, but by the whole sentence "Snow is white." We can now proceed to the apophantic level, and we see that we can *assert* the sentence "Snow is white" if and only if we are in the position to ascertain that the corresponding state of affairs holds. Clearly, it lies outside the reach of the metalanguage to ascertain this 'holding' of the state of affairs. However, there is nothing puzzling about that. Our speech is full of biconditionals relating states of affairs of different natures. For example, if I say "John is happy today if and only if his favourite team has won the football match," I am uttering a perfectly meaningful sentence in which a psychic state is related biconditionally with a sporting event. The case of attributing truth to a sentence is similar.

However, there is in the above brief presentation the delicate point which has led many philosophers to reject the theory of truth as correspondence to a 'state of affairs,' since we have said that we have to recognise in the world of referents not only an object such as snow, but also a property such as whiteness. Is this not the ontological price many are not ready to pay? Yes it is, but it may not be such a high price, if properly understood.

4.5.4 Truth-Makers

We resume here a line of thought we already hinted at in a foregoing section where we maintained that a sentence, and also a proposition, is a truth-bearer, which in itself *may* (or may not) be true, and turn out to be such only *in virtue* of something not linguistic, and even not mental, but objective and belonging to the world. The problem is now that of determining what belongs to the world, and we are normally confronted with the empiricist prejudice according to which what really exist are concrete individuals, while properties and relations, being 'general' or 'universal,' are *abstract*, and hence do not have genuine existence. However, there is absolutely no evidence nor argument in support of such a dogmatic claim, and in particular there is even no reason for maintaining that properties or relations are more abstract than individuals. Indeed, it was already clear to Aristotle[85] that there are features (namely, the individual accidents) which are *in a particular subject*

[85] See *Categories*, Chap. 2.

and are *particular* as well. For example, when we say that Socrates is white, we are referring to the particular whiteness inhering in Socrates, and which is different from the whiteness inhering in another man, or in a particular sample of snow. In this respect, there is no radical difference between admitting that a given individual substance (e.g., Socrates) exemplifies a certain general concept (e.g., 'man') and admitting that a given particular instance of white exemplifies the general property whiteness in the particular individual concerned. We may go even further, and note that sometimes we may clearly perceive, for example, a colour, without being able to determine what is the individual object *in which* the colour exists (we might, e.g., see 'something brown' without being able to discriminate whether it is a living animal or a piece of fur). Therefore, properties and relations may have, and often do have, the same perceptual accessibility that concrete substances have.

If we pass from these general considerations to the particular view developed in this work, we may even say that our criteria of referentiality are essentially bound to *attributes*, and with this term we denote properties and relations. In a certain sense, we should even say that individual substances are very seldom the object of scientific investigations since the majority of the sciences (excluding the historical sciences) do not really speak of particular individuals but—as we have already stressed—speak of 'abstract objects' which are only exemplified by concrete individual substances or things. This is the reason why we have even used an expression which is reminiscent of a traditional empiricist way of speaking when we said that "a scientific object is a bundle of attributes." This statement sounds very much like the claims by which Hume believed himself to have eliminated the notion of substance, by reducing it to a bundle of perceptions. However, in his case, this was the expression of the gratuitous epistemological dualism he shared with other representatives of modern philosophy, as we have already explained. In our case, to the extent that the scientific object is abstract, it is not a substance, but a thought-content, a *noema*, and therefore its attributes are not something which exists *in* it, but something which *constitutes* it (are parts or elements of it). When we encounter things which operationally appear to exemplify such abstract objects, these things are to be taken as individual substances *in* which these attributes are present.

This approach, which could be seen as a peculiarity of the semantics adopted on our view of scientific objectivity, has found interesting support in the doctrine of *moments* (and in their use as truth-makers) developed by a group of scholars who have resumed important investigations initiated at the beginning of the twentieth century by Husserl, Russell, Wittgenstein and Meinong, and who are developing them in the form of a rigorous ontology which is also conscious of a long (and unduly neglected) tradition in Western philosophy.[86] We shall now critically

[86] Let us mention in this connection the articles: Simons (1982) and Mulligan et al. (1984). The second of these papers contains useful references to additional literature. We should like to stress, however, that an analysis of these topics, from both an historical and a critical point of view, had already been offered in Küng (1963), which can therefore be reckoned as the first proposal for the development of an ontology of 'concrete properties' in connection with modern semantic analysis.

examine this interesting theory (which we have already briefly mentioned in note 7 of Sect. 4.4). In the terminology of these authors, moments are essentially the equivalent of traditional *accidents* or *modes*[87]; that is, they have an authentic ontological status, though they have no independent existence, but can only exist *in something*, which functions as a "fundament" for them. This something is often a substance, but can also be a set of substances (when moments are n-place relations), while moments may also be founded on other moments. Therefore, moments are particular 'objects' characterised by their ontological dependence on a fundament, and objects which are not moments are called "independent objects or substances."[88] We shall not go into this interesting doctrine in greater detail; what we have said here is sufficient to show how these moments are close to our "attributes," and how they are also close to the Galilean "affections" of which—not by chance—we have spoken at some length in Chap. 1.

Something which is also common to this doctrine and ours is that, in spite of admitting the existence of properties and relations, we are not committed to a position of "extreme realism" in what is traditionally called the problem of universals. In fact, though we have spoken of 'exemplification' (which is usually the term used for indicating the relation between a universal existing 'in itself' and concrete instances of the universal), we never endorsed such an extreme position. It is sufficient for us that the said exemplification be concretely tested through operational procedures of a *given type*. On the other hand, this 'type' allows us to avoid acceding to nominalism. The attributes are not empirically 'found' *in* substances, and then brought to some kind of generality by virtue of some similarity. On the contrary, (scientific) attributes are determined as *noemata*, and are equipped with standardised referentiality procedures which enable us to test whether they are exemplified in this or in that concrete thing. Perhaps this position (which seems to us to be rather close to the Aristotelian and Thomist attitude of "moderate" realism), is also the best suited for the proponents of the theory of moments as well.

However, we are not in total agreement with this theory, and indeed diverge on an important point. Without entering into details which would require a lengthy discussion, let us simply say that, in the view of these authors, *moments are truth-makers*. Indeed they are the truth-makers *par excellence*, and only the existence of certain kinds of 'recalcitrant' sentences suggests that also things, besides modes, may count as truth-makers. In our opinion, on the contrary, truth-makers are *states of affairs*. At first glance it might seem that here again we are in total agreement, because these authors too would accept this claim, but only because for them states of affairs are included in the class of moments.[89] We do not believe that their

[87] However, these are rediscussed and refined along the lines of the formal ontology opened by Husserl in the third of his *Logical Investigations*.

[88] See Mulligan et al. (1984), p. 294.

[89] This is concisely put by Simons as follows: "Many, perhaps all, states, events and processes involve substances, and are thus moments" (Simons 1982, p. 159).

analysis is satisfactory since it is not in virtue of the fact that we can ascertain the existence of substances, and also the existence of moments, that we are able to say whether a proposition is true; we must in addition ascertain the *inhering* of the moment in that particular substance, and this is another requirement.[90]

We are encouraged to defend this position by the fact that it reflects an obvious distinction that is often overlooked, but which is also duly stressed by several more exact authors when they analyse, for example, two possible senses of the expression "objects of thoughts" (or "objects of speech").[91] In a sense, these objects are *what we think*, in another sense they are that *about which* we think. Now, when we say (or think) that snow is white, we speak *about* snow, but we also speak *about* whiteness (this is the important achievement of the theory of

---

[90] Let us consider the example given by our authors:

> If 'This cube is white' is true, then it is true *in virtue of* the being white (the whiteness) of this cube, and if no such whiteness exists, then 'This cube is white' is false.
>
> Because the whiteness in question here is a particular dependent on the cube, and not a universal whiteness shared by all cubes, its existence does nothing to make sentences about other things' being white either true or false (Mulligan et al., op. cit., p. 297).

A minor point is that, apparently, there is a small inaccuracy in this statement since, instead of saying "shared by all cubes," the context obviously required saying "shared by all white objects," because it is not the fact of being cubic, but the fact of being or not being white, that would matter for the attribution of the moment of whiteness to a substance. This remark, however, draws our attention to a decisive point. Even though we recognise a moment as being particularised to a given substance, we cannot elevate a particularisation to the level of making the moment *dependent on that particular substance*, not only as to its *existence* (which is correct, since the substance is its basis), but also as to its *essence* or quality. This is tantamount to saying that it would be absurd to maintain that a given moment may exist *only* in a given substance. If we pushed the 'individualisation' of moments that far, we would completely trivialise their function, and would be obliged to designate them by *proper names*. On the contrary, we want to be entitled to say that whiteness has an individuality as a moment, but this in the sense that it is different from redness or blackness, and so on. However, this does not prevent this moment (and in general *all* moments) from being *aptum inesse pluribus*, so that even if the whiteness which is in Socrates does not coincide completely with the whiteness of our cube, or with the whiteness of snow, it must be in all these cases an *exemplification* of whiteness.

But now we have an inevitable consequence. If whiteness is not *in itself* the whiteness of *any* particular substance, it follows that it might well not exist in *this* cube. It might be one of the many moments which are not exemplified by our concrete cube. The contingent *fact* that this cube and whiteness are combined is the particular *state of affairs* that makes our sentence true. In this way we could even vindicate the possibility of speaking of *facts* as truth-makers, a possibility against which many contemporary philosophers would revolt, but which is acceptable to the extent that facts are understood just in this way. However, since the use of the word "fact" is open to many ambiguities, we shall refrain from using it when we specifically speak of truth-makers, or of referents of sentences, using instead the less disputed expression "state of affairs" which, in addition, has for us the technical sense explained.

[91] A significant example of this sense is provided in the first pages of Prior (1971).

moments), and *what we say* is the fact, or the state of affairs, that whiteness is a moment having its fundament in snow. In other words, even if we accept, and we can accept, that the thought-content of any sentence expresses a relation between two individuals (the individual substance and the individual moment), it is their being in the relation of foundation to attribute in the particular case in question which constitutes the novelty, the state of affairs that must be empirically ascertained, as well as constituting the specific thought-content of the sentence as distinguished from the thought-contents of the terms "snow" and "white."

As a conclusion of these reflections, we are entitled to say that a sentence cannot be true simply because it is 'asserted,' but its being true depends on the *existence of appropriate truth-makers*, which we qualify as its referents. These referents are states of affairs in which moments inhere in their fundaments, some of which may in turn be moments, but some of which must be independent objects, or substances (without any further specification as to the particular ontology of these substances).[92] Endowing attributes with an authentic ontological status (they are referents of predicates) has the advantage of showing how ill-founded is the claim that in science we may perhaps affirm the *existence* of certain things, while being unable to establish *how* they are, that is, to characterise their *properties*. This discourse ignores several points. First, that in the practice of (empirical) science we can trace things only to the extent that they exemplify properties, which is another way of putting the much more general condition that we can perceive, or in general know, a substance in no other way than by perceiving or knowing *some* of its attributes (or moments). Therefore, it is absurd to say that we may know of the existence of something without knowing (in a certain measure) what this something is (i.e., without endowing this something with attributes). Of course, we may be wrong, that is, it may happen that the substance does not posses the attributes or the moments we credit it with, but this does not eliminate the fact that, in order to correctly affirm the existence of a substance, we must know *at least some* of its attributes.

4.5.5 An Acceptable Version of the Correspondence Theory of Truth

What we have just said is another way of underscoring the 'relativity' of truth. The truth of sentences is relative to their truth-makers. At the same time, this is a way

[92] In this way we can maintain that sentences refer to their truth-makers, while in the perspective of the authors we have discussed, truth-makers are not designated by the sentences they make true (p. 303). These authors do not consider this an inconvenience, but we believe that to leave the question of the reference of a linguistic expression (i.e., a sentence) without answer is not a desirable feature of a semantic analysis. For a different but related criticism of this conception, see also Buzzoni (1995), pp. 49–53, where the expression *indirect realism* is adopted in this connection.

of recognising that such a relativity is far from implying that truth is ontologically uncommitted. Quite the contrary, relative truth entails the existence of those entities which make true propositions true. We say that in this simple position we can identify an acceptable version of the correspondence theory of truth, and we shall briefly see why it is acceptable.

First, it does not suffer from that strange prejudice we mentioned at the beginning of this section, which consists in imagining a reality 'in itself,' which should on the one hand have its intrinsic fixed structure, independent and unaffected by language and thought, while on the other be such as to be mirrored by thought and language. We shall not repeat here our criticism of this conception, and will underscore, instead, that in our perspective reality may be known (especially in science) through the *intervention* of human subjects. But this intervention results in the determination of attributes which are known as they are brought to light and, at the same time, are those actual aspects of reality which are effectively known through a particular intervention. Under different conditions, reality would manifest itself under different aspects or in the form of other attributes, but these too would be real. Therefore, a sentence which *only* contains certain predicates, which in turn denote particular attributes, refers not to an undetermined reality as such, but uniquely to that *state of affairs* which is characterised by the attributes involved and the way in which they are structured. When the sentence is true, this means that it turns out, on the basis of accepted testing procedures, that the state of affairs referred to by the sentence obtains.[93] Since the state of affairs belongs to reality (is real), we can say that in this case the proposition *corresponds* to reality, not—we repeat—to reality in general, or to The Great Fact, but to the limited portion of reality constituted by the state of affairs referred to.

What we have said already discards as inappropriate that 'pictorial' view of the correspondence theory which takes it to mean that language and thought provide a kind of reduplication of reality under the form of representation.[94] Even Frege, who was a 'realist' from many points of view, said that 'facts' are true thoughts, but explicitly included facts in the realm of sense, and not of reference. They do not belong to the world, and their being true cannot be seen as something like a correspondence with the external world either. A correspondence—Frege said—may be that of a picture with a concrete object, such as the cathedral of Cologne,

[93] We are even prepared to say that the state of affairs *exists*, since existence is an 'analogical' concept; and we can therefore say that for a substance to exist means to have an independent ontological status, that for an attribute it means to be inherent in something else, and that for a state of affairs it means to be the case.

[94] This was, in particular, the Marxist conception of knowledge as the "*Widerspiegelung*" ("mirroring") of reality; so we can say that Marxism was one of the most paradigmatic expressions of the 'rough' correspondence theory of truth. We do not underestimate, however, more refined forms of conceiving mappings, projections, or representations of reality that have been developed by several authors and that we are not going to consider here.

but nothing of this kind is imaginable for a thought, even a true thought.[95] It is therefore clear that one of the reasons for rejecting the correspondence theory may be its being conceived in this pictorial sense. However, this is by no means its most obvious sense, and in any case not the one we advocate. From what we have said, it is clear that any proposition refers only to an infinitesimally small fragment of reality, and it refers not in a pictorial way, but simply by expressing a few particular attributes which are only exemplified in certain concrete circumstances.

This 'weak' pretension of our correspondence theory also discards an unhappy impression that may be associated with the notion of correspondence, and which may be expressed as the pretension that true sentences are characterised by some kind of 'point-to-point correspondence' of their elements and internal structure with the elements and structure of reality. This was the conception of logical atomism (and partially of Wittgenstein's *Tractatus*), but was also its weakness, as well as being the reason it was abandoned by its original supporters. The gratuitous presupposition was not only (and perhaps not so much) the assumption that the internal structure of facts could be mirrored by the 'logical' structure of language, but resided in the seemingly less arbitrary assumption that there exist 'complex facts' in the same way that there exist complex propositions.[96] Our conception does not share this interpretation of correspondence, since the procedures which enable us to establish referentiality are *operational* and not *logical*. In performing them we follow, so to speak, the patterns of facts, and not the patterns of language, even when we use language in order to describe them, and even when we have used language and logical inferences to establish which sentences we want to test.

We have the impression that we are left with only one objection which is sometimes directed against the idea of the correspondence of sentences to states of affairs or to facts, the objection (which we have already briefly discussed) that one and the same true sentence implies an infinity of other sentences. Therefore, if a real state of affairs or fact should correspond to each true sentence, we would have an infinity of concrete facts for each true sentence, which is absurd.[97] Let us first consider those sentences that are logically equivalent to a particular true sentence. In this case no problem exists since (and now we recognise that it was a merit of logical atomism to have clarified this) no new attributes are created by logical relations. Logically equivalent sentences 'correspond' to the same state of affairs. As to logically non-equivalent sentences which may be implied by a true sentence, we have to stress that they do not *multiply substances* (which might be found undesirable from certain points of view), but are simply able to multiply the recognition of many states of affairs in which substances are involved. In principle, one single substance may host a potential infinity of attributes (i.e., very many 'moments' may exist *in* it), while still remaining one. Socrates may be white, a

[95] See Frege (1918).

[96] For a critical examination of this point, see Mulligan et al. (1984).

[97] We have already quoted Davidson (1969), and Keuth (1978) for a presentation of this objection.

musician, a philosopher, a husband, a master of Plato, a citizen of Athens, and so on, without producing any counterintuitive consequences. All these states of affairs, though being concrete and real (since they are concretely testable, and none can be asserted only on the basis of a logical or linguistic analysis), may be 'added' to Socrates without any ontological problems ensuing.

In conclusion, we can say that there is space for a reasonable version of the correspondence theory of truth between the two extremes, one expressing this doctrine simply as the easily acceptable (but almost empty) claim that in order for truth to be given there must be 'some kind of relation' between language and reality (i.e., that language does not speak *about* nothing), the second expressing this theory as the claim that language is able to say completely *what reality is*. The reasonable position consists in recognising, first, that we certainly speak *about* something, that this something consists of substances endowed with attributes, and that these attributes result from the encounter between our way of investigating reality and *what reality is*. This 'way of being' of reality is only partially brought to our knowledge in any cognitive situation and, moreover, only very few aspects of it are conceptualised and receive denominations in language. Therefore, all we can expect when we claim that a sentence is true is that the few attributes denoted by its terms are really exemplified in the state of affairs which they enable us to single out and empirically test. If the result of the test is positive, we are entitled to say that our sentence is *adequate to reality* (not that it is isomorphic to reality or that it pictures it), and in this notion we express the positive but limited success of our cognitive enterprise. This is, in our view, the meaning of the 'classical' theory of truth as *adequatio*, which we may accept as a possible form of the 'correspondence theory,' provided we give to this expression the more precise sense outlined here.

4.5.6 An Ontological Consequence of the Referential Nature of Truth

Let us consider a certain domain of reference, and a set of true propositions concerning this domain. It is a necessary consequence of what we have explained (i.e., of the relational nature of truth and of the referential connotation of this relation) that, *if* these propositions are true, they have *referents* to which they apply in this domain. Although this consequence is very obvious, it has a particular importance and a conspicuous force, for it enables us to use the 'reverse way' in the path of referentiality which connects referents with sentences. The direct way, which we already know, is that in which we are able to provide, for instance operationally, the immediate connection between a sentence and its referents. This allows us to conclude that the sentence is true (and we spoke actually of 'immediate truth') so that we can say that the 'existence of truth' was granted by the discovery of the referents.

Is it now possible that the 'existence of the referents' is warranted by the discovery of truth? Surely it is, at least in principle, provided we can use other tools for discovering the truth of a sentence. These tools exist, and some of them are well known and effective, such as those represented by logical deduction. If we have a sentence A of which we can say (for instance, operationally) that it is true about some domain of reference, and we can then correctly *prove* from it another sentence B, we must say that B too is true *about* the same domain of reference, and this even if B is not operationally testable. B would not be 'immediately ascertained as true,' but 'recognised as true by mediation' or 'by argument.' Still, being true, it would *necessarily* refer to some referents in the domain, owing to the structure of truth.

Radical empiricists do not usually agree with this conclusion, since it rests on attributing also to *reason*, and not only to the *senses*, the ability to provide knowledge in a proper sense. Actually, they reduce the nature of logic to a simple 'tautological transformation' of sentences in which no new knowledge is gained in the conclusion with respect to that which is already contained in the premises. In this sense we can say that only an 'analytic use' and no 'synthetic use' of reason is recognised. The traditional conception of logic was different. It was considered as the specification of those forms of argument that are *truth-preserving* and, therefore, can lead to statements that are really different from the premises (hence they say something new) and are at the same time true owing to the truth of the premises and to the *reliability* of the logical tools. In other words, radical empiricists fail to recognise that humans have an *additional cognitive instrument* in comparison with other animals, and this is *reason*.

The reliability of this instrument has been established through centuries of investigations in formal logic, though this does not entail that its use necessarily grants us truth in every concrete case. This, however, occurs also in the case of the senses. Not everything that sight attests to is a concretely existing thing (think of optical illusions or simply of images in a mirror). The reliability of the particular cognition obtained in single cases is checked by resorting to other cognitive instruments, and reason may appear among these. In order to exclude that the image in the mirror be a concrete thing, I need not 'touch' it. I could simply reflect on the absurdity of the claim that a thing can multiply itself indefinitely merely as a consequence of multiplying the number of mirrors reflecting it (such a multiplication of mirrors would not be a cause adequate to the production of new things), and try to find a rational explanation of this apparent multiplication. In this effort men have discovered (by means of *reason*) certain aspects of reality not ascertainable through the senses, that is, the laws of optical reflection, that accounts for this phenomenon. In conclusion, true knowledge can be obtained by means of several cognitive tools that are intrinsically reliable, but whose concrete application requires a kind of double-check in order to increase our 'degree of confidence' but not in order to be sure that, *if* the knowledge they provide is true, the referents of this true knowledge really exist (this consequence is embedded in the nature of truth itself).

Aside from logical deduction, other tools are produced and applied in science for obtaining true sentences, and the global task of science may now be seen again (after the specifications we have provided thus far in this study) to include, as a very conspicuous part, the effort to obtain true sentences or systems of sentences. The consequence of this is therefore that, to the extent that these efforts are successful, that is, to the extent that true sentences of a mediate character are reached in science, they necessarily imply the presence of their intended referents in the domain of discourse of the science in which they are stated.

Interestingly, such a referential commitment of truth is recognised even by scholars who are far from admitting a correspondence theory of truth, and are very cautious in admitting 'realism.' Consider, for example, these statements by Nicholas Rescher:

> Semantics constrains realism: we have no alternative but to regard as real those states of affairs claimed by the contentions we are prepared to make... As these deliberations show, we are committed to the idea *that* there is a thought-independent reality even though we are not in a position to stake any claims to definitive knowledge about *what* its nature is... Negative and regulative though the conception may be, we nevertheless require it as a tool of indispensable utility.[98]

We shall return to this kind of question in the chapter devoted to scientific truth, where we shall see that modern science provides knowledge about the nature of that reality which it investigates.

4.6 Scientific Objects are Real

We are now in the position to give a more complete answer to a question raised earlier, are scientific objects real? This question requires a clarification regarding how one is to *conceive* of 'scientific objects' and reality. We have elaborated such a clarification, and have begun to see, starting with Sect. 4.2, a progression of arguments which support an affirmative answer to this question. First, we have recognised the 'analogical' character of the concept of reality, and have stated that "real" means 'different from nothing,' leaving fully open the *kind of reality* which may be attributed to that which (in a given circumstance) has been found to be different from nothing. In such a way we easily see that scientific objects, being defined as structured sets of attributes, are abstract objects and, as such, have the ontological status of *noemata* and the corresponding kind of existence (i.e., *intentional existence*). However, when the question of the existence of scientific objects is asked, it is normally understood as the question whether they are endowed with a kind of non-mental or not purely 'intentional' existence. We have seen that this question must be clarified. If we pretend that abstract objects also exist concretely, we are making an absurd claim, since we are conflating the semantics of encoding with the semantics of exemplifying; no abstract object *concretely exists* (this is not, however, a peculiarity of scientific objects, but a feature of every general concept). What we may correctly ask, instead, is whether an abstract object is *concretely exemplified* in a particular referent or referents; and we have seen that this is the case with several scientific objects for which concrete operational criteria of referentiality are provided. In this sense we may accept the

[98] Rescher (1982), pp. 263–266.

loose expression that, at least in many cases, scientific objects are real *as referents* (instead of saying more strictly that they are exemplified by referents).

Up to this point we have achieved something, but not a great deal, as far as the domain of the natural sciences is concerned.[99] In fact, the majority of scholars in contemporary philosophy of science are ready to admit that the 'reality' of immediately 'observable' objects is not problematic (and our condition of immediate applicability of the criteria of referentiality is similar to, though more rigorous than, the criterion of observability). The problematic issue is that of attributing reality not to those 'abstract concepts' which are abstract only because they encode *general* properties and relations (concepts which can nevertheless be directly tested), but to those abstract concepts which cannot be characterised through immediately testable properties or relations (they are often called "theoretical concepts" or even "theoretical entities," when ontological distinctions are not carefully respected.) In this case, we patently cannot rely upon that guarantee which is naturally inscribed in the structure of operationality (since it seems obvious that no one can operate with 'nothing,' that no one meets, finds resistance in, is confronted with 'nothing'). Hence, the mark of reality being simply that of being different from nothing, it *follows* that objects reached by this kind of procedure are *real*.[100]

Let us note that no use of the notion of truth has been made in the foregoing reasoning. The link with reality was simply a consequence of the operational procedures securing the reference of certain predicates. However, we may also remember that, from the very moment we first spoke of such operational procedures, we have been obliged to present them as those which, in a given science, enable us to determine what its *immediately true* sentences are, to fix its data, and so on. This is why we have indifferently (but not equivocally) called them "criteria of protocollarity" or "criteria of referentiality." The rather detailed analysis which we have performed thereafter has clarified why we could and actually ought to do so. The meaning of *concepts* is usually determined in judgements,

[99] However, what we have established regarding the plurality of ontologies secured by the recognition of the 'realist' purport of the various *kinds* of criteria of referentiality is by no means elementary or trivial.

[100] Let us note that this is why there is such a general agreement about taking empirical evidence to imply what is usually called "ontological commitment" (that is, the existence of something "real" *of which* this is evidence). As we have already noted, sensory experience is an encountering, one's 'meeting' things, and not simply a formation of images. Both idealists and realists may agree about this "ontological function" of empirical concepts; idealists because, all kinds of reality being for them nothing but intentional, they have no special reason for denying reality (in this sense) to sensory evidence as well; realists because, when distinguishing between intentional reality and concrete reality, they must give at least one criterion for recognising the latter, and no better criterion seems available than that offered by sensory evidence. In conclusion, unless one is ready to claim that nothing exists, or that we know nothing (not in the sense that we do not know anything, but in the sense that what we know is nothing), one should at least admit that the referents of what we could call immediate empirical evidence are real in a full "ontological" sense. Hence, as the attributes which are operationally defined in a science have the characteristics of the immediate empirical evidence, they are referents which are real in this genuine "ontological" sense.

and this is in particular holds for scientific concepts, so that the operational procedures serve as *criteria for truth* for those sentences whose only predicates are *operational* (i.e., are predicates for which the same procedures constitute *operational definitions*).

When we then passed to the analysis of the notion of truth, we saw that it intrinsically implies a referential connotation, that is, the relation of a sentence with something extralinguistic, and we have also seen that this something may be qualified as 'reality' according to a trivialised form of the correspondence theory of truth. In doing this we found tacit help in the fact that our criteria of referentiality were at the same time criteria for truth, so that they offered a precious supplement to that aspect of the correspondence theory which is notoriously weak, its lack of a *criterion* for truth. As a matter of fact, a distinction which (though having been clearly and convincingly stated by several authors) is surprisingly overlooked by many others is that between a *definition* of truth and a *criterion* for truth. As a consequence, when one considers a certain 'theory of truth,' it often remains obscure whether this theory intends to provide a definition of or a criterion for truth. In the case of Tarski, for example, we have seen that he declared, at a certain moment, that his definition was compatible with all possible options regarding "any epistemological attitude we may have had," since his "semantic definition of truth implies nothing regarding the conditions under which a sentence such as 'Snow is white' can be asserted."[101] This simply amounts to saying that he was denying that his definition should provide a *criterion* for truth.

A clear discussion of the difference between 'definitional' and 'criterial' aspects of theories of truth is provided in Rescher (1973), beginning already in its very first pages. It is interesting that Rescher—while being the author who has proposed the probably most accurate and formally elaborate coherence theory of truth to date—at the same time explicitly recognises that the special merits of this theory reside in its constituting a powerful *criterion* for truth,[102] and admits that the correspondence theory still constitutes the best characterisation of the *meaning* of truth (despite being criterially weak).[103] The force of 'rival' theories of truth (the coherence, pragmatic, and intuitionistic theories, as Rescher calls them) appears to consist in their ability to offer criteria for the *application* of the notion of truth. It is for this reason, in particular, that Rescher correctly sees the possibility of working with various theories of truth at the same time, given that they do not conflict in the

[101] Tarski (1944), pp. 361–362.

[102] The coherence theory, Rescher says, "is designed to give (or at any rate is best construed as providing) an answer to the problem of a criterion for truth" (p. 10). "The position we shall defend supposes that coherence is not the *meaning* of truth in the context of factual claims, but its *arbiter* (to use F. H. Bradley's well-chosen word)." (Rescher 1973, p. 12).

[103] As a matter of fact, this prominence of the correspondence theory as to the determination of the pure meaning of truth has been more or less explicitly recognised by many of the most serious proponents of the coherence theory, as is sufficiently documented in the first two chapters of Rescher's book.

determination of the meaning of truth.[104] In other words, we can adopt one theory because it provides a convincing meaning of truth, and at the same time adopt another because of its superior criteria for truth. This is particularly evident when the problem arises of not losing that which may be considered the core of the correspondence theory, that is, the idea that truth entails a link with 'facts,' and cannot be reduced to a simple relation of a sentence with other sentences:

> The link from truth to factuality is not to be broken, regardless of one's preferred conception of the definitional nature of truth. Even the most ardent coherence theorist must grant, certainly not the premiss of the correspondence theory that truth *means* correspondence to the facts, but merely its consequence, that truths must correspond to the facts.[105]

We are convinced that our 'referential conception of truth,' in which referents are at the same time *identified* operationally, and considered as the *specific objects* about which sentences are expected to be (relatively) true, provides the correspondence theory (so restricted) with that *criterial* reinforcement which is lacking in its more general and imprecise formulations. In fact, the operational criteria are, exactly, *criteria*, but have the advantage of not making an appeal to something which (though

---

[104] The distinction between definitions and criteria of truth, as we have already said, is not always made in the literature, and even the meaning of "criterion" is not clear. The most usual notion of criterion is that of a *sufficient condition* (though in the philosophical-linguistic debate on the nature of criteria, they are normally considered to be neither necessary nor sufficient.) For instance, in mathematics we find an explicit definition for the convergence of a series, stating necessary and sufficient conditions for a series' being convergent. This definition, however, is often of no use for establishing *whether* a given series is convergent or not, and certain criteria have been found (e.g., Kummer's criterion). They indicate sufficient conditions for convergence that may be usefully applied in concrete cases, but which are not requested for the convergence of any series whatever. In the case of Rescher's coherence theory of truth, coherence seems rather to play the role of a *necessary* but not sufficient condition for truth, since a sentence can be considered as a good candidate for truth only if it is consistent with the system of sentences already admitted as true (necessary condition). But an additional requirement must come into play, i.e., the presence of *data* among the sentences of this system (see footnote 11).

In the case of the acceptance theory of truth, acceptance seems to play the role of a sufficient condition, not in the sense that it is sufficient to grant that the generally accepted sentence corresponds to the facts, but in the weaker sense that we have no better means for believing that it is true. In the case of the pragmatist definition, practical usefulness plays the role of a necessary and sufficient condition, since truth is displaced from the domain of theoretical to that of practical reason. (In this sense, the definition of truth as correspondence to the facts loses its representational characteristic and is reduced to a pragmatic efficacy in 'dealing' with the facts).

An interesting case is that of *instrumentalism*, a case which is related to the pragmatist view of truth (indeed, this term has been coined by John Dewey) but which has additional features. If it is understood not in the trivial sense currently attached to it (according to which it is equated with relativism and conventionalism) but in the much more refined sense presented, e.g., in Fine (1986), it maintains that what we must require of a (scientific) sentence or theory is that it be *reliable* for practical as well as for theoretical purposes. Reliability does not entail nor presuppose truth; it simply allows us to 'dispense with' truth and, in this sense, constitutes a 'deflationary' position according to which reliability is a necessary and sufficient condition not for truth, but for acceptability. The foregoing sketchy reflections indicate that even the notion of "criteria" for truth is still in need of precisions that are not fully developed in the literature.

[105] Rescher (1973), p. 27.

important and acceptable as an 'authorising' criterion, such as simple coherence or success) ignores the relation of truth to factuality. Indeed, the referentiality criteria are *criteria of factuality*; they enable us to specify a restricted, sensible and manageable *meaning* of "true to the facts" that no longer involves that reference to "The Great Fact" with which Davidson was so dissatisfied (and which is analogous to the general 'reality' of the 'rough' correspondence theory). But, in such a way, the operational procedures play the *double* role of providing the basis for a *definition* of the meaning of truth, as well as offering *criteria* for truth. Now, it was in virtue of this possibility of legitimating once again a *definition* of truth as correspondence to the facts that we could advocate (in the foregoing section) the legitimacy of two distinct claims: (a) if a state of affairs is so-and-so, then the sentence which affirms this is true; (b) but also (in the *reverse* direction), if we have reasons for claiming that a given sentence is true, for the same reasons we *must* admit that its referents exist and are as the sentence says.[106]

The purport of the second statement seems rather trivial to the extent that we consider the operational testing criteria as the *only* criteria for truth. But we do not; we maintain that truth is *in any case* a relation of sentences to referents. However, while in many cases the holding of this reference may be ascertained by the direct application of operational *criteria of referentiality* (providing 'immediate truth'), we do not exclude other criteria of truth (e.g., coherence or pragmatic criteria). This means that, *if* we should recognise, in virtue of one of these other criteria, that a sentence is true, *then* we should not only be authorised to say, but *entitled* to say (in the name of the *meaning* of truth) that this sentence has referents, where by "*referents*" we do not mean just *noemata*, but *real* referents, endowed with the same *kind* of reality as those which are reachable operationally.

This last claim may sound rather awkward, but it is not difficult to see what it would mean to say that the referent of a true sentence may not be real. It would simply mean that it may be equal to nothing, that is, be nothing, or be no referent at all. Hence, either a sentence has *no referent*, or its referent is *real*; and since true sentences must, by definition, have a referent, they cannot help having a *real* referent. Note that we could express the same idea by simply saying that, unless the referent is real, a true sentence would be true of nothing, that is, not true at all. Or, again, since a true sentence (to use the Aristotelian way of speaking with which we have agreed) must "say what is," it could not be true about nothing because, in this case, it would clearly "say what is not."[107]

---

[106] It is therefore clear that our way of indicating the correspondence between true sentences and facts is completely alien to those characterisations on which many criticisms of the correspondence theory are based, criticisms which unduly attribute to this theory the fallacy of considering language as a copy or mirroring of the world. (We have already discussed this issue; for a few additional references and examples, see Rescher 1973, pp. 8–9).

[107] Two points may be useful. The first is that we have said nothing concerning *false* sentences in order to avoid discussions regarding the 'bivalence principle,' which would take us far from our immediate concern without being of decisive help in clarifying our issue. In fact it suffices to say that *if* (under whatever form or circumstances) we are justified in stating or believing that a

The preceding remarks applying to *all* true sentences in general, it follows that not only immediately true sentences, and not only sentences which refer to empirically given objects indirectly, but also sentences which refer to so-called "theoretical entities" in science (such as, e.g., electrons), must have *real referents* (i.e., not purely intentional referents) if they are at all *true*. In other words, this means that *all* scientific objects (be they operationally or theoretically defined) are ontologically real *to the extent* that the sentences which mention them are *true*. Or,

(Footnote 107 continued)
sentence is true, *then* we are implicitly making an assertion or expressing a belief regarding the existence of some referent, unless we are expressing a purely logical truth (a case which remains outside the scope of our considerations).

The case of false sentences splits into two sub-cases. A sentence is false when it has no referent or when it says of its referent(s) something that is not the case. As is well known, some authors consider in the first sub-case the sentence to be neither true nor false. (See the famous example, "The present king of France is bald," in which one can say that the *sentence* has no referent because its descriptive term has no referent.) These authors therefore advocate (at least implicitly and perhaps unconsciously) a *referential* condition not only for truth, but even for falsity. Our conviction—which we shall not explain in full here, but which easily follows from what we maintain in the previous sections of this work—is that a more correct appreciation of the referential nature of truth should lead to considering sentences of this kind to be false, and in that way to favouring the 'bivalence principle.' This has to do with the nature of the *apophantic* discourse, which engages itself in expressing 'what is the case' or 'what is not the case,' and in which the relation to a referent is already incorporated.

A non-asserted sentence may have a sense, even though it has no referent, either because one of its terms lacks reference, or because some of the properties or relations it expresses do not hold for one or more of its individual referents. Therefore, "The present king of France is bald" belongs to the semantic discourse, until it is not uttered as a claim concerning some existing individual, in which case the non-existence of this individual leaves it meaningful, but makes it false. But also "The present president of the United States is a multiple of 7," or "Beethoven's seventh symphony is soluble in water" are sentences that are meaningful but false since, this time, they refer to existing individual objects which, for different intrinsic (in this case, categorial) reasons, cannot possess the property expressed in the sentence. That they have a sense is already granted by the fact that we can *understand* them; and it is because of this sense we can say that they cannot possibly refer to a state of affairs.

Of course the propositions expressed by these sentences have an *intentional reality* and, as such, they continue to constitute the 'denotation' of the corresponding sentence (according to our terminology) so that, in 'intentional contexts,' it may be perfectly meaningful, and even true, to say, for example: "John thinks, believes, imagines… that *p*," where *p* is one of the above sentences.

It is when we proceed to the assertion (apophantic discourse) that the referent is needed; and the asserted proposition is true if and only if the state of affairs to which it refers obtains. This requires: (a) that the referents of its individual terms exist; (b) that they may in principle possess the properties or relations which are attributed to them; and (c) that they are actually related to their attributes in the way expressed by the sentence. Failing to meet just one of these conditions makes the sentence false.

Frege, precisely because he was perfectly aware that reference is one of the basic conditions for truth, never accepted that a sentence could lack a referent; and, since in common language such sentences may actually be constructed, he saw in this fact a confirmation of the necessity of passing to an artificial language (in which no such sentences may be constructed) for elaborating rigorous logical investigations.

if one prefers, in order to discard the ontological reference of theoretical concepts in science (which was our 'problematic' issue), one must pay the price of declaring that the sentences which use these concepts are false. Unless one is ready to pay this price, one cannot eliminate the ontological commitment of theoretical sentences in science and, by extension, of scientific theories in general.[108]

We can thus draw our conclusion. It is impossible to propose an acceptable definition of truth (i.e., one which does not allow a sentence to be qualified arbitrarily as being true or false or both) without involving in it the relation of *reference*. Moreover, referents cannot be anything but *real* in the fullest ontological sense. *Therefore* (here we have at last the full justification of our claim) *scientific objects are real* in the fullest ontological sense, either because they are reached by means of operational criteria, or because they are the referents of true sentences which are recognised as true on the basis of theoretical considerations and arguments.

It is obvious that while the first warranty of reality is strong, the second is much more problematic, since it is in general problematic to establish truth outside the immediate situation of referentiality. By recognising this we are returning to the role according to which operational procedures constitute *criteria* for truth, and we are actually recognising that, in the case of empirical or 'factual' knowledge, they play the role of *fundamental criteria*. This does not only mean that they are sufficient for granting truth *immediately*, but also that it is through these criteria that truth is, so to speak, 'injected' into the discourse of the empirical sciences, whose theoretical tools would never be able *by themselves* to produce any sentence having referential purport, which would in such a way be 'specifically true in that science.' These reflections show that we can certainly 'enlarge' the domain of the sentences recognised as true beyond that of empirical sentences by using logical arguments (and in such a way we also uncover the existence of empirically non-immediately ascertainable objects), but this is possible only if some immediate truth of an operational nature is available.[109]

---

[108] We are aware that we are here making a jump, since we have said thus far that *sentences* are true or false, and now we are speaking of *theories*. Do we mean by this that theories are sentences? We have already said that this is not what we mean; and this will become even clearer in the sequel (see Chap. 6). However, we intend to underscore here the ontological commitment of theories, and this we can do via the following claim: considering that theories (even though they are not just sets of sentences) *give rise* to sets of sentences, we may say that the objects denoted by the sentences a theory obliges us to state as true must exist, unless the theory is *untenable* (be it globally, or as far as that particular aspect is concerned). This is related to the fact that, according to what we have said in the preceding note, sentences without referents must be considered false (in the sense that the non-existence of the referent is a sufficient, though not a necessary, condition for falsity).

[109] What we have just said is reminiscent of the Aristotelian and 'classical' distinction between 'immediate' and 'derived' truth, and we do not want to deny this affinity, though we do not agree with Rescher, when he qualifies it as a particular 'theory' of truth, that is, as the 'intuitionist' theory (op. cit., pp. 10–11). In fact, we have shown how 'immediate' truth does not come out of any special 'intuition,' but is rather the result of the application of intersubjective operational criteria (usually complicated and very artificial). It is also undeniable, in our view, that the immediate truths are, in a way, stable. Even if we prefer not to use Schlick's emphatic statement:

Let us also note that, while at the level of the definition of truth a sentence must be either true or false (since it cannot help but say either what is or is not the case), this does not necessarily happen with the criteria for truth. In other words, while the logic of the definition of the *meaning* of truth is classical (in the sense that it respects the 'bivalence principle'), the logic of the *criteria* for truth need not be classical, and this simply because it is not said that our criteria always enable us to recognise with certainty whether a sentence is true or not, and there may be degrees in this recognition.[110] An advantage of the criteria of protocollarity is that, in general, they are more suitable than 'theoretical' criteria for recognising with a great degree of certainty the truth or falsity of sentences. However, this should not lead us to underestimate the importance and the force of theoretical criteria for the conquest of truth. Quite the contrary, it is a fundamental persuasion of Western culture (with the exception of "radical" empiricists) that *reason* has the force of providing truth beyond the pure witness of experience. (Of course, we do not ignore that empiricists also appreciate for example the role of induction and deduction, that are operations of reason, but they are usually hesitant in attributing to these operations the capability of

(Footnote 109 continued)
"The problem of the 'basis' changes automatically into that of the unshakeable point of contact between knowledge and reality" (quoted in Rescher 1973, p. 11), we must recognise that a protocol cannot be proved wrong by other protocols. In such cases of discrepancy, we should try rather to understand and explain how the 'wrong' protocol came about, and to consider the disturbances, the imperfections of the instruments, and other conditions which, while they cannot 'annihilate' the protocol, can allow us to disregard it.

A confirmation of the impossibility of dispensing with this 'injection' of truth from the outside is also found in Rescher's coherence theory, where he correctly emphasises the indispensability of introducing *data* in order that the coherence theory not be reduced to the requirement of a purely internal relation among sentences:

> To the criticism why should coherence imply truth, we thus propose to reply: what is at issue here is not *mere* coherence, but coherence with the *data*. It is not with bare coherence as such (whatever that would be) but with data-directed coherence that a truth-making capacity enters upon the scene (op. cit., p. 65).

We completely agree with this statement but, on the other hand, we must note that, if data are to function as 'truth-makers' (as is appropriately said here), they must possess a legitimacy which transcends coherence. Rescher does not draw this conclusion since for him "[a] datum is a *truth-candidate*, a proposition to be taken not as true, but as potentially or presumably true" (p. 54). This is, in our opinion, a less satisfactory aspect of his theory, which should instead here exploit the fact (duly acknowledged by Rescher), that there exist several criteria for determining truth. This may be understood in more than one way. The most obvious is that, for example, the criteria for ascertaining historical truth are not the same as those for mathematical truth, or for physical truth. But an additional meaning is also useful. In order to take full advantage of the criteria provided by a coherence theory of truth we have to supplement them with criteria which are closer to the basic 'meaning' of truth, such as the operational criteria of referentiality. This is the position we advocate.

[110] For example, the coherence theory of truth developed by Rescher adopts a nonclassical logic, precisely because it intends to propose a criterion for truth.

fully warranting truth.) If this is actually the case, this means that we can obtain referents also by means of *rational* tools, and not only empirical ones.[111]

A vicious question surfaces: did we not begin this book precisely with the analysis of the fact that modern science has given up the claim of truth in favour of the requirement of objectivity? What does it mean therefore to try now to ground the existence of certain scientific objects on truth? Are we not becoming involved in a circle? Although these (or similar) questions may be spontaneous, they are not all of equal importance. Some of them are easy to answer. It is true that we started with the remark that modern science has *de facto* substituted the requirement of truth with that of objectivity. But a question we left open then was precisely whether an analysis of objectivity could not lead us to a rehabilitation of truth as well, which is actually the problem we are tackling now. This means, in particular, that we are using all of our analysis of objectivity to investigate the problem of truth (so no circle is involved here). And if it is undeniable, on the one hand, that we need the notion of truth to give an ontological foundation to objectivity, this does not mean that we are 'grounding' objectivity on truth. In fact this ontological commitment, as we have seen, is the *consequence* of *two* distinct factors: the structure of objectivity, and the structure of truth. We examined (and 'founded') both separately. What remains to be done is to ascertain whether they go together, and this is the essence of the question regarding the truth of scientific theories which will occupy our attention in the coming sections. But in order to present things in the most appropriate way, we shall first refine the idea of realism, and then develop a detailed analysis of the *conditions* of scientific objectivity. This should put us in the position to provide a more satisfactory account of what scientific theories are. Only after all these clarifications shall we return (in Chap. 8) to the problem of 'scientific truth.'

111 Additional considerations concerning the problem of the existence of 'unobservable' objects are developed in Agazzi (2000).

Chapter 5
Scientific Realism

5.1 Some General Preliminaries Regarding Realism

Today the expression *scientific realism* designates a broad (perhaps too broad) range of epistemological conceptions against which stands an even broader range of 'anti-realist' conceptions. We shall not survey these positions, nor enter the critical discussion concerning their merits and shortcomings. Our goal is much more practical. Since it is clear from the whole of the foregoing discourse that we advocate a form of scientific realism (because we have maintained that "scientific objects are real"), we want to make more precise and explicit *which form* of realism we advocate and at the same time give (or better recapitulate and restate) the fundamental reasons we have found for maintaining realism. Certain of these reasons are found on the level of basic epistemology (i.e. as a philosophical position that ought generally to be adopted), while others advocate realism as best capturing the nature of the epistemology *actually* presupposed by science and, at the same time *rightly* adopted by science. By saying this, we do not intend to classify our realism under one of the existing labels, nor do we intend to counter single anti-realist objections. We shall, instead, propose our view directly, as organically as possible, and discuss at the end of our presentation only those objections which seem most relevant to our position. Nevertheless, precisely because the debate concerning scientific realism is affected by a confusion deriving from the interplay of different (usually only implicitly presupposed) meanings and approaches, we find it necessary first to delineate a framework for a correct understanding of the problem of realism on a *general philosophical* level, and only then deal with more specific issues related to *scientific realism.*[1]

[1] A more detailed development of the historical considerations presented in this section can be found in Agazzi (2001). Let us stress that, owing to the specific interest we have in characterising our conception of scientific realism, we are not going to discuss other valuable conceptions of scientific realism that have been proposed rather recently, of which Niiniluoto (1999) and Psillos (1999) are significant examples. Therefore, our attention will be rather concentrated on the discuyssion of the more significant anti-realist positions.

E. Agazzi, *Scientific Objectivity and Its Contexts*, DOI: 10.1007/978-3-319-04660-0_5,

© Springer International Publishing Switzerland 2014

The debate concerning realism and anti-realism has proven not to be very fruitful because there has been very little agreement on the meaning of the fundamental concepts involved. These differences, however, are not bound to the personal idiosyncrasies of individual writers, but are rather the often unconscious consequences of a mixture of different meaning-elements that have been attached to realism through the whole course of Western philosophy. A rapid inspection of this history will help us understand certain (often implicit) features of the present debates.

5.1.1 Realism in 'Classical' Philosophy

The notion of realism has had two basic meanings in the history of Western philosophy. The first emerged in the dispute over universals of the Middle Ages, and regarded the kind of existence that can be attributed to universals such as genera and species (making it therefore an *ontological* question). In this dispute no one denied that full reality should be attributed to 'individual substances' that 'exist in themselves,' such as stones, trees, men and women, but also God, angels and devils. The issue concerned only the real existence of the 'abstract entities' such as genera and species, that we commonly denote through general concepts (and this is why this dispute is called the dispute over universals).

Given this premise, it has become customary to call *exaggerated realists* those philosophers who claimed that universals are real, that is, that they have an existence in themselves (as Plato had affirmed of his Ideas). A different position was advocated by the *conceptualists*, who affirmed that universals are simply concepts and, as such, though endowed with a reality of some kind, do not exist in themselves but only in our minds (they have only a mental reality or existence; they are *entia rationis*, in the terminology of that time). A radical position was taken by *nominalists*, who credited only individuals with real existence, and reduced universals to simple 'names' to which not even a conceptual designatum corresponds, but which are only mental (and linguistic) tools for grouping together individuals that show certain similarities. Finally, a fourth position was that of the *moderate realists*, inspired by the Aristotelian metaphysics (the most famous representative being Thomas Aquinas). According to them, universals have a double kind of existence: they do not exist in themselves but in individual substances, as their essence or form (in the technical Aristotelian sense); and they exist also in our minds as concepts.

We note that these different positions were all of a strictly *ontological* kind, and were not rooted in different *epistemological* doctrines. Indeed, all tacitly shared a common *realist epistemology* in the sense that they considered our knowledge to be knowledge of *reality* in its different forms. Moreover, all admitted that our knowledge is based both on a sensible intuition and an intellectual intuition, and differed only in the determination of the objects of the intellectual intuition. The exaggerated realists maintained that genera and species exist in themselves as immaterial substances, and can be grasped *directly* by the intellectual intuition. Moderate realists maintained that these universals can be grasped by our intellectual intuition by means of an 'abstraction' that captures the form of individual substances. The result of this abstraction is a concept that exists in the mind as an *ens rationis*. Conceptualists limited their position to this second aspect, without feeling committed to a particular doctrine regarding the status of the universals 'outside' the mind. Nominalists considered intellectual intuition to be a kind of reproduction of sensible intuition which is, admittedly, intuition only of single individual items of reality and, as a consequence, could not give to the universals any consistent *ontological* status. In conclusion, the different positions can be related to different ways of considering the role of intellectual intuition. We cannot be detained here with the details of the 'analytic ontology' that provided the ground for such discussions. We simply want to stress that a 'realist epistemology' was their common framework.

5.1.2 Realism Within Epistemological Dualism

A radical change emerged from that tacit presupposition that characterised modern philosophy (conventionally considered to have been inaugurated by Descartes) and which we have already qualified as *epistemological dualism*. As we have stressed in Sect. 1.5, this expression must not be understood in the (quite obvious) sense that we have two kinds of knowledge (such as sensible knowledge and intellectual knowledge), but in the sense that the natural aim of knowledge (that is, to know reality *as it is*), cannot be attained directly, but only by passing through another preliminary *immediate* knowledge, that of our representations or *ideas*. Therefore, the fundamental question of modern epistemology became that of determining whether or not, starting from our ideas, we can *indirectly* obtain knowledge of reality. Those who maintained that we can were qualified as *realists*, while those who maintained that we are condemned to know only our ideas were qualified as *idealists*. Therefore, "realism" has, in this context, an *epistemological* meaning. It is obvious that epistemological dualism actually contained a second tacit presupposition, namely that reality exists independently of our knowledge (ontological dualism). This presupposition can be called naturalistic, since it both reflects the common-sense conviction that there is a world 'external' to one's mind, while it at the same time constitutes a presupposition of natural science, whose spontaneous aim is to investigate and gain knowledge about the features of this 'external' world.

As a consequence, the majority of modern philosophers until the end of the eighteenth century can be qualified as *realist* in the weak ontological sense of admitting (in keeping with common sense) the existence of a *reality* 'external' to the mind, whose existence is also *independent* of the mind. But they were also *idealist* in the weak epistemological sense of affirming that we do not directly know reality but only our *ideas*. The *strong* ontological and epistemological senses of the term "realism" are that a reality *exists* independently of our knowing activity, and that it is precisely what we *attain* knowledge of in our knowing. Therefore, the question remained as to whether we can also obtain (indirect) knowledge of reality starting from our ideas, that is, whether we can from our ideas know *how* reality is, or know *what* it is.

From this second point of view, realists in the *epistemological sense* may be called those philosophers who believed themselves able to afford a positive answer to this question, and idealists those who did not see the possibility of transcending the realm of our ideas and reaching reality. Before the end of the eighteenth century only Berkeley, among the best known philosophers, expressed the full and radical idealist position in which the *ontological dependence* of things on our ideas is explicitly affirmed: *Esse est percipi* ("To be is to be perceived"). In the nineteenth century the "transcendental idealism" of Fichte, Schelling and, especially, Hegel came to the extreme conclusion that reality and thought are *ontologically identical* in the sense that reality 'reduces' to thought.

All these are well known facts, and we recall them here only to stress that they are a consequence of the (unacceptable) dualistic presupposition stressed earlier. Let us underscore the crucial difference that divides 'classical' from 'modern' epistemology (which we have already had the opportunity of discussing in an earlier section). According to classical epistemology (which in different forms goes from ancient Greek philosophy to Medieval), knowledge consists in the fact that things are *present* to the mind; according to modern epistemology, knowledge consists in the fact that things are only (at best) *represented* by the mind. Moreover, the presence of things on the classical view was not conceived in any spatial sense, and was expressed as a particular *identity*, the *intentional identity* of thought and reality. In a perception, or in an intellectual intuition, our cognitive capacities 'identify' themselves with objects, though remaining ontologically distinct from them. This ontological distinction furnishes the correct meaning of "the 'external' world," which, otherwise, would mean everything 'outside my skin.' The *representation* of modern epistemology, from this 'classical' point of view, is simply a thing's 'way of being present' to our cognitive capacities, and is ontologically dependent on both, though not produced by either. Modern epistemology, having lost the notion of the intentional identity, gives to representations the status of being direct objects of knowledge that we encounter in our mind.

Through this short sketch we certainly do not pretend to have clarified *how* this intentional identity was conceived and demonstrated in classical epistemology, and the unsatisfactory status of such a demonstration was certainly among the historical reasons for the almost sudden occurrence of 'epistemological dualism' in modern philosophy. This, however, does not prevent one from recognising the inconsistency of this dualism, an inconsistency that entailed its historical decline with the rise of transcendental idealism. Still, the 'reductive' conception of the identity of thought and reality advocated by contemporary idealism favoured the rediscovery of classical intentionality in contemporary phenomenology, as well as the efforts one finds in the philosophy of mind and in the cognitive sciences aiming at understanding in what this marvellous process (i.e. knowledge) consists, a process through which certain beings are able to 'interiorise' the external world without destroying it in order to 'assimilate' it.

From the above historical survey we can single out a few different planes on which the issue of realism can be expressed. On the *ontological plane* this question consists in determining whether certain entities that are denoted by terms of our discourse *really exist or not*, or at least to determine *what kind of reality* they have. The question whether the existence of something *depends on our knowledge* of it is also of an ontological nature. On the *epistemological plane* the question of realism consists in determining whether we are able to *know reality as it is*, and this question can give rise to different answers. One can maintain that indeed we can know at least something about reality as it is, independent of the mind; or that a reality ontologically independent of the mind exists, but we cannot know anything about it; or that we can know reality as it is precisely because it is totally or partially dependent on our mind. Finally, a third plane, which could be called *semantic*, consists in relating the question of realism to the nature of reference, and in such a way it strictly relates it to the nature of truth as well. All these are *general philosophical* positions; but they are often presupposed—consciously or otherwise—by many authors engaged in the discussion of *scientific* realism, and determine their approach.[2]

We are not interested in discussing the question of realism at its general philosophical level, but only as regards science. In order to better appreciate the current interest in the question of *scientific* realism it should be kept in mind that the dispute is very recent, since science was generally interpreted realistically until the end of the nineteenth century, and it is important to understand *why* discussions concerning the nature of science were suddenly affected by anti-realist concerns. A second historical survey is therefore appropriate.

5.1.3 Realism and Science: An Historical Overview

It is obvious and even trivial that 'classical' science (that is, *epistéme* and what is represented by equivalent concepts in the Greek and Latin traditions) was conceived in a fully realist sense, being identified with *necessarily* true and *justified* knowledge (these two requirements indicating that it was not a form of 'naive' realism). But 'modern' science too was understood in a realist way in the beginning, and a decline of this realist appreciation occurred only gradually, until a

[2] For example, those who restrict existence to observable entities are making an ontological claim (Mach, and perhaps to a certain extent, van Fraassen); others consider realism simply as the contrary of idealism (Popper), though in defending their position they may introduce arguments of an epistemological or semantic nature, as we shall see in the sequel. In particular, it must be noted that even within the logical-empiricist tradition realism was not rejected, but simply accommodated to very restrictive conditions. An analysis of this issue, and of the evolution of the 'reality principle' is offered in Tarozzi (1988).

non-realist view of modern science was advocated by Kant for very special reasons. Kant's view, however, did not really affect the general conception of science until the 'crisis' of the exact sciences that occurred at the end of the nineteenth century.[3] We shall briefly substantiate these claims.

We have seen that the principal reason for which Galileo can be considered the founder of modern science is because he explicitly declared that the condition for obtaining knowledge of "natural substances" (that is, of physical bodies) consists in giving up trying to "grasp by speculation the intimate essence" of such substances and remaining content with the knowledge of "some of their affections." This programme (as we have already noted) is still expressed in the language of classical ontology, in which *substance* was characterised as having an essence and possessing certain *accidents*, of which *affections* constituted a particular kind. All this belongs to a classical *realist ontology*.

Among the accidents of physical bodies, Galileo distinguished those that depend on the sensory abilities of the observer (colours, smells, and so on, later called "secondary qualities"), and are therefore subjective—and thus *not* "of physical bodies"—from those that are intrinsic to the body (the quantifiable and mathematisable qualities, later called "primary qualities"), which he calls, for this reason, *real accidents*. It is only with these real accidents that natural science is concerned, and it can be so concerned efficaciously by adopting mathematics as a means for describing them, thanks to *measurement*. Measurement is indeed the link between mathematics and reality. This can be expressed in terms of our notion of *operations* and is a clearly realist view of natural science that is abundantly confirmed, as we have already seen, throughout Galileo's works.

What we would like to note now is that in the Galilean assertions one finds a certain weakening of the force of *intellectual intuition* (which is no longer credited with the ability to capture the essence of things). Its role, however, remains of primary importance, since it is only thanks to an intellectual intuition that mathematical properties can be determined and described as interpretations of the empirical results of measurements, that mathematical models of physical events can be constructed (by the extrapolation of mensural thinking beyond what can actually be measured), and that idealisations of the natural phenomena can be proposed; and these are salient characteristics of Galileo's scientific method. Galileo cannot be considered an empiricist because, while declaring that natural science is based on "sensible experiences and mathematical demonstrations," he also admits that the most significant advances are made possible when "the

[3] Historical accuracy would oblige one to note that already at the beginning of the nineteenth century positivism had adopted an anti-realist conception of modern science, to which was already given fuel by Newton's (and everybody else's) inability to indicate the mechanism of gravity. Therefore one can recognise that it is fundamentally for this reason that anti-realism first gained a foothold; but it is also undeniable that the generalised decline of the realist conception of science was determined by the deep 'foundational' crisis in the exact sciences that occurred at the end of that century.

intellect does violence to the senses."[4] From this point of view we must say that the Galilean *revolution* consisted in beginning to seek quantitative mensural knowledge (knowledge of primary qualities) rather than qualitative sensory knowledge (knowledge of secondary qualities). The notion of essence could be retained, but it took a fundamentally different form than it had in Aristotle's teleological approach due to the fundamental shift in epistemology. It became, in its *nominal* guise, 'that which can be determined by measurement,' as is in keeping with our earlier quotations from Galileo.

In Galileo the term "phenomenon" does not occur, while it is frequently used by Newton. It must be clearly said, however, that Newton's concept of phenomenon is not affected by the epistemological dualism we have described. For him, phenomena are simply the "manifest" characteristics of physical events (that, however, are not just sensory qualities, but measurable primary qualities, i.e. *magnitudes)*, and are by no means "pure appearances" (as Kant would have said, considering them, after all, as secondary qualities). Newton simply places as a basic methodological requirement on what he calls "natural philosophy" or "experimental philosophy" (i.e. natural science) that one abstain from introducing any "occult qualities" which (as he says) traditional philosophers used to posit as contained in the "substantial forms" of things, in order to provide explanations of the manifest features of things. All this is well in keeping with the views of Galileo (whom Newton mentions with approval on several occasions). He also shares with Galileo the admission of a limited role to intellectual intuition, in that he too recognises the decisive importance of mathematisation in the construction of natural science. But Newton is much more clearly an empiricist, since the single general laws of physical phenomena are explicitly declared by him to be propositions obtained by 'inductive generalisations' of the phenomena, beside which possible exceptions must be carefully listed. In such a way *generality*, rather than *universality*, appears as the salient characteristic of scientific laws (on an *empiricist* conception of induction), while no ontological *necessity* is attributed to them (again in keeping with Galileo's views).[5]

Universality and necessity, on the contrary, had been considered as the characteristic features of science (*epistéme*) by the classical tradition, and had remained substantially preserved only in the view of science of the 'rationalist' representatives of modern philosophy. Therefore, it is in a way surprising that these two features were rehabilitated by Kant, who attributed them to the two paradigmatic examples of 'science' he considers in the *Critique of Pure Reason* (and in the other works of his 'critical' period). These examples are mathematics and physics (the latter being practically Newtonian physics). Therefore, he made the gigantic effort of explaining *how* these sciences could be so successful, and

---

[4] For more details and relevant quotations we refer to Sect. 1.4 of this book, and to Agazzi (1994).

[5] These general methodological principles are clearly summarised, for instance, in the *Scholium Generale* of Newton's *Principia* (Newton 1687), and, in a more elaborate form, in Question 31 of the third book of his *Opticks* (Newton 1704).

inscribed this effort in the view of the epistemological dualism that he advocated. Indeed, Kant distinguished *phenomena* from *things in themselves* and declared phenomena to be "pure appearances." However, phenomena are knowable, while things in themselves are not, phenomena being knowable because they are based on both *sensible intuitions* (the "sensible impressions") that are *passively* received by our sensible capabilities (be they those of the "external" or "internal" sense), and the a priori forms of intuition and understanding, which provide knowledge with objectivity, universality and necessity.

For Kant, the understanding is *active*, but its activity is limited to its being able to unify the content of sensible intuitions according to its own structural characteristics. These are present in every act of empirical knowledge, since they are the very *conditions of the possibility* (or the *transcendental grounds*) of such knowledge. In such a way universality and necessity were recovered for any authentic knowledge, since they are simply expressions of the fact that, on Kant's account, we cannot know anything without using the a priori forms of the understanding. The *objects* of knowledge are therefore given their *form* by the understanding, but not their *content*, which is provided by the intuition. Indeed, Kant carefully distinguishes *thinking* from *knowing*. Thinking amounts to a pure combination of concepts, while knowing requires that these concepts be applied to actually present sensible intuitions. This is why Kant is concerned with distinguishing his position from 'idealism' (which in the meantime had become the contrary of 'realism,' as we have already explained above). He qualifies his doctrine at the same time as an *empirical realism* and a *transcendental idealism*, and it seems worthwhile to analyse this position (as we shall do in the following note), since it is especially concerned with the issue of the *dependence* of reality on thought, which is still one of the main concerns of several discussions regarding scientific realism.

Realists (as they were called at that time) maintained that the existence of the objects of our knowledge does not depend on our act of knowing them; idealists maintained that it does. According to Kant, the existence of these objects does not depend solely on our act of knowing, since the *content* of empirical knowledge consists of sense-intuitions that we do not produce but passively receive, while the construction of the objects of empirical knowledge follows the conditions imposed by the categories of the understanding, and therefore depends on our intellectual a priori knowing abilities, or the "transcendental conditions" of our knowledge. This is why he calls himself an empirical realist and a transcendental idealist at the same time.[6]

---

[6] This doctrine is presented in the first edition of the *Critique of Pure Reason*, and taken up again in the *Prolegomena*. It tends, first, to free the 'idealist' from the charge of denying the *existence* of external things:

> The term "idealist" is not, therefore, to be understood as applying to those who deny the existence of external objects of the senses, but only to those who do not admit that their existence is known through immediate perception, and who therefore conclude that we can never, by way of any possible experience, be completely certain as to their reality (A368–369).

(Footnote 6 continued)
The reason we can never be completely certain of the existence of such objects is that it should be *inferred* as a causal explanation of our "inner perception," and "the inference from a given effect to a determinate cause is always uncertain, since the effect may be due to more than one cause" (ibid.).

However, while the general idealist finds himself in this position, the *transcendental idealist* is—according to Kant—much better off, since he subscribes to "the doctrine that appearances are to be regarded as being, one and all, representations only, not things in themselves, and that time and space are therefore only sensible forms of our intuition, not determinations given as existing by themselves, nor conditions of objects viewed as things in themselves" (A369). The important consequence is that such a 'transcendental idealist' can also be a *realist*: not, of course, a 'transcendental realist' who "interprets outer appearances (their reality being taken as granted) as things-in-themselves, which exist independently of us and of our sensibility" (A369). He can only be an *empirical realist*, that is, someone "who may admit the existence of matter [we should say of the 'external world'] without going outside his mere self-consciousness, or assuming anything more than the certainty of his representations" (A370). From all this the desired conclusion follows:

The transcendental idealist is, therefore, an empirical realist, and allows to matter, as appearance, a reality which does not permit of being inferred, but is immediately perceived (A371).

However, the Achilles' heel is here that the existence of matter is immediately secured just *as appearance*. There is no need to 'infer' it only because it is nothing but one of two different kinds of representation (the representations affecting the so-called "outer sense," which is itself a structure of the thinking subject):

External objects (bodies), however, are mere appearances, and are therefore nothing but a species of my representations, the objects of which are something only through these representations. Apart from them they are nothing. Thus external things exist as well as I myself, and both, indeed, upon the immediate witness of my self-consciousness. The only difference is that the representation of myself, as the thinking subject, belongs to inner sense only, while the representations which mark extended beings belong also to outer sense (A370–371).

One must admit that the consolation is rather meagre. If all of what we can be certain is the existence of our *representations*, of *appearances*, it is not the fact that we can classify certain of them under the special rubric of belonging to *our* 'outer sense' that may entitle us to be *realists* in any serious sense, that is, the sense that reality does not reduce to representation. Kant himself was certainly dissatisfied with this doctrine, which occupies several pages in the first edition but is completely dropped in the second. Here it is replaced (but in a different place and context) by the "refutation of idealism" (B274–279) which, however, is no less cumbersome and unconvincing. The basic reason for this frustration is that this doctrine depends on that *epistemological dualism* of which we have spoken several times (indeed, the pages presenting this doctrine are among the most expressive documentation of Kant's adhering to this gratuitous presupposition). Therefore, it is only by abandoning the unjustified presupposition that what we *know* are just our representations that one can be a realist in a non-vacuous sense.

We have devoted special attention to this Kantian doctrine because it has been recovered, almost literally, in Putnam's doctrine of 'internal realism.' Therefore, it is not only relevant to the contemporary debate concerning realism, but also shows features which are also those of certain positions in this debate as well.

We would like to draw two conclusions from what we have said. First, while modern natural science has remained ‘realist’ in the ontological and epistemological senses (scientists admit that physical reality has an existence in itself, independent of our investigation, and that it is endowed with certain characteristics that can be made manifest and be known by us as they are), Kant gave modern science a ‘non-realist’ interpretation (what science knows is not reality in itself, nor ontological features of reality, but a world of objects that are “pure appearances” organised according to the transcendental conditions of our understanding). In spite of this, he claimed to be, at least partially, a realist, in the sense that not everything in the objects of our knowledge ‘depends’ on us, because the sensible ‘appearances’ are only passively received by us.

We find here a very peculiar form of epistemological realism, which (as we mentioned in the foregoing note) was undoubtedly reactivated in Putnam’s “internal realism” (as Putnam himself notes). We must also repeat, however, that Kant’s solution strictly depends on adhering without question to *epistemological dualism*, postulating the unknowability of things in themselves, and reducing the whole of knowledge to something ‘internal’ to the subject. This weak point was challenged by Kant’s followers (already Jakobi had noted that “without the thing-in-itself one cannot enter criticism [that is, Kant’s “critical” philosophy], but with the thing-in-itself one cannot remain in it”), and German transcendental idealism tried to eliminate this discrepancy between reality and thinking. What remains to be seen is whether an ‘internal realism’ can be advocated without falling into epistemological dualism, a question we shall take up later.

For the moment we can note that Kant’s ‘phenomenalistic’ interpretation of science did not immediately meet with agreement. On the one hand, transcendental idealism, having eliminated phenomenalism altogether, considered natural science as a correct but still inadequate form of knowledge that must be surpassed by a philosophical understanding of reality (the romantic “philosophy of nature” that developed within the idealistic framework practically amounted to a devaluation of science). On the other hand, the rich harvest of technological applications made possible by the rapidly increasing advances of the natural sciences easily convinced the general public that science indeed provides adequate knowledge of nature, and positivism consecrated this spontaneous conviction philosophically, declaring modern science to be the unique form of adequate knowledge, and discrediting philosophy for its pretension to do so. This position implicitly expressed a realist view of science, despite the fact that positivism explicitly advocates an empiricist epistemology, but this was only the expression of the concrete success of science shown by the spectacular advances of technology. The spontaneous commonsense inference was that such a success would not be explainable if science were unable to actually know what reality is. Refinements of this pragmatic argument are to be found in the contemporary debates on realism and anti-realism under the heading of the "no miracle argument."[7]

In mathematics, this realist view began to enter a crisis in the second half of the nineteenth century. The construction of non-Euclidean geometries gradually discredited the role of mathematical intuition, showing that logically consistent geometrical theories can be constructed starting from intuitively incompatible postulates; and even when set theory seemed to provide the rock-bottom foundation for the whole of mathematics, the discovery of the antinomies in this theory destroyed the confidence that we can intuitively know even such basic 'entities' as sets. As a consequence, mathematics came to be seen as a great family of logically interconnected *hypothetico-deductive* systems expressed in a formalised axiomatic way, whose legitimacy was not afforded by their ability to describe the properties of 'mathematical objects,' but simply by their internal logical consistency.

As to physics, the realist view of Newtonian mechanics was strongly reinforced during the first half of the nineteenth century, not only thanks to the impressive mathematical developments of that same mechanics, but also by the gradual appearance of a *mathematical physics* which, concretely speaking, was nothing but an effort to express, interpret and explain the phenomena studied in the different branches of physics by means of the concepts, mathematical tools, and models provided by mechanics. Therefore, the challenge for theoretical physics was seen to consist in the elaboration of adequate 'mechanical models' for the two new branches of physics, that is, electromagnetism and thermodynamics.[8] This challenge, however, was doomed to meet with failure, for no satisfactory mechanical model could be elaborated for the electromagnetic 'ether,' and no satisfactory explanation of the second principle of thermodynamics could be provided within the framework of the kinetic theory of matter (despite the very ingenious efforts of several outstanding mathematical physicists on both problems).

The reasons for these shortcomings soon appeared to be related to the fact that physics was seriously taking its first steps in the realm of the *unobservable* (in the primitive sense of what cannot be directly perceived by our unaided sense organs). In this enterprise it was making use of powerful *idealisations* that were tacitly justified by a fundamental *presupposition*, namely, that the laws and principles of mechanics have a true *universality*, i.e. that their scope includes both the microscopic as well as the macroscopic world. Both these presuppositions were attacked by Ernst Mach when he gave his diagnosis and therapy for this crisis in physics. His fundamental philosophical thesis was a form of *radical empiricism*, according to which only sensory perceptions constitute knowledge. He did not deny a certain

---

[7] See, for example, Smart (1968), Putnam (1975, Vol. 1, p. 75), Musgrave (1988), Niiniluoto (1999, p. 197).

[8] For example, even J. C. Maxwell, in the last pages of his *Treatise on Electricity and Magnetism*, indicated as a task for future generations that of finding a mechanical description of the electromagnetic field, for which he had offered his famous equations. For details on this issue see, for instance, the large Introduction to my Italian translation of Maxwell's work (Maxwell 1972), as well as my article Agazzi (1975).

function to the intellect, but reduced it to the elaboration of general schemes that have no representational import but only a pragmatic role, in the sense that they allow us to summarise sets of similar perceptions, to make useful predictions of future perceptual situations, and also to realise concrete applications. Thus, for Mach, the intellectual constructions are simply *conventions* that can be abandoned and replaced whenever other conventions appear to be more useful. He added to this *epistemological* doctrine also an *ontological* claim: the *unobservables* are not simply unknowable, but also non-existent (indeed, he denied the existence of atoms). Therefore, we must say that he expressed a clearly anti-realist view of science that was embedded in a more general philosophical anti-realism (stemming from Berkeley, via Comte). Let us note that all the described events occurred before the creation of relativity theory and quantum mechanics.[9] They only contributed to the deepening of the crisis of classical mechanics, since they showed that many more concepts, laws, principles and methodological presuppositions of this discipline had to be deeply modified in order to satisfy the needs of the new physics.

5.1.4 The Present Characterisation of Scientific Realism (and Anti-realism)

The rather lengthy story we have sketched above was necessary in order to propose a reasonable distinction between realism and anti-realism in general, on the one hand, and specifically *scientific* realism and anti-realism on the other. Indeed, as we have already noted, in many current discussions on realism and anti-realism which allegedly concern science, we simply find more or less elaborate variants of the positions regarding knowledge in general (this is the case, e.g., even with such a famous philosopher of science as Popper). In order to understand how anti-realism could gain credit in modern science, it is useful to deepen our consideration of the crisis of the exact sciences at the turn of the twentieth century.

The reason classical mechanics could receive a realist interpretation was that it remained in keeping with the spontaneous realism of common sense (that we are far from qualifying as 'naïve realism'), to the extent that it appeared as a kind of 'prolongation' of common sense itself. Its concepts were certainly abstract, but at the same time they could be seen as 'idealisations' of concretely observable physical bodies or events. A material point could be seen as the limiting condition of a shrinking grain of sand, a physical wave as of the same nature as the waves in a pond of water, a rigid body as equivalent to a rigid iron bar, frictionless motion as similar to a perfect glass sphere moving on a perfectly horizontal ice surface, and so on. Though the concepts of classical mechanics were rigorously delineated, they remained bound to *observable* physical objects or processes, they were

[9] Mach's work on the historical development of mechanics, which substantiates his views on this issue, was actually published in 1883 (see Mach 1883).

visualisable; and this spontaneously inclined people to expect that other, not yet explicitly encoded, properties of physical objects or processes from which the idealisation had been abstracted should be exemplified as well. Unfortunately, this expectation was frustrated when models of the microworld were proposed using the idealisations derived from the observed macroworld. The way out of this difficulty, in the spirit of classical physics, would have been to find new concepts obtained via idealisation from the observation of the micro-objects, but these were unfortunately *unobservable*.

This is the frontier that separates contemporary physics from classical physics, since contemporary physics is essentially a physics of unobservable objects; and it is not by chance, as we have seen in our historical overview, that *scientific realism* began to be challenged when this frontier was encountered. Therefore, we propose to characterise the problem of scientific realism as specifically the problem of the *reality of the unobservables* proposed by scientific *theories*. The suitability of this characterisation is confirmed by the position defended by such an influential scholar as Bas van Fraassen, who accepts common-sense realism regarding the objects of everyday experience since they are accessible to observation, and denies realism regarding the unobservable entities of natural science (though it might be said that his antirealism is more epistemological than ontological). This is why a very natural way of opening our discussion of the contemporary debate on scientific realism in the next section will be, to start with, a reference to this author, whose positions have the merit of having been presented in detailed and systematic form in some fundamental books rather than in more or less important scattered papers.

5.2 The Chief Issues Concerning Scientific Realism

5.2.1 Realism and Theories

It is in a way strange that we are beginning to analyse the problem of scientific realism without having presented an explicit and articulated conception regarding scientific *theories* while, in the current discussion, the 'question of realism' is normally considered to concern scientific theories in particular (as we too take it to concern here), since certain authors, such as for instance Hacking and Cartwright, are anti-realist regarding theories while being realist regarding 'entities.' For example, van Fraassen characterises realism on the one side, and his own anti-realist position on the other, as follows:

> *Science aims to give us, in its theories, a literally true story of what the world is like; and acceptance of a scientific theory involves the belief that it is true.* This is the correct statement of scientific realism.[10]

[10] van Fraassen (1980), p. 8.

> *Science aims to give us theories which are empirically adequate; and acceptance of a theory involves as belief only that it is empirically adequate*. This is the statement of the anti-realist position I advocate; I shall call it *constructive empiricism.*[11]

In such statements truth is clearly considered to be a meaningful property of theories, both in what is taken to be the realist and the anti-realist positions. (The difference is only that the alleged realist believes this truth to be attained, while the anti-realist does not care about this requirement, not because the anti-realist denies that theories might be true, but because she is content with something less, such as empirical adequacy.) It follows that at least most, if not all, of van Fraassen's arguments developed in this work in the dispute regarding realism remain within the framework of the statement view of theories, which has been challenged for very good reasons in past decades, and lose a great deal of their force once they are deprived of this tacit presupposition, so that van Fraassen himself took up a different stand in his subsequent production.[12]

As we have already anticipated (e.g., in the last note of Sect. 4.6), we do not accept the crude statement view of theories; and in Chap. 7 we shall present a more elaborate conception, from which it emerges that theories are *not* sentences or sets of sentences, ideally replaceable by a unique long conjunction (the typical logical-empiricist view). They are rather the expression of a global *Gestalt*, which, in order to be *formulated*, must be expressed in sentential terms (this is why we do not totally reject the sentential view, as many do). However, these sentences do not express the *Gestalt* simply as a result of *logical* connections. Thus: (a) the aim of theories is far from that of telling a 'literally true story' concerning the world, but is rather of giving the most faithful depiction of a certain (partial) vision of the world under a specific point of view, usually in order to *explain*—often by indicating causal relations between the constituents of the picture—certain empirically accessible features of the world; (b) theories are therefore neither true nor false, but only more or less 'adequate' or 'tenable'; (c) nevertheless, *certain single*

---

[11] Op. cit., p. 12.

[12] Actually in van Fraassen (2008) the statement view is abandoned in favor of a representational view that the author qualifies as 'empiricist structuralism.' In the Introduction of this new book (p. 3) however, he explicitly confirms the continuity with the position defended in his book of 1980, in particular as far as the issue of realism is concerned, and he does not offer new arguments for advocating his anti-realist position (though, under a closer scrutiny, one could say that anti-realism is more definitively affirmed here). This is why we are going to concentrate our attention on the 1980 work. By these remarks I do not want to underestimate the rich harvest of novelties and interesting reflections that are offered in the 2008 work, and even the presence of significant points of contact with views that I had been defending for many years (and are also presented in the present work). For example, the non-'epistemologically dualist' conception of phenomena he defends, the contribution of experiments to theory construction, the 'pragmatic' status of scientific concepts, representations and theories, the 'indexical' component of scientific statements, representations and theories, and the role of intentionality—though on each of these points the affinities are also accompanied by significant differences that will be pointed out when they are met with in the course of this work. The reason for such differences is constituted in general by the strict radical empiricist orthodoxy to which van Fraassen intentionally sticks, as opposed to the more open attitude I adopt towards the role of reason.

sentences of a theory may be true or false, and this implies, as we have explained in the foregoing chapter, that the objects referred to in *these* sentences exist and have the properties ascribed to them (if the sentence is true), or do not exist, or do not possess these properties (if the sentence is false). Clearly, we can agree that theories do not tell a 'literally true story' about the constitution of the world, but this does not commit us to rejecting that several sentences *in* theories are true or false, nor that this has consequences for our appreciation of the real constitution of the world. In other words, theories do not have to be things of the sort that are *true* or *false* in order for us to recognise that the entities they depict exist or do not exist: to put it differently, theories are proposed as hypothetical constructs *intentionally* directed towards the world (i.e. a domain of *referents*); and if we have good reasons for accepting a theory, for the same good reasons we must accept that their referents exist. For example, Wilfrid Sellars has very simply expressed this claim without mentioning the notion of truth: "To have good reasons for espousing a theory is ipso facto to have good reasons for saying that the entities postulated by the theory really exist."[13]

Therefore, in order not to make our considerations regarding scientific realism dependent upon any particular conception of the nature of scientific theories, we have preferred to postpone such a discussion, and this has the additional advantage of not making the issue of scientific realism *preliminarily* dependent on the problem of scientific truth. Hence, without anticipating here what we are going to present more systematically later regarding the nature of theories, let us simply indicate, by means of an analogy, how theories may be 'related' to truth and describe reality without being literally true.

Consider, for example, the map of a particular city. Such a map is obviously neither a sentence nor a set of sentences; but neither is it a real picture of the city, since it does not reproduce all the details of the city, but simply those features that correspond to a certain 'point of view' (the point of view of schematically indicating the disposition of streets and squares, and sometimes also the position of certain buildings that may also be indicated iconically). All these things are represented via symbols; and the map may be enriched by means of additional symbols and become more and more *informative*. Can we say that the map is *true*? This question is meaningless if taken *literally*, since the map is neither a sentence, nor a set of sentences. Thus its not being true is not a consequence of its being only 'approximate,' but of its being an entity of a sort to which 'truth' does not apply.[14] Even if we had an aerial photograph of the city, instead of a non-detailed map, this would in any case be only a schematic representation (it would be two-dimensional, it would present the city only from the point of view of certain optical features of its streets and buildings, and so on, and would again be something of which it is not fully pertinent to speak of truth, despite its being a much better 'approximation' than the map).

---

[13] Sellars (1963), p. 97.

[14] See what we have said in Sect. 4.6 regarding the appropriate use of the adjective "true."

However, we can give a *non-literal* sense to the idea that the map is, in a way, true. The *information* one can derive from the map may be true (or false), depending on the extent to which it can be translated into *propositions* once one has learned how to interpret the map's symbols. If I correctly derive from my reading of the map, for example, that the railway station is at the corner of 7th Avenue and 25th Street, and this is what is actually the case, the *sentence* which expresses this information is *true*; and if all the items of information provided in the map are of this sort, we can say that the map is *faithful* or *accurate* (which is an acceptable way of expressing the idea that it is true, though it would certainly remain unnatural to use this adjective when speaking of a map, and in general of a 'model' of whatever concrete reality). However, if our map turns out to have certain minor inaccuracies, we shall say that it is less faithful, but still rather faithful, and *therefore* that it remains more or less *reliable*. The same is the case if we discover that it is in part incomplete (e.g., if it is not up to date). This possibility of admitting of degrees, which is very natural regarding accuracy, faithfulness, and reliability, is much more problematic if applied to truth (it is from here that the well-known difficulties of the concept of verisimilitude emanate, though they can be skilfully treated, as has been done, e.g., by Niiniluoto).

Though elementary, this analogy tells us several useful things. Not only does it show how a type of representation which it is not appropriate to qualify as true (or false) may nevertheless be 'linked' to truth, and serve the aim of expressing truth, and of describing reality (how things are). This example also indicates how a 'correspondence theory' need not be a 'pictorial theory.' The map certainly *corresponds* to the city (albeit under the restricted point of view of the disposition of its streets, squares and buildings), but is far from being a point-to-point picture of the city: it is a sort of 'idealised' model in which only certain 'essential' features are captured (obviously "essential" means those that are encoded in the particular point of view that has directed the construction of the map). The correspondence in question reduces to facts of the following sort: if the map indicates that proceeding in a certain direction along street A we shall after two blocks cross street B, this *is actually the case*. Nothing *more* than this is implied in the said correspondence (many other features of the city are out of the reach of this correspondence, which is therefore not a mirror-image of the city), but nor is anything *less* implied. In other words, correspondence may be genuine and satisfactory, even if limited to very few aspects. The important thing is that it actually holds for *these* aspects. We have already noted that information can be derived from the map thanks to a symbolisation adopted in its construction, plus the corresponding decoding of the symbols when it is used. This is tantamount to saying that it is not possible to read the map without resorting to *sense* and *noemata*, because it is on the basis of these that the map has been designed to *represent* reality, and can therefore disclose what it actually represents. In short, nothing is a *representation* in itself, but can be a representation only in virtue of an underlying *intentionality*.

The example just discussed seems to suggest that the question of realism has indeed a direct connection with theories—and coincides with the problem of the ontological purport of theories—even though it might not depend on ascribing truth to theories. However, let us not link the question of realism to the ontological purport of theories either, for the practical reason that the concept of theory is not used univocally even in the philosophy of science: theories constitute a diversified fauna, and vary widely as to their order of complexity, their intended scope, their degree of concreteness or abstractness, and so on. Therefore, while for certain very simple and 'concrete' theories we may say that they are intended to express something rather similar to a literally true story of the world, for many others, which are very close to being abstract models with only a few links to empirical tests, it would simply be absurd to advance such a claim. Moreover, one should not overlook the fact that the purpose of theories is more than simply to depict an ontology. This has obvious consequences as regards the respective ontological commitments of different theories. As we have already discussed, many theories consciously postulate the existence of abstract objects (rigid bodies, perfect gases, adiabatic transformations, and so on), and in this sense they are far from intending to literally say how the world is. Yet, if one is able to recognise the proper status and role of idealisation, it becomes clear that they have a different purpose, namely that of *causally explaining* empirical laws in which their abstract objects are exemplified, and this fully justifies their relevance as to the knowledge of the 'concrete' world; their objects have a kind of intentional or *noematic* reality, and may at best be approximated by concrete objects which sufficiently accurately instantiate the properties these abstract objects encode.[15] Therefore, while we believe that the problem of realism has significant links with the question of truth, we do not believe that we need to relate this truth to theories in order to investigate this issue.

5.2.2 The Goals of Science

With what should we then link the question of realism, if we do not relate it to the truth of theories? We propose to relate it to the *intrinsic goal* of science. We know that philosophers disagree not only as regards what the goal of science is, but even whether science has a goal. For example, Arthur Fine claims that "'the aim of science' is a chimera, conjured up in response to misplaced hermeneuticism and fear of the irrational."[16] However, we believe that some precision can make the issue sufficiently clear and less controversial. Let us not confuse the said intrinsic goal with the *proposal* or the *intention* of those who practice, promote, or use science. The first is objective, the others are subjective and variable (one person can practice science for intellectual pleasure, another to earn a livelihood, another

[15] Remarks similar to those made here are expressed in Ellis (1985), where it is stressed, in particular, that in the case of theories characterised by the aim of providing causal explanations of phenomena, it is irrational not to admit the existence of the postulated causes.

[16] Fine (1986), p. 127.

for social prestige; and similar considerations also apply to groups or institutions promoting or using science). This is not peculiar to science; it concerns *any human activity whatever*. Indeed, every human activity is characterised primarily by its intrinsic *defining goal* (even material tools are often so defined), and one is said to perform that activity if one pursues that goal as 'immediate,' even when the said activity is performed in view of other goals with respect to which the immediate goal is only 'instrumental.'[17]

Also in the case of science, an intrinsic and defining goal *must* exist. However, it cannot be detected through a sociological inquiry. (Note that, in order to perform such an inquiry, we ought to decide first—and without any cogent reason—*who* is to be asked: professional scientists, agencies promoting science, and/or the general public, which has its ideas and expectations regarding science). This internal goal must be determined on the basis of a conceptual analysis which also takes into account the history of this concept and of its effective application to specific human activities. Undeniably, as a result of this analysis, we may conclude that the defining goal of science is that of obtaining *reliable knowledge* (though the difficulties begin when one wants to make precise the concepts of reliability and of knowledge itself).

Since knowledge cannot be knowledge of nothing, it is also necessarily implicit in this statement that this knowledge is reliable if it tells us 'how things are' in the different fields in which we intend to have knowledge (for example, in the case of the natural sciences we can say that their goal is to provide knowledge and understanding of the world in physicalistic terms). This (as a consequence of our previous inquiry) is tantamount to saying that the intrinsic goal of science is to offer reliable means for attaining *truth*. One can say that on this conclusion there is a general (though often only implicit) agreement, but the differences surface precisely regarding the reliability of the *asserted* truth, since, strictly speaking, it is obvious that truth is reliable 'as such.' But one can be doubtful about the *criteria* by means of which a certain sentence can be promoted from the status of being a *belief* to being *knowledge* (which is a typical epistemological issue): strict empiricists, for instance, maintain that true beliefs are attainable only concerning sensory perceptions, or the mere obtaining of states of affairs, whereas for other epistemologies truth is not reduced to a mere *ascertaining* of single states of affairs, but includes also their *understanding* and *explanation*. These differences, as we shall see, have a profound impact on the issue of realism in general and of

[17] For example, the activity of fishing is characterised by the intrinsic goal of trying to catch fish, but this does not mean that the *primary* intention of everyone who goes fishing is to catch fish. Professional fishermen catch fish primarily for the sake of their livelihood, while anglers do it for amusement. However, it remains true that these persons actually fish *only* to the extent that their *immediate purpose* coincides with the *defining goal* of fishing. Let us also note that the pursuit of this goal remains unaffected by whether the person performing this activity actually reaches the goal. (One is fishing independently of whether one succeeds in catching any fish, or even whether there are any fish to catch, provided one *intends* to catch fish.)

scientific realism in particular. Of course, this is not incompatible with the fact that other *proposals* may closely accompany this intrinsic goal (e.g., proposals to use this knowledge to dominate nature in a Baconian sense, or proposals to better regulate our expectations of future experiences, as different forms of instrumentalism would say).[18] After several centuries in which this defining goal of science had been believed to be more or less easily reachable, we found ourselves in the twentieth century in a situation in which serious doubts were raised regarding the possibility of attaining unquestionable truth even in those domains where it had been believed to reside, that is in the 'exact' sciences. In this situation it is normal that certain other goals may be considered *sufficient* for the practice of science, since we no longer feel confident that science can attain its intrinsic goal (the situation is more or less like that in which an angler enjoys fishing even in waters where he knows that the chance of making a good catch is slight).[19]

van Fraassen's criterion of empirical adequacy, as well as other criteria, seem to mean precisely this. But here we may take advantage of the distinction between definition and criterion, which we recalled in the discussion of truth (Sect. 4.6).

---

[18] Several authors (e.g. Laudan 1977, van Fraassen 1980, Putnam 1981, Popper 1983) have pointed out the difference between the intrinsic goal or aim of science and the contingent intentions of those who are concerned with science, and have admitted that science *has* a goal. However, they do not agree as to the specification of this goal. For Laudan, the aim of science is problem-solving in a very broad sense; for van Fraassen, it is the construction of empirically adequate theories; for Putnam, it is producing a rationally acceptable representation of the world; for Popper, it is providing satisfactory explanations of what strikes us as being in need of explanation, and so on. However, I believe that all these goals implicitly presuppose, or entail, the pursuit of truth. This is even explicitly admitted by Popper (1983, p. 132). van Fraassen and Putnam do not exclude truth, but rather restrict the sense in which it may apply to science. As to Laudan, it seems undeniable—unless we limit ourselves to a purely genetic or psychological account (i.e. to acknowledging that science arises out of the spontaneous curiosity of human beings)—that what is aimed at in science is not some vague 'solution,' but the *true* solution of problems. Therefore, we believe that Laudan's proposal is not at variance with, but rather provides a useful genetic complement to the characterisation of science as a truth-seeking human activity. These are the reasons for which we feel right in suggesting the search for truth to be the aim of science. But, of course, the soundness of this claim depends on several clarifications regarding the notion of truth, many of which we have already proposed, and some of which we shall present later.

A last remark: by making the search for truth the characteristic mark of science we do not want to intend that *whatever* truth-seeking activity is a science. We want truth be not only *ascertained*, but also *understood* (according to that synergy of empiricity and logos that has been included in the definition of science since antiquity). For this reason we shall use the expression "full truth" in the rest of the present subsection.

[19] Rescher portrays this situation well:

> We realise that there is a decisive difference between what science *accomplishes* and what it *endeavours* to do. The posture of scientific realism—at any rate of a duly qualified sort—is nevertheless built into the very goal-structure of science (Rescher 1982, p. 249).

We may recognise that, while the intrinsic goal (to attain full truth) may serve to *define* science (establish the meaning of the term "science"), empirical adequacy—as well as other features—may correctly serve as *criteria* for scientificity. They are *intrinsically necessary* for the attaining of the defining requirement in a certain specific case (in fact no empirically inadequate sentence or set of sentences could be true in empirical science), and may also be considered *pragmatically sufficient*, that is, sufficient for rationally justifying one's engaging in scientific inquiry.

However, even after having defended the thesis that reaching full truth is the intrinsic goal of science, we are not committed to the claim that this is the only genuine goal, but that other goals are only indirect and must be subordinated to it. Unfortunately, it is rather common (not only in philosophy) that, when one believes oneself to have grasped a valuable point, one not only *affirms* the importance of this point, but also tends to *deny* the importance of other approaches. The consequence is that, usually, one is right in what one affirms, and wrong in what one denies. It is much more reasonable to adopt, if possible, an attitude of "*et… et*," rather than "*aut… aut*," since different viewpoints are often complementary rather than incompatible. This general remark also applies to our problem. In many human activities and institutions, a *plurality* of goals may be needed for a correct definition of their nature, particularly when they are *complex* activities. Since science is undoubtedly a complex activity, it is reasonable to admit that it is characterised through *a system of interdependent goals*, among which attaining full truth occupies a prominent position. This is why we have given the title 'The *goals* of science' to this subsection. For our purposes, it is sufficient that attaining truth be recognised as *the fundamental* among these goals.[20]

5.2.3 Science's Link with Reality

In all of the foregoing discussion we have only spoken of *truth*, while the question of realism evidently makes direct allusion to *reality*. Only adhering to some kind of correspondence theory of truth allows one to consider the two issues as being equivalent. (In effect, most of the discussions of realism that take up the question of the truth of science, tacitly entertain a correspondence theory of truth, as well as a logical empiricist conception of science.) However, we would like to return to

[20] This methodological rule of 'non-exclusiveness' applies in particular to the *instrumentalist* view of science. Instrumentalism is right in what it affirms, namely, that scientific theories must (in a broad sense) be 'reliable' (be able to account for data, make predictions, permit useful applications and so on). This does not imply, however, that other goals should be denied, that is, be considered as illusory or misleading. In particular, the search for truth and for a true description of reality cannot be dismissed as illegitimate unless sound *reasons* for such claims are convincingly advanced. Therefore, the realist can accept the instrumentalist requests, but *in addition* she maintains that other goals can legitimately be pursued by scientific inquiry.

the more direct meaning of the concept of reality, and tackle the question of realism on a more ontologically qualified level (a level, by the way, which is often present in the literature). This means returning to the more primitive question, "Does science represent reality?" and trying to answer it after the necessary clarifications.

A moment's reflection shows that the above question splits into two different questions, depending on how it is understood. The first is: "Is it *reality* that science intends to represent?"; the second is: "Does science *succeed* in representing reality?" The first interpretation essentially amounts to recognising that science certainly has some kind of descriptive intention, but that it perhaps only describes its own constructions, and not reality. The second interpretation concedes that the intention of science is to describe an independent reality, but questions its *ability* to do so. While further subdivisions are certainly possible, we shall content ourselves with these two, for they suffice, in our opinion, to frame the main positions in the realism-anti-realism controversy. According to this choice, the realist position consists in advocating at least one of the following claims: (a) science attempts to represent a reality independent of science itself, and is committed to measuring itself on the basis of its success or failure in doing so; (b) what science states is an adequate representation of this reality 'as it is.'[21] Any anti-realist position challenges at least one of these claims. What is the question really at issue? Whether science *is* realist in its orientation, or whether, if it is, it is *warranted* in being so? If only the former is the real issue, then (b) is not relevant, but unfortunately a careful inspection of the literature shows that many scholars who became convinced that science is unable to attain such warranties, were induced to 'save the honour' of science by denying that science actually *does* cultivate the dream of representing an ontologically independent reality.

The first claim is clearly less demanding than the second, but we set ourselves the task of defending both (as regards both the empirical and the theoretical aspects of science).[22] However, we do not want to oversimplify the first question, as happens when it is understood as a particularised (and trivialised) version of the old controversy between realism and idealism, according to which what is at stake is the admission of a reality the existence of which does not depend on our knowledge of it. Indeed, as we have already noted, the (often confusing) proliferation of realisms and anti-realisms depends to a large extent on the fact that the

[21] These positions are sometimes labelled "referential realism" and "truth realism" respectively (e.g., in Harré 1986, pp. 65ff.), but what we have in mind perhaps does not fully coincide with these characterisations. Therefore we prefer to avoid using any kind of terminological classification.

[22] Since this task is rather complex, and sometimes demands reference to certain issues which have already been tackled in other parts of this study, we shall once again take up some of these points in the present discussion, rather than simply refer back to them. This will sometimes involve a few repetitions, but this might not be too high a price if compared with the (practical) advantage of making this chapter more self-contained, and the (conceptual) advantage of stressing more explicitly the links between these former positions and the general question of realism.

term "realism" already had a circulation in philosophical discourse long before the question of *scientific* realism was born. It is for this reason that we offered an historical survey of these different meanings in the preceding section, not just as an erudite digression, but because several elements of such old conceptualisation are still present in the pages of certain contemporary philosophers of science.

Such an oversimplification is evident, for example, in the following statement by Mario Bunge (who, on the other hand, has defended realism with much better arguments on several other occasions): "Philosophical realism boils down to the thesis that nature exists even if it is neither perceived nor conceived."[23] Hardly anyone today would want to deny such a thesis. In this respect everyone is a realist, and realism would be reduced to a fully uninteresting platitude (perhaps only Mach dared to affirm explicitly, on a few occasions, that "bodies do not produce sensations, but complexes of elements (complexes of sensations) make up bodies."[24] Even less significant would be the defence of realism based on a victory over idealism, where idealism is depicted of as having the characterising thesis that "The world is our dream," as Popper represents idealism more than once.[25] Therefore, despite Popper's often and emphatically declaring himself to be a realist, his epistemology is hardly recognisable as a realist *philosophy* of science, since he openly admits that he has no stringent arguments for refuting idealism, even on his poor conception of it:

> From the irrefutability of idealism follows the non-demonstrability of realism, and vice versa. Both theories are non-demonstrable (and therefore synthetic) and also irrefutable: they are 'metaphysical.'
>
> But there is an all-important difference between them. Metaphysical idealism is false, and metaphysical realism is true. We do not, of course, 'know' this, in the sense in which we may know that $2 + 3 = 5$; that is to say, we do not know it in the sense of demonstrable knowledge. We also do not know it in the sense of testable 'scientific knowledge.' But this does not mean that our knowledge is unreasoned, or unreasonable. On the contrary, there is no factual knowledge which is supported by more or by stronger (even though inconclusive) arguments.[26]

This passage indicates not only that Popper is involved in a very general and not very promising debate regarding the controversy between realism and idealism, but also that his very admissions do not allow for great expectations regarding that

[23] Bunge (1989), p. 130.

[24] Mach (1872).

[25] See, for instance, Popper (1972), p. 38; (1983), p. 80.

[26] Popper (1983), pp. 82–83. The strongest of such not irrefutable arguments has been sketched by Popper e.g. in his *Objective Knowledge*. They are: the agreement of realism with common sense and with a general scientific mentality, the descriptive and referential characteristic of language, the avoidance of postulating the absurd idea that we create that which we perceive and know (Popper 1973, pp. 39–42).

which is, after all, the real issue of *scientific* realism, that is, a *justification* of the claim that science provides knowledge of reality.[27]

Let us say, on the other hand, that—conversely—certain criticisms of realism rely upon an equally unjustified portrayal of its characteristic claims. An example of such is the definition of scientific realism as the claim that theories constitute a "literally true story of what the world is like," which we have found in van Fraassen. This 'story' might well not be "the correct statement of scientific realism" he suggests it to be, in spite of his honest attempt to give realism a "minimal" formulation so that his criticism could be immune from the risk of "charging at windmills."[28] The difficulty in portraying realist positions in such a controversial context seems to appear even from Putnam's definition of the "externalist perspective," to which he wants to oppose his own "internalist perspective":

> One of these perspectives is the perspective of metaphysical realism. On this perspective, the world consists of some fixed totality of mind-independent objects. There is exactly one true and complete description of 'the way the world is.' Truth involves some form of correspondence relation between words or thought-signs and external things and sets of things. I call this perspective the *externalist* perspective, because its favorite point of view is a God's Eye point of view.[29]

The least that can be said is that this picture of a "metaphysical realism" (for which no supporters or precise references are mentioned) is so vague that probably no philosopher could be found to have subscribed to it in this extremist form.

As a matter of fact, anti-realists hardly contest the existence of a mind-independent world. But according to them science might well construct its *own* kind of reality, and know only this reality, leaving untouched the independently existing reality. On this view a reality independent of science would be alien to science, not only ontologically (it is not produced by science), but also cognitively (it is not the kind of reality that science has to do with).

This view clearly raises some non-trivial points. First, it stresses that the main problem does not concern the *existence* of things, but the possibility of *knowing* them (one aspect of epistemological realism). However, the problem of knowing things does not simply reduce to the problem of the 'adequacy' of our thinking vis-à-vis existing things—i.e. whether what science states is an adequate description of reality 'as it is' (claim (b), above)—since anti-realists of this type claim that science constructs its own 'world' or reality. Therefore, the genuinely ontological

---

[27] For a perhaps too severe, but essentially correct, appraisal of Popper's doctrine on this point, see Keuth (1978); also, Popper's realism is qualified—not illegitimately—as *fiduciary* in Harré (1986). In Buzzoni (1982) it is shown that the lack of an operational criterion frustrates Popper's efforts to provide the foundation of a consistent realism.

[28] van Fraassen (1980, p. 8). In such a way, van Fraassen expresses the methodologically sound requirement of avoiding that which Harré calls "the fallacy of high redefinition," that is, "the move by which some established metascientific concept… which has a well-understood use in scientific discourse, is redefined in such a way that there are no conditions under which it could reasonably be applied" (Harré 1986, p. 38).

[29] Putnam (1981), p. 49.

questions—concerning what is the status of this 'world,' and what is its relation to the 'external world' independent of science—do not vanish.

We must also note a circumstance which is, in a way, curious. Anti-realists not only do not usually deny the pure and simple existence of a science-independent reality, but often also concede that science actually knows this reality as far as *empirically immediately accessible* entities are concerned (we can qualify this as *realism on the empirical level*). Through this move, anti-realism (or at least a good deal of its articulations) turns out to be nothing more than a form of strict empiricism (that excludes the cognitive purport of 'theorising') endowed with an ontological endorsement of the epistemological requirements of empiricism itself. In fact, for many authors the "question of realism" only concerns the legitimacy of maintaining the existence of the 'theoretical entities' postulated by science, while such authors do not question that 'observable entities' have a real existence. We have already said something on this issue in the preceding section, and shall return to it again later.

We have been driven by this last consideration to formulation (b) of the issue, i.e. to the claim that science is able to know reality 'as it is' (of course, not in the sense that a single science or even science globally considered is able to know *the whole* of reality, but only the part of reality it intends to investigate). This means that, even if we admit that reality has an independent existence, there is no guarantee that we will be able to attribute *well determined properties* to it. This problem is clearly reminiscent of Kant's distinction between phenomena and things-in-themselves; and the separation of these two problems is already expressed in Kant's famous claim that he saw himself to be at the same time both an empirical realist and a transcendental idealist, his empirical realism implying that he admitted as unproblematic the *existence* of both phenomena and noumena, and his idealism implying that on the basis of our understanding we attribute properties to the sense-data that we acquire in sensory intuition, as we have already discussed in the preceding section. In the contemporary context, this problem may receive formulations of different degrees of complexity, which we have also explored in the foregoing sections and shall not review here, being content to recall that most of the difficulties surfacing here are the consequence of epistemological dualism, and may be settled once this doctrine has been duly criticised and overcome.

5.3 The 'Linguistic Turn' and the Question of Realism

Recent decades have seen in the philosophy of science the diffusion of a strong opposition to realism, an opposition which can be considered a development of empiricist-analytic and Popperian epistemologies (which are essentially more cognate than they are often believed to be). This can be explained rather easily. Contrary to what might be thought on the basis of the concepts used in the foregoing discussion (concepts which pertain to general epistemology), the more recent challenge to realism is no longer based on epistemology, but on the philosophy of language, and this challenge only subsequently assumed certain more familiar features within epistemology. All this is well in keeping with the 'linguistic turn,' which is one of the most characteristic marks of contemporary philosophy, and has deeply affected the philosophy of science.[30]

5.3.1 The New Face of Anti-realism

Any philosophy for which all questions reduce to an analysis of language is bound to be a form of anti-realism. In this form, anti-realism is no longer a means for rejuvenating phenomenalist positions of a more or less Kantian flavour. Rather, it insists on the (real or presumed) impossibility of our thinking, including scientific thinking, succeeding in representing reality, because this would imply the possibility of saying something outside the language. Such an anti-realism amounts to an extreme exploitation of the semantic thesis that the meaning of terms is completely dependent on the whole context within which they are uttered or written (a thesis which is often called *semantic holism*, and whose best known representative is Quine), a thesis that is (wrongly) considered equivalent to the claim that a language 'constructs' its own *objects* (understood in a *referential* sense).

The first consequence, as we have already seen (e.g. in Sect. 4.3), has been the thesis of the 'incommensurability' of scientific theories (though the incommensurability thesis does not necessarily presuppose a philosophy for which all questions reduce to an analysis of language, and in fact can even be seen as indirectly being a criticism of just such an approach, as it will be seen later). Moreover, this thesis was coupled with another: the more or less explicit presupposition that 'observational terms' established contact with reality, while it remained an open question whether the same could be said of 'theoretical terms.' This is obviously the empiricist conception of realism, but it must be noted that the

---

[30] The 'linguistic turn' may be considered as the expansion of the thesis of the 'impossibility of transcending language' according to which any investigation that apparently regards a certain subject-matter cannot avoid being an investigation of the discourse or discourses in which such a subject-matter is described or treated. Therefore, any reality whatever is always given in a language and, to express the point using a famous claim of Wittgenstein, "the limits of my language are the limits of the world." This thesis is very similar to the fundamental thesis of the 'impossibility of transcending thought' defended by idealists, according to which it is impossible to affirm something 'external' to thought, since by this very affirmation one would include this something in thought. We have discussed this point in Agazzi (1989), our discussion justifying the use of the expression "linguistic idealism" for qualifying the linguistic turn; and we have noted that this position shares with idealism not only the correct claim that reality cannot be 'separated' from language (or thought), but also the mistaken idea that reality can be 'reduced' to language (or thought). An inability to separate does not imply the lack of a distinction. This is why the anti-realism based on this reduction or 'identification' must be held to be as naive as the anti-realism based on a similar 'identification' made by idealists. Several parts of the quoted paper are reproduced in the present section.

combination of the linguistic approach to science with this empiricist conception of realism constituted the general paradigm or framework of the philosophy of science of logical empiricism, and of its development over several decades. This is why many discussions on scientific realism that were tacitly inscribed within this framework were essentially question-begging, and the fragility of this adopted framework emerged from the unsatisfactory results of the abundant analytic work done under its auspices. Indeed, once the possibility of neatly discriminating observational from theoretical terms has disappeared (since all terms are to be in a certain measure theoretical, according to the semantic holism), one cannot see what safe link with reality is guaranteed (if this link is solely granted by observational terms). Moreover, if it is true that one term may have a different meaning in two different theories, it seems unavoidable that the hypothetical reality to which such a term might refer would be different in the two cases (but this is a consequence of having conflated sense with reference, as we have discussed in Sects. 4.2 and 4.3).

This leads to two equally paradoxical consequences: either one admits that each theory 'creates' its own reality (which eliminates the idea of realism as the assertion of a reality which exists in itself independently of the science which investigates it), or one admits that realities can 'multiply' indefinitely and become the object of different theories. The second consequence would equally frustrate realist aspirations, because it would not only conflict with the idea of the existence of one reality, but would also leave us with the impossibility of knowing 'which' reality we are talking about at any one time.[31]

5.3.2 Realism and Referentiality

Within the 'linguistic' perspective the question of scientific realism can be restated as follows: the realist position maintains that scientific discourse has *actual referents*, Frege (as we have already seen in previous sections) in his essay 'On Sense and Reference' stressed the difference between the sense of a term (which is a content of thought that is 'meant' by the term), and its *referent* (which is an object constituting 'that about which' the sense in question is thought or expressed). A similar distinction has, however, been left unused just by those who, for a lengthy period of time, have occupied a pre-eminent position in elaborating theories of meaning, that is, by the mathematical logicians. As far as the interpretation of formal calculi is concerned, they have quickly embraced an *extensionalist* semantics according to which the meaning of a term is precisely the set of its

[31] Here again we refer to the analyses of these positions provided in foregoing sections, from which it has also appeared how these different claims are not equivalent, nor really logically entailed in the way sketched here. Moreover, one could also add that arguments of a sociological nature have joined the linguistic arguments just mentioned, especially after the publication of Kuhn (1962). We shall consider these additional elements later.

referents. Such an identification between sense and reference has not been made out of ignorance, but is supported by the practical necessity of conforming to the general 'philosophy' of logical formalism, according to which the symbols of a formal system have not and must not have any sense.[32] (Of course, one can ask what it was about this view that it should become an essential constituent of logical empiricism, but we cannot here make such an historical digression as would be required to answer this question.)

However, we have already noted that this extensional semantics, which seems to be the semantics most concerned to obtain referents without taking into consideration the abstract world of concepts, shows all its weakness precisely when it is used in formalising theories of empirical science, that is, theories which are *intended* to speak of a world 'external' to the language in which they are couched. The failures of extensional semantics in this field (which remain despite the many articles which continue to be published in this area trying to patch up this or that point) are a clear symptom of this essential fact: not only is it true that sense and reference cannot be equated, but also that neither of them can be eliminated, and that access to reference is guided by sense. Indeed, sense and reference must both be safeguarded if a discourse is to keep all its fundamental characteristics intact. By eliminating sense one would obtain a discourse which 'has nothing to say,' by eliminating reference one would obtain a discourse 'about nothing.' A full-fledged discourse involves the intention of saying something about something.[33]

---

[32] See Sect. 4.1 for details.

[33] As a matter of fact, the philosophy of science inspired by logical empiricism, and continued within the analytical tradition, has been characterised by a strong 'syntactic' approach that, in part, reflects the spirit of the 'formalistic' view that became dominant in mathematics and, in addition, that allowed for the application of the sophisticated tools of mathematical logic (with the skill and sense of pride entailed by the ability to master such complicated techniques). Although some interesting results were obtained using such an approach, it seems undeniable that it proved much less fruitful than its complicated machinery seemed to promise, and that it even distracted attention from philosophically central issues, in favour of rather artificial questions. This is why, in the present work, we did not follow this syntactic approach.

We are not alone in this evaluation. van Fraassen, e.g., declares: "Perhaps the worst consequence of the syntactic approach was the way it focused attention on philosophically irrelevant technical questions. It is hard not to conclude that those discussions of axiomatisability in restricted vocabularies, 'theoretical terms,' Craig's theorem, 'reduction sentences,' 'empirical language,' Ramsey and Carnap sentences, were one and all off the mark—solutions to purely self-generated problems and philosophically irrelevant. The main lesson of twentieth-century philosophy of science may well be this: no concept which is essentially language-dependent has any philosophical importance at all" (van Fraassen 1980, p. 56). We essentially agree with this judgment, and especially with its conclusion, though we think that a more balanced evaluation should be made of the formalistic approach taken as a whole (see Agazzi 1990). In any case, we want to stress that the mentioned work of van Fraassen (1980) really meant a decisive overcoming of the strongest limitations of the logical-empiricist paradigm, by introducing (with his 'constructive empiricism' a significant appreciation of the semantic dimension. (Of course, the introduction of a semantic approach in the analysis of the structure of scientific theories had already known a significant start at the end of the 1960s and beginning of the 1970s of the twentieth century, in the works of Suppes and Sneed, later developed by Stegmüller and his

At this point it may be clear why we have suggested identifying (within the context of the philosophy of language) the position of scientific realism with that which attributes referents to scientific language. On the one hand, we must say that *without realism* one cannot give the *referents* of a language, and this precisely because a referent is an extralinguistic object to which the particular language under consideration 'refers' to as other than itself. In the case of scientific language, therefore, if one does not admit the existence of a reality different from the pure 'language game' constituted by that language, one cannot attribute to it the ability to refer to something, but only, at most, the ability to proceed according to the rules of the game internal to the language itself. Furthermore, if we interpret scientific language as simply a language game that is internally coherent and conducted according to rules accepted by a given community of speakers, but *without referential* purposes or possibilities, then we shall never be able to hold a realist position towards science, because we shall already have accepted that it does not intend to talk about a reality distinct from its own language. These two theses, which imply each other, are therefore logically equivalent, and we are justified in saying that the thesis of the referentiality of scientific language is the expression of the thesis of scientific realism when one moves from the epistemological level to that of the philosophy of language.

Some may be dissatisfied with these arguments, suggesting that a true realist would not be happy simply to maintain that scientific language 'refers to' something different from itself. What more might be demanded is that this something be reality and not, for instance, pure illusion, a mere intellectual construction, or even just the private world of one's sense perceptions. The objection has weight, especially because it invites us to specify what one can and cannot hope to establish about scientific realism by remaining *within* the philosophy of language. Clearly, within this philosophy, one will not be able to say a lot about the 'kind of reality' to which the referents belong, and this for the good reason that this is not a linguistic problem. We have therefore no difficulty in acknowledging that the question of scientific realism is not *wholly* soluble within the philosophy of language. Nevertheless, the foundation of the referentiality of scientific language is a

(Footnote 33 continued)
collaborators of the so-called 'structuralist school,' of which we have already had the opportunity of speaking in Sect. 3.2 of this work. One can also mention Agazzi (1976) as a contribution in this direction.) Yet van Fraassen still remained partially prisoner of the linguistic approach and especially of radical empiricism. An overcoming of the first limitation is attained in his other fundamental work (van Fraassen 2008), where the notion of 'scientific representation' entails abandoning the previous approach to philosophy of science in terms of theories and their truth (where truth was replaced by the notion of 'empirical adequacy'), and the new perspective of an 'empiricist structuralism' is presented, in which theories are conceived as systems of models and representations to which the predicate 'true' does not apply. The rather curious consequence is that, whereas van Fraassen's first view attributed to science at least a rough empirical realism, the second view is much more radically anti-realist, since there is no possibility of finding, in the different levels of models proposed by the author, a 'bottom level' of which it could be said that it 'represents' the external world.

necessary condition for the establishment of the thesis of realism (since it provides a great number of essential ingredients for this foundation), and this fact justifies the attention that we would now like to devote to it.[34]

5.3.3 Symptoms of Referentiality

It is difficult to deny that the language of science is intended to be referential. One need only consider the attitude of the majority of practising scientists, who share what has been called "spontaneous realism." On the basis of this, they *intend*, first, to devote themselves to describing and understanding some sector of reality (and not simply to creating mere intellectual constructions or to developing some complex language game); second, they *believe* themselves to be doing something of this sort; finally, very many of them *believe* that science can succeed in this enterprise (others may be more sceptical and occupy intermediate positions). Of course, what scientists say, think and believe is not in itself sufficient to determine what science really is, but it cannot be considered irrelevant either.[35] We can also add (last but not least) that the general attitude modern society has gradually developed concerning science is precisely that science is a trustworthy and reliable (probably even the most trustworthy and reliable) tool produced by humankind for *knowing and understanding reality* and operating with it. Therefore, it is uncontentious (a) that science has a referential *intention*; it is however something else again to assert (b) that it *succeeds* in constructing a referential discourse; and finally still another to clarify (c) the *type of reference* which scientific discourse can have. This, of course, is not yet sufficient for knowledge and understanding, since it remains to be investigated 'how good' a discourse it is regarding an independent reality, and we shall come soon to this additional issue.

---

[34] What we have said explains why, for example, a phenomenalist could also accept our thesis and say that scientific discourse is not a simple language game, that it must have referents, but that these are only phenomena, and not things as they *really are*, or things in themselves. The essence of the views of van Fraassen (1980) and Putnam, for example, could be included in this line of thought. This means that they actually share some of the features of realists, in the sense explained here. The fact that they do not accept this qualification (at least in full) clearly depends on certain additional requirements they impute to realism, to which they are not prepared to subscribe. The examination of these requirements will occupy us soon, but what we have seen is already significant because it explains why the most recent debates concerning realism have increasingly abandoned the orbit of philosophy of language in which this discussion was couched for many years, and take into account more traditional non-linguistic factors, such as scientific progress, the success of science, the aims of theories, and so on.

[35] As expressed by Shapere, whose statement we fully endorse: "According to a widely cited slogan, the philosopher of science must pay attention to what scientists 'do' rather than to what they say. I believe, however, that we must attend to both, though of course with a great deal of critical awareness" (Shapere 1984, pp. XXXVI–XXXVII).

We should now like to touch briefly on the second point, contenting ourselves with an elementary, but fundamental, remark: it is one of the most characteristic and uncontroversial marks of the *empirical* sciences that *certain assertions cannot be accepted as true even though they are endowed with meaning*. This concerns in particular those assertions which are rejected by empirical evidence. Given that we are dealing with assertions endowed with *meaning*, one cannot say that they are rejected because they do not correspond to the *rules of the language game* of the particular science in which they occur, but because there is a *non-linguistic* condition which prevents their acceptance.

One could object that in this case too the paradigm of the language game is present, because a rule common to all language games which characterise the experimental sciences is precisely to establish that all sentences which describe direct experimental results can, or even must, be accepted, while sentences which are irreconcilable with sentences describing such results must be rejected. Despite appearances, this objection is very weak because it ignores the fact that a rule of this kind rests on a *non-linguistic* condition, such as that of taking into account operations and observations of a concrete nature which concern the sphere of 'doing something,' rather than that of 'saying something.'

In the case of the rule of accepting sentences which describe experimental results, and of rejecting sentences which contradict them, it would not only be naive but even misleading to ignore that this rule has been introduced in science because experimental results have always been credited with the role of being the direct view of 'reality' with which science concerns itself. If we want to describe the situation as it is, we ought therefore to say: if there exists a reality which is endowed with its own structure, it is clearly not possible 'to say anything and everything' about it, because certain propositions which refer to it will turn out to be false, as they indicate *what it is not*. Hence, the fact that in the experimental sciences certain propositions can be forbidden—because certain conditions of referentiality (experimental results) are opposed to them—is already an important symptom of the fact that these propositions speak of reality.

5.3.4 Semantic and Apophantic Discourse

The arguments so far presented bring us back to the distinction already made in this work (see especially Sect. 4.4) between semantic discourse and apophantic or declarative discourse. The former limits itself to 'signifying,' while the latter 'asserts' (i.e., affirms or denies). The establishment of the sense of terms does not imply asserting or denying in a literal sense, but rather a more general 'saying,' to which in particular the dimension of truth and falsity is alien. The semantic discourse, too, uses declaratory or descriptive sentences, for instance in definitions. But why do we then say that definitions are not true or false, despite their consisting of descriptive sentences? This question has given rise to much discussion in the past, and has in particular added heat to the disputes on the difference between nominal and real definitions that we have already considered.

The only way to escape unambiguously from misunderstandings appears to be this: the semantic discourse is neither true nor false (and within it definitions in particular are neither true nor false) because it is *non-referential.* As soon as we give it a referential direction, it transforms itself into apophantic or declarative discourse (this is the case with what are termed 'real definitions,' which are sentences purporting to be true of real objects, attributing particular properties to them). As is clear, it is not the form, but the *intention* of a discourse that makes it semantic or apophantic. In the case of apophantic discourse, its intention is to state that some state of affairs (expressed semantically by the meaning of a certain statement) obtains, or is the case. But if the discourse is, for example, in the form of a question, its intention is not to 'state that,' but to 'inquire whether' the state of affairs obtains, and the discourse is therefore no longer apophantic, though it must keep a semantic dimension (i.e. have a meaning) which 'serves' not an apophantic, but, let us say, an 'inquisitive' purpose.

Our conclusions allow us now to understand clearly what it would mean to deny referential import to empirical sentences and theories. It would mean reducing them to the level of semantic discourse, to solely being instruments for establishing meaning. Someone might perhaps find this perspective acceptable, but it has the serious fault of not explaining the difference between the empirical and purely formal sciences. Granting that we can legitimately say regarding the latter (albeit with caution) that they contextually give sense to their own terms, we cannot say the same of the empirical sciences, because in them the presence of empirical data introduces something that spills over the boundaries of the pure and simple linguistic context.[36]

For this reason we must say that the empirical sciences appear as discourses of an apophantic or declarative nature. The establishment of an apophantic discourse is characterised by the fact that reference emerges together with sense, and furthermore does so in such a way as not to be independent of sense. In fact, as we have repeated several times, the search for referents requires a non-linguistic activity which in many cases (especially in that of the sciences) is even of a clearly 'practical' type, such as instrument manipulation, observation in suitably created conditions, and so on. This activity therefore consists in *exploring the world,* and not in *exploring language*. However, it is no less true that this exploration of the world in search of referents takes place on the basis of sense; otherwise, we would not be able to *recognise* the referent when we meet it. Here is the solution to the paradox already stated by Plato, according to which one can only know what one knows already. The point really is that we know a referent only because in being *acquainted* with it we recognise in it the attributes expressed in the sense with which we began our search, and thereby know *that* it has those attributes. But the referent was not already known to us before we met it (we did not know *that* it had

[36] See the introductory considerations of Agazzi (1976) for more details.

those attributes until we were *acquainted* with it). When the referent is traced in this way, several properties can be 'asserted or denied' about it, and in this way true or false sentences can be produced. The apophantic logos is therefore that in which the *notion of truth*, directly linked to that of reference, is established, as we have already discussed.

5.3.5 The Excessive Claims of Contextualism

For the sake of brevity, the already mentioned approach, according to which each term adopts a sense which is totally determined by the context within which it is set, is here called *contextualism*. From this it follows that, for contextualism, any term set in different scientific theories will have a different sense in each of them. The result, as we have already seen, is the thesis of the incommensurability of scientific theories, the non-existence of true progress in science, and the impossibility of referring theories to a common reality. These consequences are unavoidable only on the view inspired by the philosophy of language approach we are considering here, and are symptoms of its weakness, whereas they are avoidable on other approaches that we will consider later. For the moment let us note that, faced with this situation, it is more than legitimate to ask why the comparison of theories should take place on the basis of their senses, and not instead on that of their referents. After all, it was the traditional conviction of scientists and epistemologists that two rival theories, which spoke about the same reality, could be compared in the sense that one could be found to be false and the other true about that reality, if they were asserting not only *different* things, but *incompatible* things. This remark is very important because it expresses the idea that one theory can be better than another even though its sense cannot be 'compared' with that of the other (the two senses are simply 'different'). This is so if, on the basis of experimental test, it can be presumed to be true, while the other, on the same basis, must be declared false.[37] Why can we not continue to adopt a similar view? One can say, because it is a point of view founded on realism, which we today reject. This answer, however, clearly cannot serve as a reason for *justifying* the rejection of realism. Different reasons have to be offered which, in particular, hit the nodal point of what we have called the traditional discussion, that is, the thesis that theories with different senses can deal with the *same* referents.[38]

The relation between the determinacy of sense and the identifiability of referents is not as strict as might appear at first sight. In the first place, some indeterminacy of sense is compatible with the possibility of identifying referents. One

[37] Here again we speak of true or false theories, following a common way of speaking, without recalling what we have already said in Sect. 5.2.1 about the legitimacy of this extendend use of the notion of truth.

[38] This is also a central point in Dilworth (2008).

can successfully identify a referent even if only *some* of the semantic features are determined, as long as those features are the ones *the linguistic community has agreed to use* to identify the referent. For instance, whales were once classified as fish and are today classified as mammals, so that the sense of the term "whale" has undoubtedly changed. Should we then say that the referents too have changed, that is, that what we call whales today are not the same animals as those we used to call whales? Not at all; in fact, there exist a sufficient number of properties of whales (for instance, their morphological properties) which enable us to identify whales, and which remain the same even today. Similarly, in different theories of empirical science groups of characteristics can exist which remain unchanged even within two different contexts, and can be used to trace the referents and to recognise that they are the same.

From what has been said several times above, we can infer that these characteristics will be the ones linked to empirical evidence or, more precisely, the ones which are connected with determining the operations (e.g. of measurement) which are *materially the same* in both theories. When this happens, we shall be able to say that the referents are the same; and we could then proceed to the comparison, even without denying that all the terms of the two theories, because of the influence of their different theoretical contexts, receive more or less different senses. In other words, a 'referential part' of the sense of certain terms exists which is not sensitive to contextual variance because it is linked with that *extra linguistic* and operational component which characterises empirical science.[39]

It may even be that *different* terms with rather different senses end up denoting the same referents, despite their being located in different theories. One can think of Dalton's "atom" and of Avogadro's "molecule." Despite being different terms they are characterised in two different theories through a series of common properties, among which are some of an experimental nature, such as that of being the smallest particles in a gas the combination of which possesses all the chemical

[39] This point has already been presented in greater detail in Sect. 3.3, and corresponds to a 'minimal' condition of stability, which is sufficient for our purposes. However we would at least like to mention a more elaborate doctrine developed in several papers by Shapere which convincingly accounts for the historical change of meaning (and even of reference) of scientific concepts, without entailing incommensurability, provided one considers the actual *reasons* that have determined the piecemeal change in question:

> The idea of 'chain-of-reasoning-connections' disposes of the problems of "incommensurability" which have been the source of so many relativistic and sceptical views of science during the last two decades.... Later ideas in science are often *rational descendants* of earlier ones, even if they abandon a great deal, or even all, of what was in those earlier ideas. (Shapere 1984, pp. XXXVII–XXXIII)

The requirement of the stability of reference has been notoriously investigated in the theories of 'direct reference' and 'causal theories of reference' elaborated by Putnam (1975) and Kripke (1980). Since they are well known and, moreover, are of a general character and less related to the problem of 'scientific' realism, we believe we can dispense with presenting them here.

properties of that gas. This can be enough to maintain that the two theories concern the same referents, while attributing partially different characteristics to them, up to the point that one theory can be false and the other true with respect to these. (As regards our example, Dalton asserted of his atoms the erroneous thesis that they were indivisible using the chemical techniques of the time, while Avogadro did not maintain this.) The moral of this story is, therefore, that even in those cases in which there is only an imperfect possibility of translation of one concept into another (Avogadro's 'molecule' only imperfectly 'translates' Dalton's 'atom'), the possibility of finding a common referent may not be compromised.[40]

This same thesis can also be reiterated by inverse reasoning, that is, by showing that *scrutability*—to use Quine's expression—does not necessarily imply "determinacy of translation," that is, the complete homogeneity of sense in two different contexts. In order to illustrate this, instead of the example of a common name, such as "whale," "atom," "molecule," we shall use a proper name, the referent of which is a well-determined individual, rather than a class of individuals (that is, an extension). This also has the purpose of emphasising how there is no substantial difference between proper names and common names from our present point of view (i.e., referents can be individuals or natural kinds, and in science they are in particular natural kinds).

For example, for a certain number of people uninformed about the history of philosophy, the proper name "René Descartes" might be associated with two senses, such as the founding of analytical geometry and dying in Stockholm on 11 February 1650 at the court of Queen Kristina of Sweden. For another group of people with knowledge of philosophy, but without mathematical knowledge, the same term might be associated with the sense of being the author of the *Discourse on Method* and of having died in Stockholm on 11 February 1650 at the court of Queen Kristina of Sweden. The two senses are clearly different, but the referent can be univocally determined through the description of the date and of the circumstances of death, up to the point that we can very well think that two interlocutors are in perfect agreement in attributing to the name *René Descartes* the same referent, despite each attributing to it a different sense. Identity of referent does not therefore imply identity of sense, even if it can be a basis for the search for such an additional common ground (in our example the two interlocutors can reach the point of exchanging their respective information about Descartes, and so in the end associating with the name the same—enlarged—sense as well as the same reference). The senses, however, need not even overlap to have commonality of reference; it is enough to mention here that this case occurs when "complementary" descriptions, in the sense of Bohr's principle of complementarity, are provided about the same referents within different scientific theories.

The two interlocutors have been able to determine that the object of their references was the same despite the difference of sense because, within the two

[40] For a good analysis of this topic from the point of view of the philosophy of language, see Smith (1981), pp. 106ff.

significant contexts, there was at least one common aspect of the respective senses which was *independent of the senses of other aspects* (i.e., the date and the circumstances of Descartes' death do not depend, in this case, on the fact that he was a philosopher rather than mathematician). Only if there existed a reason to consider the two characteristics *incompatible*, could we have doubts about the identity of the referent—but even in this case there exists a double possibility of a way out.

For example, it could be the case that, after deeper examination, we should conclude that, because of an extraordinary coincidence, two men have existed with the same name, and died on the same day under the same conditions. But it is much more likely that one would ultimately consider incorrect the attribution to our referent of either the one or the other of the two allegedly incompatible properties, as might be revealed for instance in admitting that the referent of "René Descartes" either did not write the *Discourse* or did not invent analytical geometry. As one can see, far from being inscrutable, the referent is generally delimited with reasonable confidence. In our example this is due to the rather unproblematic situation of determining the date and circumstances of the death of a famous person. In more usual cases the referent is captured through a restricted portion of the meaning that is related with familiar operations, and it is the referent itself which is firmly held onto up to the point that the referent guides the decision as to whether senses which are attributed to it are admissible or not.

Certainly, as has been seen in this case too, it must be so that *certain* unambiguously shared senses exist, so that the referent can be identified. In our fictitious example, these were the senses of the common language terms used for describing the date and circumstances of the death of Descartes, and in science they are the senses available within a certain community in order to understand what kind of instruments, operations, and readings can be intersubjectively accepted. Once the referent has been found, it is the referent itself which guides the choice of other predicates. In the case of empirical theories, we repeat, the best candidates to be employed as providing a sense *independent* of the context and suitable for guiding the search for referents are the operational predicates linked to the execution of empirical controls.[41]

---

[41] It may be useful to note that there is a difference between our position and that presented by Dilworth (2008): he suggests that sameness of operations is only a *criterion* for sameness of referent. Sameness of referent—according to him—is *determined* by the intentions of the persons applying the conceptual scheme or theory in question. This not negligible difference depends on the fact that Dilworth strictly links *reference* with the personal *intention* of the subject using a term, and does not consider it (as we do) to be a semantic constituent of the meaning of a concept. This is why his "perspectivist" conception of science, though having many points of contact with our view, has certain points of difference as well. The main reason for this is probably the fact that we are investigating scientific *objectivity* and, therefore, leave out of consideration the subjective factors. In particular, Dilworth can easily admit that incompatible attributes can be assigned to the same referent (because two different subjects can 'refer' to the same object though attributing it incompatible properties), whereas for us the referent is intersubjectively identified within a certain linguistic community by means of a certain term, and for this reason cannot be characterised by incompatible attributes.

Let us conclude. It is true that the sense of a term always depends on the conceptual context within which the term is used, provided that this dependence is understood in a genuine semantic sense and not as being algorithmic. This is because not all terms (in an empirical context) are *logically interconnected* in the sense of being *interdefined*. Therefore, in certain cases, terms which one could call 'free' can reappear in other contexts while continuing to remain free. There they can, thanks to their liberty, guide the search for common referents of the discourse in which they occur. For example, the term "light beam" occurs with two different senses in a corpuscular and in a wave theory of light. Yet it keeps a certain independence from these contexts, and scientists belonging to the two rival schools can use light beams for common experiments of reflection, refraction, diffraction, and so on. As a result, one can accept the 'contextualist' approach to meaning without reaching the extremist consequences to which, for reasons that are not essential to its internal coherence, this approach has itself led. Ultimately, it is a question of not losing that degree of common sense which allows us to understand that the identity of 'that of which one speaks' does not require the identity of 'that which is said about it,' but only the compatibility of the different predications.[42]

---

[42] Such a very obvious fact has been stressed by those authors who have been aware of the importance of not reducing the whole of meaning either to sense or to reference. Let us simply quote two examples. Harré says:

> We must routinely distinguish the business of establishing that something exists from the ever open possibilities for further research into what it is that exists. A referent can persist as the focus of empirical research and as the subject of predication even through recategorisations of a rather drastic sort. We can maintain our focus on an existent while our researches into its nature lead us to abandon every statement we once thought true of it, except that it exists and that its nature is such as to secure it a place in some referential grid (Harré 1986, p. 66).

And Putnam, in his paper 'Explanation and Reference' (in his 1975, II, pp. 196–214) attacks the alleged dependence of reference on sense and context ("concepts which are not strictly true of anything may yet refer to something; and concepts in different theories may refer to the same thing") and, moreover, recognises that, though the intension of a concept might not be well determined, elements having referential relevance may occur in it:

> I said before that different speakers use the word "electricity" without there being a discernible 'intension' that they all share. If an 'intension' is anything like a necessary and sufficient condition, then I think that this is right. But it does not follow that there are no ideas about electricity which are in some way linguistically associated with the word. Just as the idea that tigers are striped is linguistically associated with the word "tiger," so it seems to me that some idea that "electricity" (i.e. electric charge or charges) is capable of flow or motion is linguistically associated with "electricity" (II, p. 200).

It is interesting to note that, in the same paper, Putnam strongly attacks the philosophy of science of logical empiricism, charging it with "idealism."

5.4 The Ontology of Scientific Realism

For reasons of clarity we explicitly call the attention of the reader to the double meaning that "scientific object" has had in the present work. Initially it had the meaning of something 'clipped out' of things according to a certain point of view (therefore, it had an implicit ontological-referential sense). After the recognition of the 'functional' and 'relativised' nature of the notion of thing (Sect. 4.1.6), and the analysis of intension and extension, the notions of 'abstract object'—encoding certain properties—and 'concrete object'—exemplifying certain properties—(Sect. 4.1.2) were defined. We recognised then that any science necessarily studies abstract objects, but with the intention of knowing an extra mental reality to which it 'refers,' and in which it intends to find 'concrete objects' that are 'referents' exemplifying its abstract objects (Sect. 4.2). In such a way a *scientific object* is, in a first primary sense, an *abstract object*, but we are entitled to recover our initial more intuitive sense by saying that a thing 'becomes a *scientific object*' when it becomes the *referent* of a certain science. As a consequence, the question regarding the reality of scientific objects splits into two sub-questions: are they real simply as abstract objects (as *noemata*) or are they also real as referents? This was the issue analysed in Sect. 4.6, where it appeared that we could avoid this splitting of the attribution of reality by instead saying (more rigorously) that a scientific object is 'real as referent,' that it is 'exemplified by referents.' Nevertheless, we easily recognise that such a simplification is possible only if one shares our analysis in terms of encoding and exemplifying; and, in order to avoid such a prerequisite, we shall go on in our discourse by adopting the usual way of speaking, according to which, when the question of realism is raised, it regards the existence of objects understood as extra-mental (or extralinguistic) referents of scientific concepts, or sentences, or theories. This clarification, by the way, is a justification for the rather detailed semiotic discussions we have developed in the foregoing sections, and for the (seemingly redundant) references to those analyses that we might perform in what follows.

We have already argued for realism's being a precondition for reference, in that reference—being of an extralinguistic nature—already has characteristics which are attributed to 'reality' in the context of the philosophy of language (though in a sense that is not univocally understood by different authors). We have also noted, however, that one cannot merely on the basis of the philosophy of language pretend to specify which type of extralinguistic reality the objects of reference, i.e. referents, have. Before continuing we should like to stress once more that one is obliged to admit that something is *real* if one has already admitted that it is *different from nothing*. Clearly, from this point of view even a dream, a mathematical calculation, or something imaginary must be considered real because, despite being different from what we call concrete and material reality, such entities are also different from nothing, as is evident from the fact that we can describe them, moreover in a way that is true or false (I can assert that I have dreamt of a white horse while I really dreamt of a black cat). Therefore, we shall say that these various types of thing differ, not in the *fact* of existing, but in their *way* of existing. (A house exists physically in such a way as to be capable of being perceived with the senses and to be operationally used for living in, while its image exists as a mental entity at different levels, and according to various modalities.) It would therefore be arbitrary to maintain that only things which belong to a single and well-determined kind of reality (that is, to physical reality) are real. The notion of reality—as we have already stressed in Sect. 4.2—is not a 'univocal' but an 'analogical' concept, as already stressed by Aristotle. Therefore, we want to repeat explicitly that, for us, *reality, existence* and *being* are equivalent, since we say that something is real if and only if it is different from nothing, and nothing itself is simply the contrary of being, which in turn is understood as the simple fact of existing. We are aware that these notions have been characterised and mutually related in various ways in ontology, but our choice simplifies several issues, as it has already done in the preceding parts of this work, and will continue to do in the sequel.

5.4.1 Reference and Reality

If, on the one hand, we should not be trapped in the pretence that reality is only of one type, neither can we use the distinction between the fact of existing and the mode of existing as a crowbar to back the claim that one cannot avoid being a realist in all possible situations. How then can one avoid misunderstandings in this direction? We shall describe a discourse as realist if it *intends* to speak of a reality *of a certain type*, and succeeds in this intention.[43] Therefore, a discourse which speaks of physical reality is conceived as realist only if we can claim that it actually succeeds in speaking about this reality and not, instead, only about conceptual images of it. However, a discourse which proposes to deal with dreams or hallucinations will be truly realist precisely if it succeeds in its aims, independently of the fact that dreams and hallucinations are not physical objects. On the other hand, this discourse would not be considered realist if it succeeded only in talking about physical situations which accompany dreams and hallucinations, such as

[43] If we wanted to be extremely scrupulous we should say that a discourse in itself does not 'intend' anything, but only the speaker who uses this discourse intends to say something. This is in a way correct, but we have already explained when we spoke of the possible differences between sentences, propositions and statements, and more extensively when we spoke of judgments and of apophantic logos, that we tacitly (and obviously) understand that the discourses we consider are stated by persons. But, on the other hand, we can also tacitly presuppose that there is concretely a broad consensus regarding the 'intensions' of the linguistic expressions belonging to a certain natural or disciplinary language, a consensus that dispenses one with the need of knowing the subjective individual intentionality of every single speaker. This is what actually happens when we read a book, a paper, or consult a dictionary, and even more strictly when we have to do with a disciplinary language (which is, after all, the subject matter of our investigation). Let us simply refer to the general discussion we have devoted to *intersubjectivity*.

electrical or chemical states of the brain (therefore, physicalist reductionism is far from being a guarantee of realism; it is rather its negation).

What has been said so far using the notion of 'type of reality' or of 'way of existing' can be better specified in terms of *properties* and of *criteria of referentiality*. To say that not all of what exists has the same type of reality ultimately means acknowledging that various entities possess different properties, and that we are able to refer to different types of entity depending on our ability to access their properties, which we use, in this way, as *points of reference*: they are the ontological target of those "points of view" of which we have often spoken in the preceding sections of this work. It is enough to look back on what was said where we spoke of sense as a 'guide' in tracing referents. It is easy to see that that discourse was nothing but a description of how properties serve to identify referents precisely because properties are attributed to referents in an intentional act of the subject (the subject's "point of view"). However, it is no less essential to acknowledge that these points of view spring not only from the subject, but from the meeting between the subject and reality, as is clear from an example.

For instance, a toothache is as *real* than anything ever is (it is enough to think about the profound difference between the *being* and the non-being of a toothache for the suffering subject). However, a toothache does not have a colour, a mass, a location in space, a shape nor many other properties which would allow us to qualify it as a physical entity. In this case, the criterion of referentiality is only a subjective state of pain which is sufficient for us to state that it *exists*, but despite the greatest effort, we could never succeed in attributing a colour or other such properties to pain. Therefore, it is not within the capacity of the subject to attribute properties to things at its pleasure. On the contrary, while, for example, we can attribute a colour, a mass and a shape to a leaf, we cannot say that it is odd or even, monosyllabic or polysyllabic, introvert or extrovert, given that these are properties which serve to qualify and identify other 'types' of entities.

Our reasoning has so far been developed on the level of everyday language, but it can be extended without difficulty (indeed, in an even simpler way) to the case of scientific discourse. As we have by now repeated at length, each scientific discipline presents itself as a discourse which has an intentional relationship to reality from a certain 'viewpoint,' that is, it sets itself the task of investigating only certain aspects or *properties* of reality. Because of this, it selects for its language a certain circumscribed number of *predicates* and, for the purpose of succeeding in its referential effort, associates them with some standardised *operations* which we can call indifferently "criteria of objectification," "criteria of protocollarity," or "criteria of referentiality." These operations 'clip out' specific objects (i.e. referents) of a given science within the vast sphere of reality. Moreover, the operations do not apply to nothing, but to already identified referents ('things' of daily experience within a particular historically determined community), which furthermore are subjected to empirical and not purely linguistic and intellectual manipulation. Therefore, objects (referents) arise which cannot avoid being *real* as well.

One might wonder whether the *properties* attributed to these objects are or are not real, but at this point the deep-seated naivety of this question should be clear, given that in any particular science only those entities which have such properties are acknowledged as objects, so that, for that given science, an object is nothing but the structured set of the properties which can be operationally attributed to a thing precisely because they are operationally *referred* to it and not only *thought* about it. (We have already seen in the foregoing chapter how this discourse can be extended also to theoretically defined objects, via the notion of truth, but we are not interested in this question here.)[44] This, evidently, does not exclude the possibility that a certain determined referent possess other properties as well, which can be investigated by other sciences, or which can even be the object of non-scientific discourse. Using the way of speaking adopted in this book, we can say that, when one clips out an 'object' from a 'thing,' one leaves out of consideration a great variety of aspects of this thing. This means that the *referent* one is reaching, though being 'encountered' by means of certain operational procedures, is much richer than the bundle of operationally defined characteristics or attributes that those procedures are able to demonstrate and 'sum up' in the object. This does not mean, however, that this same referent cannot be further investigated by means of other criteria of referentiality and become in such a way the subject-matter (the object) of other objectification procedures.

Our position could be expressed by saying that there is a *distinction* (but not a *separation*) between the realm of objectivity and that of reality in this precise sense: the domain of objectivity is always much more *restricted* than the domain of reality (do not forget that, according to our definitions, reality coincides with existence, and therefore encompasses the total domain of being), and it can never be brought to coincide with it. Indeed any objectification depends on a point of view *within another point of view* (that is, the broader point of view in which 'things' are given, which is in itself 'contingent' upon a certain historical situation and never encompasses 'the whole' of reality). This must not be understood, however, as if there were secluded parts of reality perpetually immune to any objectification. On the contrary, there is no part of reality which may be thought of as not being able *in principle* to undergo objectification (such a claim would be a concealed form of *epistemological dualism*). Awareness of this fact enables us to obtain a deeper understanding of certain claims we have already made in the context of our discussion of truth. What is dangerous in contemporary science is not to maintain that scientific propositions are *true*, but to pretend that they are *complete* in the sense that they tell 'the whole truth' about reality, that they apply to the whole of existence. Indeed, if the propositions of science faithfully describe *objects*, and these are, according to a correct realist viewpoint, part of *reality*, the propositions must be recognised as being true, since truth is nothing but the

---

[44] The reader can easily see that the content of this section is to some extent a recapitulation of certain results presented in greater detail especially in Sects. 4.2 and 4.3. See also Agazzi (1997b). We believe that these repetitions are justified due to their making this chapter on scientific realism self-contained.

property of a correct statement about reality. However, scientific statements cannot pretend to be *complete*, since they always leave substantial portions of reality out of consideration; and they can become false if they pretend to cover this part of reality as well.[45]

The above discourse may sound convincing to the extent that we apply it to directly accessible entities, since the concrete operations enable us to put our hands on the 'real object' while admittedly 'clipping out' only a delimited number of some 'thing's' aspects or properties. The situation seems much more problematic however when we have to do with *theoretical entities*. In this case it seems more reasonable to separate the reality of properties from the reality of objects.

Let us try to sketch the reasoning supporting such a view. If we consider, for example, an electron, there are certain measurements that we can perform in order to attribute to it, let us say, a mass, a charge, a spin, and so on. It is also helpful to speak of such measurements as expressing *properties* of 'some object' because it helps our mind to synthesise them. But, as a matter of fact, all we can do is perform these measurements—there is no moment when we are actually acquainted with the object, i.e. the electron. Why should we then be authorised to speak of it as something really existing without perceptual evidence for its existence? This seemingly reasonable argument is actually involved in the old superstition of epistemological dualism which, in this case, consists in conceiving of the electron as a kind of 'substance' that lies 'behind' its properties, and which is such that we never encounter *it*, while we are able to encounter its *properties*. If one thinks this way, however, one conceives of the electron as a 'thing' and not as an 'object,' and one has removed oneself from physics by this very fact. If we instead conceive of the electron as an object, it must be conceived of not as something to which properties are attached, but as something which is *constituted* by these properties. An object is to be considered as the 'structured' totality of the objectively affirmable properties and not as a mysterious substratum of these properties. This might sound as a Humean positivism, but it is not, since we do not maintain that such properties are exclusively our perceptions: they are ontological aspects of reality, and may even be perceptually unattainable.

Let us note that in everyday experience we behave in a similar way. When I maintain that there 'really' is a piece of paper on my desk, I can do this because there exist a certain number of properties, perceived by my eyes and by other sense receptors, that give me evidence for this being the case. But the piece of paper is

[45] This kind of realism can be accepted even by those who feel inclined towards a certain 'phenomenalism,' to employ an expression we have already met. In fact, if such a phenomenalism does not want to be confused with the incorrect thesis that we only know phenomena but not reality, it must identify phenomena with what we previously called the *aspects* of reality that are considered as objects by a particular science. In this sense every scientist can or must be a phenomenalist as far as he is aware that only certain features of reality are treated by his science but, once this correct admission is made, he must also recognise that these phenomena are part of reality, and thus he must confess to being a realist. For this reason it seems advisable not to speak of phenomenalism, since such a word is frequently used to designate the doctrine that we can only know certain *appearances* and not reality as it actually is.

actually nothing more than *the structured totality* of its properties, and we all know that it possesses several other properties that are not directly accessible through perception (such as, e.g. its chemical properties). In the case of this example, the presence of a certain amount of properties directly accessible by observation was sufficient to catch the referent, but this is not essential. An object is a complex *structured* reality, as we have pointed out, and there is no *reason* to pretend that *all* the properties that go into this structure be observationally testable. (We leave out of consideration here the fact that the *synthesis* of the properties into one object is not just another property, but the result of conceptualisation, as we have hinted at in Sect. 2.7.) This amounts to saying that the substance, essence or form (according to characterisations offered during the history of Western philosophy) are metaphysical or ontological principles introduced in order to understand principally the unity of the multiplicity and the permanence under change, and therefore are not themselves additional properties that should be 'attributed' to an entity. In order not to engage in such traditional discussions we have preferred to use the perhaps more neutral notion of a 'structured' set of properties, where the notion of structure here plays the role of the above-mentioned principles. In the case of some objects it may be that none of the properties attributed to them is empirically testable. In such a case we are nevertheless obliged to admit the existence of such objects for theoretical *reasons*, as we have already explained in our previous discourse regarding truth. We are even more obliged if it is possible for us to derive from the existence of this object, in a logically cogent way, certain previously unobserved features that we actually observe in conformity with our prescriptions. In other words, as we have already discussed, this purely theoretically admitted object must be recognised as real unless we have *reasons* for admitting that our theory is false. We are not entitled to impose on *ontology* certain limitations based on *epistemological* tenets devoid of solid argument.

The above clarifications show how fragile a particular distinction is that is sometimes proposed in the debate about realism, the distinction between a *realism of properties* and a *realism of entities*. This distinction is useless since, in science, entities (objects) are (as we have maintained) nothing but *structured sets* of properties. Traditional ontology was aware of this profound connection between properties and ontological identity. Scholastic philosophy, for example, admitted that, on the one hand, a substance cannot be *equated* with its accidents while, on the other hand, we can know a substance only by *knowing* its accidents (and this was not epistemological dualism since it was recognised that we know the substance, though only partially, through some of its accidents). From this awareness followed a principle, *talia sunt subjecta qualia permittuntur a praedicatis suis* (subjects are such as they are permitted to be by their predicates), where one can see that the entities are determined by their properties or attributes.[46]

[46] This criticism also applies to those scholars, such as Ian Hacking, who are ready to admit a 'realism of entities' but not a 'realism of theories' (see, e.g., Hacking 1983), because theories are nothing other than the way of expressing the properties of the entities they admit, i.e., of saying 'what these entities are,' and without this one would not be able to say what the entities are whose

Within the framework presented here, the 'realist' import of scientific *applications* acquires relevance, though not in the grossly pragmatist sense according to which success is the best guarantee of truth. Rather, the realist import of scientific applications becomes relevant in the more rigorous sense according to which if we succeed in *operating* on reality, letting ourselves be guided by a science, it follows that this science has picked out some actual properties of reality, and that it has reached an interpretation of reality that is correct at least up to a certain point. We shall see in the sequel that from here stems an even more decisive role for *technology* as a support for scientific realism.

5.4.2 Realism and the Possibility of Error

Standing against what we have been saying so far is the fact that no scientific theory is ever certain of its own truth, and that, moreover, the history of science attests to the continuous change of theories which, it seems, one could interpret either as an uninterrupted series of 'falsifications' or as the indication of a lack of reference. It is often said that it is not by chance that the crisis of scientific realism at the beginning of the twentieth century was a consequence of the discovery of the falsity of Newtonian mechanics.

In this accumulation of arguments several different aspects are present which it is necessary to distinguish. First, the falsity of a theory (admitting, for the sake of argument, this way of speaking) can in certain cases indicate that the theory is without reference, but in other cases it cannot. Second, it is a question of seeing whether the cases in which it is said that a theory has been falsified are really cases of falsification, or more simply instances of a change of reference.

Let us discuss the first point and see how, in certain cases, the falsity of a theory implies the acknowledgement of the non-existence of its referents. An example

(Footnote 46 continued)
existence one admits. To be fair, we must recognise that Hacking is pointing (more implicitly than explicitly) to that requirement of operationality which, as we have seen, is fundamental for the *referentiality* discourse and for the claims of existence. In fact he maintains that even unobservable entities can be credited with existence if they can be used (to put it briefly) in the building and functioning of some scientific instrument or concrete process, because this fact shows that they are endowed with some causal ability, which indicates that they are 'real.' We note that this indicates that in scientific discourse and practice they have become so familiar that they can be taken for granted and 'used,' and this (to use our terminology) amounts to their being considered as *things*. This does not eliminate the fact, however, that they have been made identifiable through a long process of determining their *properties*. For instance, the electron (to use Hacking's example) was submitted to experimental manipulations in order to establish its charge, to determine that it can be diffracted, that scattered electrons conserve their total energy, that they can be filtered, etc., and in such a way they could result (thanks to considerable *theoretical* work) in that kind of entity that could be *concretely* 'sprayed' in the Fairbank attempt to detect free quarks. Their 'reality' as entities cannot be separated from a realist interpretation of the theory determining what kind of entities they are.

relating to a singular term can be that of the stories that for centuries were narrated concerning Hermes Trismegistus, considered to be the author of those writings known as the *Corpus Hermeticum*. Successive criticism has shown that this figure, in whom very serious Renaissance scholars such as Marsilio Ficino still believed, has never existed, and that the *Corpus* was written during the age of Imperial Rome by philosophers of neo-Platonic inspiration who invented the existence of this scholar, roughly coeval with Moses, for the purpose of adding credence to their doctrines. The falsity of the theory therefore coincides, in this case, with the elimination of its alleged referent. A certain analogy exists with the phlogiston theory, once adopted as the basis of the incipient science of chemistry and today abandoned. In this case, too, one can say that the falsification consisted in discovering that the term "phlogiston" does not have a reference. However, one could be slightly more tolerant, and maintain that we actually use another name in referring to certain gaseous products (for instance hydrogen) which can be seen to emanate from particular chemical reactions and that once were covered by the term "phlogiston."[47]

However it must not be taken for granted that the falsification of a theory implicitly denies the existence of its referents. In the case of the Ptolemaic and Copernican theories, for instance, one can claim that the referents remain the same (earth, sun, planets), and that the latter has shown that certain assertions of the Ptolemaic theory relating to the stationary position of the earth, rather than of the sun, in the planetary system, are false (We simplify this discourse a little, realising that it could use certain refinements). We can say that this is, after all, the most common situation, which well corresponds to the fact that in general a discourse can be said to be false when it 'succeeds in its reference' but 'says' of its referents that they have properties which they do not have.

When one concentrates specifically on studying scientific theories, much greater importance is acquired by the cases in which the purported falsifications must be interpreted neither as an elimination of their respective referents nor as a discovery of false assertions made about these referents, but as a *change* of referents. We easily appreciate this possibility when we bear in mind that the referents of every scientific theory are usually 'clipped out' from 'things' (that is—in the most common cases—from the referents of common sense) through standardised, precise and limited operations. From this it immediately follows that, if the set of these operations changes, the operational meaning, that is, the 'referential meaning' of certain basic terms changes, and with it so too change the *objects* to which theories refer. One can for instance read the transition from classical to quantum mechanics in this way. But it is then clear that two 'rival' theories can both remain true, each obviously *about* its own objects (or referents as

[47] For the example of phlogiston see the discussion in Smith (1981), pp. 112ff. Also Poincaré once made a remark regarding the alleged elimination of the referents of old theories: "Barely 15 years ago, was there anything more ridiculous, more quaintly old-fashioned, than the fluids of Coulomb? And yet, here they are reappearing under the name of *electrons*" (Poincaré 1902, p. 164).

one may prefer), so that they are not really rival. In this way we shall in fact have broadened the range of known truths, as the new ones take their place beside the old, and do not replace them.

Note that in this perspective one can explain what Popper tries to express with his unsatisfactory theory of verisimilitude, according to which there exists *truth in itself*, which is intrinsically unattainable, despite the fact that successive theories more and more closely approximate it in an endless asymptotic process. The misunderstanding here lies in having reified truth, so that the cognitive enterprise is not thought of as a process which aims at knowing reality, but at knowing truth. Now, while there is nothing strange in stating that the enterprise of knowing reality can be an ideally infinite task since each set of *true* pieces of knowledge about it only picks out partial aspects of it, at the same time it seems absurd to say that we are certain of approaching truth even if we have no possibility of taking knowledge of reality as a term of comparison, to assess whether we have really come closer to it.

These last considerations allow us to acknowledge in the realist position the most solid basis for talking of the cognitive progress of science. It may consist in the elimination of errors equivalent to showing the non-existence of purported referents, or in the elimination of earlier erroneous assertions about referents which are preserved—in the technical sense in which the new theory retains the same 'objects' as the preceding one. One can also think of cases in which certain everyday referents are successively 'objectified' through operational predicates which are totally or partly different. We shall then say that the different theories permit one to increase true knowledge about these referents by emphasising different *aspects* of them. When the diversification of operational criteria is such as to leave doubts about the fact that the everyday referents are still the same, we shall certainly talk about incomparable (or "incommensurable," but incommensurable on an empirical-operational basis and not only on a semantic-contextual basis) theories, and scientific progress will consist in having brought to light new *objects* of knowledge.[48]

In all these cases it will be perfectly legitimate to speak of cognitive progress even in a *cumulative* sense, meaning that either one knows more and better about the same referents, or one knows more because new referents have been discovered. Both truth and error contribute to this progress in the way sketched above, and this fact justifies the common conviction, which is also that of the scientific community, according to which human knowledge, even if fallible, nevertheless proceeds in the discovery of what is true (that is, in the acquisition of true propositions) insofar as it has ever greater success in describing and understanding the structure of reality.

---

[48] We shall provide more details on this point when we speak of scientific progress.

5.5 Arguments for Anti-realism

Having presented the features of the form of realism that we are advocating, we now intend to outline some of the main anti-realist arguments, and assess them from our point of view. Before beginning this presentation, let us note that anti-realism seems in many cases to be an emotional rather than a rationally grounded attitude in a way which is reminiscent of the allergic repulsion many scholars manifested against metaphysics at the time of logical empiricism (and perhaps also today). In this latter regard Tarski once noted, "one gets the impression that the term 'metaphysical' has lost any objective meaning, and is merely used as a kind of professional philosophical invective."[49] If one considers the efforts of certain contemporary philosophers *not* to be classified as realists, despite the objective content of their doctrines, one comes more or less to the same conclusion with regard to the term "realism."[50] One must recognise, however, that sometimes realists also show the same emotional attitude.[51]

---

49 Tarski (1944), p. 363.

50 Here is a paradigmatic example. The opening lines of Fine (1984) present a strong anti-realist declamation:

> Realism is dead. Its death was announced by the neopositivists.... Its death was hastened by the debates over the interpretation of quantum theory .... Its death was certified, finally, as the last two generations of scientists turned their backs on realism and have managed, nevertheless, to do science successfully without it" (op. cit., p. 83).

However, when one considers what Fine's own position is—which he calls the "natural ontological attitude," which he claims to be equidistant from realism and instrumentalism—one finds that it consists in a so-called "homely line of argument" thanks to which "it is possible to accept the evidence of one's senses and to accept, *in the same way*, the confirmed results of science" (p. 95). The consequence is that even the most classical theoretical constructs of scientific theories are to be taken as existing:

> I have similar confidence in the system of 'check, double-check, triple-check' of scientific investigation, as well as the other safeguards built into the institutions of science. So, if the scientists tell me that there are molecules, and atoms, and y/J particles, and who knows maybe even quarks, then so be it. I trust them and, thus, must accept that there really are such things, with their attendant properties and relations (op. cit., p. 95).

How is one to distinguish this position from the usual realist positions (and indeed the most committed ones) is hard to say. The difference may be seen in the kind of argument adopted (quite a weak one that relies on a fiduciary trust in the scientific community), and in the almost unintelligible affirmation, according to which true realists "add onto this core position" something like "a desk-thumping, foot-stamping shout of 'Really!'" (p. 97). This is certainly no *argument* against realism, but rather the expression of an emotional defence against being considered a realist. By our remarks, however, we do not exclude that, by a deeper scrutiny of Fine's position, one might perhaps conclude that he is ready to accept a form of 'empirical realism,' but we are not interested in this additional point.

51 An example is Popper. We have already noted that he recognises himself to be unable to provide conclusive arguments in favour of realism. However, he says:

Those forms of anti-realism whose efficacy mainly rests upon a misrepresentation of realism also incline in a similar direction, and in such a way obtain a victory against a straw man.[52] This is not necessarily done intentionally, and may be the consequence of the fact that realist positions are often formulated rather vaguely. The remedy, however, is not to construct an allegedly clear and unambiguous formulation of realism which turns out to be an *overstatement*, and then attack it (as has been done). Rather, one should critically scrutinise those explicit formulations that certain realists actually have proposed (which are certainly not impossible to find). The following is, in general, the fair way of rationally discussing a thesis T: if one wants to *defend* T, it is usually a good strategy to try to justify a *stronger* thesis T_s from which T follows, while if one wants to *reject* T, the correct strategy is to reject a *weaker* thesis T_w from which the refutation of T will follow *a fortiori*. Unfortunately, discussions on realism often adopt the unfair (and ineffective) reverse strategy.

Among the formulations of realism which we can qualify as spurious (and, therefore, as such that they do not represent serious targets of anti-realist criticisms), that which turns out to be a particular form of *epistemological dualism* is paradigmatic. According to this spurious formulation, the realist sets science the task of investigating 'a World out there' that is both ontologically and epistemologically independent of us, maintaining that science is capable of describing this world in a set of true sentences. Clearly, in such a picture, realism appears involved in inextricable difficulties, which include the problem of comparing that which we think we know with the truth, or the problem of securing 'access' to the external world, or of disentangling the 'reciprocity' of our intervention in the world and the intrinsic features of the world.[53] As already stressed, all these are

---

(Footnote 51 continued)

> This robust and mainly implicit realism which permeates the *Logic of Scientific Discovery* is one of its aspects in which I take some pride. It is also one of its aspects which links it with this *Postscript*, each volume of which attacks one or another of the subjectivist, or idealist, approaches to knowledge (Popper 1983, p. 81).

Clearly, philosophical positions are not something of which one is to be or not to be proud, and Popper is here using the language of someone who has fought for years a long battle for a noble cause rather than the language appropriate to an 'objective' philosophical discussion.

[52] This fact is also discussed in the first chapter of Dilworth (2007). As examples of misrepresentations of realism that work as presupposition for the defense of certain forms of antirealism we can mention those of van Fraassen and Putnam already considered in Sect. 5.2.3.

[53] Let us give a few examples of these 'dualistic' portrayals of realism. Ellis says:

> We can investigate nature and develop a theoretical understanding of the world, but we cannot compare what we think we know with the truth to see how well we are doing. (Ellis 1985, p. 69).

Also Fine, though beginning his presentation of realism in a seemingly neutral way (1986, p. 150), imperceptibly shifts to a neatly dualistic misrepresentation of it:

ways of implicitly and gratuitously assuming that we know *our representations* (in this case they are the 'scientific images') of the world, and not the world itself *through* our representations. If one does not fall victim to this fantasy, one can understand that our ways of observing and even operating on the world are precisely our ways of having *access* to it, and uncover certain of its features under these conditions, while being aware, at the same time, that these cannot be features of the world *cognitively independent* of our knowledge, precisely because they are what we want to *know* of the world.

Rather often the strategies in the debate between realists and anti-realists follow the path of placing the burden of proof in the opposite camp. Though rather common in the whole history of philosophy, this strategy is not really decisive, since even successful criticism of *one* of the opposed positions only in special cases (i.e. in the case of direct negation) provides conclusive reasons *in favour* of the other. This is why we shall first be interested essentially in the analysis of those arguments which are put forward *in favour* of anti-realism, and only later shall we examine certain arguments which are put forward *against* realism.

5.5.1 The Radical-Empiricism Argument

Anti-realism is very short on 'positive' arguments (i.e. arguments that do not reduce to a shifting of the burden), but this is not surprising. Every 'anti-' position is doomed to be in this situation, since its being against something does not imply its being in favour of a positive alternative. The most common way of presenting a positive characterisation of anti-realism (i.e., of saying what scientific statements, concepts, theories *are* from an anti-realist point of view), seems to reduce to the fallacious picture of epistemological dualism. Scientific statements, concepts and theories are *our representations*, and we really know only these representations. Owing to the fallacy of epistemological dualism, this is a poor argument indeed (it is not even an argument, but just a prejudice). This is why we have to look for less radical forms of anti-realism, that is, forms which admit at least a partial ability on the part of science to describe the real world.

As a matter of fact, this is what happens at present. Most of the anti-realist positions reduce to denying ontological reference to *theoretical* concepts, or constructs, of science, while admitting as unproblematic such a reference for the

(Footnote 53 continued)

The problem is one of access. The correspondence relation would map true statements (let us say) to states of affairs (let us say). But if we want to compare a statement with its corresponding state of affairs, how do we proceed? How do we get at a state of affairs when that is to be understood, realist-style, as a feature of the *World?*... The difficulty is that whatever we observe, or, more generously, whatever we causally interact with, is certainly not independent of us. This is the problem of *reciprocity* (Fine 1986, p. 151).

observational concepts. Why? It is not difficult to find here a *radicalisation* of the empiricist attitude, that is, a transition from the easily admissible thesis (a) that every existence claim about the world must explicitly *be linked* with sense experience, not only to the already more controversial thesis (b) that such claims must *ultimately rest* on sense experience, but even to the extreme thesis (c) that every existing entity must be *directly ascertainable* by sense experience.[54]

This is the position explicitly advocated, for example, by Bas van Fraassen.[55] It has the merit of being explicit and clear; however, it is hardly convincing, not only because it has the inconvenience, for example, of excluding from the domain of reality such unobservable entities as dinosaurs, no less than a lot of 'familiar' microscopic objects (such as bacteria, viruses and macromolecules) which we really 'observe' by means of instruments, but especially for a reason of principle, that is, because it makes ontology dependent on an *epistemological* criterion without any argument to support such a move. In other words, it is straightforward to admit that whatever is directly observable is real (though many precisions should be added even to this statement), but one does not see why whatever is real must be directly observable as well. For example, certain people are devoid of sight, and therefore cannot see, let us say, stars, while we know, on the other hand, that stars exist. This indicates that their existence does not depend on their being visible In fact, it would not help to say that they are visible to most of us, because it is the fact that they can exist, even if they are not visible *for somebody*, which already indicates that existence and visibility are not *the same thing*.

This remark is generalisable. It is arbitrary to make the existence of scientific objects dependent on special epistemological features (or prejudices). In fact, many doubts as to the existence of scientifically characterised entities rely upon difficulties in 'visualising' them, or of reducing their features, properties, ways of

---

[54] Indeed, as we have shown in our historical reconstruction (Sect. 5.1), modern science was realist (with few exceptions) until the end of the nineteenth century, and anti-realist positions emerged when natural science began to be typically concerned with *unobservables*. For this reason we have maintained that the most suitable way of characterising the issue of *scientific* realism, as distinct from general *philosophical* realism, is that of considering *the reality of unobservables* as its central problem.

[55] For van Fraassen, this actual and concrete observability by unaided human senses is a defining characteristic of what is real, so that, for example, even certain entities which are theoretical according to the usual standards in philosophy of science (e.g., stars or galaxies which cannot be directly observed *de facto*), may be considered real for him, since they could in principle be approached and seen directly, let us say, by astronauts in a spaceship. (For a characterisation of this position see, for example, van Fraassen 1980, pp. 13–19.) In order to be completely fair one should recognise that van Fraassen's position is more sophisticated, in the sense that he does not flatly say that what is empirically unobservable does not exist, but that we are not logically compelled to *believe* that unobservable entities exist. This is, however, a rather unclear jump from an ontological and epistemological plane to an epistemic one. Indeed there are no logically cogent reasons that compel one to *believe* that a certain sensory perceived thing really exist. From our side, we have tried to offer, in the foregoing parts of this work, certain logically cogent reasons for admitting the existence of unobservable entities, based on the analysis of the referential nature of truth.

operating, to familiar patterns. This attitude is arbitrary, and we have already argued against the requirement of 'visualisation' as a mark of scientific objectivity.[56] We could even go further and say that failing to find a convincing ontological model does not imply anti-realism, but simply that we are still unable to encompass in a fully satisfactory framework the realities we *have reason* to admit.[57]

Our last statement provides an answer to the rejoinder that the radical empiricist could put forward against our observations: what criterion of reality could we accept besides direct unaided sense observation? Our answer is (in addition to a defence of indirect or instrumental observation, which we shall leave aside for the moment), the existence of *rational arguments that provide truth.* This answer is not unexpected after what we have said in the foregoing chapter, but here we should like to point out that for those entities which are not only directly unobservable, but are determined *only* by means of theoretical constructs, the *only* warrant for their existence is the *truth* of the sentences which describe them and this brings us back to the 'truth' (in the non-literal sense already explained) of the theories where they are postulated. This explains why a radical-empiricist anti-realism must also reject the possibility that theories be true, as van Fraassen, for example, indeed does.

5.5.2 The Negation of Truth for Theories

We do not intend to consider, for the moment, strictly instrumentalist positions, and shall limit our discussion to those approaches which accept a 'cognitive goal' for science. This acceptance, as we have seen, cannot really make sense if it is not also expressed as the admission that science aims to attain truth. But then differences easily emerge as to *what* may be credited with being true in science. At least starting with Mach, a classical position has been that of admitting the possibility of being true to those scientific statements which are observational in the strict perceptual sense, while denying this possibility to those parts of science which go beyond perceptual observation, and may be regarded as *theories*. In brief, theories are not true (or false). Let us explicitly stress that the position we are advocating in this work, though it asserts that theories *are not* properly characterised as being

---

[56] See Sect. 2.8 of this work. This thesis is well defended in McMullin (1984, pp. 10–15) where it is significantly said that "imaginability must not be made the test for ontology. The realist claim is that the scientist is discovering the structures of the world; it is not required in addition that these structures be imaginable in the categories of the macroworld" (ibid. p. 14). In keeping with this correct remark, McMullin shows that the anti-intuitive conclusions of quantum physics did not have an anti-realist meaning even for Bohr, since Bohr was simply maintaining that what can be inferred about the world "is entirely at odds with what the classical world view would have led one to expect" (p. 12).

[57] More on this point will be said in the discussion regarding 'phenomenological theories.'

true or false, makes this assertion on the grounds that we *do not* accept the *statement view of theories*, which is the only view which would make *full* sense of attributing truth or falsity to them. But we are maintaining that theories—while being non-sentential gestaltic representations of a certain domain of objects—permit, and indeed impose, the formulation of certain *propositions* which are either true or false, in spite of their not being the record of sense perceptions.[58]

The Machian position may therefore be correctly expressed by saying that in science we have true sentences (the ones describing perceptual observations), and theoretical sentences (the ones overstepping the level of such observations, and constituting 'theories'). Theoretical sentences are neither true nor false, but must simply harmonise with the perceptually true sentences, and may do so in several different ways, so that their choice remains very open (instrumentalism and conventionalism creep in). The most elaborated version of Machism is van Fraassen's constructive empiricism, that is "most elaborated" for two reasons: first, because Mach's view is a form of *phenomenalism* (since, as we have seen, he considers objects to be only sets of perceptions), while van Fraassen's is a form of *common-sense realism* (since he himself says at the beginning of his work that he accepts the reality of those things that are directly observable, as everyone does in ordinary life).[59] The second reason is that van Fraassen's view has the advantage of a terminology for characterising the status of theoretical sentences on this approach, namely as being or not being *empirically adequate*. This choice of terminology is interesting, since it retains the central feature of the classical concept of truth (the notion of *adequacy* or *adequatio*), but limits it to strictly empirical evidence: "a theory is empirically adequate exactly if what it says about the observable things

---

[58] A position that denies the possibility of speaking of truth in the case of theories is advocated by Dilworth: "Scientific theories are seen not to be entities of the sort which are either true or false, but to be structures which are more or less applicable depending on the results of certain measurements." In this case no radical empiricism dictates such a claim, but the general view that theories are not linguistic entities (and for this reason it is correct to say that they are not the kind of entities of which truth or falsity can be predicated). In fact Dilworth defends a view of scientific theories as applied *models* that are conceived with the purpose of explaining (rather than merely describing) concrete reality, a view that we also share in part and that we call a *Gestalt view* (in keeping, by the way, with the intuitive basis of Dilworth's view, namely his Gestalt Model), of which we have spoken several times in this work. The difference, however, is that Dilworth advances his view as an alternative to the statement view (and develops it so as to provide an account of both theory conflict and scientific progress), whereas we defend it as a *complement* to the statement view, because we maintain that theories entail the *linguistic expression* of the content of their corresponding Gestalt, which is obtained by means of a certain number of sentences. Therefore, we are entitled to speak of the truth of theories in an 'analogical' sense, that is, by relating their global truth to the truth of their explicitly formulated sentences in a suitably specifiable way.

[59] Let us note that we do not qualify common sense realism as 'naive realism' since we attribute to common sense a much deeper meaning and significance than what is often meant by many philosophers. The relations between common sense and scientific knowledge are by no means trivial, and we refer to Agazzi (2002) for a discussion of this issue.

and events in this world, is true—exactly if it 'saves the phenomena'."[60] However, one must note that this solution is much less satisfactory than it might first appear, for it leaves us with no means of qualifying those sentences which were problematic. We are told (perhaps more implicitly than explicitly) that a theory is empirically adequate if its *observational* sentences are true, but nothing is said about the status of its *theoretical* sentences. In other words, the *semantics* of the theoretical sentences remains fully undetermined (this is, probably, something van Fraassen is ready to accept, since his interest seems to be only epistemological; but on the other hand it does not seem satisfactory for a philosophy of science to admit that the semantics of a considerable amount of scientific statements remains incomplete only for epistemological reasons).

It is therefore no wonder that van Fraassen (like Mach and many other phenomenalists in both past and contemporary science) essentially advocate an epistemic and pragmatic type of discourse. Instead of clarifying what theories *are*, he tries to say what it means to *accept* rather than *believe* a theory, and how this fact may have consequences as regards our relying on a theory for purposes of practical behaviour or intellectual investigation (this, by the way, gives an instrumentalist connotation to his approach—as van Fraassen recognises).

We do not wish to discuss the soundness of this distinction between accepting a theory and believing it. According to van Fraassen, believing entails admitting the truth, while accepting does not: we accept a theory because we believe that it is empirically adequate, but then we do not believe in the truth of its theoretical parts, being simply content with its other 'virtues,' such as its explanatory power and its ability to increase our control of reality, these attributes, however, not entailing truth. This distinction could perhaps be briefly expressed by saying that by accepting a theory we are committed to using it and working with it as though it were true, but remaining agnostic as to whether it really is true. The reason we do not discuss the meaning of this distinction is that we see no sound reason for excluding the assignment of truth to the theoretical sentences of a science.

In fact, van Fraassen tries to give such a reason by noting that "we can have evidence for the truth of a theory only via evidential support for its empirical adequacy," so that no warrant for the belief in the truth of a theory could be

[60] van Fraassen (1980), p. 12. We want to note an interesting consideration made by Dilworth, that points out an implicit extension of van Fraassen adequacy also to unobservable entities: "what a theory referring to unobservables 'says about observable things and events' is that they take the form they do as a consequence of the nature and behaviour of the unobservables. So his own manner of expression suggests that a theory's being 'empirically adequate' implies its also being 'theoretically adequate,' i.e. that it correctly depict what is unobservable." (Dilworth 2007, p. 38). In van Fraassen (2008) the notion of "appearance" is introduced in addition to that of "phenomenon," and appearances are defined as the result of measurements. We do not find this peculiar convention (which we would attribute to the radical empiricism of the author for reasons that we do not explore here), useful, and we have additional reasons for not using it, since our theory of objectivity, as will be more and more clear, intends to analogically concern all the sciences, and not only physics and, in particular, is totally independent of measurement. On this point see Agazzi (1978e).

stronger than the simple warrant for belief provided by the evidence in favour of its empirical adequacy. Therefore, every additional belief is "supererogatory."[61] This argument is more elegant than convincing. Its elegance resides in advocating a principle of parsimony (why should we accept more beliefs than those which are strictly necessary), but it is doubtful that this principle is always wise. For example, if we have reasons for believing in the truth of a set of sentences (hypotheses), and we can logically deduce from them other sentences (theorems), it is clear, from the one side, that the reasons for believing the theorems cannot be *stronger* than the reasons for believing the hypotheses but, on the other side, nobody would deny that we are not only entitled, but even obliged, to believe in the truth of the theorems as well, unless we reject logic.[62] The moral of this story is that reasons must exist for *denying* the truth and justifiability of theoretical sentences, and not only for admitting them, and these reasons are not provided by the anti-realist arguments we have analysed. Moreover, these arguments suffer the serious drawback of leaving unanswered the question of what the semantic status of theoretical sentences is, and this is certainly not a minor defect in a philosophy of science. The attempt to circumvent this question by saying what we do when we 'accept' such sentences does not eliminate this difficulty, as we have seen.

5.5.3 Explanation and Truth

Similar remarks apply when we consider other requirements which are usually demanded of theories and which, in anti-realist positions such as that of van Fraassen, are regarded simply as 'virtues' not entailing belief and truth. We shall consider only one example—explanatory efficacy—which is, however, very significant, since (according to most presentations in the philosophy of science) the specific task of theories is to provide explanations of phenomena. Separating explanatory efficiency from the truth of the explanatory hypotheses is not peculiar to van Fraassen, though he discusses this problem at length.[63] The arguments by which this separation is defended are elaborate, but they have a common basic feature. They show that we do not require *more* than empirical adequacy whenever we accept or prefer an explanation conceived of in the standard way, whereby a

---

[61] van Fraassen (1985), p. 255.

[62] A more detailed reasoning might be appropriate here. Our application of the parsimony principle is indeed correct only in the case that the conclusions are *necessarily entailed* by the assumptions. In the case of expansive inferences (that are typical for the introduction of theoretical constructs in science) however, this might not be the case. Therefore, considering van Fraassen's position, we should rather object that one does not see any justification in the fact that the same logico-inferential procedures, whose legitimacy and soundness he admits as far as they produce inferences whose conclusions do not overstep the phenomenal domain, suddenly become unreliable once this domain is overstepped. This kind of criticism can be found in Alai (2010).

[63] See, for example, Fine (1984), p. 89.

theory permits the *logically correct derivation* of the explananda from the hypotheses constituting the theory.

All this may be true, but—we submit—it does not capture the genuine meaning of "explanation." We do not wish to refer here to several critical examinations which have tried to show that correct logical inference is too meagre a requirement since explanation must include, for example, ontologically more robust features such as causal links.[64] We would simply like to ask an elementary, but direct, question: would we honestly accept, as an explanation of a fact, an hypothesis which we know to be *false*, but which accidentally happens to be such that we can logically deduce this fact from it? It seems that no one would honestly say "yes." In other words, it is, so to speak, inscribed in the concept of explanation—since the time when it was first expressed through the notion of 'giving the reason' (*lógon didónai*) in Greek philosophy—that explanation needs *both* logical cogency *and* the truth of the premises. If we were told, "This hypothesis is false, but it explains your facts," we would simply think that we are being mocked, and not being provided with any explanation at all. Therefore, to accept an explanation means to believe that the hypotheses it employs are true, even though we could not be *certain* that they are true, or—if we prefer—even if we do not *know* (but only suppose and believe) that they are true.[65]

From what we have said, it also appears that the distinction between accepting and believing a theory is, after all, not as useful as van Fraassen imagines. In fact, it may make some sense from a pragmatic point of view, but from a *cognitive* point of view one cannot see how one could accept a theory without believing that its theoretical *sentences* are true. (We have already said that if a theory is something *more* than just a system of sentences, one can accept it without holding it to be true in a strict sense.) Often we can only provisionally *adopt* or *use* a theory about

[64] Let us quote a significant statement by McMullin: "Theory explains by suggesting what might bring about the explananda. It postulates entities, processes, relations, themselves unobserved, that are held to be causally responsible for the empirical regularities to be explained" (McMullin 1984b, p. 210). This view is also central to the conception of explanation accurately presented in Dilworth (2007). Similar considerations are expressed by Rescher:

> In attempting answers to our questions about how things stand in the world, science offers (or at any rate, both *endeavors* and *purports* to offer) information about the world. The theory of sub-atomic matter is unquestionably a 'mere theory,' but it could not help us to explain those all too real atomic explosions if it is not a theory about real substances. If I hypothesise a robber to account for the missing jewelry, it is not a hypothetical robber that I envision but a perfectly real one. Similarly, if I theorise an alpha particle to account for that photographic track, it is a perfectly real physical item I hypothesise and not a hypothetical one. Only real objects can produce real effects .... The theoretical entities of science are introduced not for their own interest but for a utilitarian mission, to furnish the materials of causal explanation for the real comportment of real things. (Rescher 1987, p. 38).

[65] Arguments in favour of including truth among the requirements of genuine explanatory efficacy are expressed, for example, in Leplin (1984a, p. 212) and McMullin (1984a, p. 29).

which we are highly doubtful, and which we know to include certain unsatisfactory aspects; but in these cases it would be inappropriate to say that we *accept* the theory.

Looking more closely at van Fraassen's position, however, we can find an implicit but powerful reason for his refusal to pass from acceptance, to belief, in the case of theories, this reason being the well-known allergy to metaphysics that has been so endemic to radical empiricism. This allergy transformed the epistemological requirement of observability into an ontological requirement about what deserves to be credited with existence.

Empirical adequacy goes far beyond what we can know at any given time. (All the results of measurement are not in. They will never all be in; and in any case, we cannot measure *everything* that is measurable.) Nevertheless there is a difference: the assertion of empirical adequacy is a great deal weaker than the assertion of truth, and the imposed limit of 'acceptance' delivers us from metaphysics.[66]

Why we should be "delivered" from metaphysics remains here a rather emotional question, but we can try to understand it by considering what are, according to van Fraassen, the goals of science from an empiricist point of view (otherwise van Fraassen's argument *for* empiricism would *presuppose* empiricism).

To be an empiricist is to withhold belief in anything that goes beyond the actual observable phenomena, and to recognise no objective modality in nature. To develop an empiricist account of science is to depict it as involving a search for truth only about the empirical world, about what is actual and observable.[67]

The above depiction is simply a synthesis of what is developed in detail in van Fraassen's work. His 'constructive empiricism' may be regarded as an aprioristic limitation on the aims of science dictated by the double tenet that: (1) the interest of science is not that of ascertaining how things are but simply of producing a model in which all (and only) observable features of the world can be accommodated; (2) unobservable entities are 'metaphysical' ballast that we must avoid. The first tenet is mistaken since it misconceives the genuine aim of achieving knowledge generally, of which scientific knowledge is only a particular form (as van Fraassen correctly maintains). We always want to know how things are, and we certainly admit that observable features are the first source of information about this; yet humans have also always been interested in knowing whether unobservable entities really exist, and this owing to the exigencies of *logos* that demands a *reason* for what is ascertained, a reason that in many cases amounts to the indication of a *cause*. Now, a *general* characteristic of our knowing activity (that we find also in everyday life) is that *in order to explain (find the reason for) what we 'see,' we look for something we do not 'see.'* It is simply a coherent application of this general characteristic that the first question regarding

[66] van Fraassen (1980, p. 69). A fully developed metaphysical alternative to logico-linguistic philosophy of science (partially different from the one that is being developed in the present work) is offered in Dilworth (2007).

[67] van Fraassen, op. cit., pp. 202–203.

unobservables and scientific knowledge is whether what is observed is *caused* by the unobservables. It seems that too many philosophers of science who have dealt with this issue have missed this point. To deny a priori such a possibility is a genuinely metaphysical claim (since it expresses a statement about 'what really exists').

Observability and non-observability are only *epistemological* characteristics; they do not necessarily imply an *ontological* difference. To be sure, many metaphysical doctrines admit the existence of 'supernatural entities' (as an ontological qualification), but this does not necessarily imply that such entities are always unobservable. For example, Christians believe that Jesus was really God, despite his also being visible thanks to his 'incarnation.' In the case of the sciences, the unobservable entities are not believed to belong to a supernatural reality but simply to the reality of mundane physical things. They simply overstep the limits of our sense perception—and this is why their existence can be stated only *by argument* and not by observation *alone*. As a consequence, their existence is less *certain* than that of observable entities, but this is an *epistemic* (not an 'epistemological') disadvantage which should not be confused with an ontological difference.

A realist understanding of science captures whatever an extreme empiricist understanding does, but without *unduly* disregarding the requirements of *logos*. That this disregard is arbitrary also results from the surprising limitation of the notion of truth advocated by van Fraassen. He suggests that scientific theories aim to be *true*, but true only of observable entities. Why could they not also be true in their theoretical part? Only because of the unproved tenet that their unobservable referents do not exist. But in such a case they would be simply false. van Fraassen does not come to this conclusion, and he must leave the theoretical statements of science in a vague limbo: they are perhaps neither true nor false. But this is incompatible with their being declarative statements which we may not know with *certainty* to be true or false. This, however, is an *epistemic* situation that does not affect their *semantic* and *referential* status.

The very valuable features of constructive empiricism can be suitably incorporated within a correct realist approach that consciously limits the truth of the propositions of a given science to its specific and *empirically circumscribed* domain of objects. It is incorrect to say that, for a realist, "a theory cannot be true unless it can be *extended* consistently, without correction, to all of nature" (p. 86). Quite the contrary, as we have explained at length, a non-naive realist maintains that a theory is true of those aspects of nature that constitute the *specific domain of objects* of that theory, and empirical adequacy is a fundamental prerequisite for this to happen (and actually for the investigation of the whole of nature) without exhausting the *possibilities* of knowledge we might have in the whole of this domain.

A couple of final remarks. As we noted in the foregoing discussion regarding empirical adequacy, denying the transition from explanatory efficacy to truth seems to correspond to the wise adoption of a criterion of parsimony, but it is questionable whether this is a *rational* parsimony since it results in *misrepresenting* the concept of explanation altogether. Moreover, it is subject to the same objection we have already considered: if we do not find any objection to assigning truth to the logical consequences of a set of sentences which we admit to be true, why should we object to assigning an (*epistemologically* less warranted) truth also to the hypotheses from which these sentences may be derived?[68]

Furthermore, in philosophy of science no less than in everyday life, it may happen that by being parsimonious in one domain we become prodigal in others. This seems to be the case with van Fraassen's approach. After all, by permitting us to believe observational statements and to qualify them as true, while asking us to refrain from believing theories and remain content with only 'accepting' them, he saves something (i.e., the commitment to truth for theoretical sentences, and to existence for unobservable objects). However, he is obliged to *duplicate* the methodology of science by allowing all the usual machinery of explanation, confirmation, and similar 'virtues' to operate in favour of belief, truth and ontological commitment when it comes to observable phenomena and objects, but admitting only acceptance, simple empirical adequacy (not truth), and ontological agnosticism when this contingent border is overstepped. What is the *justification* of this 'inflationist' policy? There is obviously no other reason than the *radical empiricist* dogma already discussed at the beginning of Sect. 5.5.1.[69]

We may conclude these reflections with a general remark: Ockham's razor (*entia non sunt multiplicanda praeter necessitatem*) is certainly a wise intellectual principle, but it also admits of a 'counterpart' which is no less wise (*entia non sunt diminuenda praeter necessitatem*). The conjunction of these two principles says that we must have *good reasons* both for introducing and for denying entities, properties, and so on. Anti-realist arguments, after all, are not very sound from either of these points of view. They seem to eliminate 'without necessity' perfectly reasonable features of scientific investigation; they introduce without reason implausible redundancies; and, finally, they leave unanswered certain important questions (e.g., what is the semantic status of theoretical concepts and sentences). We shall now consider whether certain arguments in favour of realism are immune to these defects.

---

[68] This point is also made in Glymour (1984), pp. 188–189.

[69] This issue is presented with much efficacy in Fine (1986), pp. 167–168:

> Despite the uniformity of practice, however, constructive empiricism does feel a need to multiply its interpretation of that practice (going here for acceptance, and only there for belief). That need goes against its deflationist promise. It is generated only by the prior commitment to empiricist epistemology.... What positive arguments or reasons connect the two, providing the grounds to multiply interpretations of the inferential practice? The answer is that the constructive empiricist has no argument. It goes its inflationist way in order to prop up empiricist epistemology. There is no other (or better) reason that supports its chosen path (p. 168).

5.6 Realism and the Success of Science

There is a commonplace argument in favour of realism when doubts about the ability of science to describe the world as it really is are advanced: how could science be so successful, both in predictions and practical applications, if it did *not* correctly describe and explain the world? Perhaps because this argument is so simple and widespread in everyday discourse, philosophers have usually paid little attention to it, looking for more elaborate and 'critical' approaches. (Exceptions exist however; for example, this kind of reasoning was used against extreme conventionalism at the beginning of the twentieth century by scholars such as Poincaré.) In more recent years, this argument has been resumed with particular emphasis, and is symptomatic of the fact that the discussion concerning scientific realism has tended to abandon the restricted arena of philosophy of language (for the success of science is certainly not something which can be dealt with within a linguistic framework). The most popular formulation of this argument is Putnam's claim that only realism allows us to rationally explain the success of science in the sense that, without realism, such a success should be considered a miracle:

> The positive argument for realism is that it is the only philosophy that doesn't make the success of science a miracle. That terms in mature scientific theories typically refer (this formulation is due to Richard), that the theories accepted in a mature science are typically approximately true, that the same term can refer to the same thing even when it occurs in different theories—these statements are viewed by the scientific realist not as necessary truths but as part of the only scientific explanation of the success of science, and hence as a part of any adequate scientific description of science and its relations to its objects.[70]

So many different claims are squeezed into this passage that it is no wonder that the thesis it expresses has been challenged from several sides. (What does "the success of science" mean; when is a theory *mature*; what is *approximate truth*; is permanence of referents necessary for realism; is realism itself a kind of scientific hypothesis?, and so on.) This is a good example of how even elementary issues may become very complicated once they fall into the hands of philosophers. However, we are not going to follow the details of these discussions, but shall content ourselves with a few remarks regarding certain more delicate issues.

5.6.1 *In What Does the Success of Science Consist?*

The first problem comes from the meaning that must be associated with the expression "the success of science." Some authors consider this success as consisting more or less in *scientific progress*, and are therefore interested in defending the thesis that there actually is progress in the Popperian sense (according to which a theory T′ supersedes a theory T because T′ can deductively explain what T could explain, as well as phenomena incompatible with T, and not explainable by T)

[70] Putnam (1975), p. 73.

against, for example, the criticism coming from the 'incommensurability' thesis.[71] Note however that even on a deductivist approach it is not necessary to consider the success of science to consist in progress of a *cumulative* sort.[72] Indeed, the 'predictive' success of science is already a *fact*, impressive in and of itself independently of any sort of accumulation. Therefore, the real question is simply that of seeing whether this progress can be captured by deductivism. Since prediction is *systematic*, when it is the result of a derivation from assumptions about the existence of certain (unobservable) entities, it cannot be attributed to *chance* but (unless it is a rationally inexplicable miracle) its success must depend on the *truth* of these assumptions, and in such a way one is returned to the problem of how this truth can be established.

Against the plausibility of this reasoning it has been alleged that there have existed *false theories* which were successful as regards their predictions. Larry Laudan has in particular been active in developing this line.[73] However, it is not certain that his long list of examples must be interpreted in such a sense, for the following reasons. First, as we have said several times, it is not fully appropriate to consider theories as true or false, but rather as more or less adequate. Now, it may well happen that a particular theory which turns out not to be adequate from several points of view and is therefore replaced by another, remains *partially* adequate from certain points of view; and this is enough to afford an understanding of its predictive success. This success depends on those parts of the theory which are adequate.[74] In particular (to be explained more extensively later), a theory

---

[71] A deductivist attempt to account for scientific progress may be found in Boyd (1984), where, among other things, a defence of the significance of crucial experiments, even in the presence of a kind of theory-dependence, is presented (p. 59), and a defence is provided of the claim that the technological realisations of science cannot be paradigm-dependent (p. 60).

[72] This means that one would not be disturbed even if what Fine says were true: "The plausibility of the explanandum (that the conscientious practice of science leads to abundant instrumental success) is an artefact of our historical perspective" (Fine 1986, p. 152).

[73] Consider Laudan's very strong statement:

> For every highly successful theory in the past of science which we now believe to be a genuinely referring theory, one could find half a dozen once successful theories which we now regard as substantially non-referring (Laudan 1984, p. 212).

This affirmation should be backed by the historical studies the author has presented elsewhere (e.g., in Laudan 1977), but it is debatable that the results of these studies really justify this claim, as we shall see.

[74] Here our example of the map may provide a useful (though only partial) analogy. If we have a good map of a city, printed a few years ago, and which then proved very accurate in describing the city centre, we may still use it even if we know that in the meanwhile new suburbs have developed, which are either not described, or only very imperfectly described. This means that we can still rely on the accuracy of our map as far as the city centre is concerned, and use it for finding streets and squares which we had not visited or seen at the time when we first bought the map, while we are to be very careful about those parts of the city which are new. With respect to them, it is more than probable that our map will not help us, and that we need a new one. That this

contains much more than that which is strictly *objective*, since it provides a *Gestalt* where different factors of representability, even of visualisability, are admitted, and which affords a framework for the strictly objective features of its discourse. It is therefore rather natural that the referents of a theory be conceived of with many of these non-objective and redundant features, and that as a result they may not exist *as such referents*. But this does not mean that the referents of the theory do not exist *at all*. They exist, but they have properties which are not captured by the intuitive picture which was used to imagine them (the existence of different models of the atom at the beginning of the twentieth century could be an example in point). Only when the postulated referents are characterised through properties which actually play a logical (and not just a psychological) role in explanation and, especially, in prediction, can they be credited with a solid ontological status.[75] Second (and this has also been noted earlier), theory change often corresponds to a change of *objects*, and this may imply that we apply new criteria of referentiality to things already studied by other means, a fact that may induce us to believe that the old referents did not exist, while they simply do not show up in our new objectification.[76]

(Footnote 74 continued)
new map also contains an accurate description of the city centre might be an advantage, but it is not necessary. We could use both maps, according to our concrete needs.

[75] In McMullin (1984) there is an interesting discussion of the features which must characterise a theoretical construction describing 'structural' properties of a scientific domain in order for it to be a reliable candidate for ontological reference. In particular, 'fertility' is a desired virtue (see pp. 30–34).

[76] For this reason we must consider the following statement of Laudan as mistaken (a statement which, however, is indicative of a typical misunderstanding in the interpretation of the historical succession of theories):

> After all, presumably many theories which we believe to be false (e.g., Newtonian mechanics, thermodynamics, wave optics) were—and still are—highly successful across a broad range of applications (Laudan 1984, p. 228).

In fact, we must say that Newtonian mechanics, thermodynamics, wave optics are *still true*, not 'in general,' but *of their objects*, that is, of those aspects of reality which may be investigated by resorting to their criteria of referentiality. Of course, it is true that absolute space, for instance, does not exist, but this was not the kind of 'entity' which was implied in the predictions of Newtonian mechanics. It was a 'redundancy' which Newton had introduced for certain metaphysical reasons. A confirmation of this unsatisfactory approach, which fails to take into account the 'relativity' of scientific truth to the 'objects' of a theory, is given by Laudan a few pages later:

> It is well known that statistical mechanics has yet to capture the irreversibility of macrothermodynamics as a genuine limiting case. Classical continuum mechanics has not yet been reduced to quantum mechanics or relativity. Contemporary field theory has yet to replicate the classical thesis that physical laws are invariant under reflection in space (p. 238).

This is presented as an objection to the (allegedly) realist claim that new theories *must* include old theories as limiting cases. However, this is not a requirement of realism (despite certain realists' perhaps having made it one), since in certain cases it *may* happen that a new theory includes the old as a limiting case (when the objects remain the same, as we have

What we have just said indicates that *preservation of the referents* is not a necessary requirement for realism in the sense that, while it is to be admitted that the referents of an earlier theory are not 'annihilated' when a later theory is accepted, this does not mean that they must *remain referents of the new theory as well.* On this delicate point (which we believe ourselves to have made clear thanks to our theory of scientific objectivity) Putnam also appears to be mistaken, when he says:

> If one believes that the terms of T_1 do have referents,... then it will be a constraint on T_2... that T_2 must have the property that *from its standpoint* one can assign referents to the terms of T_1 (Putnam 1984, p.143).

In this claim it is not difficult to see the confusion between things and objects, of which we have often spoken. As a matter of fact, we are not able, for example, to describe and explain the macroscopic properties of a body in terms of quantum physics, but this is not what we *should* be able to do (though it might perhaps be *desirable* that some results in this direction could be obtained). For such a description we must use criteria of referentiality proportioned to this task, such as those of classical mechanics and many other disciplines that explore macroscopic *things from their standpoint.* On the other hand, when the criteria of referentiality remain the same, we have the right to give importance to the persistence of the referent *in spite of theoretical changes*. This is clear in our approach, and is also presupposed in what Putnam says about the electron remaining the same across different theories (pp. 145–146). However, while we can secure this constancy of reference because we attribute it to extra-theoretical *operational* criteria, Putnam gets into trouble, since for him existence is *intratheoretic* (p. 149), and this really makes his 'internal realism' rather problematic. Indeed, any form of realism should at least let *existence* be theory-independent, while allowing that characterisations of the existing referents are theory-dependent.[77]

After the precisions proposed here, it is possible to consider under a better light the notion of 'approximation to truth' which is used by several realists, but which suffers from ambiguities. Laudan is right in protesting that "Unfortunately, few of the writers of whom I am aware have defined what it means for a statement or theory to be 'approximately true'."[78] But the confusion already arises when he speaks indifferently of the truth of a statement and the truth of a theory. Without anticipating here what we shall present later, we shall say that a statement may be

(Footnote 76 continued)
already explained), but such cases are by far the less frequent. In general, we have (as in the examples quoted by Laudan) a *change of objects*, and therefore we have truly referring theories which do not preserve the old reference, not because they 'eliminate' it or show it to be non-existent, but because they investigate *a new one*.

[77] We are not going to deepen the discussion of Putnam's internal realism here. (See especially Chap. 3: 'Two Philosophical Perspectives,' pp. 49–74 of Putnam 1981.) We have provided such a discussion in Agazzi (2001).

[78] Laudan (1984), p. 229.

true (or false), but then it is simply true or false, and it is not appropriate to say that it is 'approximately true.' (At this point, however, we do not want to enter into the discussion of the 'degrees of accuracy' that characterise the different kinds of referential procedures, a point we shall approach later.) On the other hand, a theory is strictly speaking neither true nor false; and this is why it may be said, in a certain non-literal way, to be approximately true. This, as we have already explained, would be better expressed by saying that a theory may be more or less *accurate*, in the sense that the information it contains gives rise to several true sentences in the most *relevant* cases, while giving rise to false sentences only in a reduced number of rather marginal cases. As a consequence, successive theories may represent an increase in 'accuracy,' and this may be presented (in a non-literal sense) as progress in acquiring increasing 'degrees of truth,' as Leplin, for example, maintains.[79]

This approach also has the advantage of rendering fallibilism more interesting from the point of view of knowledge-acquisition in science. The more usual Popperian way of seeing the question is that the elimination of errors already constitutes an approximation to truth, which is in a way correct, but only provided that what we substitute for the error is not itself a *worse error*. Now the Popperian theory of verisimilitude, apart from its many other weak points, fails to take this problem into account. According to the view advocated in the present work, however, the notion of accuracy permits us to elaborate a different picture of theory-change in which degrees and improvements are possible. In particular, the opening of new points of view (new objectifications), the technological availability of more refined criteria of referentiality, and so on are clearly part of this improved approximation, which is fully compatible with the possibility (and actually the continual occurrence) of errors.[80]

A last question which may be discussed concerns the issue (concealed in Putnam's quotation made at the beginning of this section) whether scientific realism could itself have the status of being a scientific hypothesis.[81] Several discussions have been devoted to this topic, since this thesis seems to involve circularity, or to suffer from the difficulties of abduction, and so on. We do not want to enter into this discussion here, and content ourselves with referring to Boyd's analysis, which we consider satisfactory, in virtue of which it can be shown that "all that is claimed is that the instrumental reliability of the methodology of mature science depends upon the development of a theoretical tradition that embodies approximate knowledge of unobservable as well as observable phenomena."[82] This remark is an (essentially empiricist) interpretation of the success

[79] Leplin (1984a), p. 202.

[80] For a deepening of this discussion we suggest taking account of the treatment of the relative acceptability of scientific theories, as well as of the criticism of Popper's 'verisimilitude,' contained in Dilworth (2008).

[81] For a good discussion of this issue see Alai (2012).

[82] Boyd (1984), p. 77.

of science taken as an empirically ascertainable fact, and we are therefore justified in taking it as a basis for interpreting scientific methodology in a way which supports the realist view (in this case, the ontological purport of theoretical constructs).

5.6.2 The Special Relevance of Technology to the Issue of Scientific Realism

According to the conception presented in this work, science provides *objective knowledge* thanks to its determining its objects and obtaining intersubjective agreement by means of *operational procedures* that consist in the competent use of specialised *instruments*. In addition, these operational procedures grant science its referential import and justify its claim to provide knowledge of reality. The mention of operations, instruments, concrete procedures, and ascertainable results brings to mind certain fundamental features of technology, and suggests that there exists at least a deep kinship between science and technology.[83] Therefore, one might wonder why only very few and brief hints regarding technology have been made up to this point. The reason is that the said kinship certainly exists (and we shall see this in a moment), but it must not be pushed so far as to become an identification. If this happens, several confusions inevitably arise in different fields. Unfortunately, such a confusing identification has become widespread. For the general public, science and technology are usually considered to be one and the same. For example, if one asks a normally cultivated person what the most salient advancements in *science* have been in the last century, one will very likely be told that they are such things as radio, television, atomic energy, the human conquest of the moon, and organ transplants, but it is unlikely that relativity theory, quantum mechanics, or the discovery of the structure and function of DNA will appear in the list. This indicates that *technological realisations* are perceived as great *scientific achievements* much more than the achievements that deserve to be considered milestones in the advancement of science proper. However, this is not just the unreflecting view of the general public. Very cultivated people also share it; and several philosophers have argued in favour of this identification (for which the neologism "technoscience" has also been coined).

Precisely because we are convinced that the said identification of science and technology is confusing and misleading, we have carefully avoided speaking of technology in a discussion that has been concerned with many delicate and complex issues regarding the *cognitive purport* of science, a discourse in which a premature involvement of technology would only have damaged the clarity of the analysis. As we have stressed on several occasions, *distinction* must not entail *separation*. But distinctions are necessary when one wants to understand complex

[83] See, e.g. Queraltó (1999).

realities. In the present case, the distinction between science and technology comes from the consideration of their *defining goals* (in the sense already discussed in Sect. 5.2.2). The specific goal of science is the acquisition of *knowledge*, the specific goal of technology is the *production* of efficacious *tools* for attaining a wide spectrum of particular practical goals. This distinction does not imply separation for the following reason: on the one hand, in order to attain its goal, science requires the use of several tools (indeed, highly sophisticated tools in contemporary science); on the other hand, technology can realise the production of more and more efficient tools by exploiting the advancements of scientific knowledge. This interplay determines such a dense network of feedback loops that we can say that contemporary science and technology are 'consubstantial,' though not identical.[84]

Something of a deepening of this realisation is useful in view of certain consequences we shall derive regarding the question of scientific realism.[85] We shall introduce our analysis by distinguishing between the meanings of two English terms that are often considered synonymous, that is, "technique" and "technology." The first is used to denote the particular 'art of doing' that characterises the performance of a given activity (e.g. pictorial technique, textile-making technique, skiing technique, and so on) but we can also use this term to denote the whole of such practical skills or 'efficacious ways of acting' (more or less as we do when we speak of atomic technology or biotechnology, as well as of technology in a general sense).

Technique has always existed, since it consists in humans' constantly developing special skills and efficacious forms of action in order to satisfy a wide range of needs. The advancements of technique were produced by the accumulation of empirically based improvements in existing practices, of the results of trial-and-error efforts to solve certain problems, and of occasional fortunate findings. Greek thinkers (especially Plato and Aristotle) have coined the term "*téchne*" to denote an efficacious way of acting which is also accompanied by the knowledge of the *reasons* or *causes* of its efficacy. In such a way the nature of *techne* was analogous to that of *epistéme* (or science), because both required the satisfaction of a *theoretical* foundation capable of providing the *why*. In the case of *epistéme*, knowing *that* a certain proposition is true had to be backed by knowing *why* it is true; in the case of *téchne*, knowing *that* a certain procedure is efficacious had to be backed by knowing *why* it is efficacious. Therefore, we must recognise that the Greek notion of *téchne* overstepped the borders of simple 'technique' by requiring in addition a substantial contribution of theoretical reflection, whose foundations were looked for in general principles of a philosophical nature.

With the creation of modern natural science (and its non-philosophical features which we treated at the beginning of this work), the bulk of theoretical knowledge from which the justifications for the success of technical practices were drawn became the variegated domain of the limited but objective knowledge obtained in

---

[84] Buzzoni advocates an even more radical connection between the scientific and the technological dimensions (see Buzzoni 1995 and 1997).

[85] For greater detail, see Chap. 4 of Agazzi (1992), as well as Agazzi (1999).

the different sciences. We propose to call "technology" this new face of technique that consisted, on the one hand, in requesting a 'theoretical supplement' to pure technical efficacy (in keeping with the spirit of classical *téchne*) and, on the other, in taking *science* as the source of such theoretical justification. This amounts to characterising technology (fundamentally, though not exclusively) as *applied science*. This characterisation allows us to see that distinguishing technique from technology is not all that strange. Indeed, many cultures have existed that developed an advanced technique in the absence of a significant science, but even in those cultures (such as the Western culture) in which science has been powerfully promoted, it is possible to write a history of technique quite independently of the history of science, since even today there are sectors in which technical skills and know-how progress according to an internal empirical dynamics and accumulation of successful practices without one really knowing *why* they are successful.

Therefore, it seems correct to see a 'bifurcation' in the history of technique. After the creation of modern natural science, a new branch grows on the old science-independent trunk, this is technology, which aims at pursuing the traditional goals of technique by *applying scientific knowledge*.[86] The most important novelty of technology does not appear, however, from the pure and simple fact of being applied science. Humans have always wanted to make use of their knowledge for the satisfaction of their needs and goals, and have therefore applied this knowledge in those domains where it seemed useful. Therefore, it is not particularly noteworthy that, when the possibility of acquiring, by means of new methods, a much better knowledge of nature, people expressed the wish, and even the certitude, that such knowledge would greatly contribute to the improvement of the human condition. This is usually presented as the Baconian view of science, but is actually to be found in several other authors such as, for example, the 'rationalist' Descartes. The idea that, by uncovering the 'secrets' of nature, human beings would be able to put nature itself at their service, has inspired magic and astrology for centuries; and it is not by chance that magic, astrology, and alchemy have flourished for at least a century (in the Renaissance) along with the new natural science and not at variance with it.

The real novelty of technology lies in the fact that the application of scientific knowledge it typically realised was the construction of *machines*. Machines were not unknown in the tradition of technique, but they were rather simple and, moreover, they were a kind of amplification of the typical product of technique,

[86] A linguistic precisation may be advisable. The distinction between technique and technology proposed here sounds rather natural in many modern languages where the corresponding words are present, whereas it may sound a little peculiar in English. Indeed, in this language the word "technology" is normally applied throughout the history of humankind, such that e.g. the development of ancient tools constitutes instances of technological development. Therefore it is certainly possible to cover all the cases one wants to cover using the English terms "technique" and "technology" in their usual senses, and qualifying them when necessary. Nevertheless we considered it meaningful to elaborate this a little in our conventional definition of technology, not only for purely analytic purposes, but also because it helps in understanding the spirit of modern civilisation, as will also appear from further considerations we are going to propose later.

that is, of *tools*. A machine is the assemblage of several tools that render possible or more efficacious the realisation of a certain goal. Modern machines are different: they are *designed or projected* in advance, as the concrete application of knowledge provided by a given science or set of sciences. We shall call them "technological machines" (it would sound a little strange, but not arbitrary, to call them "scientific machines"; and we note that *how* a technological machine functions and *why* it functions that way are known *before* the machine itself is concretely constructed, and not as a result of empirical trial and error. Such machines are concrete realisations of abstract models in which causal relations are specified.

This, in particular, explains why machines could quickly convert themselves into models for interpreting and explaining natural phenomena. In a machine nothing is hidden or mysterious, since the machine has been constructed according to a design in which every detail has its purpose. Therefore, if we are able to interpret a certain natural system 'as a machine' we have completely understood the system constituting the machine, and think we have correctly explained its features. Of course, strictly speaking, it doesn't follow that, from understanding how machines work, I shall consider myself to understand how things that *are not* machines work by conceiving of them on analogy with machines. The reason the machine analogy is so ubiquitous in science is because the functioning of machines transparently embodies the various principles and laws of science, in particular that of physical contiguity. This is why interpretations of animals and also of the human body as mechanical machines were proposed already at the beginning of the developments of mechanics, and other types of machines (chemical, thermodynamic, electrical, cybernetic, and so on) were proposed later with a similar purpose, along with the intellectual prestige obtained by new branches of science.

Let us examine the fact that a technological machine is the concrete realisation of an abstract model more closely. Using a terminology adopted in our preceding semantic discussions, we can say that the machine *exemplifies* the properties and functions *encoded* in its model; and this is literally true not only because the machine is a *concrete individual thing*, but also because the exemplification (as we have noted) admits of degrees of approximation and accuracy. In fact, the creation of a well-functioning machine usually cannot be accomplished on a first try, but requires several refinements or concrete adjustments that are not imposed by 'mistakes' in the design but by particular features of the materials employed. (Hence the creation of functioning 'technological' machines requires *both* 'theoretical' and empirical or scientific and practical input.) At the end of this work, the machine functions 'as it ought to,' which means that its functioning is a *state of affairs* that corresponds to the *predictions* contained in its *project* or design.

Let us compare this with the case of a non-technological machine. In such a case, we should rather say that the concretely constructed machine does or does not function in conformity with our *expectations*; and if it does not, we do not know whether this failure is due to some accidental imperfections of the material conditions or to some *reason* of a theoretical nature, a reason entailing that a machine of such design *could not possibly function* in the expected way. (For example, for centuries people had known that they could lift water by means of pumps and had increased their efficiency be means of technical improvements, however, they soon realised that they were unable to lift water up to a level greater than approximately 10 m from the level of the water. Only with the creation of hydrostatics in the seventeenth century could this impossibility be explained giving theoretical reasons independent of any practical improvement of the pump.) Without a theoretically designed project it is impossible to say how the machine *ought to function*. The project, however, does not properly formulate any prediction. The predictions 'contained' in the project actually are the predictions made by the *scientific theories* which have permitted the proposal of the complex *noema* that constitutes the project, and contains not only prescriptions as to the way of realising the *structure* of the machine but also as to its *functioning*. This functioning is something that *happens*; it is a *state of affairs* that constitutes a *confirmation* of the theories used in projecting the machine.

From a strictly logical point of view, this confirmation is similar to those provided by scientific experiments that satisfy the 'predictions' of a certain theory. It has been correctly stressed in the literature that prediction plays a stronger role than explanation in supporting the acceptance of a theory, since explanations are correct deductions of already known true sentences (in which the hypotheses can be selected and accommodated in order to allow such a deduction to obtain), while predictions (on the Deductive Model) are logical consequences of what are admittedly *hypotheses*, and are not granted to be true in advance.

The outcome of an experiment, however, though it strictly speaking constitutes a state of affairs, has the features of an isolated *event* and, moreover, concerns the confirmation of a single hypothesis. (The thesis that not single hypotheses but whole theories are confirmed or rejected by experimental tests is a rather speculative counterpart of 'semantic holism,' which holds in principle, but plays a small role in actual scientific practice.)

The correct functioning of a machine, on the other hand, is a permanent state of affairs, a 'fact' that is a confirmation of a whole system of scientific theories. Therefore, its force as regards the truth of the theories involved is much greater than that of an experimental confirmation obtained in the rarefied atmosphere of a laboratory in which an artificial and highly sophisticated experimental set-up has been constructed. If we now shift our attention from the consideration of a single machine to the whole world of technology, we must admit that we are in the presence of a gigantic and irrefutable confirmation of the truth of our scientific theories and of the *realist* purport of science.

The specific 'criterion of reality' for scientific objects that technology introduces (in addition to those of operational referentiality and truth by argument, already discussed) is the fact that technology *makes use* of such objects, and it is obviously not possible to make use of something that does not *exist*. More precisely, technology consists in the production and use of *things*, that is, of those entities whose reality is not controversial, not in a vague or general sense but as far as they are *referents* that *instantiate* in their attributes the properties of certain scientific objects.

This is the exact meaning of the qualification of technology as *applied science.* To apply means in this case to use, and the use is made within certain contexts of reality to which science is being applied. This is only possible, however, if the particular aspect of reality to which a given scientific theory is applied can be spoken about *truly* by the discourse of that science, that is, if the things it encompasses possess attributes that are adequately treated in that discourse. But there is more: since technology does not consist in the pure production of artefacts, but in the realisation of *efficient* artefacts and procedures, it follows that in technology those attributes of reality are exploited that are capable of showing a *causal effect* on the world.

This reflection suggests a couple of considerations. The extreme empiricist and positivist tradition led many people to eliminate the notion of cause from science, and to replace it with the 'metaphysically uncommitted' mathematical notion of functional relation, or the notion of logical connection. As a consequence, scientific explanation has also been portrayed as a hypothetico-deductive chain in which the sentences describing the explananda are logical consequences of the hypotheses, without suggesting that the properties of the objects mentioned in the hypotheses could be the cause of the properties appearing in the explananda. We have already mentioned certain criticisms that have been formulated against this too-narrow logico-linguistic conception of scientific explanation, and we are now in the position to vindicate the rights of a truly *causal explanation* on the basis of the evidence in favour of causal links that emerges from technology.

A second point concerns the existence of unobservable scientific objects. We have maintained that their existence must be admitted in virtue of the truth of the theories in which they appear as *logical reasons* for asserting the existence of observable entities. This is why we have already explicitly stressed that *not all scientific objects* are the results of operations. We can now add that, if we consider without prejudice what is actually done in scientific theories, we see that observable features of certain concrete objects are explained as *causal consequences* of the properties of unobservable objects, and this entails that such objects have the *same kind of existence* as the observable ones. Indeed, concepts can logically 'produce' only other concepts; statements can linguistically 'produce' only other statements; neither can produce materially observable *facts in the world.*

We venture to propose a plausible characterisation of a *mature science* which we found mentioned in a vague way in a passage from Putnam. A mature science is a science that has given rise to a significant technology. This means, for example, that we can provisionally admit certain theories that are 'empirically adequate,' without admitting their truth as van Fraassen says, until we have significant *predictions* confirming them. This fact (especially in conjunction with other 'virtues' discussed in the literature) already justifies attributing *truth* and *ontological reference* to them, but the existence of *technological applications* is the last decisive step that assures that they have been able to adequately treat *those aspects of reality they intended to treat.*

These last words are very important. They underline the fact that technological success does not eliminate the *partial* or *limited scope* of scientific theories. The fact that we can use classical mechanics in creating many machines or for sending rockets into space certainly means that this mechanics is *true of its objects* and therefore 'tells a true story' about *certain aspects of reality*. This can also be expressed by saying that this theory is *partially true* of reality, but only if we mean that it does not speak about the totality of the attributes of reality, and that, consequently, it can speak properly only of such referents that possess these attributes. In other words, it is not correct to say that this mechanics is true regarding *the whole of reality* because other aspects of reality exist that must be accounted for by means of other theories which, in turn, can be used as a basis for different technologies.

Let us also revisit the issue of *theory-ladenness* that has for decades been the cornerstone of many anti-realist positions. Considered from the point of view of technology, it shows all the poverty of its linguistic one-sidedness. Indeed, what is more theory-laden than a technological machine or a sophisticated technological process? Several scientific theories, often of great complexity and involving many abstract concepts, are brought together in order to design one single machine, so that the machine is literally theory-soaked and theory-dependent. However, this does not prevent the machine from being *real* and usable, and this not *despite* the heavily theoretical context that has determined its design, but *thanks to* this context, which has permitted us to endow the machine with all those properties that allow it to function and be useful. The *existence* of the machine is an *onto-logical* feature that is independent of any theory, and the machine can be seen and used also by persons that are unaware of the theories that have presided over its realisation.

In conclusion, let us note that technology belongs to the *world of life*, not only because it consists in the production of an enormous quantity of concrete artefacts, but also in the production of a no less impressive number of procedures that deeply affect our daily lives and even our habits and customs. Certain philosophers (such as Husserl and Heidegger, and their followers) have strongly criticised science for having separated humans from the *world of life* (*Lebenswelt*), and for having created an artificial framework of ideas that prevents us from having a genuine contact with being, with reality. We do not deny that an 'absolutisation' of science (that we call "*scientism*") can have such an undesirable effect, but we must also recognise that science has not lost contact with the world of life, and technology is the most tangible evidence of this fact.

Interpreting technology simply as the expression of an insatiable desire to 'dominate' reality is a gratuitous misrepresentation of its nature that stems from a prejudicial attitude of 'theoreticism,' for which intellectual contemplation is the only genuine approach to being. We agree that a proper place must also be given to contemplation, to art and poetry, and to moral and religious feelings, in order to obtain a full understanding of reality. We cannot underestimate the fact, however, that our approach to reality consists primarily in *having to do* with it, that is, in a vast display of *life practices* within which we form our concepts, our expectations, and our worldviews, and in which we try to prosper. This is why (as we have already noted) scientific concepts are also bound to basic *operations* that belong to life practices; and this is why scientific knowledge ‘returns’ to the world of life by *permitting* an increase and improvement of our life practices through technology. This, on the other hand, does not at all mean that such an improvement results in a real *global improvement* of the human condition: in order to attain this goal other considerations regarding the *orientation* of human praxis are needed, considerations that cannot come from science and technology, as we shall see at the end of this work. Such an involvement of science in the concreteness of life-practices does not tarnish its intellectual purity, but simply prevents us from the mistake of *separating* its cognitive aspect from the rest of its actual *reality*, whose complex nature reflects itself also in its cognitive dimension. We shall take this wider horizon of science into consideration in the next chapter.

Chapter 6
The Contexts of Objectivity

6.1 The Historical Determinateness of Scientific Objectivity

6.1.1 The 'Historical A Priori' of Science

In what we have said up to now science has been essentially (though not exclusively) considered as a system of sentences. Such an approach, however, although being useful for clarifying several aspects of the problems posed by the task of understanding science, is limited and insufficient in other respects. We shall now begin to tackle questions which oblige us to overstep the limits of this approach.

When we were led to consider the conditions and presuppositions of scientific objectivity, we did not find them by means of any *logical* analysis of sentences, for they were not conditions or presuppositions for stating a sentence within a certain scientific context, but were rather conditions for the existence of the context itself. We must also say, however, that this kind of investigation cannot be of a *methodological* nature either, for a methodological inquiry involves an examination of the links existing between scientific sentences and the said conditions by illustrating, for instance, their operational articulation (and, in general, the way they enter into the shaping of the patterns of a certain scientific objectification), but does not question the conditions themselves.

To focus more closely on our problem, let us recall what we have already said rather briefly about the general structure of any objectification. This is only possible because we can consider as 'given' and unproblematic not only many entities to which we refer ('things') but also many concrete and abstract tools for defining the operational predicates necessary to speak 'about' them. We have already noted that this feature of 'givenness' and of being unproblematic is only relative, in the sense that it can be problematised, but 'outside' the given objectification. So, bacteria may be taken for granted in pharmacology, which *presupposes* their existence and properties, and tries to elaborate strategies for controlling them. But this does not prevent another science, such as microbiology, from problematising bacteria and their properties. The same can be repeated in the case of instruments,

E. Agazzi, *Scientific Objectivity and Its Contexts*, DOI: 10.1007/978-3-319-04660-0_6,
© Springer International Publishing Switzerland 2014

as we have already noted when we spoke of the telescope being unproblematic in astronomy and problematisable in optics, or when we noted that differential equations are taken for granted in mathematical physics but can be studied and questioned in the context of the differential calculus, and so on.

These considerations, however, seem simply to stress that *theories* always presuppose *theories*. Certainly, this expresses the 'conditioned' status of every theory, but sees it to consist in a dependence not on something *extralinguistic* but on other linguistic constructions; and this would push us to conclude that, after all, logical analysis is enough to account for this dependence. This remark is only seemingly correct for, in order to reflect the actual situation exactly, it should be formulated differently, that is, theories usually presuppose the *results* of other theories; and this is a difference of no little moment, for these *results* are not to be taken as *sentences* but as *referents*. We should also point out, in addition, that one also finds at work (usually in an implicit way) in all these theoretical constructions certain fundamental *principles* of an ontological or metaphysical nature, but we shall leave this aspect out of consideration for the moment.

This way of speaking is explicitly adopted here in order to underscore the different position of theories under the different aspects involved. If we assume (as we have already done occasionally, and shall do more explicitly in the sequel) that the *specific* task of theories is to *explain* facts (and laws), and that this task is fulfilled by *hypothetically* proposing models, conjectures, hypotheses to be tested, we easily see that those alleged 'theories' which serve as prerequisites for any objectification are *not* characterised by such features. Indeed, they constitute reliable *information*, that is, something which is not there in order to explain anything but simply as a knowledge of 'how things really are'; and, because of that, this knowledge is not taken hypothetically but categorically.

Let us stress that this does not imply any dogmatism, but rather expresses the fact that no acquisition of knowledge can start without presupposing a ground of already established knowledge, and this means that these two *roles* are absolutely indispensable. The knowledge or information taken for granted is simply that which, at a *given* stage of the development of science, appears to be free from any reasonable, concrete and specific doubt, without meaning that it is to be understood as untouchable and indisputable in itself. If good *reasons* should appear for questioning this knowledge, it could be submitted to correction and even rejected; and such reasons might sometimes be imposed by knowledge obtained within a certain discipline, or sometimes from arguments elaborated outside the discipline.

Given these clarifications, we are confident that we have explained why even accepted theories may become 'referents' in the process of constructing a new objectification equipped with its own theory (in the proper sense). Old theories are taken for granted and *used*, they are not 'investigated,' and the scientist *avails himself* of them much in the same way as he avails himself of instruments, laboratories, facilities and so on. This in particular means that these old theories will never be assumed as hypotheses on which new theories *logically* depend. They belong to the 'factual' conditions for the emergence of a new field of inquiry (and are in this specific and restricted sense *a priori* with respect to this field), with all its conceptual, operational and theoretical features. It is especially because of these characteristics that the situation of historical determinateness which we are describing differs from the 'background knowledge' of which Popper speaks, and which—not being 'historicised'—can hardly escape an infinite regression (in fact *must* lead to an infinite regression, since Popper does not admit stable scientific results which can be assumed to provide reliable background information).

This situation may be illustrated by means of an analogy. A biologist using an electronic microscope is normally ignorant of the theory of this instrument, and would be completely lost if he should have to repair it, not for lack of technical skill, but for lack of knowledge about how it is made, why it is so made, and so on. In a similar way, a pharmacist working on the preparation of a drug on the basis of clinical tests need not know on which theoretical and empirical basis biologists established the existence of bacteria, their metabolism, the possibility of their inducing illness in man, and so on. He simply *assumes* the existence of such referents as a result of microbiology and *applies* his methods to them.

However, apart from the considerations suggested by these examples, the question becomes clear in its conceptual basis if one considers the extreme logical consequences of the thesis that theories *always* presuppose other theories. Indeed, it immediately gives rise to an infinite regression which renders the existence of theories impossible, for, in order for this process to be started, a *first* theory must be postulated, while, if the above claim is correct, this too must presuppose some other theory and hence it can *not* be the *first*. This simply means that this regression must stop somewhere with a theory (or set of theories) which have no other theories as prerequisites. One way of finding this basis, which is frequently proposed as sound and almost obvious, is indeed quite artificial and unrealistic. It is the way according to which we go back by analysing more and more elementary theories, until we reach the level of common sense, and here our knowledge is to find its starting point by resorting to certain elementary predicates which are related to immediate perceptions such as 'red,' 'round,' and so on.

In the case of scientific discourse, this alleged logical reconstruction (some of the most convinced supporters of which we find among the early neo-positivists) is pure fiction, for there is no mythical starting point of any concrete theory in a perceptual realm of virgin predicates, but every scientific discourse only takes shape starting from a pre-existing *cultural* background, that not only contains pre-existing objectifications but also a display of *intellectual frames* that we can qualify as *principles*, many of which are of such a general scope that they also constitute the structure of *common sense* understood in a deep, non-trivial way.

Moreover, in the case of everyday discourse a similar condition also holds, for the results of the different objectifications flow into the realm of common discourse through a process of progressive acculturation, and constitute a kind of stratified body, the most superficial and perhaps not completely solidified layers of which are constituted by the most recently accepted objectifications, while the oldest ones represent the bulk of currently accepted and widely spread knowledge; and the most ancestral ones may fragmentarily appear in some strange symbiosis with the more recent. (Consider, e.g., the fact that we usually say, and think, that the sun rises in the morning, while at the same time we know, and think in other contexts, that it is relatively at rest with respect to the fixed stars, while the earth revolves around it, rotating on its axis; and, lastly, we might know and think, in still another context, that the sun travels with the solar system and our galaxy in the cosmos.) In conclusion, what we inevitably meet when we try to analyse actual knowing procedures or systems of knowledge is a certain 'intersubjective basis' in which both common sense and scientific theories have their roots, and which does not go back, both historically and epistemologically, as far as many seem to believe.

In order to stress the fact that this basis constitutes the prerequisite and the condition of all scientific knowledge, we shall qualify it as *a priori*; but, on the other hand, in order to stress that it must not be confused with certain structural conditions of our thinking or knowing capacities, we shall call it *synthetic a priori*. Note that the similarity to Kant is only superficial, since for Kant both "synthetic" and "*a priori*" are adjectives (predicated of the noun "judgment"), while for us the term "*a priori*" is a substantive and indicates the complex of concrete and actually existing situations (ontology) and information (epistemology) which enable scientific knowledge to grow, while the adjective "synthetic" explicitly stresses the concreteness, the 'givenness' of these conditions that are the result of empirical inquiry and of an accumulation and correction of knowledge. One might think that, to this end, we should have used, perhaps, the term "empirical" and have spoken of an *empirical a priori*; but this, apart from a few advantages bound to the fact that this *a priori* includes referents, operations, and so on, which are of an empirical nature, has clear disadvantages as well, since it could lead us to overlook the fact that many non-empirical components also enter into this *a priori*, such as mathematical tools, accepted doctrines, general sentences, and models of particular realities, as we shall see more clearly in the sequel. (In short, our 'synthetic' does not mean 'empirical' as in the case of Kant.)

In any case, we need not worry about terminology because, if calling this *a priori* "synthetic" is useful in order to stress its epistemological status, we shall not stick to it, as we can immediately qualify it on the basis of a more informative characterisation, by calling it the *historical a priori*. We shall therefore say that all scientific knowledge and every scientific theory is based upon an historical *a priori*.

It is also clear that this *same* historical *a priori* also affects common sense and everyday discourse as well as personal experience. Yet we do not claim that it plays an equally important role for them. Hence it is possible, on the one hand, that, in order to understand and justify the knowledge possessed by single individuals, reference to the Husserlian *Lebenswelt* and the tools of a transcendental phenomenology could be of predominant importance. In the case of science, on the other hand, we are convinced that such tools are superfluous and might even lead us in the wrong direction, more or less as though we were invited to use a magnifying glass to look at and describe the house in front of us. We would surely perform our task better by looking at it with the naked eye. The same may be said of science. It can be constructed only after human knowledge has become considerably complex, embedding many preceding processes. The consideration of the historical *a priori* seems therefore the appropriate level at which an investigation of the foundational conditions of science may be located, other levels being better suited to investigate various kinds of pre-scientific knowledge which, although being implicit in it, are not specific to it. (Note, e.g., that it is only at this level that an account may be given of the fact that science cannot be an individual enterprise, a fact which cannot be discovered either by a logical or by an epistemological or phenomenological investigation.)

Let us now proceed to a brief survey of this historical *a priori*, adding details to the picture of it at which we have already hinted. There are some components of this *a priori* which we may conventionally call "material," and which are constituted by the different kinds of concretely identifiable 'things,' accessible and manipulable at a certain historically determined time. These things include the domain of what we call the natural world, but also (and for scientific inquiry to an even more important degree) all possible kinds of artefacts and products of human skill and technology. Besides this, there are also institutions, organisations, libraries, laboratories, financing channels which belong to the 'material' part of the historical *a priori* and which directly affect science, while other existing social features might concern it less directly. Recent 'sociological' philosophy of science has overemphasised this social–historical context, losing sight of the even more essential epistemological, intellectual and ontological aspects of science itself, but this does not mean that one can ignore or underestimate this context.

There is then what we may call the mental *a priori*, which contains accepted scientific theories, received tenets of the most different kinds, mathematical and logical apparatus, world outlooks, basic metaphysical conceptions, individual and social ideals, personal commitments, ideologies, and so on. This mental a priori is less restricted than the material a priori in the sense that it is essentially the same for the whole of the scientific community living in a certain historical epoch, but is still historically characterised. Not all of these ingredients are of equal relevance for the construction of scientific theories, but we shall leave the possibility open to take any or all of them into consideration if it should prove useful or necessary for us to do so.[1]

[1] An excellent and original characterisation of what we here call the *mental a priori* has been provided in a series of papers by Dudley Shapere. After having convincingly argued that no knowledge-seeking enterprise can take place without relying upon a background of beliefs, he has shown how this does not imply relativism or scepticism, or simply the elimination of objectivity from science. For science has elaborated criteria for checking the soundness of these beliefs, criteria which consist mainly in the beliefs being 'successful,' 'doubt-free,' and 'relevant' to the specific domain involved. The system of such beliefs constitutes what Shapere calls the background knowledge or background information of scientific research, not so much 'in general,' but regarding all of its concrete items. The difference between this position and Popper's reference to a 'background knowledge' is mainly to be found first of all in the much more precise and detailed expression of that notion, as well as in the fact that this background knowledge is explicitly recognised with its characteristics of soundness and reliability, which are alien to Popperian falsificationism. The distance from Kuhn's and Feyerabend's conceptions of paradigm, paradigm-change, and implicit arbitrariness in the choice and acceptance of them is even greater.

How the material component of the historical *a priori* may work in determining the shaping of scientific theories is rather clear in itself and does not require a special explanation after what we have already been saying about how scientific objects are constructed by means of standardised operational procedures—procedures which obviously presuppose the existence of instruments, laboratories, and adequate technological development. It is easy to enlarge this immediate image of the material conditions of scientific objectivity somewhat so as to include in it socio-institutional aspects for, if it is true that a certain objectification cannot be established unless we have at our disposal, for instance, an instrument which costs one million dollars, it is also clear that the availability of these funds somehow enters into the conditions of that objectification. We shall in any case say at once (and we shall return to this point later with a detailed analysis) that this part of the material conditions has little to do with the 'noetic' or cognitive side of objectivity proper.

As for the mental part, the performance of some of its components is clearly analogous to that of the material components. For example, it is clear that the establishment of the General Theory of Relativity was possible thanks to the existence of the tensor calculus in mathematics no less than the Special Theory benefited from the existence of the sophisticated optical instruments employed in the Michelson-Morley experiment. The influence of other components of this 'mental part' is less immediate—such components as, for instance, general worldviews or metaphysical conceptions. Yet their influence is effective and important, as has been stressed by several scholars who, in recent times, have particularly related their positions in the philosophy of science to an analysis of what happens in the history of science. In any case, we might venture to say that in the presentations of these scholars (let us mention only Thomas Kuhn as a good

---

(Footnote 1 continued)
In fact Shapere's detailed historical analyses show the deep *rationality* which guides the retaining and changing of those beliefs (a rationality, however, which does not coincide with formal deduction). In such a way he can convincingly escape the useless complications of 'theory-ladenness':

> But worries stemming from the 'loading' of background beliefs into what counts as subject-matter and observation are also dispelled. For we see that it is not just *any* background belief that can be used in shaping the course of science. There are conditions governing what can be so used, conditions which have been arrived at in the course of inquiry and which have become ever more stringent with the development of science. The scientific enterprise is thus a process of building, or rather of coming more and more to be able to build, on the best beliefs available. Insofar as science is able to proceed in the light of its best beliefs, its arguments and alteration are rational. The relativism into which Kuhn's views collapsed is thus escaped, even while all aspects of science are left open, in principle, to revision or rejection (Shapere 1984, p. XXV).

A valuable paper, in which these ideas are applied to the analysis of a concrete example, is Shapere (1982).

example) what appears reasonably clear is *that* such an influence exists; but it is not sufficiently explained *how* it works, and *why* it has to occur.[2]

We believe, on the other hand, that the doctrine of scientific objectivity proposed in this book provides a natural explanation both of the 'how' and of the 'why' just mentioned. For we have stressed that scientific objects are constituted by considering things 'from a certain viewpoint'; and this means that, in considering them, we decide to limit our attention to only those attributes which we single out as *relevant*, in order to characterise these things 'as we see them' under that specific viewpoint.

6.1.2 Relevance and Interpretations

We have called into play the category of *relevance*, which is extremely delicate due to an intrinsic ambiguity it contains, being involved, in common use, both with practical and with cognitive contexts. We shall discard from present considerations the practical aspect, and shall therefore not consider the sense of "relevant" according to which, for example, we decide to take into account simply the calorie content of a piece of bread because we are *interested* in not becoming overweight by eating calorie-rich foods. We shall limit our attention, instead, to the *cognitive* relevance according to which a certain feature is considered decisive, in order to express what we *mean* by a certain approach to reality. It is clear, therefore, that such a relevance depends on the whole image, on the whole approach under which reality is being considered, and which we could express by a single and better word by calling it a *Gestalt* (as we have already done on several occasions). This technical term of cognitive psychology expresses fairly well, in our opinion, the central aspect of our notion of 'viewpoint.'[3] Indeed, it entails on the one hand the idea of relevance (expressed in the distinction between *figure* and *background* which gives rise to different *Gestalten*, depending on which parts of an image are

[2] This, on the contrary, is no longer the case with scholars who have gone much deeper into the details of this interplay, as has, for example, Rom Harré (see especially Harré 1970). Also well-known scholars such as A. Koyré and L. Fleck have earlier produced notable work along this line. Regarding this necessary interplay between philosophy of science and history and sociology of science let us express a very simple but significant remark: science is a *cultural reality*, and not a *natural reality* (in the sense that it is a product of human activity and not simply the outcome of physical processes). Therefore, it should be obvious that the study of this reality belongs to the domain of the "sciences of culture" and not of the "sciences of nature" (to use a famous distinction). In other words, physics investigates the features of a certain domain of nature, whereas investigating the features of physics amounts to describing and understanding the characteristics of a complex cultural reality, a task that, by itself, requires the contribution of many approaches, besides that of logic and methodology, typical of different "human sciences." This fact is explicitly recognised, e.g., also in Stepin (2005), p. 46.

[3] The suitability of such a terminological choice is confirmed by the fact that it has been adopted as the Gestalt Model of scientific theories proposed in Dilworth (2008).

considered to belong to the figure and which to the background); and on the other hand it hints at the fact that the relevant elements give rise to a 'unity' of a given kind, and receive their relevance precisely from the role they play in that unity.

Let us note that this explains how cognitive relevance is independent of practical relevance. For example, it may be true that we consider bread only from the viewpoint of its calorie content because we are 'interested' in avoiding becoming overweight. But this does not imply that the correctly relevant cognitive patterns, in order to satisfy this interest, are in the consideration of the calorie content rather than of the total mass of the bread. That calories and not mass are causally relevant to becoming overweight is *knowledge* that comes from scientific investigation.

Now the problem becomes that of understanding how these *Gestalten* come about; and it is here that general worldviews, *Weltanschauungen*, and also quite special metaphysical conceptions about particular sectors of reality, play their role. They suggest the main lines of the possible *Gestalten*, and sometimes they can favour the assumptions of some of those lines and make it difficult to accept others. The most often discussed example in this connection is that of the metaphysical worldview that supported Ptolemaic astronomy, and which produced resistance to the acceptance of the Copernican system (which in turn was favoured by an alternative worldview).

But these metaphysical *a priori* need not be expressed through detailed 'images' or 'pictures' of reality. Sometimes they may simply rely on some rather abstract metaphysical principles such as those concerning the possibility of time being finite or infinite, of the universe not having a precise origin in time, and so on, which may be clearly shown to orient the most significant theories in contemporary cosmology.[4]

[4] For a good discussion of this issue see Hübner (1978). Popper too has expressed a positive appreciation of the role of metaphysical views in producing fruitful "conjectures" that can be expanded into scientific theories, and within the Popperian school, J. Agassi in particular has stressed the importance of "influential metaphysics" in the development of science (see Agassi 1974, 1981). One could note, however, that the notion of metaphysics adopted by these authors was a rather poor one, essentially amounting to a 'general worldview.' Much more significant had been in the past the relation between metaphysical principles (understood in the strong ontological sense of tradition), and the construction of scientific theories proposed by William Whewell (1847), who explicitly criticised the positivist opposition to metaphysics advocated by Comte. Such an explicit link of metaphysical general principles with modern science, however, did not find significant prosecution until the work by Dilworth (2007), a whole book that is the most highly elaborated view along the lines anticipated by Whewell. We shall return to these issues in Chap. 10. An exception to what we are saying here is constituted by Émile Meyerson, who investigated the presence and function of such universal principles as that of the regularity of nature, of causality, and of identity in the construction of science, both from an epistemological and an historical point of view. One should note, however, that according to his approach, such principles are of a psychological rather than of a metaphysical nature and that, in addition, he was not in keeping with the mainstream French philosophy of his time, so that (in spite of the admiration of certain important philosophers and scientists), his work did not receive the positive evaluation it deserves, and is still rather little known today.

If we want to schematise the process just hinted at, we might say that at any historical moment there is a complex background of metaphysical perspectives (in the sense that they intend to express the fundamental features of "reality as such"—to use the classical Aristotelian characterisation of metaphysics), partially unquestioned, partially conflicting, partially still in a state of formation. These yield a display of *Gestalten* concerning the most various fields of reality, and determine, accordingly, sets of attributes which are considered relevant in order to characterise these *Gestalten*. It is at this level that the rise of a scientific theory might be seen as the effort of making a certain *Gestalt* explicit in all its potentialities, and this poses two simultaneous, but logically subordinate, problems, one of concept formation, and one of theory construction. (To be more precise, the general metaphysical principles determine the *discipline* or *science*; *theories* are constructed and compete *within* a particular discipline, i.e. they accept the *same* principles, but make explicit a particular *Gestalt* within the framework of these principles.) The problem of concept formation (as we have seen) is solved by sorting out operational procedures which are able to fix the meaning of some of the relevant attributes, and in such a way to fix the *domain of referents* of the theory (referents—as we have explained at length—that are not only empirically or operationally attainable, but can also be theoretical). The problem of theory construction appears then as the effort to 'speak about' the referents in an adequate way.

For example, the act of singling out as 'fundamental' certain quantifiable predicates such as mass, length, time, and force has clearly been the consequence of the pre-scientific conception, represented by the mechanistic worldview which approached the whole of reality from the point of view of matter and motion. The efforts to make this *Gestalt* precise led first to some inadequate pictorial representations such as those contained in the treatises of Descartes, which lacked the operational anchorage necessary for giving rise to objectivity, but eventually led to the operationally-based Newtonian mechanics. Needless to say, the success of a certain scientific objectification works as feedback in favour of the *Gestalt* which promoted it, and this is why Cartesianism was superseded by Newtonianism in France itself during the eighteenth century.[5]

Taking this approach, one can see, for instance, how the different schools which were fighting one another so intensely in the field of psychology until not so long ago were actually expressing different pre-theoretical outlooks on the domain of psychological reality, or on the idea of what is scientific. These outlooks thus determined the choice of related protocollarity criteria (such as the pure observation of behaviour, rather than introspection, or the administration of tests) and in such a way psychologists, reproaching one another for not being scientifically

[5] Through this process of feedback, science 'internalises' even its general metaphysical presuppositions, and discriminates between different metaphysical patterns, as is shown, for example, in the already quoted works of Shapere. However, in the sequel we shall consider this internalisation of metaphysical presuppositions again—an internalisation which may lead to scientism if not kept under control.

correct, did not realise that they were simply handling different objects in the different schools. The conflicts came from the fact of having arbitrarily advanced the metaphysical pretention that the *particular* perspective of a single school was able to exhaust the *whole* of psychic reality.

We shall only hint at two points which find their natural place here, but which will receive more detailed consideration later. The first is that in the above analysis it is implicit that reality must be 'understood' in a certain way before becoming the object of scientific inquiry proper; and this understanding has the character of a global *interpretation* of the relevant attributes of reality. The subsequent 'explanation' always occurs, then, within the context of a certain interpretation. As one sees, the two famous categories of *explanation* and *understanding*, of *Erklären* and *Verstehen*, which are often advocated in order to claim the irreducible opposition between the 'natural sciences' and the 'human sciences' (*Naturwissenschaften* and *Geisteswissenschaften*), are being considered here (of course, giving them a meaning not identical with that present in that old dispute). It appears that these epistemological categories are, on the contrary, related, and that, in particular, understanding also has to do with the natural sciences. This fact will offer us useful reasons for reflection, as will the strongly related fact that a *hermeneutic* moment is contained in every scientific enterprise (where "hermeneutic" strictly means "related with interpretation," and does not imply reference to a particular philosophical doctrine so denominated).

Our second point is that it has become fashionable today to speak of the dependence of scientific objectivity on social, ideological and political conditions. It seems to us that the only place in which these claims may find a correct and not a trivial or partisan treatment is exactly at the level of the historical *a priori* which we are presenting. We shall therefore return to this issue when we tackle the problem of the so-called neutrality of science.

In concluding our considerations we could say that, as regards the problem of scientific objectivity, the historical *a priori* presents itself as a system of given gestaltisations, of available information, and of given operational 'possibilities' of translating them into the objective referents of scientific theories. In this way, historical determinateness appears as something different from conventionalism, necessitarianism, and arbitrariness. It is different from conventionalism because conventionalism would claim that the fact that scientists reach a certain agreement about their way of objectifying reality is only a consequence of decisions and choices they freely make.[6] This, as we have seen, is not at all the case, since the availability of particular gestaltisations does not depend on the wills of scientists. Nor does the scientists' being embedded in a particular cultural background which spontaneously leads them towards certain worldviews, nor their having definite technological and/or intellectual tools at their disposal, depend on their wills. The intersubjective agreement is therefore the result of a complex system of interactions which certainly leave room for decisions and choices, but within rather

[6] This, as we have seen, was affirmed at least to some extent even by Popper (cf. note 4 above).

narrow limits, and, in any case, not to such an extent that the objectification can be made dependent *only*, or even *mainly*, on these decisions and choices.

On the other hand, the opposite extreme, which would see historical determinateness as a form of necessity, is also untenable. It is clearly so, first of all, because 'matters of fact' are always contingent in the proper sense (unless one is a confirmed follower of Spinoza or Hegel, but one would still have to *prove* the contrary); and this excludes the strictest sense of logical and ontological necessity. But this is so also because historical determinateness, as we explained above, represents a system of given possibilities, more than a system of determined structures. We could say that the historical system admits of a great number of degrees of freedom in the relations among its 'fixed' components, and that this explains the possibility of making choices and decisions in it. In any case, owing to the 'fixed' components (which we could call the *internal constraints* of the system), these choices cannot be arbitrary, and this is why scientific treatments present themselves as reliable even to people who are actually unable to understand their specific content.[7]

An important consequence which we can easily derive from the above-sketched features of historical determinateness is that it leaves complete freedom to the *creativity* of the scientist. Indeed, it entails that this creativity is a necessary ingredient of science. The reason is the following. If the historical determinateness consists in displaying many possibilities given by the *presence* of certain conditions, it is clear that we need something *in addition* in order for these different possibilities (or, more exactly, *some* and *only some* of them) to be actualised. This is the task for the scientist's free creativity. In other words, we cannot confine ourselves to saying that, in order for the general theory of relativity to exist, historical conditions such as the existence of the tensor calculus or good optical instruments for testing it were needed; we must also add that Einstein was needed as well! This sounds very obvious, but it tends to be neglected by several present conceptions of science, especially by some which are proposed within the domain of dialectical materialism.

As a matter of fact, many features of our general conception of the historical determinateness of scientific objectivity should also be acceptable to a dialectical (or historical) materialist, possibly together with an effort to reduce the role of the mental *a priori*. But one reason that a consistent historical materialism would not be able to fully account for science is that it could not *explain* the role of free creativity. It follows that it might be able to account for many (true or alleged) possibilities determined by social and economic conditions, in order for a certain scientific objectification to take place, but the *individual use* of these possibilities, made by the single scientist in order to construct a theory, cannot be accounted for along this path.

---

[7] It is by correctly underscoring this fact that Shapere has illustrated how *rational* scientific choices are, and that Harré (particularly in Harré 1986), has insisted on the *reliability* of science, and on its being essentially an enterprise based on a "moral order" of trust.

We would like to conclude this section with a couple of considerations concerning the relationship between *contingency* in the sense presented here, and a *necessity* which is still inherent in scientific constructions in general. The meaning of this contingency is now fully clear. It consists in the choice of the criteria of relevance (in which even personal, social and ideological factors may play a role) as well as in the actual availability and in the choice made of operational tools for giving shape to criteria of objectification, or of certain intellectual tools, as explained above. Yet, *within* this contingency, as we have remarked earlier, a kind of substantial necessity exists, once the discourse realised under these contingent conditions is open. Or, in other words, once one of the open *possibilities* is actualised, the degrees of freedom within the selected possibility are (though not totally eliminated) drastically reduced, and we come very close to a kind of necessity. We could express the same conception by saying that it is not 'necessary' to be concerned with this or that domain of objects; but once the choice of objects is made, what we can say about them necessarily becomes limited. We shall use this remark later in order to see how a sound objectivity in science is compatible with the fact that the *Gestalten* and the viewpoints which generate them can be questioned.

6.2 The Historical Dimension of Science

6.2.1 Why Was it so Difficult to 'Historicise' Science?

What we have said in the preceding section leads us to propose some suggestions, not so much regarding circumscribed issues such as that of the nature of scientific objectivity and its conditions, but rather regarding science in general, that is, regarding that marvellous phenomenon of human civilisation of which the production of objective knowledge can be seen as the major aim, but which also includes many other features that are likely to remain unnoticed or underestimated if one's attention is focused only (or chiefly) on epistemological or methodological problems. If we consider human civilisation globally, we can see it as something which develops through history in an organic way in the sense that all its aspects are interconnected and participate in the historical development in a triple sense: (a) they are themselves subject to an 'internal' historical change; (b) they are 'made by history' in the sense that their 'internal' history is to a certain extent determined by the context of the general historical course; (c) they are 'factors of history' in the sense that they contribute to the making of the general historical course. Because of this intimate interconnection, we feel that nothing 'human' is irrelevant to the understanding of man's history and civilisation.

Furthermore, to the extent that we become aware that a deeper understanding of humankind and its civilisation implies an acquaintance with its history as well, we realise that this is also true of civilisation's various components. This claim, however, must be correctly understood: (a) it means that knowledge of the past may often help a 'genetic' understanding of the present; (b) it may also mean that the present can sometimes be better understood via certain analogies with the past; (c) and it may mean, on the contrary, that an understanding of the present can be made easier by considering its differences with the past. These aspects still do not exhaust the reasons for an historical interest, since they remain confined to a limiting presupposition, that is, that the study of the past has an interest *because* it is instrumental to the understanding of the present. A genuine historical interest, however, does not take shape unless the past is considered interesting in itself, so that our primary aim should be that of understanding the past independently of the advantages (if any) that we could then draw for a better understanding of the present. If we ask 'why should we be interested in the past?,' the most convincing answer could be 'because it belongs to us'; and in this sense doing history is a kind of testimony or homage that humankind pays to itself. This is why a clear symptom of the fact that something is perceived as constituting a highly valuable component of human civilisation is its being conceived of historically according to all the dimensions of historicity outlined above.

When it comes to science, it appears highly questionable that it be commonly perceived according to the 'historical consciousness' just mentioned, despite the great development of studies in the history of the sciences which have taken place in recent decades, and also despite the increasing use of historical discussions made in works devoted to the philosophy of science. We shall first try to understand why the situation is such, and then provide some critical reflections.

Let us begin with some descriptive remarks. When we consider the 'common way' of seeing, appreciating and understanding the different forms and manifestations of human civilisation, we become aware of a rather surprising fact. We are ready and spontaneously inclined to place them in an historical perspective and consequently to judge them according to an 'historical consciousness,' with practically only one exception: that of science. No one finds it difficult to admit that the poetry of Homer, Virgil, Dante, Goethe or Baudelaire has been authentic, attaining the heights of absolute value and even at times being of a hardly attainable perfection, all the while recognising that to understand this poetry, to appreciate its excellence, and to penetrate its meaning, the effort must be made to put it within its historical context (and ideally to put *oneself* within that context) rather than to judge it according to the modes and forms of the poetry of our own time. What we have just said regarding poetry also applies to music, the fine arts, philosophy, law, social and political institutions, ethical concepts, religions and customs. In the case of science, such an historical consciousness is almost entirely lacking, even among cultivated people. The history of the sciences is not normally a part of the store of knowledge of these people; but this situation, far from being the cause of such a lack of historical consciousness, is rather its consequence. This happens because we are unconsciously persuaded that science is not, properly speaking, an historical phenomenon; we have the impression that it has not had a real history.

This statement calls for some clarification. In fact, all those who have received what is known as a basic education and have been encouraged to study some mathematics and science are familiar with the theorems of Pythagoras, Thales and Euclid, the Cartesian co-ordinates, Archimedes' principle, Newton's laws, and so on. They have probably heard the Ptolemaic and Copernican systems of the world mentioned, and Darwin's theory of natural selection (and naturally they also know that these persons belong to a more or less remote past). This suffices to show us that there is a certain 'past' in science; but this is not enough to prove that there is a 'history' proper. For these different names are linked to some isolated 'discovery' that finds its place in an exposition organised according to logical, systematic, didactic or other criteria, but which falls outside an historical interest in a proper sense. These names play an essentially mnemonic role that facilitates the reference to a certain statement whose meaning and value are entirely determined by the place it occupies within a contemporary scientific discipline. Hence it is not difficult to understand the most common view of science: it is seen as something that only has a present (it could be defined as the present state of our knowledge), while its past no longer belongs to it since, if there was something in this past that deserved to be saved, it is already incorporated in the present (it is therefore still present). The rest has been forgotten and is no longer of interest or importance.

Certain stereotypes easily appear in such a perspective. We believe that science properly speaking only appeared in a very recent epoch, having been preceded by some isolated and almost accidental discoveries that were made fortuitously within an intellectual context that was still confused and primitive, with the sole exception, perhaps, of Greek mathematics, for which we gladly recognise a high degree of logical rigour. Moreover, the brevity of this course, which should make the undertaking of an historical reconstruction easier, does not arouse the necessary intellectual interest precisely because such a history would be reduced to a sort of catalogue of 'truths' and 'errors,' the former of no use because they are already collected and preserved in present-day science, and the latter just because they are errors. We clearly see, then, why in the case of science the tendency is to judge the past in the light of the present and, in any case, to 'rid' ourselves of it, in contrast to what happens with other forms of cultural expression.

What is the reason for such an astonishing difference? A first answer to this question (or a first part of a possible answer) may come from the consideration of the historical period during which the 'sense of history' developed as a fundamental constituent of European culture, that is, the Romantic Age, dominated by an historicism that was expressed in the thought of its philosophers as well as in the work of its artists, men of letters and, especially, its historians. They no longer limited themselves to political history, but engaged in the reconstruction and interpretation of all forms of human culture.

For these people history was essential not only as a recognition of facts; it appeared as a request for the conscious understanding of any issue to which it offered a new dimension, beyond the simple rational analysis in which the men of the Enlightenment had placed all their hopes. History became an interrogation of the past that nourishes the present, a dialogue of the present and the past that also contains inspiration for the future. It was a great reservoir of sense that could play the role of a revelation and a promise. For this reason, we recognise the greatness of each century, and the task of the historian is to glean from each period an interior knowledge that grasps it as it really was.

However, parallel to this historicist atmosphere, another line of thought developed in the same decades of the nineteenth century, namely the positivist movement that continued in its way the 'progressive' movement of the philosophy of the preceding century. August Comte proposed that for each area of reality that man attempts to know, three steps are involved: the theological, the metaphysical, and the positive. The last corresponds to the moment in which a certain branch of knowledge becomes science and, in that way, achieves its definitive maturity.

At first glance, we should say that here we have an historical comprehension of the development of human knowledge. But we soon see that in reality the movement of history is brought under an *a priori* interpretation, abstract and arbitrary, which judges the past in an essentially negative way. It is considered to be immature, having an inadequate view of things that should be surpassed and forgotten once the light of scientific knowledge liberates us from the gropings and phantoms of other more primitive forms of knowledge.

The affinity of this position with Hegel's concept of a history of thought that culminates in his own system, or with the Marxist concept of a history destined to end with the coming of the classless society guaranteed by the dictatorship of the proletariat, strikes us immediately. Hegelianism was questioned and superseded even before the death of its creator, while Marxism, having situated the advent of a classless society in an indeterminate future, could give itself a practically unlimited waiting period. But compared to these, positivism had the advantage of declaring itself the champion of science when science was already celebrating its triumphs (the more spectacular of which it would continue to celebrate for a long time to come). Under these conditions, it is not surprising that European culture let itself be convinced that science has no authentic history, that what went before was not the course of its internal evolution but rather, so to speak, a kind of prehistory of no real importance (with the additional prescription, made by many scholars, that we need to continue fighting against metaphysics and religion, in order to prevent them from stopping the progress of science).

It could be objected that Comte's theses, though too schematic, emphasised a fact that meets with general agreement, namely that the sciences of nature began their progress in the seventeenth century, while many 'human sciences' began theirs only in the nineteenth century. This does not imply, however, that they do not have a history. It is only that their history is more recent. For there to be a history, it suffices that there be change and growth in time, and this is certainly the case with the sciences.

We reply that the simple setting down of a past and a future, of a change and a growth in time, is not enough to establish an historical consciousness. This implies (as we have seen) that there is an actual interest in understanding past events, with the meaning they had when they occurred. Furthermore it suggests that we can recognise the true value of past events, independently of whether they still retain a meaning and a value today (which would be different), while probably helping us to better understand certain events of our own time (usually thanks to the differences rather than to the similarities). As we have pointed out, this has become normal for almost all manifestations of culture except for science; and we formulated the hypothesis that, for science, the obstacle to its inclusion in historical consciousness is a result of the fact that it has been a prisoner of the anti-historical approach of positivism precisely during that period in which Romanticism approached all other forms of human culture historically.

However, there are certainly other reasons for the exclusion of science from historical consciousness that are worth exploring. One of the most important is found in the task that is spontaneously assigned to science, that of giving us a faithful and *objective* picture of the different sectors of reality it deals with, without mixing it up with interpretations, judgements and evaluations. In the case of the arts, philosophy, law, religion, and customs, we usually think that these areas are expressions of human subjectivity, that they represent the products of the creativity of man, the richness of his sentiments and intuitions, and we are thus spontaneously inclined to evaluate them for what they are in themselves, to see them as testimony to an inexhaustible wealth that is always of interest to us, that may inspire and at times guide us, but which usually attracts us in a spontaneous way because it satisfies our *subjective* feelings. In short, what makes these manifestations of the past interesting to us is precisely the presence of those elements of subjectivity that—on the contrary—we insist be *eliminated* from the discourse of the sciences.

The multitude and variety of expressions that appear to us a fecundity and richness in those areas create in the case of science an intolerable situation that must be avoided in order to arrive at a univocal image of reality: that which reproduces its true structure, that is one and only one, and with regard to which no liberty may be taken. Consequently—according to the most common way of thinking—a scientific proposition has only two possible destinies: it is either true (because it describes reality as it is), and then becomes an enduring part of the patrimony of science (or, at least, is kept in science to the extent to which it can be believed to be true), or it is false, and will be rejected as soon as it is recognised as such. The historical moment when a true proposition was formulated does not influence its value. The fact that it was discovered a century earlier or later does not change its position or its meaning within science (at most it may enhance the prestige of that century).

In this perspective it is clear that no proposition or scientific theory has any historical meaning. It is a-historical and, consequently, all science is deprived of an historical dimension. A history of science is not impossible in this perspective, but reduces to the presentation of that catalogue of truths and errors we have already mentioned. This would have a documentary interest and would respond to a desire for erudition, but would not constitute an historical perspective proper. We have used the conditional tense here, but it must be said that most of the histories of science that have been published up until recently are of that type.

Since the first years of the twentieth century, the idea that science provides 'true' and unshakeable knowledge about nature has entered into the profound crisis we have already discussed. However, this has not contributed to its acquiring a dimension of historicity, because the conception that has gained ground has been that of science's having an essentially pragmatic value (further facilitated by the great successes of technology seen as 'applied science'). In the framework of such a conception there is still less interest in taking into account the scientific propositions and theories of the past. If they were abandoned because they were no longer of use, there is no reason for returning to them. A science conceived as the repertory of knowledge that is useful to us *now* obviously offers no conceptual space for attributing a significance to a form of knowledge now become useless and 'obsolete.'

These are some of the reasons that explain the lack of historical perspective that still today characterises the most common way of conceiving of the sciences.

6.2.2 The 'Historical Consciousness' of Science

We have spoken of a 'common way' of seeing science, but this should not be considered as a reference to the opinions of uncultivated people. The a-historical perspective outlined above is widespread among scientists, and is well reflected in the philosophy of science that dominated until the 1960s (and probably still dominates today). The logical empiricists and Popper not only had in common the reduction of science to a set of 'theories' which were essentially systems of logically interdependent sentences, but moreover both proposed as 'science' a very abstract linguistic construction which could at best be seen as a rough schematisation of *contemporary* science (and even, in particular, of mathematical physics). Furthermore, this way of viewing science was a-historical not only because it absolutised the present while discrediting the past, but also because the present itself was by no means considered in its actual historical features. Contemporary science as it was actually practised was seldom analysed in it, and it is astonishing how few scientific facts, concepts, laws, and theories are sporadically presented by these authors, how ultra-elementary and roughly described they are, and how the very same 'examples' are used to support opposite claims.

This reflects the above-sketched idea of science being either the accumulation of a-historical truths (logical empiricism) or the elimination of errors (Popper). Therefore it is not surprising that the one camp maintained that Galileo's mechanics was absorbed in Newton's, and this in Einstein's as limiting cases (the truths being preserved), while the other said that Galileo's mechanics was disproved by Newton's, and this by Einstein's (errors being eliminated). But what matters more is that these authors explicitly refused to consider 'external' conditions to be relevant for the analysis of science and the understanding of its way of acquiring knowledge.

The situation has only apparently changed with the works of Kuhn, Lakatos, Feyerabend and many others who gave space to considerations of examples taken from the history of science in elaborating their views in the philosophy of science. Indeed, one cannot overlook that the aim of these authors was that of discussing a theoretical, and not an historical problem, the problem of theory change. Historical evidence was proposed with the view of disproving certain philosophical tenets (and this certainly underscored the fact that philosophy of science cannot be pursued with respect to a more or less imaginary or overly idealised schematisation of science, but must take actual science into consideration). If one carefully considers these analyses, however, one finds that they incline towards a form of sociological relativism rather than towards an appreciation of the historicity of science (and actually it is rather transparent that the specifically epistemological background of these authors was still the traditional one: Kuhn and Feyerabend were much in keeping with analytic philosophy of science, while Lakatos was a Popperian). Moreover, the majority of the historical examples were selected with the aim of defending or rejecting specific theses in the philosophy of science, and they were given a—usually debatable—interpretation capable of performing this function. All this often amounts to using history of science as a kind of 'toolbox' from which one takes the most useful weapons for defending one's epistemological claims while remaining quite far from a genuine 'historical consciousness' of science, which is to be found only in few contemporary authors whom we have already mentioned. But if what we have advocated in the discussion concerning the 'historical determinateness' of scientific objectivity is sound, we must conclude that the appreciation of a genuine 'historical dimension' of science (in the full sense presented at the beginning of this section) is an indispensable part of any understanding of science, and also of any philosophy of science. Let us see how this awareness could develop from existing and already-admitted requirements.

The first might be the already-mentioned claim that philosophy of science cannot be a discourse that 'floats' in the world of ideas without effective contact with the real world, but must demonstrate that its models of scientificity correspond to an acceptable degree to the world of science as it is, and not as it is imagined or postulated to be. In other words, the philosophy of science is obliged to exercise an 'empirical check' upon itself, and such a check can only be performed by considering a history of the sciences that does not claim that the forms of science (past or present) that do not correspond to the model are not really scientific.

A consideration of this kind is not easy because the models set up by analytical philosophy of science are not satisfied by a good number of present sciences nor, for even stronger reasons, by the sciences of the past. The only exit from this impasse is to recognise the existence of different *hermeneutic* contexts (or criteria of intelligibility) that characterise the great variety of contemporary scientific disciplines and direct their choice of empirical criteria, the empirical and theoretical concepts they employ, and the models of explanation they adopt. By doing this, we are already beginning to 'historicise' science (that is, we are putting it into its 'contemporary' historical context). To this must be added—though the step is no longer difficult—the awareness of an historical dimension that also extends into the past. If we complete this awareness, we open up the possibility of understanding the sciences of different ages according to the hermeneutic contexts that characterised them. We can glimpse the ties between the sciences and philosophy and consider how these endeavours have interpreted certain conceptions of the world and of man, and how they have contributed to their modification and evolution. In short, the sciences and philosophy can be inserted as vital constituents in the development of human civilisation.

We certainly do not wish to suggest that philosophy of science must be resolved within the history of science. We only affirm that a philosophical understanding of science as a construction of human thought, and of the different sciences as the articulation of this construction aiming at knowing certain sectors or aspects of reality, cannot do without an awareness and examination of the ideas and ways of conceptualising that have determined the elaboration of scientific theories in history. In saying this, we also hope to avoid the impression that a philosophical comprehension of the sciences must 'submerge' them in their historical and social contexts. Such a dissolution would be incorrect as far as the pure and simple history of the sciences is concerned, and even more so as regards their philosophy.

The philosopher of science as such does not need to reconstruct the genetic pathways that have led to the formulation of certain scientific ideas or principles, or to the construction of certain instruments. However, he is obliged to pay attention to such ideas, principles and material or mental techniques when they actually become a framework for the construction of a given scientific theory (namely, by furnishing the hermeneutic context for the selection of its basic concepts, or the ensemble of the logical and mathematical constructions that underlie the theoretical architecture of a discipline).

Thanks to this historical awareness, the philosophy of science will rid itself of a certain number of false problems that greatly disturb its present discussions, and that have led to attitudes whose effect has been that of discrediting the philosophy of science itself. We refer especially to the aforementioned question of the comparability of scientific theories, and the possibility of admitting progress in scientific knowledge. We have seen that some philosophies currently in vogue deny the possibility of establishing an effective comparison between scientific theories, and hence of establishing a cognitive preference among rival theories which recognises *progress* when one theory is replaced by another. In such a way the common belief according to which, thanks to the development of the sciences, our present knowledge is quantitatively and qualitatively better than that of our predecessors, is considered naive and unjustified. This general belief is also shared by scientists and, as a consequence, they become inclined to regard with suspicion a philosophy of science that seems unable to justify such a very fundamental conviction they have. Formerly, scientists had a certain sympathy for philosophy of science because, while still considering it rather schematic and distant from the problems they dealt with in their daily work, they saw in this philosophy a serious attempt to understand the structure of scientific thought. Today this sympathy and this interest are gradually declining. This is then the ironic situation: the philosophers of the logical empiricist tradition ended up by studying an imaginary science, owing to a lack of historical sensitivity and knowledge, while many philosophers of science who constantly refer to the history of the sciences in their written works present an image of science that is equally artificial and is rejected both by laypersons and scientists. How can this be explained?

The answer is not difficult. The thesis of the 'non-comparability' of scientific theories was born, as we have seen, within a logico-linguistic approach that led to the tenet of the 'theory-ladenness' of empirical terms. And it is in the light of this tenet that the advocates of the non-comparability thesis often wished to read the history of science, in order to compel it to give them support. If we try to be more attentive to the true lesson of history, we can receive a much richer impression of the past. First, we can ascertain that many 'fundamental' hermeneutic frameworks exist that have often inspired scientific conceptions throughout history. Each of them lasted only for a certain time, but the hermeneutic framework that inspired them proved capable of leading to other scientific conceptions in later epochs (thus these frameworks have an historical stability that goes beyond the precariousness of their specialised 'concretisations'). For example, the atomist and continuity intuitions regarding the physical world, the conceptions of potential and actual infinity in mathematics, the role of chance and necessity in natural phenomena, the paradigms of final and efficient causality as models of intelligibility in the various domains of knowledge, are only some of the basic conceptions we could cite, not to mention more specific examples, such as the different ways of conceiving of space and time. The history of the sciences shows us how these general conceptions recur and evolve in a way comparable to that of certain intuitions of Plato and Aristotle in the history of philosophy, or that of the idea of democracy in the history of political institutions, or the forms of property rights throughout the history of legal systems.

In conclusion, just as the philosophy of law was nourished by the history of law, and political philosophy and the philosophy of art were nourished by knowledge of the history of politics and art, so the philosophy of the sciences finds in the history of the sciences concrete realisations on which to reflect, since they are stages in the maturation of science which, while being consigned to the past, have a meaning for the present, even though they are not directly utilisable. Just as we study Dante, Beethoven or Roman law, not with the intention of writing poetry in Dante's style, or producing Beethovian compositions, or introducing into our legal systems the forms of Roman public law, but with the intention of having a better understanding of the nature of poetry, music and law (and even of cultivating our poetic, musical and legal sensitivity), a more than superficial knowledge of the history of the sciences greatly helps our comprehension of this fundamental dimension of human civilisation, and may even be profitable for our scientific education.

In addition, we must not underestimate the importance of the scientific results that have been achieved in the past and that maintain a validity that we may consider as definitively established. What we must avoid is to 'flatten' them in the simplistic view that claims that everything of value in the science of the past is preserved in the science of today. The most correct way to evaluate these results is to understand that, throughout its history, science has made several of those ‘cuts’ out of reality we have often mentioned and that, within them, many objective truths have been established. These are lastingly acquired for the patrimony of human knowledge (this is why, for example, within the ‘cut’ corresponding to elementary geometry, the theorems of Euclid’s *Elements* remain valid within the limited domain of its ‘objectification’ and continue to be studied). We must not lose sight of the fact that the introduction of new domains of objectification does not depend solely on the constitution of new hermeneutic frameworks, but also on that ‘background knowledge’ of which we spoke in the preceding section, and which can be seen as the historical ‘accumulation of the results’ of science. In such a way we concretely recognise the intrinsic value of accumulation in science, even in those parts of it which we no longer use at present.

A comparison may help us to better understand this point. When in a specialised museum we look at the scientific instruments that have served the researchers of past centuries, we are often astonished at the beauty, perfection and fine workmanship of these tools. They retain all their intrinsic value, their ingenuity of design, even though we should never think of using them today, because the instruments of today are more accurate. On a closer look, however, we see that this improved accuracy is linked partly to pure and simple technical progress and partly to the fact that we have gone on to study other domains of research in which the older instruments are no longer of use (the pragmatic side of science is thus involved here). But if we wish to fully understand the science of a certain age we are obliged to take its instrumentation into account, and even at times use these instruments to repeat the observations and experiments that were then possible.

Only by again reaching this level of consideration can we return to science all the spiritual and cultural value that is its due. Just as we can admire Roman law, or Michelangelo’s statues, without thinking of being able to ‘use’ them for the concrete requirements of our age, but at the same time feeling that they belong to us as a living part of our history, in the same way we must adopt a similar attitude towards the history of science. In addition, only in this way can we justify an intuitive conviction that we all share, namely, that geniuses such as Euclid, Archimedes, Galileo, Newton and Maxwell are on a scale of grandeur that goes beyond that of many of our Nobel laureates, and that they have contributed to the building of our civilisation to a degree that is not inferior to that of the great geniuses of the arts, letters, philosophy, law and religion. Through such a realisation we can hope that contemporary science may play its role in the building of our culture, a role that is partly lacking today precisely because we have too often considered it as a simple collection of provisional knowledge that is of interest and is meaningful only as far as it is practically useful.

6.3 The Hermeneutic Dimension of Science

We are now going to develop in a more detailed way certain remarks concerning the 'gestaltic' preconditions of scientific theories which we have already mentioned occasionally, particularly in the last sections.[8] We shall see that these preconditions correspond to a specialisation, in the case of science, of the general notion of a 'way of representing reality' when this notion is applied to actual—and therefore particular—cognitive situations. What makes these situations concrete is the fact that the knowing subject meets a great deal of data through their individually affecting her sense receptors in space and time. These data, however, are immediately unified, in a spontaneous and unconscious manner according to a certain 'way of representing them' that already organises them under a kind of 'general' structure. Such a fact has never escaped the best investigators of our knowing activity, from Plato and Aristotle to Kant and the psychologists of the *Gestalt* school.

These last have stressed the difference between sensations and perception, the latter being an organisation of sense data which is not univocally contained in them, but rather corresponds to a 'shaping' of them according to a pattern superimposed by the unconscious but effective activity of the knowing subject. Pictures which may be seen as representing very different objects (such as a vase or two opposing human profiles) depending on what in the picture is perceived as being the background or foreground are well known. No less famous, and widely discussed in recent philosophy of science, is the example used by Wittgenstein of a picture which may be seen as representing the head of a rabbit or of a duck depending on the stress laid upon certain details.[9]

In all these cases it emerges that 'seeing' must be analysed into a 'seeing that' and a 'seeing as.' The 'seeing that' may be interpreted as a taking into account the single (and, so to speak, factual and isolated) components of a given cognitive situation, while the 'seeing as' corresponds to the way of unifying these components into an actual representation. The most interesting fact, however, is that, contrary to what might spontaneously be thought, 'seeing as' comes before 'seeing that,' since the constitutive elements are identified by *analysing* the representation, and this is tantamount to recognising that no representation exists which is not already a certain *way of representing*, and it is because of this that one and the same thing can be seen completely differently depending on which *Gestalt* one

[8] In this section the points of affinity (that are numerous and significant throughout the whole of this work) and the points of difference with regard to the considerations developed in Dilworth's very valuable *Scientific Progress* (2008) become particularly clear. The affinities lie in Dilworth's Perspectivist conception of science and his Gestalt Model, while the differences depend on our recognition of the correct (though partial) appreciation of the linguistic aspect of theories expressed by the sentential view that, in particular, has led us to try to overcome certain conclusions obtained by authors endorsing that view, by remaining *within* their approach, as will be seen in detail in Sect. 7.2.

[9] By the way, this is the figure Dilworth uses in elaborating his Gestalt Model.

uses. It might also be shown that even 'seeing that' is a form of, or a lower degree of, a certain 'seeing as,' but we are not interested in this analysis for the moment. We want explicitly to point out, however, that in this 'seeing as …' or 'seeing as thus or so' resides the justification of the transition from the pure gestaltic or *representational* aspect of a theory to its explicit *linguistic formulation*. Indeed, as Nelson Goodman has convincingly demonstrated in his classical work (Goodman 1968), even a picture has a denotational and a predicative function, since representing something 'as thus or so' amounts to attributing it a property, and in this sense a picture plays the role of a *statement*. Therefore, the formulation of a theory as a system of sentences is only the linguistic counterpart of its gestaltic nature.

What we want to underline is rather that, as we have already noted, a certain measure of generality is already present at the level of sensory knowledge, owing to the presence of the 'form,' i.e. of the *Gestalt*. Indeed, this form or *Gestalt* is a 'model' for the organisation of the data which the knowing subject already possesses, and to which he 'brings back' the data. If the subject did not already possess the *forms* of the vase, the human profile, the rabbit and the duck (to remain with the mentioned examples), she could never see the pictures which are presented to her *as* a vase, a profile, a rabbit or a duck. This is the ancient discovery of Plato that 'to know is to recognise.' Leaving aside the problem of the origin of such a form, let us underline that the form has a general character, since it has already served to unify several other complexes of data; and this is why it may function as a unifying instrument for the new data we are now confronted with. The age-old view is that sensory knowledge is confined to the particular, while the universal is the privilege of intellectual knowledge. But this view should be replaced (or at least complemented) by a view that is more 'continuistic' in its suggesting that at *every* stage of our knowledge we are in the presence of a relation between the particular and the universal. This is so since what is universal at a certain level functions as particular at a higher level, that is, at a level where a further *unification* occurs. Contemporary philosophy has already advanced along this path, as it has generally replaced the dichotomy particular-universal with that of data-construction (let us simply mention the Husserlian treatment of this issue).

In the above discussion we have once used the term "model" in referring to the unifying form or *Gestalt*. We believe that the notion of model, despite its occurrence in several guises in contemporary philosophy of science, has not yet been fully investigated from the point of view of its *hermeneutic* function, which we shall stress here. This function will turn out to be essentially that of providing a 'way of representing' a given field of inquiry and, by that, it will appear very close to the concept of scientific theory. In such a way the concept of model will be seen as an intellectual tool much more significant than the rather trivial device for 'visualising' complicated situations which it is often claimed to be.

6.3.1 Explaining, Understanding and Unifying

No consensus exists in the philosophy of science as to what a scientific theory is. Yet a rather general agreement still exists in admitting that one of the fundamental *tasks* of a scientific theory is that of explaining data.[10] But what does it mean "to explain"? Philosophers of science have quickly discarded any psychological meaning of this notion, by eliminating its common and spontaneous interpretation according to which to explain means to 'bring back' what is unclear and unknown to something which is already clear and known. They have privileged instead a more philosophical and technical notion of explanation according to which it consists in providing the 'why' of what is empirically evident. Moreover—and this was a decisive step—the ostension of such a 'why' has been identified with performing a *deduction* of the empirical data from sufficient hypotheses. But deduction is a typical *logical* procedure that applies to *statements*, so that in the last analysis scientific theories have been considered as systems of hypothetical statements from which factual statements describing data can be correctly deduced. This is the famous *statement view* of theories, encompassing the *deductive-nomological* model of scientific explanation which has dominated empiricist-analytical philosophy of science, and which was perfectly in keeping (as we have already noted) with the 'linguistic turn' that has characterised most of the philosophy of the twentieth century.

This view has been strongly criticised in recent decades, and it seems to have been discredited by now. Our opinion, however, is that this view is of greater merit than its opponents are ready to admit, since it enables us to analyse the logical and linguistic *aspects* of scientific knowledge, aspects which certainly cannot be underestimated, though they are not such as to *exhaust* the nature of scientific theories, as we shall see. It seems more sensible, in other words, to remark that this conception should not be made *absolute*, in order not to overlook *other* aspects of scientific explanation, on which the logical and linguistic ones may even depend.

We claim indeed that explanation constitutes—at the *intellectual* level—a component of that process of *unification* that we have already seen to operate at the level of sensory knowledge, and more precisely of perception. It is not arbitrary to

[10] In Harré (1964), for example, a distinction is drawn between a weak sense of "theory" according to which theories simply *connect* observable relations expressed in terms of descriptive concepts (he calls them "reticular theories," see pp. 9ff.). And a strong sense according to which theories *explain* the observed regularities by resorting to conceptually more complex structures in which different theoretical concepts occur, and from which the observed regularities causally follow (he calls them "explanatory theories": see pp. 18ff.). One might note that this distinction goes back at least to Poincaré. Other meanings and tasks could be mentioned, among which one can also find a diminutive one (not uncommon among experimental scientists) according to which the term "theory" indicates a conceptual scaffolding provisionally admitted with the hope of its being replaced with more 'solid' knowledge. This sense however is not altogether derogatory, since it hints at the very important role that a conceptual pre-comprehension plays in empirical investigation as well, and is in keeping with the hermeneutic outlook we are presenting here. We only prefer to limit the meaning of "theory" such that theories are tools for scientific *explanation*.

maintain—and cognitive psychology seems to confirm this claim very clearly—that knowing is in a broad sense unifying, and that, in particular, "to think is to unify" according to a famous statement of Kant who, however, expresses a view that is present in the whole history of philosophy. Now, since the co-ordination of empirical data within the context of a coherent deductive net certainly is a form of unification which respects the fundamental intellectual law of *logical coherence*, it is clear that scientific explanation, as it is 'currently' understood, is a particular way of satisfying the fundamental requirement of unification which characterises any knowledge whatever. The question, however, remains open whether this is the *whole* role of explanation, because it is certainly correct to ask whether explanation (understood in the restricted sense considered thus far) exhausts the whole horizon of intellectual unification and, in addition, whether it has something to do *more* than simply unifying (that is, whether it must *provide reasons* different from pure hypothetical-deductive inferences but, for example, of a *causal* nature).

Apart from, and in a certain sense prior to, *explanation*, the intellectual phenomenon of comprehension or *understanding* must obtain. Unfortunately, the relations between these two concepts are obscure for opposite reasons. On the one hand, their use in ordinary language makes them practically synonymous and interchangeable; on the other, they have been used to indicate cognitive processes of a very different nature during the controversy that developed among certain famous scholars around the beginning of the twentieth century concerning the vindication of the 'scientific' character of the historical and social sciences. Some of them have seen 'explanation' (*Erklären*) (which they conceived as consisting in subsumption under a law or laws) as being the task and the methodology typical of the natural sciences, while identifying the provision of 'understanding' (*Verstehen*) with the task and methodology of those sciences which were sometimes called 'of the Spirit' (*Geisteswissenschaften*), 'of culture' (*Kulturwissenschaften*), 'social–historical' and, more recently and in a much broader sense, 'human.'[11]

It would be wrong to qualify that distinction as arbitrary or confusing. If it is considered within the precise historical debate where it was advocated, it was rather useful and legitimate. Yet it is much less useful and clarifying if we remain faithful to it outside that particular context, as though it expresses a final and clear-cut separation. Therefore it is much better not to contrast 'explanation' and 'understanding' but rather to see them as distinct, but interconnected, moments of the intellectual knowing activity. Their relationship might be summarised in the following elementary claim, "it is not possible to explain something which has not been understood."

This claim expresses a certain conceptual antecedence of understanding with regard to explanation in the sense that explanation is to develop within the horizon of understanding; and one might venture to say that, after having understood *how*

---

[11] It would lead us too far afield to analyse this controversy, in which scholars such as Droysen, Dilthey, Simmel, Windelband, Rickert, Croce, Collingwood and Weber have expressed different views. Let us simply refer to von Wright (1971) for a contemporary discussion of this topic, which also gives some historical hints to the debates of the past.

something is, one proceeds to trying to explain *why* it is so. Yet we should not want to create misconceptions concerning this 'antecedence' which must not be conceived of in a temporal but—as we have stressed—in a conceptual sense. In this sense understanding continuously accompanies explanation, and the latter may be seen as a constant deepening of the former. Moreover it is by no means a passive deepening, for it may happen that explanation meets with such difficulties that a substantial modification of the frame of reference provided by the original understanding is required, and this is tantamount to saying that something that appears seriously inexplicable might turn out also to be not understandable. In this sense it is also correct to maintain that the result of a successful explanation is that we finally 'understand' our subject matter. It would be naive to take this consideration as contradicting our first statement that understanding comes 'before' explanation: we have already pointed out that this is not a temporal precedence, and now we can more properly point out that in this case (as in numberless others) we are in the presence of a feedback-loop in which beginning and end are relativised. These considerations will become clearer in the sequel. For the moment we simply note that—if things are as we have claimed—understanding and explanation *must* enter with equal right into both the natural and the human sciences.

What has been discussed has direct relevance to the notion of *model.* To make oneself a model of a certain domain of reality means to *understand* it, in the sense of obtaining a unified representation of the domain which transcends the sphere of pure sensory experience in unifying it. This unification is a second-level gestaltisation which presupposes a first-level gestaltisation, and which provides a representation whose traits are not sensorially perceived but *thought.* Now, as we have found it reasonable to recognise at the root of explanation the presence of a fundamental intellectual procedure—that of *deductive argumentation*—so it seems reasonable to recognise at the root of understanding the presence of a no less fundamental intellectual procedure, that of *interpretation.* Therefore, if we agree to call the dimension of deductive argument *logical*, we can agree to call the dimension of interpretation *hermeneutic* (so long as it is recognised that we are using the term "hermeneutic" in its strictly etymological sense, without implicit involvement in additional connotations bound to particular philosophical doctrines of today); and we can conclude that in every scientific discourse both logical and hermeneutic dimensions are present, and that the latter 'presides' over the former. If we now recall that a decisive role has been recognised for the model in the construction of the 'understanding' interpretation, we can also conclude that a non-eliminable *hermeneutic function* is attributed to the model by this fact.

6.3.2 The Hermeneutic, Heuristic and Analogic Function of the Model

The above specifications enable us to see why the hermeneutic function of the model cannot be confused with a rather trivial requirement of 'visualisation.'

Indeed, we can even allow ourselves to use the terminology of 'vision' without becoming involved in undesirable equivocations, for we have spoken of the problems of "seeing as" and "seeing that," two expressions which have only accidentally to do with the material images produced by the sense of sight, since they point rather to an intellectual 'seeing,' to an *insight* which occurs also with respect to the most abstract entities. (In this sense we commonly—and correctly—say that the mathematician is able to 'see' the properties of such entities as numbers, etc.).

In this deeper optics nothing is lost of the heuristic and analogic value of the models that has so frequently been stressed in the specialised literature, since we are here in the position of understanding the real *reason* for such very important functions. Indeed, also in those cases in which, in order to investigate a given domain of objects, a certain model of that domain is constructed within a different and already known structure of objects, the meaning of this procedure does not reduce to the mere returning by analogy to something which is already familiar to us in order to make the task of our investigation psychologically easier (the analogic function). Nor does this procedure reduce to the—certainly not negligible—exploitation of the structural analogy between the two domains of objects with the aim of hypothetically transferring to the still unknown field the homologue of what is known inside the already explored field (the heuristic function). These are very respectable pragmatic motivations. They do not clarify, however, *why*, among the many possible models, it has occurred to us to use exactly that one. The reason is the following: at a given moment it has appeared to us that that model was able to provide us with the *insight*, the conceptual point of view, the gestaltisation that was necessary for *understanding* our domain of objects. This gestaltisation—as we have stressed—is a 'seeing as' which, precisely because it relies upon an analogy, becomes a 'seeing as though' and, as far as the analogy retains its validity, provides us with the heuristic stimulation to directly look in the unknown field for further possible confirmations of that 'seeing as.' However, while the analogic purport and the heuristic impulse of a model disappear when the initially ascertained analogies are no longer of use, the model's hermeneutic function may continue to hold for much longer, that is, until the gestaltisation contained in the model—that serves to *directly* unify the data of the domain of objects under investigation—enters a crisis owing to *reasons internal* to that very domain of objects. How this may happen will be seen later. We may note by the way that what has been said makes more precise the often advocated claim that a model is the 'preamble of a theory.'[12] As we shall see, a theory consists in a system of statements explicitly expressing the main lines of the structure contained

[12] Beside some classical works devoted to the concept of model in science, such as Black (1962) and Hesse (1966), we refer to Dilworth (2008), where the concept of model plays a very important role and is conceived in a way that has several affinities with ours. Dilworth's conception of the role of models in scientific theorising is further developed in his (2007). More details regarding our conception (with particular stress laid upon the *structural* nature of models) may be found in Agazzi (1969, especially Chap. 9), and Agazzi (1986 and 1987).

in the model, in such a way that the model itself can be put to test by the inference of single precise statements that can be empirically tested.

6.3.3 The Model and the Construction of the Domain of Objects

The considerations concerning the hermeneutic dimension of science, and the special meaning that the notion of model receives within this perspective, constitute a clarification and enrichment of the general view concerning the nature of scientific objectivity to which this book is devoted. We shall now provide some evidence of this fact and, at the same time, indicate how this approach affects our way of conceiving of laws and theories.

The basic claim of the conception of scientific objectivity advocated in this work is that every science considers reality from a specific 'point of view' to which some operational criteria of referentiality are associated for the purpose of providing an empirical check on what is said about reality from that point of view. It is clear that the said points of view already express a first gestaltic—and hence hermeneutic—orientation, which therefore lies at the root of the very *construction* of scientific objects. This is simply an initial moment that presides over the identification of the *data* of any actual science. It also shows, however, how any datum is itself a construction from a certain point of view—a first gestaltic unit. Such data, when they occur with a certain uniformity, are further unified in formulations which are sometimes called empirical generalisations, and which are usually said to be obtained by induction. But the inductivists overlook that 'seeing' these uniformities is by no means something univocal and automatic, for it is equivalent to discovering a new *Gestalt*, a new unification, and therefore a new model which inscribes itself within the original gestaltisation and enriches it. Many scientific *laws* are precisely generalisations of this kind.

But laws themselves soon begin to appear as data that demand to be inscribed in a wider *Gestalt* or model, and the conceptualisation of this model itself requires the intervention of a hermeneutic moment. This is the birth of a *theory*, which thus coincides—at its initial moment—with the proposal of a model, understood as a global vision, as a 'way of representing' the whole of the data and the already discovered laws. But the possibilities of 'understanding' a given set of data within a certain unification are—as we have stressed several times—multiple; and each unification adds something of its own to the data, which gives rise to the problem of checking whether this 'something' is more or less arbitrary. This is tantamount to saying that the theory must be tested. But how can this actually be done?

The answer is that, in order to test the theory, we must make it explicit and *formulate it linguistically*; and this corresponds to translating its intuitive content, its global insight, into a finite set of *hypotheses* that are intended to transcribe, so to speak, its most salient features. It is interesting to note that at this stage a kind of detachment of the theory from its model takes place, since the theory, owing to the very fact of its becoming the linguistic translation of the model, actually *refers* to it, and only through it also refers to the objects which, as we know, are included in it.

In this very brief presentation we can find a satisfactory interpretation of one of the several senses in which the concept of model is used, the sense implicit in the statement that sometimes a model acts as 'a substitute for a theory' (rather than as a simple 'preamble of a theory,' as we have recognised above). A typical case of this kind is Bohr's model of the atom, which is actually a theory of the atom. Indeed, if we wanted to present Bohr's *theory* of the atom we could do nothing other than present his *model*, so that Bohr's theory is in the last analysis the theory of Bohr's model. If we conceive of models as no longer simply having the subsidiary function of 'illustrating' theories (a function which *some* models sometimes have), but as entities that even 'construct' the domain of objects of theories, and from which theories themselves genetically derive, it is only a matter of consistency to say that a theory is the theory of its model. Admittedly, one may remain doubtful about our way of distinguishing models and theories, and also concerning how we can say that a theory is a theory of some model. Both the theory and the model, even if they are conceived of as being distinct, are *of reality*, as is determined by the intention of the theorist (and, in an intersubjective way, of the scientific community adopting that theory). This perplexity is, in a way, spontaneous, but we have already met an almost identical situation when we noted that the expression "scientific object" receives a double meaning: in a first (and more proper) sense the scientific object is an *abstract object*, a *noema*, an *intellectual construction* that *encodes* a certain number of attributes. In a second sense, the *objects* of a given science or theory are the *referents* that *exemplify* the abstract object and belong to that extralinguistic and extramental 'reality' in which they are 'clipped out' of 'things' by means of operational *referential* procedures (revisit Sect. 4.6 for details). In the same way we can say that in certain cases a theory is the theory of a model (that is, of an abstract object) though its intended *referents* are entities different from the model itself, but hopefully *exemplifying* what is described in the model. In other cases, the model is only a heuristic preparation of a theory that, after an initial step, may consist of a much broader display of concepts than those originally suggested by the model. Let us stress, by the way, that both models and theories, in the empirical sciences, are proposed *hypothetically* and, therefore, must be *tested* according to the criteria of referentiality specific to the science where they are proposed (For a detailed discussion of these topics we refer to Chap. 9 of Agazzi 1969).

A theory—and we have seen the reasons for this—is here being presented as a linguistic construction, and this returns to the 'statement view' of theories its title of legitimacy. But this restitution is only possible because the hypotheses of which a theory is constituted are neither the result of problematic inductions (as Popper's criticism has correctly stressed), nor the result of "bold conjectures" (as Popper too hastily pretends). We can say that the weakness of the statement view resides in its insufficient elaboration of the 'context of discovery' lying behind a theory, a context where the hermeneutic dimension, the conceptualisation of 'global' models, plays a vital role, of which the representatives of logico-linguistic philosophy of science have been too little aware. Indeed, both inductivists and Popperians limit themselves to looking for the source of *single* hypotheses and are unable to propose persuasive solutions with regard to such hypotheses because both ignore the hermeneutic and 'global' dimension that makes of the hypotheses something which is 'dug out' of the theory. In this sense the theory is very different from empirical laws, as we have briefly said above. This very same weakness is also reflected—as we shall see immediately—in the shortcomings exhibited by the statement view in the context of justification.

6.3.4 The Life and Death of a Theory

A model is linguistically translated into a theory in order to be tested; and at this stage one must say that those features which the supporters of the statement view have elaborated really work for a certain while. New statements describing data (usually in the form of predictions) are looked for, and it is checked whether the data actually found are compatible with the hypotheses. As long as the agreement with the data obtains, the theory is safe, otherwise it is 'falsified.' The said compatibility is exclusively seen in terms of logical deducibility, and this is inevitable since one does not see what other explicit relation might be established among statements.

This is a very simple pattern, but it is far from satisfactory. Apart from the fact that no confirmation of a theory can be final—which is clear from a strictly logical point of view—there is also the fact that falsification too is not final, while this is *not* justifiable from a purely logical point of view. We can then understand that Popper has always tried to maintain the decisive role of falsification.[13] In his earlier work he tended to claim that, if even one single sentence incompatible with the data is deduced from a hypothesis, the hypothesis is falsified and should therefore be dismissed. The history of science shows that this does not actually happen in most cases, and this fact may still be explained from a logical point of view by noting that in the deduction of the 'refuting' statement *several* parts of the theory are used—besides the hypothesis which is being tested—so that the conclusion should rather be that the theory as a whole has been falsified (the Duhem-Quine thesis). But then the theory should be dismissed; and this is historically even less usual than the dismissal of a single hypothesis. Popper has given a pragmatic explanation of this kind of methodological laxity: one does not dismiss a falsified theory unless a replacement theory is available. Among Popper's disciples, Lakatos has developed a rather skilful and celebrated form of 'sophisticated

[13] Popper, in his eagerness to distance himself from verificationism, claims that *no* scientific claims can be verified—a clumsy thing for him to do, since it implies that neither can any scientific claim be falsified, since falsification requires the verification of the negation of what is being claimed. Aiming at the enemy, he shot himself in the foot.

falsificationism,' intended to replace Popper's alleged 'naive falsificationism,' but without being fully convincing.

The reasons for these inadequacies may be clear from our perspective. If the hypotheses of a theory constitutes a system of statements aimed at expressing a given *Gestalt*—i.e. a model—linguistically, each statement is, so to speak, the description of an isolated detail of that *model*, a detail, however, that finds its proper location and 'representational role' *in* the model. Now it is not only true that having referentially confirmed some single details does not warrant the acceptability of the image as a whole (as the thesis of the non-ultimate character of confirmation correctly states). But it is no less true that having discovered that a certain detail fails to have a referential counterpart does not require the rejection of the image as a whole. This rejection will depend in part upon the quantity of the details which have been found to be 'unfaithful,' but especially upon their strategic *importance* or relevance in the context of the whole representation. No justification of this difference of importance of single details in the model can be poured into the *logical* relations among the statements describing them. (Once more we are confronted with a *hermeneutic* question.) This is why the statement view turns out also to be weak in the context of justification. It is unable to account for the lack of an ultimate role for falsification which, on the contrary, is understandable within a gestaltic view of theories.

At this point we can see how the hermeneutic force of the model is the decisive factor that supports the theory at every moment, and determines its life or death; and, reciprocally, the hermeneutic force of the model is also the decisive factor with regard to how the theory contributes to the enrichment of the model itself. If it is true—as we have maintained here—that a theory is always a theory of its model, it follows that whatever the theory permits us to discover in its domain of referents (e.g., thanks to the predictions it provides) contributes to the enrichment of the model which initially suggested the theory. But this can happen because, in the very act of predicting, the theory is still supported and pushed by its model. Indeed, predictions do not 'gush' from the hypotheses by a spontaneous deductive force. They have to be 'imagined,' that is, they must be extracted from further *insights*, from further projections of the intellectual 'seeing' that are nourished by the hermeneutic and heuristic potentialities of the model.

From what has been said it is also clear when a theory ceases to be valid, which is when the hermeneutic force of its model is no longer effective. This may happen because the successive gestaltisations are incompatible with those that had originally oriented the construction of the domain of objects of the model. This happens in practice when too many serious incompatibilities occur between hypotheses and empirical data. But it may also happen because new gestaltisations of the same objects offer themselves in a more convincing way, that is, because a new and more efficient model surfaces. The emergence of a new model 'reorients' all the data of the field; and the reasons for its greater 'persuasiveness' generally consist in certain aspects of reality becoming 'visible' which before were hidden. This brief sketch will be integrated with other considerations when we come to speak of theory change and theory comparison. Let us note, incidentally, that this hermeneutic view of theories allows for a 'translation' of those notions of incommensurability, incomparability, and incompatibility that in the current literature are usually bound to the statement view of theories. They can be expressed as different degrees of discrepancy between gestaltisations that may occur either as 'modifications' or as 'replacements' of a given *Gestalt*. When a real '*Gestalt*-switch' takes place, the new theory is incompatible with the preceding one, and this also in the absence of any meaning variance.

6.3.5 Science and Interpretation[14]

In the foregoing parts of this section we have advocated a view in which science has a hermeneutic dimension, mainly by elaborating the notions of understanding and model, and it might appear surprising that we made scant use of the notion of 'interpretation,' which is certainly the central concept of hermeneutics. The reason for our doing so was essentially practical. We preferred to work with concepts that already have an accepted circulation and have been rather carefully analysed in the critical literature concerning the sciences, rather than use the concept of interpretation that occurs rather in the critical literature concerning the humanities and the arts. Yet it will not be difficult to show how interpretation is the basic intellectual activity which gives rise to understanding and models in the sense presented above.

We shall do this by first introducing a particular analogy of science with art. Analogies of this kind have been proposed in the past and at present, and they are mainly meant to underline the elements of 'creativity' and/or 'beauty' which are present in science. Without minimising these aspects, we are going to consider here another one, that for which art is 'interpretation'; and, in order to avoid useless digressions, we shall chose a particular example in which the artist is clearly meant to be an 'interpreter,' that is, the example of a musician interpreting a musical score. This means that, rather than the two most popular images of scientific activity—that of the construction of a great building to which every scientist contributes a new brick, or that of the progressive exploration of an unknown continent—we prefer to see the work of science as comparable with that of 'interpreting' a musical composition. The useful perspectives disclosed by this comparison will induce us to briefly revisit certain central claims of our discussion under the new light coming from the correct appreciation of the role of interpretation in science, a role which will fully justify the 'hermeneutic dimension' which we are advocating for it.

We have said that the activity of science cannot be conceived of as the progressive, systematic and complete discovery of an unknown country, but is rather like the interpretation of a complex musical score. What is typical of this artistic

[14] The content of this subsection has largely been taken from Agazzi (1985b).

activity (in comparison with other more 'creative' artistic productions) is the fact that the notes are all precisely 'given,' and the indications for their execution are also largely furnished by the composer. Nevertheless we find it normal that each performer gives a personal 'interpretation' of a score that remains the same for all. We are even willing to recognise that there may be two, three or four interpretations that sometimes differ significantly from one another but each of which appears to us to be superb and 'faithful' to the score. Obviously, here it is not a question of the natural disposition that we have to admire the originality of the artist and the strength of his creativity, since in the case in question he is not allowed to produce something new (namely, a new musical composition), but to acquaint us with something that is already accomplished; and if he seems to us to have gone beyond the limits of faithfulness to the score in an excess of originality in his interpretation, we judge his efforts as being of poor quality.

It must be pointed out that the said 'fidelity' does not consist in a so-called correspondence to 'what the composer meant' (as some affirm), but consists rather in the discovery of expressive features that are 'objectively' in the score, while requiring the intervention of the interpreter's aesthetic sense in order to be revealed. These features then have a bipolar status: they are born, so to speak, from the encounter between a 'point of view,' an intuition on the part of the interpreter on the one hand, and the concrete structure of the composition on the other, so that without the intervention of the interpreter they would never be revealed. But we realise that this way of describing the situation is still insufficient to express the substance of the phenomenon in question, since the said features were not there like diamonds hidden beneath the ground, thus already formed and only waiting to be brought to light. They only 'potentially' exist (never more than in this case do we perceive the pertinence of this Aristotelian notion), like the infinite cuts that can be made in a solid body, each of which is entirely determined as soon as the plan for dissection is chosen, but beforehand is only a possibility. With a musical composition we may thus have a scrupulous performer without talent who limits himself to a faultless reading of the notes of the score, but we may also have different interpreters who, while respecting the minimal demand to play 'without errors,' and in addition do not betray the sense of the composition 'as a whole' (which is much more difficult to define), give us more or less interesting 'cuts' of the content of the work.

The form of the dynamics of scientific knowledge is much more similar to this model of artistic interpretation than may be suspected. In fact, the concrete reality of things faces us like a musical score, and the knowledge of its intrinsic richness requires the employment of many interpretations, because it is also a wealth of 'cuts' that cannot pass from potentiality to act without the intervention of a 'point of view' that reveals them. We must be careful, however, not to reduce this intervention of the interpreter to the level of a transcendental structure of the forms of consciousness. If this structure exists (and it seems difficult to deny its existence, even though we cannot effectively determine what it consists of), it is something that ideally intervenes 'before' the cuts we have mentioned.

To use our example, this structure is the one that allows us to read the notes of the score correctly, and that finds its equivalent in the knowledge of reality that we have at the level of ordinary experience. Just as the musical interpretation begins with the correct reading of the score and always remains with the obligation not to betray it, so the different sciences begin with ordinary experience and are obliged not to contradict it (therefore, they all use and presuppose the transcendental conditions of our knowledge). However, they are distinguished by their going beyond the level of ordinary experience, and this occurs because each of them has a particular 'point of view' on reality. This point of view has not the characteristics of a structural *a priori* of pure reason, or of a transcendental condition of knowledge as such, but of an 'interpretation.' Consequently, each science effects its own 'cuts' within the reality of 'things' according to its own point of view, and attempts to develop all its particular potential. What we have here called 'point of view' may be denoted by more technical philosophical terms such as 'criteria of intelligibility,' provided this terminology is not understood as pointing to the plane of pure transcendental conditions but specifically to what is *added* to this plane.

In the foregoing parts of this work we have presented the claim that scientific knowledge is constituted by the creation of domains of 'objects' that result from 'cuts' from the 'things' of ordinary experience, domains obtained with the help of concepts, and supported by operational procedures. Therefore, we shall not repeat these theses here, but dwell upon the character of *interpretation* that belongs to such a constitution of the domain of scientific objectivity.

Unfortunately, the notion of interpretation has been thought to be totally foreign to the exact sciences, since there is the tendency to tie it to the idea of a basic uncertainty, as well as to the idea of a double subjectivity. In fact, we usually see interpretation as the work of a subject who tries to understand the undeclared intentions of another subject and, because of this, can never get beyond a basic state of uncertainty and guessing. From this comes the aspiration to make of the exact sciences a discourse in which 'interpretations' are put aside in order to hold to the prudence of naked 'descriptions' and, at most, to the logical and empirically guaranteed solidity of 'explanations.' What this perspective misses is that the aim and result of an interpretation are quite simply the production of a 'comprehension' of something, and not necessarily the conjectural reconstruction of the intentions of a certain subject. (Of course, we do not deny that in certain sciences the interpretation must consist in the reconstruction of the intentions, ideas, and purposes of a subject.)

At this point we are in the position to be able to evaluate the ingenuousness of the positivist (and neopositivist) conception of a science limited to directly reflecting the intrinsic structure of the real by means of a scrupulously neutral use of purely sensory experience and the formal and tautological transformations of logic (including mathematics). While acknowledging that the tools of scientific knowledge must remain empiricity and logos, we must admit that these are never pure, but are inscribed and nuanced according to the contexts of interpretation within which they operate.

It is useful for us now to return to our analogy of musical interpretation. The amateur, the performer without talent who is only able to read the score correctly, represents, as we said above, the stage of 'common knowledge,' while the true interpreter goes beyond this stage to give a rendition of the composition based on a certain 'interpretation,' which corresponds to the stage in science of the construction of a theory. But here our analogy exhausts its usefulness because there is an essential difference between the interpretation of a musical score and what happens in the sciences, since in the first case the global viewpoint involved is eminently individual and subjective while the interpretative frame that determines the intellectual space of a discipline or scientific theory is of a supra-individual and, more precisely, historical nature (while containing certain elements of individual genius).

In other words, the ideas and criteria of intelligibility that determine the interpretative context, within which arise—through a process of specialisation—the empirical and theoretical concepts of a scientific discipline, are the expression of an historically determined cultural milieu; and, even when one person has co-ordinated these elements in a new synthesis, this cannot provide a new scientific theory until is accepted by a community (at least by what is called the 'scientific community'). We must simply recall that the operational procedures that permit the empirical 'concretisation' of the viewpoints thus achieved are also bound to the historical context (since they are generally offered by the technical possibilities available at a certain period within a certain epoch). The same is true of scientific explanations, since they generally respond to what in a given epoch is considered to be a good explanation, a rigorous argument, and also to the technical means (belonging to what we may call the 'techniques of reason,' such as systems of logic and mathematical theories) available to effectively present the desired explanations. In Sect. 6.1 we have called this set of conditions the 'historical determinateness' of scientific objectivity, and the remarks that have just been made certainly suffice to justify the affirmation that science has the same status of historicity as do other manifestations of the human mind. This is because historical factors powerfully condition (without strictly necessitating) human interpretations of reality.

Before going further into this subject, let us use the preceding clarifications to dissipate certain equivocations circulating among those philosophers of science who have disregarded some of the factors we have discussed.

It should be stressed that the context of interpretation, the interpretative frame we have mentioned, has a global and holistic nature, is still mainly undetermined, and can develop in several directions, depending on the particular concepts it presents and the operational procedures that are associated with some of them. For this reason, it would be inexact to confuse it with a 'theoretical context' in the proper sense. The correct name we must use to designate it is rather that of a 'hermeneutic context' and, as we have seen, it is at the same time pre-empirical and pre-theoretical with regard to the specialised type of empiricity and theoreticity appearing in the sciences. We thus return in conclusion to the basic claim presented in this section, namely, that every science has a *hermeneutic* dimension that cannot be eliminated (be it a 'natural' or a 'human' science) since it is indispensable to its constitution, and is not to be confused with its *theoretical* dimension.[15]

6.3.6 Interpretations and Data

One consequence of this realisation is that the data of experience serving as an empirical basis in the sciences can never have an 'absolute' value. This correct affirmation (which in relatively recent epistemology has been stressed by Popper and his followers) has been construed in the mistaken affirmation according to which even empirical or observation statements are affected by an indispensable and vague coefficient of *theoreticity*. This is the famous thesis of the 'theory-ladenness' of any scientific statement, which we have already discussed in various contexts, and whose most striking consequence has been the thesis of the impossibility of objectively comparing rival scientific theories and thus of evaluating the 'progress' of knowledge in the sciences. This fallacious solution of a real problem comes from having misunderstood the precise sense of the non-absolute nature of the data of experience in science. We may synthesise this by saying that the data in question are never 'pure' facts but always 'interpreted'

---

[15] The considerations presented in this section have significant affinities with the positions advocated by the "idealisation" school of Nowak and Krajewski, and with Dilworth's Perspectivist conception of science as presented in his (2008), as we have already mentioned. Furthermore, our hermeneutic dimension corresponds to Dilworth's *relative a priori* as developed in his (2007), which consists of particular metaphysical principles taken to constitute the core of modern science. Less immediate, but equally interesting, are the affinities with the epistemological views elaborated by Hans Lenk, an author who has devoted many works to the study of hermeneutics proper and who has, in particular, elaborated a fully-fledged theory of what he calls a "schema-interpretation" or "structural model" approach to knowledge in general and to scientific knowledge in particular. Within this view it is possible to overarch the traditional split between the natural and social sciences as well as the humanities, since all these disciplines structure their fields and objects according to the activation of schemata or structural models by using procedures for establishing, stabilising and activating schemata and models as cognitive constructs in order to shape their respective apprehensions of the world. Schema interpretation admits of levels of categorisation according to the variability of the respective schemata. In such a way Lenk has developed a hierarchy of levels of interpretation consisting of six different levels, the last of which includes epistemological and philosophical as well as methodological interpretations of a meta-character. Therefore, anything that is conceivable appears as "interpretation-dependent" or "interpretation-laden" without being "theory-laden" in the most widespread sense of this expression. Though we believe that Lenk's efforts constitute a very significant contribution to epistemology and philosophy of science produced within the framework of hermeneutic philosophy, we shall not devote a specific analysis to it owing to the already mentioned fact that our work is intended to remain rather close to the analytic spirit that prevails in contemporary philosophy of science. We thus limit ourselves to inviting the interested reader to become acquainted with the most recent publications of this author, such as Lenk (1993, 1995, 1995a and 2000).

facts. However this does not mean that they are always and necessarily interpreted by the means offered by the scientific theory *within which* they act as an 'empirical basis.' On the contrary, they are necessarily interpreted inside the hermeneutic context within which the theory in question is totally inscribed. Only a small part of their significance is enriched by the specific theoretical context into which they enter.

The difficulty with the problem is that any concept in a scientific theory is bound to all the others by the logical ties that determine its specific meaning in a global manner, so that its meaning also 'depends' on those of all the others, in particular on those of the theoretical terms as well. All this (as we have already stressed) cannot be denied, but it does not keep us from distinguishing that part of the meaning that an empirical term draws from the simple 'hermeneutic context' in which it is placed (and which belongs to it even *before* it is used in some particular theory), from that 'supplementary' and variable part that is added as a contribution of the 'theoretical context' (or contexts) of which it may become a part. This possibility of distinguishing exists and (as we have already seen in detail) rests on the fact that the terms acting as the constitutive elements of the empirical statements are directly attached to the operational procedures mentioned. What we are able to add now to the previous analysis of the *intensional* constituents of the meaning of scientific concepts is that they include a *hermeneutic component* as well. In the special case of *operational concepts*, we have already remarked that their meanings include a *referential* 'stable core,' besides the *theoretical* variable part. We now must add that they also include a *hermeneutic* part enveloping both of them. Let us stress however that these components, though interrelated, still retain a certain independence, as has been shown here.

To synthesise the results of the preceding analyses, we must say that a 'critique of scientific reason' cannot be limited to a critique of pure reason (which could only suffice to specify the transcendental conditions of knowledge in general), nor be integrated by a critique of linguistic reason (which would confine us within the paradox of 'theory-ladenness'). It must still be completed by a critique of hermeneutic reason and of historical reason because, in the construction of the sciences, reason operates not only according to the conditions of its intrinsic functioning but also in the context of a certain linguistic and hermeneutic *a priori* that bears the signs of historical determinateness.

One more remark seems necessary. We do not wish to give the impression that this flood of *a priori* conditions implies that scientific theories are a sort of necessary and predetermined result of this complex structure. Actually, there is a feedback loop among all these elements. The internal dynamics of experience and logical argumentation may lead us to the abandonment of one theory for another within the same hermeneutic context, but it may also be the case that the dynamics of the different sciences leads to a modification of a certain hermeneutic context, and sometimes to the revision of certain *a priori* presuppositions of a linguistic or even of a transcendental order. The history of science proves this; and far from denying the historical nature of the sciences, this confirms it, since once more the dynamics of the various sciences show themselves to have the same kind of effect on the evolution of the human mind and culture as other manifestations of it.[16]

[16] Though they perhaps accord with a different spirit, the claims made in these last remarks are well in keeping with views expressed and convincingly defended by Shapere in several of his writings.

Chapter 7
Corollaries in the Philosophy of Science

7.1 Laws, Hypotheses, Theories and Experiments

After having advanced a good deal in the presentation of the basic features of our general conception of science (including the presentation of its hermeneutic dimension), we shall here say something about certain central notions that have been the object of numberless analyses and discussions in traditional philosophy of science. We shall consider them only as far as they seem at least in part to be readjusted within the perspective advocated in this work. Therefore, we are not going to mention or discuss the many aspects of their traditional presentation that we do not consider affected by our way of conceiving of science. Such notions are in particular those of (scientific) *law*, *hypothesis*, *theory*, and *experiment*, that we have already mentioned in the foregoing parts of this work but which we shall consider in a more systematic way now.

The least one can say is that considerable differences still exist in the meanings attributed to these concepts, and in the interpretation of their mutual relationships, despite the great work of clarification and standardisation that has been done in nearly one century of methodological reflection on science. Just to give an example, let us begin by sketching how these notions are correlated in the usual presentations of the so called 'experimental method.' Within a given field of investigation—it is said—the experimental scientist begins with a systematic *observation* of facts from which he tentatively derives a generalisation which is considered as an *hypothesis*. In conformity with this hypothesis, an *experiment* is then designed in which, if some factual conditions are realised, a certain effect should be observed. If this effect is actually manifest, and if the same procedure gives the same result in a sufficient number of significant and differentiated cases, the hypothesis is promoted to the level of a *law*. If several such laws are discovered in which some common concepts occur, and which may be connected by particular logical links, they constitute a *theory*. In this perspective, "hypothesis" clearly has the meaning of 'working hypothesis,' and is seen as a provisional step which will hopefully lead to a law, while no qualitative difference is made between laws and theories, the latter being seen as essentially the same thing as collections of laws.

E. Agazzi, *Scientific Objectivity and Its Contexts*, DOI: 10.1007/978-3-319-04660-0_7,
© Springer International Publishing Switzerland 2014

We could say that in this sketch the accent is put on what is sometimes called the "context of discovery."

A different characterisation is found in the most typical descriptions of scientific explanation contained in almost all standard books of philosophy of science directly or indirectly inspired by logical-empiricist philosophy. This happens because, within this perspective, the context of discovery is intentionally left out of consideration, while all the attention is focused on the "context of explanation." Explanation in turn is conceived as a logical deduction of an *explanandum* from an *explanans*. If the *explanandum* is an individual event, the *explanans* consists at least of a general 'law,' in conjunction with suitable particular conditions (this is the covering-law model of scientific explanation). The statements expressing laws and initial conditions constitute the set of assumptions, premises or 'hypotheses' of the deduction. But there are also explanations of laws on this view, and they consist in deductions in which more general laws constitute the premises or hypotheses of the *explanans*, while a less general law is the *explanandum*. No decisive difference is therefore to be found between laws and hypotheses on this view, and since the hypotheses constitute the theory, no essential distinction is made between laws and theories either. At most, the attempt has been made to distinguish between 'experimental laws' and the laws which more specifically constitute the theory by claiming that experimental laws have an observational character. However, apart from the fact that this would not eliminate the basic identification of laws and theories in general, it has been noted, for instance, by such an authoritative scholar as Ernest Nagel, that:

> It is doubtful whether a rigorously precise sense can be usefully assigned to the word "observable"; and to the extent that the distinction between experimental laws and theories is based on a contrast between what is observable and what is not, the distinction is patently not a sharp one. In any event, no precise criterion for distinguishing between experimental laws and theories is available, and none will be proposed here.[1]

The reason for the difficulties found in giving a univocal characterisation of the notions of law, hypothesis and theory—which have already emerged from the two examples presented here—depends to a large extent on the various uses that are made of these notions in the different sciences, and even in one single science such as, for example, physics. Certain laws concern a single system (e.g., Kepler's laws), others are more or less immediate empirical generalisations. Others (perhaps the majority) express functional relations between measurable quantities, while others express statistical relations between collections of events. In the presence of this variety, several efforts have been made with the view of finding some basic requirement which could characterise a statement as a law. It has been noted, for instance, that it is not enough that this statement be general. It must also be 'necessary' and not 'accidental,' though its necessity is not to be identified with the logical necessity of mathematical sentences. Such a *physical* necessity should intuitively correspond to the idea of a property intrinsic to real things, or some

---

[1] Nagel (1961), p. 83.

regularity existing in nature, or some causal connection between events. But these are 'metaphysical' features which are hardly translatable into linguistic conditions. Even the remark that scientific laws permit the formulation of 'contrary-to-fact conditionals' has been of little help, for no logically satisfactory theory of these conditionals seems to exist which does not implicitly refer to the very notion of physical necessity or lawfulness; and this clearly implies circularity.

In the foregoing section we noted that the meaning of "theory" is itself not firmly established. Sometimes theory is spoken of when the intention is to indicate a restricted number of hypotheses (or even a single hypothesis) that are proposed for the explanation of a few facts, while one is at the same time aware that this 'theory' is incompatible with other known facts or with other better-established theories. In this spirit a theory is a pragmatic and even conventional working tool for concretely proceeding in an inquiry. Sometimes one also speaks of 'phenomenological theories,' where one is satisfied with establishing certain functional mathematical relations (often very artificial) that allow one to connect certain phenomena with others without the possibility of equipping these mathematical links with any physical interpretation.

Despite all this, we can say that the most generally accepted sense of "theory," at least as it is codified in the specialised literature of the philosophy of science, is that of something which is proposed in order to provide an *explanation*. In particular, this sense is perfectly in keeping with the general concept of science which has developed in Western civilisation where, as we have seen, science has always been thought of as providing an intellectual form of knowledge able to indicate the *reasons* for its claims.

7.1.1 Laws and Hypotheses

It is by considering their specific aim or *intention* that we can obtain some clarification concerning the distinction between laws, hypotheses and theories, a distinction that cannot be established on the basis of purely *linguistic* or logical criteria. A simple but efficient way of drawing this distinction is to say that the formulations of laws are intended (by science) to express *that* something is the case, or to express 'how things are,' while hypotheses (taken in a strict sense) and theories try to say *why* something is the case. In this sense one could also say that laws have a 'factual' character (they are so to speak 'general facts of nature') while hypotheses and theories have a 'conceptual' and 'conjectural' character (to be rigorous, one should say that hypotheses have a 'factual-conjectural' character, i.e. they are educated 'guesses' about empirical states of affairs). Since explanation is in a general sense the explanation of facts, it is straightforward to say that one of the main tasks of hypotheses and theories is to explain laws. We have said "one of the main tasks" in order not to overlook that a theory may also be used to explain single events (a feature which enables us to retain the notion of theory, that is, of explanatory intellectual construction, also in the case of many 'human' sciences where the subject-matter may be constituted by single events, objects and states of affairs). It is certain, however, that in the case of the natural sciences the isolated event is practically of no interest, and in any case no theory would be proposed, but at most applied to explain it. In these sciences theories are proposed for explaining classes of events which regularly occur under suitably determined circumstances, and they are nothing but the referents of empirical laws. Hence the actual 'facts' of natural science are the empirical laws (even Popper's "basic statements," owing to the way they must be established through repeated inter-subjective testing, are the expressions of empirical laws, though he does not use this term to denote them).

This distinction between laws and theories is by no means new, but belongs to a tradition that was rather well established and was then interrupted with the advent of the 'linguistic turn' in the philosophy of science. To see this one can simply refer to Campbell's *Physics: The Elements* (1920), from which we shall now quote a few clear and illuminating statements, which are all the more significant since they speak of laws and hypotheses as propositions (which means that this view is reconcilable with at least some kind of statement view of theories, though Campbell was totally alien to this philosophical trend). Indeed, Campbell says:

> Laws are propositions asserting relations which can be established by experiments or observation…; the relations asserted, if not always the same, have always a common feature which may be described as "uniformity of association" (pp. 38–39).

Another interesting remark is the following: "it is to single propositions which can be stated adequately in a single grammatical sentence or mathematical equation that the term 'law' is confined" (p. 44). The interest of this remark consists in the fact that Campbell does not confine the expression of a law to being a mathematical equation, so that if we follow Campbell we should even be able to assimilate the logical empiricist way of expressing empirical laws, that is, as general sentences of the form (x) (Px→Qx). It has been correctly underlined that these generalisations are very poor examples of actual physical laws (which are hardly comparable to statements such as "all ravens are black"). In any case, even such rudimentary examples may be retained, since they are "single grammatical sentences" whose aim is that of "asserting relations which can be established by experiment or observation." The fact that these sentences have the "grammatical" form of a conditional *does not make them hypotheses,* since this depends on the intention and not on the form of the sentence, and the intention is here clearly that of *affirming* something and not of *supposing* something.

A deep aspect of Campbell's conception of science consists in how he sees laws and theories as being related. Not only are theories proposed for the purpose of explaining laws, but laws themselves cannot be recognised as such unless they are explained within a theory:

> When we have got our laws we want to explain them by theories; and so again we reject from consideration any propositions, however much they may resemble in their structure those which we accept as laws, which refuse to fit in with our theories, the form of which is dictated chiefly by preconceived ideas of what a theory should be (p. 98).

This view offers a very natural solution to the problem which we mentioned above, and which has challenged all efforts at solution by means of logical and linguistic tools alone, namely the problem of distinguishing a law-like statement from one expressing a simple 'factual uniformity' of a contingent nature. 'Contrary-to-fact conditionals' cannot do this, while the possibility of inserting the statement in a theory explaining it can, because this tells us *why* the 'uniformity of association' occurs, eliminating in such a way the suspicion that it be only the result of chance.

Coming now to the way theories may provide explanations of laws, Campbell explicitly indicates that this happens by means of the formulation of *hypotheses*. According to him, hypotheses are characterised by their containing some "ideas" specific to the theory (they correspond rather well to what have later been called "theoretical concepts"), and a "dictionary" (anticipating Carnap's notion of "correspondence rules") relating these ideas to "concepts" (corresponding rather well to the later notion of "observational concepts") which occur in the formulation of laws. "A theory," he says "is said to explain certain laws if it is these laws which are implied by the propositions concerning the hypothetical ideas" (p. 123).[2]

Considering such clear and well-justified distinctions, one may be somewhat shocked by Popper's saying:

> In any hypothetical deductive system, these less universal statements are themselves still strictly universal statements in the sense here understood. Thus they too must have the character of *hypotheses*—a fact which has often been overlooked in the case of low-level universal statements (Popper 1959, p. 75).

What are here called "low-level universal statements" are practically the 'empirical laws,' as may be seen from the following statement in which it is said that "for example, a falsifying hypothesis can be of very low level of universality (obtained as it were, by generalising the individual co-ordinates of a result of observation)" (ibid., p. 87). Popper's thinking here is completely within the confines of the Deductive Model. He would take a statement such as 'this is copper' to be a "low-level universal statement" since something's being copper presupposes the holding of particular laws. However, this whole line of reasoning on Popper's part is an ill-advised attempt to discredit verificationism, the problem being that it discredits falsificationism to the same degree. What is amazing is that Popper here considers the lack of recognition of the hypothetical nature of

---

[2] For greater detail, see Campbell (1920), pp. 122–123. In particular, it will be seen that Campbell explicitly admits that "a theory is a connected set of propositions" (p. 122), and that he has no difficulty in speaking of the truth of theories:

> The theory is said to be true if propositions concerning the hypothetical ideas, deduced from the hypotheses, are found, according to the dictionary, to imply propositions concerning the concepts which are true, that is to imply laws; for all true propositions concerning concepts are laws" (p. 123).

attempted expressions of empirical laws as something that has been *overlooked*, while in fact he is himself overlooking (or perhaps simply ignoring) that a variety of arguments existed in the tradition for not conflating hypotheses and laws. A spontaneous question is how could this important distinction become lost (since not only Popper but the whole logical empiricist philosophy of science has usually disregarded it, even in those cases in which the term "law" has been preserved).[3]

The answer to this question seems to consist in the fact that a superposition and finally a confusion occurred concerning two distinct meanings which are commonly associated with the idea of something's being hypothetical. In one sense, when we say that something is hypothetical we mean that it is *not certain*, and we refer in this way to an *epistemic* feature regarding our knowledge of a particular state of affairs. In a second sense, we say that a certain statement is hypothetical when it is *assumed* or presupposed as an hypothesis for the sake of a given argument, and we refer in this way to a *logical* feature or function of the statement which is independent of its being certain or uncertain, or true or false. In the standard descriptions of the experimental method, the term "hypothesis" is often used in the first sense, that is, in the sense of "working hypothesis," of a guess concerning a given state of affairs, and this very *same* statement is promoted to the status of a *law* when (thanks to the positive result of the experimental tests) we believe that the uncertainty has been removed. Let us note two things. First, that this 'hypothesis' has not been used to explain anything, and even less *proposed* with this end in view (it is almost ridiculous to claim that "all ravens are black" (in conjunction with "this is a raven") might explain *why* a given raven is black; and it is even more ridiculous to say that we arrived at this generalisation in order to explain why this particular raven is black). Second, that a genuine hypothesis in the logical sense does not change its role as a consequence of changing its epistemic status (if a sentence is assumed as an hypothesis for proving a theorem, it continues being an hypothesis for that theorem even if we were to establish that it is of a very high degree of certitude within the mathematical theory to which it belongs). Therefore, the working hypothesis which has become a law is by no means an hypothesis in the logical sense (that we might express with a more precise term "premise").

Popper's move, which is perfectly consistent with the spirit his 'fallibilism,' is to reject every possible positive certainty from science, including that which was traditionally ascribed to empirical laws, and this could be expressed by saying that

[3] In many cases the term "law" has been used to denote hypotheses of a medium-level generality, while hypotheses of low-level generality have been called "empirical generalisations," and hypotheses of high-level generality "principles." According to this hierarchical distribution, *explanation* is seen as a logical deductive chain descending from the principles to the empirical generalisations, while *meaning* is seen as going up-stream from the empirical statements to the laws and the principles. A particularly clear presentation of this view may be found in Braithwaite (1953). However there have also been empiricist philosophers who did not underestimate this difference (see, e.g., Chap. 5 of Nagel 1961, who refers back to Campbell; this issue concerns the distinction between laws and *theories*).

these laws are doomed, after all, to perpetually remain at the stage of 'hypotheses' (i.e., of uncertain statements). But how can this claim be supported? By denying that a universal statement may be considered true on the basis of a great number of positive instantiations, since there is no logically valid inference from the particular to the universal. This is true, but it presupposes that the *only* way to establish truth is by logical deduction from true statements (except in the case of 'protocol statements'), and here lies Popper's well-known aprioristic opposition to induction. However, this is not the point which interests us here. The point is rather that, after having qualified empirical laws as 'hypotheses' on the basis of this epistemic character, they are then put on an absolutely equal footing with the other, genuine hypotheses (which Popper significantly calls "conjectures"), which are introduced as theories in order to *explain* facts and generalisations (i.e., in order to explain, if possible, *why* all ravens are black, rather than simply stating *that* they are so).

We have spoken of Popper, but similar considerations apply to his antagonists, the logical empiricists. Where Popper eliminated *laws* in practice by reducing everything to hypotheses or 'conjectures' (which are suitable for explaining facts, according to the Deductive Model), the empiricists eliminated *hypotheses* in practice by seeing all scientific sentences as being generalisations of varying scope which may be obtained by induction (i.e., as sharing the basic features of laws). It is clear that both positions are one-sided. To see things clearly one must be able to recognise that *two* different aims are present in science, that of ascertaining, of stating, of describing, and that of understanding and explaining. The discovery of laws belongs to the first aspect, and includes *empirical* as well as *theoretical* laws (i.e., laws concerning the behaviour of entities that are postulated by the theory and are not immediately testable but must at any rate be empirically confirmed indirectly). Discovered laws are expressed as being more or less certain, but as nevertheless *being the case*. They are *stated*, and are expected to be further *confirmed*; and it is intended that from such laws we can try to obtain *predictions* (while they are at the same time open to correction or even rejection). Discovered laws are not something which is *assumed* in order to explain something else.

On the other hand, *hypotheses* are explicitly assumed with the view of explaining what is known. They are created rather than discovered; and while laws usually survive the demise of the hypotheses or theories proposed for explaining them, hypotheses and theories are much more subject to change. This, however, is rather natural since they are intrinsically *suppositions*. Reducing laws to hypotheses would amount to claiming that all the work of science reduces to *supposing*, ignoring that an equally important and perhaps more expanded part of this work consists in *ascertaining* and *discovering*. This of course does not prevent one or more laws from playing a role in a concrete explanation, as part of the set of *logical* assumptions which are used in it. However, laws are used as 'tools' in the explanation in the sense that it is the theory which suggests which laws to choose, and how to coordinate them in order to explain the particular fact seen 'through' them. Moreover, we are not claiming that the respective domains of laws and hypotheses (especially as far as theoretical laws are concerned) cannot overlap to a certain extent, or be interconnected.

Nevertheless, they should not be confused, and their distinction might also be of help in appreciating the difference which (despite their *formal* affinity) exists between explanation, confirmation, and prediction. Explanation relies primarily upon hypotheses and only secondarily on laws, while confirmation and prediction (or retrodiction) are essentially grounded on laws and only indirectly concern the hypotheses of the theory. From this point of view we can more easily understand that logical empiricists, having practically concentrated completely on the category of laws, have attributed a greater weight to induction, verification, and prediction (which play a decisive role in the discovery of laws, but not in the invention of hypotheses and theories), while Popper and his followers have seen in science almost only hypotheses, and in such a way have underestimated confirmation and induction, and given all the weight to falsification and the free creation of conjectures.

7.1.2 Theories

What we have said accounts for the two different ways of conceiving of theories and theory construction which have been predominant (though with several variants) in twentieth century philosophy of science, the logical-empiricist and the Popperian. Both share a common core (a common view with two different emphases), both being based on what we have termed the Deductive Model, characterised by the following features: (a) theories are proposed in science in order to fulfil the task of *explanation*; (b) theories are sets of *propositions* or statements (statement view); (c) hence explanation consists in a logical *deduction* of certain statements (*explananda*) from other statements (*explanantia*) of a higher level of generality; (d) theories must in any case cope with *empirical* evidence. Because of this common core this conception is synonymously called in the literature the "covering-law" model of explanation, the "deductive-nomological" model of explanation, and the "Popper-Hempel" model of scientific explanation, a way of speaking which, rather than expressing an historical record, manifests the agreement of the two schools of thought, of which these two names are among those of the best-known proponents.

The agreement, however, no longer exists as regards the issue of theory construction and, more generally, the relation between theory and experience. The contrast might be expressed by saying that the first school is characterised by a strict 'empiricism' which leads to conceiving of a theory as a body of laws obtained by a process of acquiring generalisations of increasing scope based on induction (i.e., by giving the widest possible application to the conceptual scheme of the 'experimental method'), while Popper's view is typically 'rationalist' in the sense that hypotheses and theories are said to germinate from the conjecturing force of the mind, and empirical evidence is used only to test them.

It seems legitimate to say that the empiricist view is weaker than Popper's as far as its ability to characterise the specific role of theories is concerned. Indeed, if a theory has to explain experience, it is hardly thinkable that this could really be done by using experience itself as a basis for the explanation, that is, essentially, by generalising experience. If it does not overstep experience, explanation seems condemned (at least in the last analysis) to the almost tautological triviality of explaining that this particular raven is black because all ravens are black; while, if to explain is to give *reasons*, we expect these reasons to be of a *different kind* with respect to the *explanandum* (and in the case of physical facts these reasons should involve reference to a cause or causes). As we have already stressed, inductive methodology is very useful to help us discover laws, but not for inventing hypotheses; and if a theory must contain hypotheses in particular, the empiricist doctrine does not sufficiently account for theory construction.

This weakness has other symptoms as well. For instance, the empiricist tradition has usually qualified theoretical concepts as simply being those which are not 'observational,' i.e. in the poorest possible (and most ineffective) way, while theoretical concepts (as we have already noted elsewhere) are introduced with the view to theorising. In other words, theoretical terms (which for the empiricist view, going back to Comte, are anomalies in science) are introduced in postulating the existence of unobserved entities, properties, or processes which are intended to explain the behaviour of what we are able to observe.

The Popperian view, on the contrary, captures the intrinsic dynamics of explanation much better, in its postulating something which is not directly offered in experience but is needed for its being intellectually understood. However, what remains less satisfactory for this view is how Popper's 'conjectures' are to be linked to empirical evidence. This problem does not simply regard the origin of the conjectures (a question that Popper relegates to psychology) but also their intrinsic ability to 'concern' the realm of experience. There must be something 'about' which we are conjecturing. To say, as Popper does, that even his 'basic statements' are hypothetical, implies that we are constantly hypothesising about hypotheses; and this is no more satisfactory than the *opposite* claim that we use laws (i.e. facts) to explain laws (i.e. other facts) as the empiricists say. The correct solution, as we have noted above, seems to be that we introduce hypotheses (i.e. conjectures) to explain facts (i.e., states of affairs established beyond reasonable doubt, though, like every claim, subject to the possibility of error).

The above considerations amount to recognising that scientific knowledge, like every form of human knowledge, walks on *two* legs, experience *and* reason, or empiricity *and* logos, which are certainly interconnected but ought not be confused. The reason for the confusion *may* be (but *need* not be) the fact that we formulate the content of experience no less than the content of our conjectures and reasonings in the form of *sentences*, and sentences *as such* do not say anything concerning their intention, their origin, or their justification, since they are simply linguistic entities. Therefore, in any science, we should distinguish what we consider to be 'factual claims' (which we can call the realm of laws) from what we intend to be 'explanatory claims' (which we would call the realm of theories). But it is no less true that this distinction cannot be grounded on the simple analysis of the linguistic form of the sentences belonging to the two sectors. Therefore, from the moment we decide to consider a science *only* from a linguistic point of view, the two domains merge; and it is significant that in current philosophy of science the term "theory" is often used to indicate *all* of the sentences of a science (so that the initial claim that theories are constructed to provide explanations is practically forgotten).[4]

The view advocated in this book, on the contrary, safeguards this distinction, and provides grounds for it. Predicates related to operational criteria of referentiality circumscribe in a science the sector of experience, and allow for the formulation of empirical laws which can be established and tested independently of the theory (or of the theories) which try to explain them. On the other hand, what we said already in Chap. 2 concerning how scientific objects emerge within a given 'point of view,' and developed further in the section concerning the hermeneutic dimension of science, clarifies how and why the sectors of experience and theory are in fact interrelated: both of them have a common root in a given *Gestalt*.

This is again something which fully escapes the statement view of science in general, and of theories in particular, not because this additional awareness is incompatible with it, but because it is unduly neglected. That the statement view is not incompatible with this general awareness may be seen by referring again to Campbell. He certainly recognises that laws and hypotheses are propositions, and correctly stresses that the character of theoreticity is bound neither to a higher degree of 'complication,' nor to a greater distance from experience, but "involves another step in the development of ideas" (op. cit., p. 121). But then he goes on to claim that "in order that a theory may be valuable it must have a second characteristic; it must display an analogy" (ibid., p. 129). A careful reading of the context in which this claim is made (and in particular of the considerations that "analogies are not 'aids' to the establishment of theories; they are an utterly essential part of theories, without which theories would be completely valueless and unworthy of the name" (ibid., p. 129) clearly shows that Campbell attributes to theories much more than the kind of unity which might come from logical connection. He actually means that they should provide a 'representation' of their

[4] It is for this reason that in logical-empiricist philosophy of science we often find the declaration that, for epistemological investigations, a theory can be equated with the logical conjunction of all its sentences. This view is also shared by Popper who, besides equating laws and theories, says that "scientific theories are universal statements" (see Popper 1959, p. 59 and, in general, Chap. 3 of this work).

field; and in such a way he is rather close to what we have said when speaking of a *Gestalt* and a model as preconditions for the existence of a theory.[5]

Leaving Campbell aside, we can say that a more satisfactory view of scientific theories, which distinguishes (without disconnecting) them from laws, and accounts for their explanatory task and power, seems to be attainable through the considerations we have presented in Sects. 6.3.1–6.3.3 of this work, which we shall not repeat here.

There is an important point that we have only indirectly mentioned occasionally in the preceding sections, and which is seldom investigated in the literature; it is the fact that the construction of a theory, while necessarily keeping the 'domain of objects' of a science stable, has the effect of enlarging the number of referents which are admitted into this domain. Indeed, as we have seen, while empirical laws concern the 'immediate referents' of a science (those which can be identified through the criteria of protocollarity or objectification), a theory usually postulates

---

[5] Owing to the interest of this point, we gladly quote a relevant passage from Campbell *in extenso*:

> *The importance of the analogy*. We see then that the class of physical theories of which the theory of gases is a type has two characteristics. First they are of the form which has been described, consisting of an hypothesis and a dictionary; if they are to be true, they must be such that laws which are actually found to be true by observation can be deduced from the hypothesis by means of logical reasoning combined with translation through the dictionary. But in order that a theory may be valuable it must have a second characteristic; it must display an analogy. The propositions of the hypothesis must be analogous to some known laws.
>
> This manner of expressing the formal constitution of a theory is probably not familiar to most readers, but there is nothing new in the suggestion that analogy with laws plays an important part in the development of theories. No systematic writer on the principles of science is in the least inclined to overlook the intimate connection between analogy and theories or hypotheses. Nevertheless it seems to me that most of them have seriously misunderstood the position. They speak of analogies as 'aids' to the formations of hypotheses (by which they usually mean what I have termed theories) and to the general progress of science. But in the view which is urged here analogies are not 'aids' to the establishment of theories; they are an utterly essential part of theories, without which theories would be completely valueless and unworthy of the name. It is often suggested that the analogy leads to the formulation of the theory, but that once the theory is formulated the analogy has served its purpose and may be removed and forgotten. Such a suggestion is absolutely false and perniciously misleading. If physical science were a purely logical science, if its object were to establish a set of propositions all true and all logically connected but characterised by no other feature, then possibly this view might be correct. Once the theory was established and shown to lead by purely logical deduction to the laws to be explained, then certainly the analogy might be abandoned as having no further significance. But, if this were true, there would never have been any need for the analogy to be introduced. Any fool can invent a logically satisfactory theory to explain any law (Campbell 1920, p. 129).

other entities and speaks of their properties and processes in order to explain the laws. But in such a way these entities could hardly be denied a right of citizenship among the referents of the science involved. How could we in fact seriously accept that a theory speaks all the time of these entities through its 'theoretical' concepts and at the same time deny that these entities have a referential status comparable to (though not identical with) that of the theory's 'immediate referents'? It is clear, on the one hand, that we are again led to the issue of realism already discussed in Chap. 5. But the point is interesting in itself, and from a simply intuitive point of view. If we are convinced that science has offered humankind the possibility of knowing a great deal of things which were previously unknown, this means that our set of referents has actually greatly expanded in comparison with its ancient borders. The question now is, how can we be justified in admitting new referents to the 'domain of referents' of a science? The discussion of this issue will lead us to a view of *experiments* in science which is rather unconventional.

7.1.3 Two Kinds of Dependence Between Experiments and Theories[6]

According to the traditional view, experiments are designed to put theories to the test, and "theories" fully deserve this honourific denomination only if the test is successful. Of course, this does not eliminate the hypothetical character of theories even after favourable tests, but points out a strict interdependence between theories and experiments. On the one hand, we must say that experiments depend *genetically* and *logically* on theories since they are conceived and designed as explicit questions formulated by means of the concepts of a theory and with a view to testing it. On the other hand, we must say that a theory depends on experiments as far as its *legitimacy* is concerned, for the negative result of relevant experiments would imply the elimination of the theory in question, thereby affecting its very *existence*.[7]

On this line of reasoning we find a certain ambivalence in the notion of theory which is directly reminiscent of the distinction between 'hypotheses' and 'laws' in

[6] The content of this subsection has also appeared as a paper Agazzi (1988e). The specific contribution of experiments to the construction of theories has also been pointed out (though in a more delimited role) by van Fraassen (2008, pp. 111–113).

[7] From what we are saying it is clear that we are using the notion of experiment in a very technical and restricted sense. In the common way of speaking this notion is actually used more loosely, especially in its adjectival form. So, for example, the 'experimental scientist' is often said to be one who spends his time doing research in a laboratory and collecting data, rather than proposing interpretations or mathematical models of these data as the 'theoretical scientist' does. In this sense experiments cover the whole area of empirical observation independently of their being intended to test some particular hypothesis or theory. Without denying the legitimacy of this broader use, we shall confine ourselves to the more restricted one just mentioned.

the standard way of presenting the experimental method. For "theory" is understood first as meaning generically what is conceptually elaborated beyond the level of *experience*, and has rather the character of a hypothetical construction (in keeping with the well-known distinction between 'observational' and 'theoretical' concepts). But in that case a theory is understood to be a fully fledged intellectual picture of reality which can be accepted only if it accords well with reality, and the successful outcome of experiments is meant to provide this warranty. If we follow this bifurcation of meanings, we can say that experiments depend on theories in the first sense while theories in the second sense depend on experiments. In this way our investigation might be closed with a rather obvious but trivial result. However, it is possible to follow the line of thought suggested by the above bifurcation of meanings in order to find a nontrivial characterisation of the role of experiments. This we shall do by briefly restating some of the fundamental views advocated in this work in a slightly different form, better suitable to our purpose.

A theory (as we have seen) develops out of a 'point of view,' which isolates certain aspects of reality, aspects that are *intended* to be the object of a particular investigation. This means that these aspects have already been singled out by means of concepts which enable us to state *what* we intend to study, concepts which have to be equipped with suitable operational criteria by means of which we can *refer* to reality under the specific point of view we have adopted. The combined presence of these concepts and operational criteria amounts to the determination of the 'domain of objects' of the inquiry, and at the same time provides us with an initial quantity of empirical evidence concerning these objects, that is, with a good deal of *data* given by the immediate application of the said criteria to reality. This is what we have above called the level of *experience*. We may also call it the level of *observation*, provided we do not consider observation to be a kind of passive 'looking at things.' It is rather a kind of 'looking into things,' in the sense that it is performed within the framework of a given 'point of view,' and involves the concrete operational manipulation of physical tools as well as a reliance upon a rich display of available knowledge which teaches us 'how' to observe in order to reach 'what' we intend to observe.

Theory enters as an attempt to describe and understand the domain of objects so identified, by utilising all the available information that is concretely constituted by the collected data. This attempt cannot avoid being one of interpreting and explaining the data; and this implies that it has two aspects, the elaboration of concepts and the formulation of statements by means of these concepts. (These two aspects, however, should not be considered as successive 'steps,' for concepts are always elaborated in mutual connection with other concepts, and this gives rise to statements in which their meanings take shape.) Along these lines, the construction of a theory may be seen as an expansion of the concepts which initially expressed the specific 'point of view' on reality (and were contained in an implicit way in that *Gestalt* or model of which we spoke in the foregoing section). This expansion depends in part on the implicit potentialities of these concepts and in part on the information contained in the empirical statements describing the available data. All this amounts to the explicit creation of new concepts and new statements that deserve to be called 'theoretical' not because they are opposed to 'observational' concepts and statements (as was supposed according to the old but untenable dichotomy), but because they exist for the sake of understanding and explaining the data. In conclusion, we could say that the purpose of a theory is to provide a 'conceptual space'—a frame of rationality or a condition of intelligibility—which as such belongs to the level of *meaning and intension*, while still needing to be connected with additional *referents* for that part of the conceptual framework which was not already linked originally with operations.

7.1.4 The Referential Task of Experiments

To provide a reference for the meanings expressed in the sentences of a theory is the specific task of experiments. This is why any relevant experiment is conceived and designed under the strict supervision of the theory of which it should express a *referential logical consequence*. Its being a logical consequence of the theory makes it dependent upon it; but its being referential makes it dependent upon the operational criteria and, in this respect, independent of the theory. It is true, as we have stressed several times, that these criteria themselves depend on the specific concepts which constitute the particular 'point of view' adopted, and usually also depend on many laws and concepts belonging to the background knowledge that provides the context of the 'historical determinateness' of the objectification involved. But *these* concepts do not belong to the theory. They are common prerequisites both of the theory and of the operational criteria, and work as guidelines for the determination of the 'domain of objects' of the discipline to which the theory belongs. Once this is clear we can see that if the experimental test is successful (in the sense that the proposition logically derived from the theory turns out to be immediately true according to the admitted operational criteria of referentiality), it provides (at least to a certain extent) the theory with a *guarantee of reference*; if it is not successful, it indicates (at least to a certain extent) that the theory has *no reference*.

Some refinements are needed concerning the referential purport of experiments. The most immediate, but misleading, idea associated with this claim could be that an experiment puts us into direct contact with the referents of the theory, in the sense that it enables us to *observe* them. Such a case is not totally excluded in the preliminary steps in which simple conjectures of elementary theories are put to the test. However, such instances are rather exceptional and not very interesting. In the case of all mature natural sciences, experiments do not allow us to directly observe the intended referents of a theory, that is, those objects which the theory is 'speaking about' (such as, e.g., elementary particles). Here the experiments simply provide us with a "guarantee of reference," which is usually indirect but reliable, as we have explained in the long discussion devoted to the 'reality of the unobservables.'

A very simple example may help clarify the issue. A photograph which I see is for me a reliable guarantee of reference concerning the existence of the person or building of which it is a photo, although I do not actually see the person or building but simply their photograph. Why? Because in normal circumstances no photograph can be taken of something which does not exist and, moreover, a photograph normally reproduces a good deal of optically detectable features of an object; further, we have in principle alternative means of checking the 'veracity' of the photograph. Of course, the possibility exists that 'false' photos be created, or that a 'true' photo be taken of an object under so peculiar conditions that it gives a misleading image of it. However, a careful inspection of the photo and the circumstances in which it was taken, and especially a comparison of it with other photos taken by other persons in different circumstances (and perhaps consideration of information coming from other sources such as letters or records) may increase the strength of the 'guarantee of referentiality' up to the point of making it a practical certainty. It is clear even from this elementary example that a not negligible number of presuppositions are involved in every referential claim which goes beyond the mere witnessing of the immediate evidence. Some of them concern matters of fact, others are fragments of causal explanations, and others are sketches of 'theories' in a broad sense.

In the case of the exact sciences the situation is analogous to that of everyday life, with some complications and some simplifications. The complications depend on the fact that the search for referential guarantees (i.e., the design of experiments) involves a very sophisticated theoretical, mathematical, and technological machinery, together with all the skills related to it. The simplification consists in the fact that only a few aspects of reality are involved and, therefore, the direction in which the research has to be done is better determined and the margins of possible equivocation are more restricted. After these refinements it should be clear in which sense it is appropriate to say that experiments have a referential purport: they show those empirical features which we expect to be consequences of certain properties of our referents, and which we can understand only as such consequences.[8]

---

[8] The example of the photograph has been proposed on purpose, since it reminds us that in everyday life, and in most sciences and other domains of knowledge, the greatest part of our claims and beliefs of referentiality are made (and moreover with a fully 'realistic' intention) only on the basis of 'indirect but reliable' guarantees of reference. For example, no one doubts the existence of Napoleon, in spite of the fact that no one living today has ever seen him. This is so because we can rely upon a continuity of historical tradition which goes back in a reliable way to the records of those who saw Napoleon, and because many documents still exist that testify to his existence. But in the case of very remote persons or events we must reconstruct or infer their existence solely on the basis of documentary evidence that is presently accessible to us, and which must be correctly interpreted and explained.

Similar reasoning applies when we accept certain existential claims of the theory of natural selection in biology, or when we accept statements concerning the internal composition of the earth or certain stars, and so on. It should be clear therefore that if difficulties are put forth concerning the referentiality of theoretical concepts in the natural sciences, these cannot be justified by saying that the guarantees are only indirect, or that a theory is always falsifiable since such objections are equally possible in very many cases in everyday life. The real reason seems therefore to reside in the preconceived closure regarding the admission of *any* kind of reality which does not share the familiar features of commonly observed things. But this—different from a reason—is simply an obscurantist prejudice.

Let us stress that, each experiment being by necessity a very limited logical consequence of the theory, its guarantee of reference or disproof thereof can only concern a restricted part of the theory. In other words, where one positive result ensures that at least certain parts of the theory are endowed with reference, many positive results ensure that several parts are—though in the latter case it still remains uncertain whether the whole theory enjoys such a guarantee. Symmetrically, one must say that one or more negative results prove that some aspects of the theory lack reference, but cannot exclude that other aspects have it. This way of looking at things seems to account in a more satisfactory manner for the well-known facts that a finite number of 'confirmations' are not sufficient for calling a theory true, while neither are a finite number of 'falsifications' really sufficient for declaring it false. These facts are usually explained by considering scientific theories as equivalent to a *logical conjunction* of propositions, a point of view which has proved unsatisfactory in several respects. According to the approach advocated here, a theory may rather be seen as a more or less faithful representation of its intended referents. It can never be perfectly faithful (for reasons we have already discussed when speaking of 'encoding' and 'exemplifying'), but this does not imply its rejection so long as it is considered sufficiently faithful. Experiments exist precisely in order to make such a decision possible.

7.1.5 Other Aspects of the Interdependence of Theories and Experiments

Given the above clarifications, we may try to characterise the role of theory and experiment by means of certain other distinctions, for example by saying that theories describe 'possible worlds' in which the known phenomena (i.e. the data) may occur, while experiments try to single out the 'real world' to which they belong. This they do by ideally discarding all of those possible worlds which, though they have a meaning, do not have a reference within the *intended domain of objects*. If we consider things in this way we can better appreciate the fact that theories are always intrinsically hypothetical. This feature (as we have noted) is often interpreted as expressing the fact that we can never be *certain* of their truth. But it seems better to recognise that theories' being hypothetical means that they express *possibilities*, and that they are intrinsically unable to do more than that because the task of providing a transition from possibility to reality belongs to something else, that is, to experiments.

Another way of expressing this is to say that theories can only determine the formulation of *questions* and cannot provide the *answers* to these questions, which are expected to come from the experiments. By saying this, it is clear and beyond doubt that the answer 'depends' on the question, but only in so far as its *meaning* is concerned (this meaning also providing an indication of how to determine the answer). Whether the answer be positive or negative, however, does not depend on the question, but must rely on other sources. In our case these are constituted by the operational criteria which, in performing the experiments, bring us into contact with the *intended referents* of the theory, as we have seen.[9] Using a terminology which we have already adopted, we may say that theories determine the constitution of the *semantic logos* (i.e. the realm of concepts and intensions), while experiments are oriented to the construction of the *apophantic logos* (i.e. the realm in which something is affirmed or denied of some subject-matter that constitutes the reference of the discourse).

We want explicitly to avoid, however, that our last claims be understood in an extremist sense which might equate them with the doctrine of the theory-ladenness of all scientific sentences, which we have already criticised. This is why we now want to restrict the purport of the statement that the theory 'provides the meaning' of the question which the experiment tries to answer. This statement is to be understood as follows: the theory provides the *significance* of the question, its *point*, its purpose and reason, besides providing the question with a global meaning resulting from its position in the general context of the theory. But it is not as though the question would be *meaningless*, that is, *not understandable*, without the theory. This cannot be the case simply because, in order to be performed, an experimental test must be entirely describable in terms of operational concepts and their meanings which, as we have already remarked several times, do not depend on the theory. This is why we need the theory in order to design the experiment (hence the experiment depends 'genetically' on the theory, as we have already said), but not in order to justify it. Once the experiment has been performed, it assumes an independent existence and is in no need of help, simply because it has the same character as the *data* which are the indisputable basis a theory is challenged to account for, and which it cannot modify or dispense with.

7.1.6 A Hermeneutic Approach to Experiments

Let us conclude by indicating how the hermeneutic approach to theories proposed in this work provides a good understanding of the reciprocal relations between theories and experiments. Our claim has been that the 'wholeness' of a theory has to do with a *hermeneutic* level which is connected to the *semantic* and the *referential* levels in different ways. Once we start investigating reality from a certain 'point of view,' we are already applying a particular *interpretation* (hermeneutic

[9] Considerations of this kind are also expressed in Buzzoni (1986) and (1987). A remark similar to this has been expressed by Harré concerning a broader context, namely that of the kinds of things that one is prepared to consider real:

> The general conceptual scheme we adopt determines the kinds of things, properties and processes we are prepared to admit, but what there actually is will be found out by investigations such as turning over stones and analysing the electric impulses in the circuits of radio telescopes" (Harré 1964, p. 50).

level) to it, an interpretation which guides our choice of concepts (semantic level) and our devising of operational testing procedures (referential level). This is so because we begin our inquiry by referring to things and 'seeing them in a certain manner,' that is, having in mind a certain *Gestalt*, or image, or model of them. Then we try to make this model explicit by expressing it in terms of propositions; and these are the *hypotheses* which in their *unity* constitute the theory. However, since the model is always a 'seeing as' (which necessarily oversteps the crude evidence of 'seeing that'), it contains elements which can be correctly conceived, and therefore gives rise to *meaningful* sentences (semantic level), though these sentences might not correspond to actual features of our objects (referential level). This is why we try to check these additional features by means of experiments.

Every time we (even by chance) uncover new facts which are in keeping with our *Gestalt*, or model, we strengthen and enrich it through further details; and this also happens when, through performing experiments, we check whether all the details of the model are correct. Therefore, the outcome of the experiments *in any case* implies a modification of the model—either in the sense that it is enriched (when the outcome is positive), or in the sense that it has to be altered (when the outcome is negative). The extent of an alteration depends on the *relevance* of the details disproved by the negative result; and, again, this relevance cannot be appreciated on the basis of either logical features (from a logical point of view every proposition is only either true or false), or semantic features (even disproved sentences remain meaningful). An appreciation of this relevance is only possible on a *hermeneutic* basis, that is, by considering to what extent the unexpected result challenges the whole picture, the whole *Gestalt* or model. It may be that the result simply implies some correction that leaves the substance of the model untouched; or it may be that it demands a complete *reorientation* of the interpretation and, thus, the replacement of the old model by a new one. But this is possible because the experiment has a referential purport, which enables it to break the 'hermeneutic circle' instead of remaining prisoner to it.[10] This, by the way, is also the reason why experiments usually provide means for comparing theories. Theories are compared on the basis of *referents* and not *meanings*, and it is possible (indeed, it is very common) that we are able to refer to the same objects in spite of interpreting them in different ways (hermeneutic level) and thinking of them through different concepts (semantic level).

Once these points are clear, we can understand in what sense it is possible and even necessary to claim that theories depend (in a certain sense) on experiments. This is simply a consequence of the fact that theories, although being structural descriptions of 'possible worlds,' are constructed with a view to being descriptions or representations of the 'real world,' at least in the following weak sense. The

---

[10] This discourse is obviously rather close to other treatments in which theories are also likened to *Gestalten*, though it is importantly different from them, since it does not entirely reject the statement view of theories, but simply reduces it to its legitimate limits. A development of the *Gestalt* approach in the form of the Gestalt Model and the Perspectivist conception of science, with detailed applications to the study of concrete examples, is to be found in Dilworth (2008).

possible world which a given theory describes must include the features of the domain of objects the theory is about, and this entails not only the empirically known features but also those which are as yet unknown, but which should exist according to the model. In order to satisfy this requirement, a theory has to undergo certain tests of referentiality concerning these additional features; that is, it has to submit itself to the judgement of experiments which, apart from supporting or weakening its referentiality claims, have the immediate effect of increasing the amount of empirical data it is *obliged to account for*.

This is the reason why, without pretending that a theory should be eliminated simply on being confronted with some negative experimental results, we must say that neither can it be considered satisfactory so long as these difficulties exist, and that it is expected to undergo the necessary corrections in order possibly to overcome them. But this is also the reason why theories are perpetually subject to change as a result of experiments. Since every experiment brings in something *new*, its interpretation produces a further elaboration of the model which again contains something overstepping the immediate evidence, and hence calls for further experiments, in an interplay which progresses indefinitely without ever being circular.

7.2 Theory Change and Progress in Science

The question of theory change and theory comparison has arisen several times in the course of this work, but as a collateral consequence of more general discussions. We would therefore like to devote an explicit treatment to it in which we avail ourselves of several results obtained in the preceding sections. In particular, we shall further develop the thesis that theory comparison relies upon the *referential* aspect of theories that remains substantially unaffected even if we pay due attention to the logico-semantic aspects and to the historical and hermeneutic dimensions we have recognised and analysed.[11]

The problem of why a theory should be accepted or abandoned has been among the most debated in the literature of recent years, but it is also one in which much confusion has been caused by the interference and the uncontrolled exchange of different planes of discussion. Indeed, this issue may be discussed on a factual plane; it may be approached from a psychological or a sociological point of view; it involves epistemic attitudes; it includes logical aspects; and it also has a significant pragmatic component. Unfortunately, it often happens that different scholars lay stress on one single plane as if it were the only one which matters, trying to discredit the other approaches. What such authors are actually able to do is to show the limitations of some of the other approaches (if each were to be taken as an all-explaining tool), but they are not fair enough to recognise those aspects of the problem which are rather well accounted for by the other approaches; and,

[11] The content of this section was published, with very minor differences, in Agazzi (1985).

moreover, they seem unaware of the shortcomings of their own approach, if it too is taken as all-explaining. An advantage of our coming discussion will be that we certainly consider the question of scientific progress in connection (indeed, in strict connection) with the question of theory change, but without having to answer the question 'why' theory change has occurred. It will be enough for our purposes to see *that* it has occurred, and to examine the situation which results in science by virtue of this change. The answer to the 'why question' will be studied only after this clarification.

7.2.1 The Notion of Progress

It has been typical of our Western culture to believe that the situation emerging from change in history is the manifestation of progress, and the concept of progress has been endowed with a rich variety of shades of meaning, the analysis of which we must leave to the historians of ideas. Indeed, owing to the thematic delimitation we have introduced here, we shall simply be concerned with the idea of progress as applied to scientific *knowledge*; and this means for us, as we have already explained, to knowledge consisting of ascertaining, understanding and explanation, that is, knowledge organised in the form of scientific theories. Under this restriction, the problem of scientific progress seems to become very simple, as everybody seems ready to accept that scientific progress may perhaps be questionable if we consider all aspects of science (and especially those which concern its social impact), though its cognitive side seems to progress in a very patent and undeniable sense.

This first impression must be treated with caution, however, when we consider, as we have accepted to do, scientific knowledge as it is expressed in scientific theories. What is really evident is that in every discipline there is a succession of theories in time—that a change occurs—but one which does not by itself imply progress. In order for change to be considered as progress, the factual ascertaining of its having occurred must be accompanied by a value-judgment of some sort, enabling us to claim that the new situation is better than the old. The difficulty lies precisely in the determination of this 'better.' Let us note that this difficulty does not vanish when we enter the domain of knowledge (which might seem to have dispensed with this kind of question, being, as is usually said, value-free). Indeed, it is precisely when we consider scientific knowledge that we are very naturally inclined to share the opinion that we know 'more' and 'better' than people belonging to older generations. Now, even the exact determination of this 'more' is problematic for, under its seemingly purely quantitative categorisation, it actually covers *much* more, that is, it includes the idea of a 'selective' process consisting at least in the elimination of *old errors*, and the continual addition of *new truths*. The idea of 'better' is then more or less vaguely understood in terms of the attaining of deeper insight, or of better correlation between different truths; and its expression is perhaps given by the familiar image of science as a great and complex building to the erection of which every single scientist contributes by adding his own brick, while every generation of scientists contributes by adding something like a new floor with beautiful rooms. (An image we have already mentioned with scepticism in a foregoing section). Roughly speaking, there is some kind of consensus in the appreciation of the development of scientific knowledge not simply as change but rather as *a linearly and cumulatively progressive process.*

7.2.2 The Deductive Model of Scientific Progress

This common-sense idea became very naturally the tacit guideline for those people who first elaborated the conception that scientific knowledge actually consists in the construction of scientific theories, and who gave a 'classical' characterisation of scientific theory which has remained in force for a long time. According to this view—which was mainly developed by the logical empiricist philosophy of science beginning with the Vienna Circle and continuing with analytic philosophy—scientific theories are systems of logically-connected *sentences* that have the task of explaining observed facts. In order to reach this goal, sentences were divided into two classes, those which are purely descriptive of individual observed facts (the *explananda*), and those which are introduced as hypotheses from which formal *deductive* chains could be started that would issue in the factual sentences.

This very familiar picture may be seen as expressing at the same time the statement view of theories and the deductive-nomological model of scientific explanation. It is also understandable that these two patterns should be adopted when the problem of theory change came to be investigated within the said tradition, and that they should also be used in order to show when and why theory change occurs, and how this may be interpreted as progress.

However, these scholars remained unaware of some subtle but effective differences which manifest themselves when relations between different theories are considered, rather than relations within a single theory. Indeed, if a theory is essentially conceived of as a system of sentences (which, as we have seen, is by no means so naive and old-fashioned as many believe today), it seems appropriate to consider the relation between the sentences which make up the theory as being *logical* in nature, as few other kinds of relations might be seen *between sentences*, and such relations would certainly not be very significant from the point of view of expressing knowledge. Furthermore, the most typical and best known logical relations between sentences are surely the *deductive* ones, by means of which both logical consequences and logical contradictions become explicit.

What is questionable, however, is whether a model which is suitable for expressing the relationships within a system of sentences can be correctly extended to express the relationships among those whole systems of sentences that are scientific theories, when globally understood. The logical empiricists, and also several philosophers of science who cannot be classified among them, seem to have too hastily accepted the legitimacy of such an extension. This has ultimately led to a serious impasse, the cause of which has been seen by many to lie in the statement view of theories but which, if better understood, is rather to be identified with the generalisation of the Deductive Model to theories as wholes, rather than limiting it simply to the expression of relations between sentences.[12] Even if we accept theories to be systems of sentences, we cannot claim either that theories themselves *are* sentences, nor that their mutual relations may be formally reduced to those existing among their sentences, although the consideration of these relations may be of interest in several respects. We have already discussed in the preceding sections how a theory's being global depends on its being the linguistic expression of a particular *Gestalt*, and how the results of testing single sentences can affect the theory and the model. Therefore, we shall not repeat those considerations here. Our intention is to clarify how, even within a substantially linguistic consideration of theories, the problem of theory change and theory comparison can receive a satisfactory solution.

Let us now briefly sketch how the Deductive Model has been applied in order to account for theory change, and especially in order to interpret that change as progress. What we are going to say is so well known that we shall confine ourselves to very few indications. According to the model, the transition from an existing theory T to a succeeding theory T' could be seen as legitimate and implying progress mainly in two ways. The first involves considering the new theory T' as including the old theory T as a sub-theory (which means, in short, that the axioms or the general laws of T may be *proved* as theorems of T'). Even this minimal condition of a syntactic nature, however, reveals itself to be too demanding, for it further implies that the primitive notions of T be *definable* in terms of the primitive notions of T', and the actual fulfilment of these requirements shows that such a change, by which progress is seen as an 'embedding,' actually occurs very seldom. Let us stress, however, that this first idea constitutes the most consistent application of the Deductive Model capable of preserving the global character of theories. In fact, one can say that the underlying hope was that the old theory, as a whole, could become a proper part of the new theory, and that this would have been a real relation between systems and not between their elements (as is the set-theoretic relation of inclusion if compared with that of membership). The failure of this hope was simply the indication that such an 'inclusion' can hardly been interpreted as logical dependence or deducibility.

This kind of relation between theories being patently too optimistic, a second, less engaging one was proposed that tries formally to express the familiar and common-sense opinion according to which a theory T' should supersede a theory T if T' is able to explain all the facts which T was able to explain as well as some facts that it was unable to explain. This second application of the Deductive Model is indeed rather different from the first, as it no longer implies a kind of relation between theories considered as wholes, but tries to compare theories simply by

[12] For a presentation of the Deductive Model and its use to express the central notions of both logical empiricism and Popperianism, including theory change, see Dilworth (2008).

considering their deductive power with regard to single sentences. If we indicate by e any empirical sentence belonging to T and by e' any empirical sentence belonging to T', the two applications of the Deductive Model can be presented as follows:

(A) : $T' \vdash T$
(B) : $(\forall e)[(T \vdash e) \Longrightarrow (T' \vdash e)]$ and $\exists e'[(T' \vdash e')$ and *not* $(T \vdash e')]$.

Let us note that by writing that *not* $(T \vdash e')$, that is, by saying that e' is 'not deducible' from T, we actually include two different *subcases*, the one in which the *negation of* e' would be deducible from T, and the other in which neither e' nor its negation e' are deducible from T. Because of this double possibility the variant (B) was also accepted by people such as Popper, who could not accept an intertheoretic relation of the kind (A). Indeed, according to his view, a new theory T' supersedes an old theory T because it emerges from a falsification of T, and this means that T' must be able to explain an empirical fact which goes *against* T (and which does not simply lie *outside* T); that is, T' is to 'explain' by logical subsumption an empirical fact that is the *negation* of something T has predicted. In such a way, T and T' are seen as *incompatible* (and this excludes any possibility of there being a sub-theory relation). But the variant (B) may still be accepted in the sense of its first subcase, as we noted above. However, owing to the incompatibility of T and T', it was necessary for Popper to say that T' must be able to explain *in its own new way,* and according to *its own point of view,* the *explananda* with respect to which T was already successful, as well as new *explananda* with respect to which T was not successful and even falsified.

Popper developed this general idea further, introducing criteria for measuring the 'truth content' of different theories in order to establish the superiority of T' over T, as well as proposing his theory of *verisimilitude* or approximation to truth. Popper shared with the logical empiricist approach two essential tenets. The first was the conception of theories as deductive systems intended to provide an explanation of facts in the sense described above; the second was that the comparison of theories, and thus the problem of justifying and interpreting theory change, was to be approached according to a deductive model focused on the relation of logical deducibility between the hypotheses of the theories and some single sentences belonging to them.

It was precisely this Deductive Model that was later recognised as inadequate to account for scientific progress; and its crisis was generally interpreted as a refutation of the 'sentential view' of theories as such. We are of the opinion that the question of the sentential view of theories requires a different kind of discussion, and we have already indicated the form such a discussion should take.

7.2.3 Theory-Ladenness and Incomparability

The Popperian conception contained a point which eventually led to a complete revision of the entire question. Indeed, when Popper said that T' must be able to account for the *explananda* of T in its own new way and according to its own new point of view, he was (inadvertently) admitting by this very fact that a *change of meaning* occurs when we pass from T to T'. But then scheme (B) above can no longer be preserved, for it would be a logical mistake to overlook this change of meaning in a formal deduction. In other words, it is a mistake to assume that T and T' are able to explain *the same explanandum e*, because "*e*" means two *different* things when it is considered as a sentence of T and when it is considered as a sentence of T'. This is certainly not Popper's position, but it was easily seen as an 'unperceived' implication of his doctrine by those philosophers of science who had so fully adhered to the linguistic turn as to have also embraced the 'semantic holism' of Quine. As we have already seen, this holism maintains that no concept in a theory has an independent meaning, its meaning always being *context-dependent* (or *theory-laden*, as it became customary to say sometime later). The consequence is that two different theories cannot really be compared in terms of the Deductive Model: they are *incommensurable*. It is also well known that this incommensurability or incomparability was almost automatically interpreted as an impossibility of speaking of progress in science in any correct sense (i.e. in the sense allowed within the Deductive Model) for, in order to speak of progress, as something different from pure change, we need *comparison*.[13]

It is superfluous to present here the very well known alternative views that people such as Lakatos, Kuhn, and Feyerabend have proposed for interpreting theory change. We need only note that the consequence of their criticism was a dismissing of the Deductive Model and, more generally, the denial of the idea that inter-theoretic relations may correctly be understood as *logical* relations at all.

[13] We are not interested in discussing here why the notions of incommensurability, incompatibility and incomparability are *not* equivalent. Let us simply note that they may be equivalent within the perspective of the Deductive Model, but not outside this model. For instance, Kuhn and Feyerabend speak frequently of incommensurability, but for Kuhn incommensurability does not imply incomparability (moreover, only certain theories were considered incommensurable by Kuhn and Feyerabend—and not all the same ones by both). There has been a great deal of confusion stemming from the equation of incommensurability and incomparability, chiefly due to the fact that these notions are very vague and hardly defined in an explicit way, such that, e.g. two theries that are considered incommensurable for logical-semantic reasons can be compared on the basis of other pragmatic criteria. Furthermore, how could incompatibility be equivalent to incommensurability or incomparability in any case? Indeed the most spontaneous sense of incompatible concerns logically opposed sentences, but incompatibility can also be applied to global *Gestalten* or paradigms. Kuhn and Feyerabend, for example, sometimes affirm that they are distinct, but often treat them as practically equivalent, and for neither of them is incommensurability to exclude incompatibility. This is probably due to the fact that in their writings there is a conflation of views related to the linguistic approach (typically the thesis of 'meaning variance'), and of views more related to the *Gestalt* approach (typically the idea of a *Gestalt*-switch). It seems that Dilworth is right when he maintains that for Kuhn and Feyerabend (despite what they sometimes say) incommensurability involves a more fundamental change than that implied by meaning variance, and that this change can better be grasped by thinking of it as analogous to a *Gestalt*-switch phenomenon. This interpretation is well presented in Dilworth (2008 p. 64), a book where several excellent analyses of different approaches to the problem of theory comparison are offered, including thoroughgoing criticisms of the logical empiricist, Popperian and Sneedian approaches.

This is so whether they be relations of logical consequence and compatibility or of contradiction and incompatibility, since both presuppose the condition of semantic uniformity which does not hold, owing to the context-dependence of the meanings involved. A more general consequence of this crisis has been the dismissal of the conception that scientific theories are systems of sentences or statements (because, after all, the relationships between statements cannot help being of a logical character in the last analysis). Along this line, a non-statement view of scientific theories (which was able, nevertheless, to preserve all the features of formal rigour which are traditional in the professional philosophy of science) was developed especially by people such as Suppes, Adams and Sneed, the last of whom also proposed new ways for accounting for theory change. However we are not going to discuss these conceptions here.[14]

One main purpose of this brief historical account was to show the variety of perspectives involved in this succession of approaches and models which undoubtedly leaves us with a feeling of uneasiness. We cannot help recognising that the criticisms addressed to some older views are pertinent, but we also feel that those views contained some legitimate claims which do not appear to be sufficiently accounted for in the new approaches, despite the conceptual advance they entail in other respects. We may say that the older and the newer conceptions are all rather one-sided in the sense that they correctly capture some features of scientific theories and of scientific change, but emphasise them in such a way as to disregard or at least underestimate other no less correct and essential features.

7.2.4 The One-Sidedness of the Examined Positions

What we would like to say can be illustrated by means of some simple examples. Let us consider, for instance, the discredit thrown on the statement view of theories. This appears legitimate to the extent that it expresses a reaction against the purely linguistic view of science typical of the logical empiricist school, and in this sense it correctly stresses that scientific theories are not *just* systems of statements (as we have seen). But this correct perception cannot prevent us from recognising that scientific theories are certainly *also* systems of statements, since scientific knowledge cannot help involving *knowledge that has been formulated*, and this necessarily implies the use of sentences. In addition, we must be able to recognise what a theory '*says*' in order to discuss it, to test it, to obtain predictions from it, and so on.

When we mentioned this question above, we said that the dismissal of the statement view of theories probably resulted from criticisms of the Deductive

[14] For a brief but rigorous account of this position, and of its relation to most of the other positions mentioned here, see in particular Stegmüller (1979). For more detail see Agazzi (1981b); for a critique, see Dilworth (2008), Chap. 11.

Model of scientific theories, and this might give the impression that the statement view could perhaps be preserved, provided we dismiss the Deductive Model. Yet this conviction would be wrong, for we must recognise (unless we advocate a purely instrumentalist view of science) that one of the most specific aims of scientific theories is that of providing rational understanding and consistently harmonising what is factually known within a certain domain of research. But this is *rather* satisfactorily expressed by the notion of 'scientific explanation' such as it has been codified, but certainly not invented, by the logical empiricists, and this explanation is usually expressed in the form of a logical deduction. The reason why this conception was only 'rather' satisfactory is that logical empiricists totally ignored (owing to their allergy to 'metaphysical' notions) the concept of cause, and in such a way their conception is that of systematisation rather than explanation, as recognised already by Duhem. But, on the other hand, we know that for them the philosophy of science was considered as an analysis of the *language* of science, and the concept of cause oversteps the linguistic horizon. It follows that a *complete* dismissal of the Deductive Model would imply that this concept of scientific explanation be dismissed as well, and in the best of cases be replaced by an essentially different and more adequate alternative.[15] However, after having recognised the fundamental importance of the deductive aspect of scientific theories—which corresponds to the fundamental role that *logos* plays in every explanation understood as a 'giving the reasons why'—we must remain able to perceive that this role is not all-embracing, and that experimental evidence, for example, plays a no less important role in science.

We are immediately confronted with the difficulties of the complex relations existing between the empirical and theoretical aspects of science (in particular the inability to separate them distinctly, as we have seen in the discussion of the role of experiments in the preceding section). If this fact already suggests that we must not overemphasise deduction within theories, even more caution has to be adopted when we take into consideration relations between different theories; and we have already noted that the Deductive Model shows its weakness especially here. Still it seems to us that the best known among the alternative approaches which have been proposed have their own drawbacks. In order to see this matter more clearly, some distinct aspects of theory change must be analysed.

(i) If two theories T and T' *exist* as a matter of fact, the problem of comparing them from a purely logical point of view is legitimate and has even received considerable attention in mathematical logic, where it has been shown when, how, and to what extent such a comparison is possible, if we understand it as something to be expressed by means of formal deductive tools, or by means of model-theoretic approaches. Yet this problem has very little to do with the question of theory *change*, and this for two distinct reasons. First, because it is not said that T' emerged as a kind of modification or rejection of T (they might

[15] In Dilworth (2007, Chap. 6) is proposed what the author intends to constitute such an alternative—the Principle-Theory-Law model of scientific explanation.

well be two rival theories existing at the same time, as has so often been the case in the history of science). Second, because claiming that theory *change* is essentially or primarily a question of logic means a transition to a different problem. This problem is that of investigating the *reasons* or the *motives* for the said change; and while it is perfectly correct to study which deductive or, more generally, which logical *relations* exist between two theories, it would be at least very debatable to maintain that the one has replaced the other *because* of certain logical imperatives.

(ii) The second problem just mentioned might be characterised as the question *when* and *why* theory change occurs, and here we have many divergent answers. It must be stressed, however, that this variety of answers and their discord are, at least to a great extent, the consequence of two different attitudes which may be adopted, both of which are legitimate, provided they do not claim to be the only legitimate attitudes. We should call them, respectively, the *descriptive* and the *normative* approaches. According to the descriptive approach, our question has to be answered by a careful historical investigation; and from this it certainly appears that theory change has occurred in science in connection with and because of a great variety of factors, among which logical rigour and respect for empirical evidence often play a marginal role. Kuhn, and several other scholars who have placed great importance on the historical aspect of science, have laid so much stress on this aspect that they almost completely rejected the thesis that logical and empirical requirements have something to do with the evolution of science. On the other hand, those scholars who adopt the 'normative' attitude are ready to admit that the actual historical development of science shows many examples of a kind of 'laxist' divergence from what ought to be the correct *scientific* behaviour in abandoning one theory and accepting another in its place. They claim, however, that this fact should not prevent us from recognising that the authentic and *legitimate* ground for justifying and evaluating scientific change is, and especially *ought to be*, that provided by a convergence of purely logical and empirical requirements.

It would be a waste of time to try to give reasons for the one or the other of these two attitudes, simply because they are both right to a certain extent, while neither can account for the phenomenon of scientific change in its entirety. Logical and empirical requirements *alone* are surely not sufficient to determine theory change. This is not so because scientists are laxist in their actual praxis but because these requirements are never fully *cogent*, as has been shown *ad abundantiam* in the discussions of recent years. On the other hand, all the other factors, be they of a socio-cultural, philosophical, metaphysical, or pragmatic nature, though being of great importance in 'priming' and orienting scientific change, are not sufficient to make it happen, unless logical and empirical requirements are satisfied at least in the final stage of the process. We might even venture to say that the extralogical conditions are more relevant in the case of 'macrochanges' (involving transitions between large and very general theories), while logical and empirical considerations play a more important role in the case of 'microchanges' (involving restricted theories which remain within the frameworks of the more general ones).

(iii) A third set of questions arises when we consider the situation *after* the change has occurred, and these are independent of the possible reasons which eventually led to the change. Now we have two distinct theories T and T', and our problem has at least three aspects: what are the relations between T and T'; do these relations provide a possibility of comparison; does this comparison open the possibility of speaking of progress? Unfortunately, most of the discussions of the past years suffer from a confusion of these distinct questions, as is shown by its being the case that, first, quite frequently the inability of certain relations to provide criteria of comparison was taken as the impossibility of comparing theories at all; second, the incompatibility of theories was often confused with their incomparability; third, continuity was confused with cumulativity; and finally, the lack of cumulativity was confused with a lack of progress.

7.2.5 A New Approach to the Distinction Between Observational and Theoretical Concepts

Before trying to disentangle this skein, we would like to indicate a feature which is responsible, in our opinion, for many difficulties occurring in the evaluation of the role played by logic and experience in the analysis of scientific change. This feature is particularly visible in the traditional problem concerning the relation between observational and theoretical concepts in science. We have already presented a detailed examination of this problem in the first chapters of this work, but it is useful to recall certain essential points in the present discussion. The accepted presupposition was, at the beginning, twofold: the two classes of concepts are mutually independent, and the empirical evidence is univocal while the logical constructs which make up theories are of very different kinds. The first presupposition was gradually destroyed through the well known discussions which eventually led to the admission that all concepts in science are theory-laden and that, in such a way, no absolute separation may be drawn between observational and theoretical terms. *As a consequence*, it was claimed that the same variability which characterises theoretical contexts also affects empirical evidence, so that this evidence could no longer be considered as an independent basis for comparing theories.

To our mind, the real weak point was the second assumption, in the sense that empirical evidence *in science* is not at all univocal, but is already subject to rich variability *in itself* and independently of the theoretical framework. But this relativity of empirical evidence, precisely because it does not depend on the theoretical context in question, gives us the chance to restrict the impact of the theory-ladenness thesis, and may grant the possibility of theory comparison. This relativity of empirical evidence (as we have repeatedly stated in this work) is expressed by the fact that in every scientific theory it is not the empirical evidence 'in general' which plays a role but only very restricted and *specific* empirical evidence—that which results from standardised operational procedures of observation and measurement.

These procedures are certainly bound to some *non-empirical* context (the context which allowed for the design and use of the instruments, the context depending on the 'historical' and the 'hermeneutic' dimensions of science); but this is *not* the *theoretical* context of the theory where these instruments are used as tools for providing the empirical evidence. For this reason we have abandoned the terminology of *observational* terms and adopted instead the terminology of *operational* terms. The consequence is that the distinction between operational and theoretical terms is not absolute since it is relative to the particular theory considered. However, this relativity does not imply that the distinction not be clear and unambiguous *within any single theory*. We have already stressed that, in such a way, we can avoid a *total* relativity in the meaning of *all* scientific concepts. It may be admitted that one and the same concept change its meaning *at least to a certain extent* when passing from one theory to another, but this does not imply such a radical modification of meaning as would prevent us from comparing theories even in terms of the Deductive Model. This means that incommensurability is not a necessary consequence of the correctly stressed existence of a certain semantic relativity of scientific concepts (see Sect. 3.1 for details).[16]

An important feature of this approach is that it enables us to recover a great part of the statement view of theories without reducing it to a purely logical feature. Indeed we consider different scientific theories as systems of sentences which are intended to speak truly about *a specific domain of objects*; and we already know how this domain of objects is determined. According to an analysis developed at length in Sect. 3.1, the *meaning* of the predicates occurring in a given scientific theory may be conceived as follows: in the case of the operational predicates part of this meaning is directly bound to their defining operations, and may be called its "referential part." Besides this, there is another part which depends on the context of the whole theory, and which comes from the net of logical relationships that link the basic predicates reciprocally and with the other theoretical predicates (we shall call this their "contextual part"). In the case of the theoretical predicates, their entire meaning depends on the context of the theory, being influenced in particular by the logical relations existing with the operational as well as the other theoretical predicates. Therefore, they are endowed with only a 'contextual meaning.'

This remark is important, as it enables us to see that part of the meaning of the *operational* or *basic predicates* is not context-dependent (or 'theory-laden'). It follows that, if we can consider sentences which are constituted *entirely* of

[16] The use of "operational" instead of "observational" (in the sense we have proposed) has also the advantage of recognising that operational concepts and terms can also be unproblematically predicated of the referents of theoretical terms. In a certain sense they even *must* be so predicated. For example, to the electron a mass and a charge are attributed that are (or may be) operational concepts in the theory but are attributed to the electron in a non-operational way, that is, by indirect measurement or calculation.

operational predicates, we can restrict our attention to that part of their meaning which only depends on the operational meaning of their predicates, that is, which simply expresses their 'referential meaning.' It is certainly easy to recognise that "meaning" has been understood *intensionally* in the above. But this is not a feature peculiar just to our discourse, since all talk about the context-dependence or theory-ladenness of the meaning of concepts necessarily refers to the intensional aspect of this meaning. What we explicitly add to this feature is emphasis on the fact that referring to reality by means of certain standard operations *belongs to the intension* of some concepts, and represents a part of this intension which is not sensitive to the rest of it. This claim seems to us very well grounded in the concrete analysis of any operational concept. On the other hand, it should be superfluous to remark that the fact that measuring instruments and the standardised way of using them depend, generally speaking, on 'other theories' does no harm. Indeed, these 'other theories' belong to what we have called the 'background knowledge' available at a given historical time, and therefore do not affect the pure operational character of the referential meaning *within* the theory involved.[17]

7.2.6 Theory Comparison

With this premise borne in mind, we can proceed to consider the possibility of comparing theories. Let us assume that two theories contain exactly the *same* operational predicates, defined by means of the *same* operations. In this case the two theories speak about the *same* domain of objects and we can try to look for a totally operational sentence which, for instance, is a logical consequence of the hypotheses admitted by the first theory while being rejected by the second. Being *totally operational*, this sentence may be tested by means of the operational testing procedures equally admitted by both theories; and the result of the test will

[17] A last remark, just to dissolve possible residual perplexities: the meaning of the operational terms depends, in some way, on laws, since in the majority of cases (especially in physics) such terms express magnitudes whose measurement, which is performed by means of the operations, presupposes the validity of the laws upon whose application the instrument is grounded. We must note, however, that the truth of a law in which the occurrence of certain operational terms among which a 'uniform' relation is recognised to hold does not depend (or may not depend) on the laws that enter into the definition of the operational terms. This is so because, for example, several different laws often exist that lead to the same result, and can provide the basis for designing different measurement procedures leading to the same numerical result. Of course, as we have noted on other occasions, we must carefully investigate when this change of procedures may be understood as a kind of 'extension' of a given domain of objects, and when it entails abandoning such a domain. Interesting considerations on this point have been expressed by Campbell (1920, pp. 39–42) who, however, does not—in our opinion—sufficiently stress that the laws implicit in the attribution of an operational meaning to certain terms belong to the historical context, to the 'background knowledge,' and not to the theory that is intended to speak about the new objects. It is through not making this distinction that the doctrine of the 'theory-ladenness' of all scientific terms has found its way.

authorise a discriminating judgment in favour of the one or the other. In this case we should say that the two theories are *comparable* and prove to be *incompatible*. In the event that the basic predicates are the same, and no operational sentence may be found that supports one theory and disproves the other, we should say that, as far as we were able to determine, the two theories are *comparable* and *compatible*.

Let us now consider the case where two theories contain operational predicates which are not *completely* identical. According to our approach we shall simply say that they do not speak about the same objects, and *because* of that they are to be considered as *incomparable* or *incommensurable,* by resorting to our criteria.[18] But this situation, as we have seen, is not the only one possible, contrary to what is maintained nowadays by so many philosophers of science. Moreover, this incomparability is *not* the consequence of a meaning variance depending on the logical contexts of the respective theories, but of the *referential* meaning of certain concepts—a meaning which is context independent. In addition, we may also say that incommensurability is in a way a matter of degree, for it may happen that two theories, though incommensurable in a very strict sense, share many operational predicates which might allow for at least some comparison of certain of their sentences.

7.2.7 A Legitimate Sense of Cumulative Scientific Progress

We are now in the position to understand why it is legitimate to speak of scientific progress even in a cumulative sense. This is so because in the history of science there are several theories which have established a rich set of true sentences about certain specific domains of objects, and this truth is never destroyed by the fact that other theories have proposed new systems of true sentences about new domains of objects. Quite the contrary, the new truths remain *together with* the old ones and *complement* them. The result is that the global amount of scientific *knowledge* is increased in the sense that ever more aspects of reality become known as a result of this proliferation of *viewpoints*, leading to a proliferation of domains of investigation. In addition, scientific *laws* (which are different from theories, as we

---

[18] Let us note that we are intentionally remaining here within the *linguistic approach* to theories, and we now see that on this approach one cannot have *both* incommensurability *and* theory conflict. On the other hand it is known that, for instance, both Kuhnand Feyerabend assume that incommensurable theories can sometimes conflict, and this is carefully explained also on the Perspectivist conception presented by Dilworth (2008). The reader of the present work also knows that we too are not reducing theories to *pure* linguistic constructions, and that we have also outlined a discourse regarding theory change and theory comparison that is not of a logical–linguistic nature. Still, what we are trying to show in this subsection is that *even* within a linguistic approach (i.e. the statement view of theories) incommensurability is not a correct thesis to maintain.

have explained in Sect. 7.1) remain valid within their domain of application independently of the theories that deductively explain them.

It may be clear by now why this progress cannot be understood as a purely logical fact. Indeed, we have seen that theory change very often means the opening of a new domain of inquiry, that is, the investigation of a new *domain of reference*, and this is by no means a matter of pure logic, since it necessarily involves, on the one hand, the invention of new 'perspectives'—and these are not a logical consequence, generalisation, or particularisation of already existing ideas. As regards this issue, let us simply refer to the section devoted to the hermeneutic dimension of science, where we explained how different theories are linguistic explications of but one global scientific interpretative model. Moreover, these theories or perspectives, in order to become effective, must be endowed with appropriate *operational support*, and this again oversteps the boundaries of logic.

Why must we say that this progress is not necessarily *linear*? Because the different domains of objects are very seldom, so to speak, embedded in one another. This may sometimes be the case, when the basic predicates of a certain discipline prove capable of totally expressing the concepts of another discipline, which in such a way becomes a sub-discipline of the first. But this is very rare indeed. The most general situation is that the different domains of objects are essentially separated. However, this situation too must be seen *cum grano salis*. In several cases we can see that the respective domains are not totally disconnected, that they admit of a certain overlapping, and that there are some borderline problems which may be investigated by the tools of two disciplines or of two theories within the same discipline. This was probably the case with classical and quantum mechanics. The problems that led to the creation of quantum mechanics were at the beginning such that a treatment of them in terms of classical mechanics was not completely impossible, though at the price of several *ad hoc* adjustments. This was the indication that classical mechanics had, so to speak, reached its limits, and that its ability to master the class of new phenomena which were being discovered was exhausted. The perception of this difficulty led to the opening of new viewpoints which ultimately developed into the creation of a new theory, with new operational procedures, and hence with new objects of its own (micro-objects as distinct from macro-objects).

We could also say that science admits of both linear and non-linear progress. The first occurs within a given theory when no change occurs in its domain of objects. The second occurs when the adoption of a new theory also implies a transition to a new domain of objects. What is interesting here is that both forms of theory change are compatible with scientific truth and with the idea that progress in science *also* means a kind of accumulation of truth. In the first case this accumulation concerns truth about the same domain of objects (as we have already noted, classical mechanics is still a domain of research where new truths are being discovered). In the second case, the accumulation of truth means the discovery of new truths inside new domains of objects. Once one has understood that truth is *relative* (in the sense of being relative to the objects investigated), one can understand theory change, be it continuous or discontinuous, or be it in terms of commensurability or incommensurability, as not compelling us to give up the idea of scientific truth. Instead of discussing in detail all the different possibilities of combination between comparability and incomparability, compatibility and incompatibility, linear, non-linear, and cumulative progress, we prefer to present a few diagrams (one of which has already been presented Sect. 3.2) in which the situation can be captured in a more articulated and synoptic way.

7.2.8 Diagrams

1—The meaning of operational concepts in two different theories T and T': (Fig. 7.1).

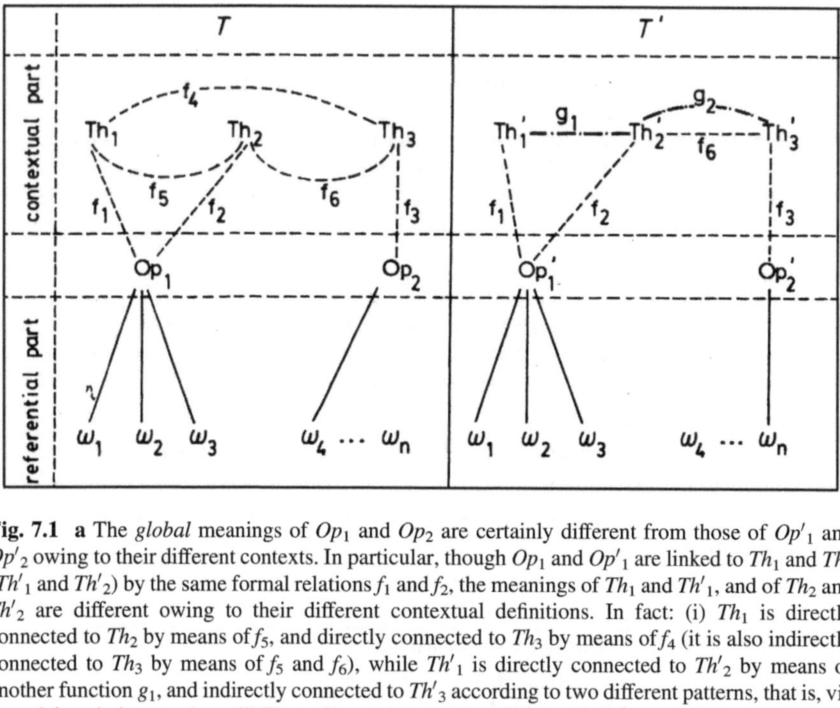

Fig. 7.1 **a** The *global* meanings of Op_1 and Op_2 are certainly different from those of Op'_1 and Op'_2 owing to their different contexts. In particular, though Op_1 and Op'_1 are linked to Th_1 and Th_2 (Th'_1 and Th'_2) by the same formal relations f_1 and f_2, the meanings of Th_1 and Th'_1, and of Th_2 and Th'_2 are different owing to their different contextual definitions. In fact: (i) Th_1 is directly connected to Th_2 by means of f_5, and directly connected to Th_3 by means of f_4 (it is also indirectly connected to Th_3 by means of f_5 and f_6), while Th'_1 is directly connected to Th'_2 by means of another function g_1, and indirectly connected to Th'_3 according to two different patterns, that is, via g_1 and f_6 and via g_1 and g_2. (ii) The referential meanings of Op_2 and Op'_2 are different, as they are bound to two different operations ω_4 and ω_n But this difference in meaning also affects Th_3 and Th'_3 by virtue of the functional relation f_3, so that the meanings of Th_3 and Th'_3 are really different; and this fact affects the whole theoretical context of **T** and **T**′. **b** As to the *referential meanings* of Op_1, Op_2, Op'_1, and Op'_2, some distinction has to be made: (i) Op_1 and Op'_1 have the same referential meaning, since they are directly related to the same operations ω_1, ω_2, and ω_3. (ii) Op_2 and Op'_2 have different referential meanings, as they are related to two different operations, ω_4 and ω_n. **c** If we now remember the strict relationship between the operations and the *objects* of theories, it is easy to see that this has automatic counterparts in the referential meanings of their operational concepts. The resulting possibilities are sketched in the following diagram

2—Relations between the domains of objects of two theories T and T': (Fig. 7.2).

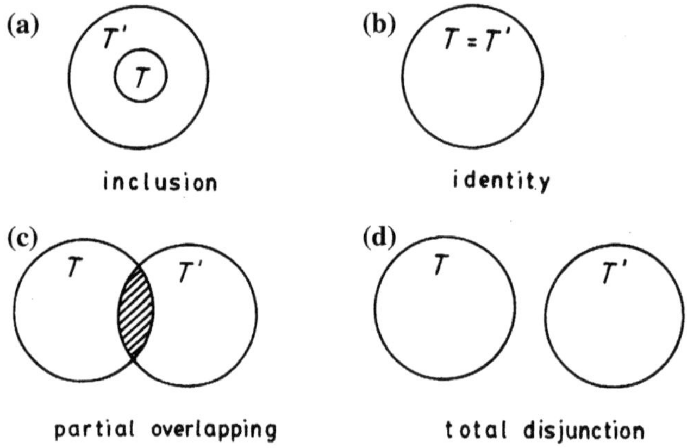

Fig. 7.2 **a** This is the case when *all* the operational concepts of ***T*** also appear with the *same referential meaning* in ***T'*** while ***T'*** contains other operational concepts of its own. We shall call the two theories *locally comparable*. **b** In this case *all* operational concepts in ***T*** and **T'** have the *same referential meaning*. Therefore, ***T*** and ***T'*** deal with the same domain of objects and are *fully comparable*. **c** Here ***T*** and ***T'*** have in common at least *some* operational concepts with the *same referential meaning*, while other operational concepts (even if they are labelled with the same *name*) actually have *different referential meanings*. This is in particular the case presented in the previous diagram. We shall call the two theories *partially comparable*. **d** In this case all the operational concepts have *different referential meanings* in ***T*** and ***T'***, and this means that they deal with fully different domains objects. We call them *incomparable* (or *incommensurable*)

As is clear, incommensurability is by no means the only possible case, but just one among four possibilities.

3—Comparability and incomparability versus compatibility and incompatibility.

The four cases examined correspond to:

(a) **T** and **T'** are *comparable*	A. They are also compatible: *(i)* Case (2a): **T'** is an *extension* of **T** *(ii)* Case (2b): **T** and **T'** are *complementary* *(iii)* Case (2c): **T** and **T'** are *partially compatible* (existence of some *correspondence principle*)
	B. They are *incompatible*: *(i)* **T'** falsifies **T** and embeds the objects of **T** in a broader but not radically different domain of objects *(ii)* **T'** falsifies **T** and retains the same domain of objects
(b) **T** and **T'** are *incomparable*: (case(2d))	*(i)* they are *neither compatible nor incompatible* *(ii)* **T'** does *not falsify* **T** *(iii)* **T** and **T'** are both simultaneously 'true' *of their respective objects*
It follows:	

4—Types of progress involved:

(a) Case 3a Ai: **T** and **T**′ are comparable, compatible, and **T T**′: continuous linear and cumulative progress.
(b) Case 3a Aii: **T** and **T**′ are comparable, compatible and complementary: continuous, non-linear, cumulative progress.
(c) Case 3a Aiii: **T** and **T**′ are partially comparable and partially compatible: discontinuous, non-linear and cumulative progress.
(d) Case 3a B: **T**′ falsifies **T** and replaces it: discontinuous non-cumulative progress.
(e) Case 3b: **T** and **T**′ are not comparable and neither is falsified: discontinuous cumulative progress.

Comments
It may be said that the typical logical empiricist position was limited to our case 4a; the Popperian doctrine was limited to case 4d; the 'incommensurability' doctrine of Kuhn and Feyerabend was limited to case 4e. However Kuhn and Feyerabend introduced non-logical criteria for comparing theories, so that for them incommensurability does not entail incomparability. Yet they failed to recognise here the possibility of progress, owing to a lack of a clear and effective distinction between incompatibility and incomparability.

7.2.9 Final Remarks

Let us conclude by saying something more about the statement view of scientific theories. As can easily be seen from the whole of our considerations, we share the opinion that scientific theories are *fundamentally* (although not exclusively) systems of sentences, provided we do not forget that they are also and necessarily systems of sentences *about* some domain of *intended objects*. To disregard or to deny this fact is a serious flaw which, in particular, makes it impossible to explain why, after all, scientific theories and in general scientific knowledge is presented in papers and books that cannot help but be composed of sentences (and this independently of the already considered major reason that scientific knowledge must be explicitly formulated, which again necessarily implies the use of sentences).

On the other hand, this does not mean that we have to accept the Deductive Model for understanding scientific change. The reason is that such a model is limited to the consideration of the possible *syntactical* relations between theories, relations which actually play a rather negligible role in scientific change. But if we concentrate our attention on the *semantic* (meaning change) and *referential* (change of objects) properties of these systems of sentences, a satisfactory understanding theory change may possibly be obtained. As we have seen, this is so because the possibility of standardising the access *to the referents* (thanks to the scientific operational procedures) provides every science with a certain 'stability of meaning,' at least as far as some of its concepts are concerned. In this sense our approach may be seen as a vindication of the '*empiria* of science' after many years in which the '*logic* of science' has been so dominant.

This vindication of the empirical aspect of science, however, must not be confused with the old approach according to which the empirical burden was totally placed upon *observations*. This was too weak a ground, as the long and largely sterile debate on the difference between observational and theoretical terms has shown. Our choice in favour of *operations* (with its methodological specifications which have been fully spelled out in the preceding parts of this book) enables us to take into full consideration, on the one hand, the fact that experience is always performed from some 'viewpoint,' within a certain '*Gestalt*,' as well as, on the other hand, the fact that, owing to its being anchored in certain *referents*, it provides us with a 'stable core' of meaning which, among other things, allows us to speak again of 'scientific progress' in a consciously recognised multiplicity of forms and meanings.

The historical, cultural, and sociological considerations widely exposed by the representatives of the so-called post-empiricist philosophy of science in recent decades are not in our perspective dismissed as being mistaken. They are actually included in the 'historical determinateness' and in the 'hermeneutic dimension' of science, of which we have spoken at length, and which are sufficient to criticise the Deductive Model as applied to scientific change, without underestimating either the fundamental role of empirical evidence, nor the due role of logical consistency. Moreover, while recognising that the empirical evidence is shaped, interpreted, and oriented by many factors of a non-empirical nature, we have also recognised that this evidence retains an independent force that not only retroacts in feedback loops on the whole theory and its inspiring model, but can even lead to their rejection.

Chapter 8
Scientific Truth Revisited

8.1 Specific Issues and Objections Regarding Scientific Truth

In Sects. 4.4 and 4.5 we studied the notion of truth from a general point of view and derived certain applications to the domain of science. The issue of *scientific* truth, however, was not thematically addressed because, in those sections, we were essentially concerned with the characterisation of the truth of *sentences* while, when people deny truth in science, they often speak of the truth of *theories*. In subsequent sections we have seen that there is a legitimate sense in which one can also speak of the truth of theories, and this amounts to recognising that the concept of truth is *analogical* (more or less in the same sense that the concept of reality is analogical), so that it would be arbitrary to reject as spurious the very common use that leads us to speak of 'true theories' or 'false theories' (a use which sometimes gives rise even to the extreme claim that only whole theories and not single sentences may be said to be true or false). In addition, it seems natural to admit that, if truth is a property of single sentences, it might also apply to *sets* of sentences. This claim is straightforward if one such set is conceived to be the result of linking the single sentences of the set by means of logical operators, since in this case the result is again a sentence whose truth-value is determined by the truth-values of its components via the truth-functional definitions of the logical operators.

The statement view of theories actually maintains that scientific theories are essentially sets of sentences, connected by formal logical links, so that it would be legitimate (at least in principle and for purely theoretical investigations) to consider a theory as a unique sentence of considerable length. This view is practically rejected now but, as we have explained, we believe that this rejection has gone too far, since it should rather be recognised that scientific theories are not *just* sets of sentences, but are *also* that. We have also seen, however, that connections linking these sentences together are *not* reducible to formal logical links so that, in particular, a theory could by no means be equated with the logical conjunction of all its admitted sentences.

E. Agazzi, *Scientific Objectivity and Its Contexts*, DOI: 10.1007/978-3-319-04660-0_8,
© Springer International Publishing Switzerland 2014

Once these clarifications have been made, it makes sense to speak of the truth or falsity of a theory in an 'analogical' sense based on two facts. First, that theories actually have a propositional aspect; second, that they can be more or less 'faithful,' 'accurate,' or *adequate* to the description of the domain of objects they try to interpret in making a given *Gestalt* explicit, and this notion of 'adequacy' is precisely the one used in the classical definition of truth.

Since the presentation of all these details required a rather broad discourse, we have been obliged to postpone to this point the *specific* consideration of the issue of *scientific* truth, especially because we must take into consideration the most important *objections* to this truth, and such objections are sometimes addressed to the truth of theories and sometimes to the truth of sentences. At this point we can treat these two types of arguments on the same footing.

8.1.1 Science can Only Approximate Truth and Never Attain Truth Itself

This statement has an entire legion of supporters who favour it for different reasons. It seems to express, first, an attitude of wisdom and modesty which should characterise every human intellectual enterprise, including science. To many it expresses the limits and finitude of the human mind confronted with the mystery of reality. For others it is simply the conclusion we must be ready to accept from the fact that no scientific theory has survived indefinitely so that, if we want to avoid the desolate scepticism which would lead us to say that science (and human knowledge in general) is always wrong, the most we can attain is an 'approximation to truth.' We can be confident, however, of being able to attain this approximation for essentially two reasons. The first is that not all of what seemed true in discarded theories perishes with them, but some of it is preserved and receives a new justification in subsequent theories. The second is that we are at least able to find and reject our errors, and this untiring elimination of falsities suggests by itself an approximation to truth.

This complex of reasons has found a kind of synthetic approach and an effort towards systematic expression in Popper's philosophy of science, and especially in the doctrine of *verisimilitude*, which he has equipped with a technical and formal apparatus. Popper's intention was to provide the notion of verisimilitude with a kind of analogy to Tarski's definition of truth, and his theory may be seen to rest on three foundational doctrines. One is the rehabilitation of the correspondence theory of truth; another is the theory of the autonomous subsistence of a 'third world' of conceptual entities independent of the world of empirical reality and of mental states (this point, however, is only weakly related to his notion of verisimilitude); and the third is the technical definition of verisimilitude (provided either in purely logical or in logico-probabilistic terms), accompanied by a metric for the alleged approximation to truth.

All these points have been the target of increasingly destructive criticism, which began with formal and technical flaws found in the definition of verisimilitude itself, continued with the finding of weak points and inconsistencies in the doctrine of the third world, and finished with attacks on the correspondence theory of truth. It is not the aim of this book either to attack or to defend anyone's doctrines. We shall therefore not further criticise an enterprise which we judge to have been a generous (though not fully satisfactory) effort at shaping a non-dogmatic but still engaging, severe and constructive conception of science; nor shall we report other people's criticisms. What we want to bring to light is rather a basic presupposition which is common to Popper's and others' doctrines of the 'approximation to truth,' and which is in our opinion untenable. The criticism of this presupposition will implicitly explain Popper's (but not only Popper's) failures, and will at the same time provide us with some indication of a more appropriate way to satisfy the legitimate intellectual requirements at the root of the 'approximation' doctrine.

The presupposition we are thinking of is the 'substantival conception of truth' of which we have spoken in Sect. 4.4.1, distinguishing it there from the 'adjectival conception.' In that discussion we showed that the adjectival conception is the more appropriate for considering truth in the context of science, without pretending that the substantival conception is wrong or misleading in other contexts. We shall now see why the substantival view has certain features that may induce people to adhere to it spontaneously.

The substantival use powerfully induces one to consider truth as a substance in the classical sense, that is, as something subsisting in itself. Moreover, if we consider the use of 'truths' in the plural, we must even say that truth appears as a domain composed of several individual entities so that, for instance, when we speak of "the truths discovered by Newton," we could actually draw up a finite list of them and refer to them individually one after the other. The combined effect of these two facts (the substantialisation and the pluralisation of truth) almost irresistibly leads one to conceive of truth more or less as a collective singular, that is, as the referent of a singular noun which designates a set of individual entities, but which may also be used on occasion to denote members of the set.

An example of this kind of noun is the substantive "man." When we say that man has created science and the arts, man must be freed of insecurity and poverty, and so on, we use "man" in such a way as means 'humankind' (i.e. as a collective singular). But we can equally well use the same term in order to denote generic single individuals and say "I saw a man in the street," or "there were three men in that car." Finally, a kind of amorphous status can be envisaged, as when we speak of the dignity of man, man's position in the universe, the importance of man's life, and so on, where "man" is understood neither as referring to humankind, nor as a single individual man, but rather as a universal essence which we do not claim to be purely conceptual (because it would not make sense to attribute dignity, position in the universe, life, and so on to a concept), and to which we might tend to recognise a more or less 'subsistent' existence according to our ontological inclinations.

The question regarding 'truth' is quite similar, and the substantival use of this notion may induce us to conceive of truth either as a kind of domain containing an indefinitely large number of single truths (like humankind being constituted of individual persons), or as a kind of universal essence which is distinct from every particular truth, though participating in all of them (as every person participates in the universal essence of man).

From this picture it easily follows that, owing to the fact that our investigations can at best assemble a limited number of single truths, we shall always remain distant from 'the truth,' whether it be conceived of as the sum of single truths, or as the universal essence which cannot be exhausted by the knowledge of a finite number of its instances. On the other hand, as the progress of human research has put us in possession of an increasing number of single truths, we may claim that we are in the process of an indefinite 'approximation to the truth'; and this itinerary may be seen either in the form of an 'infinite task' (if we incline towards the 'collective' conception of truth as an indefinitely large domain of individual truths), or in the form of a 'regulative ideal' (if we incline towards the conception of truth as a 'universal essence,' the depth of which we shall never finish sounding).

Some Popperians might be happy to see in the above-sketched conception a rather faithful and sympathetic portrayal of Popper's theory of verisimilitude, including also the 'third world' theory and the notion of approximation to truth depicted both in the form of an approximation to the whole content of the domain *T* of all true sentences of a theory (the 'collective' conception of truth) and in the form of a 'regulative ideal' for science (the 'essentialistic' conception of truth). That this is so is undeniable,[1] but it would be too much to attribute to Popper a doctrine which is much older than his philosophy, and has been advocated by a number of famous philosophers before his time. Indeed, we can say that this doctrine is the core of Platonism. If one wanted to reproach Popper for something, one might perhaps ironically remark that, after having attacked essentialism for most of his life and having presented Plato as the bad conscience of Western civilisation, he has finished by advocating an ontology which is a good model of Platonic essentialism.[2]

---

[1] For example, the notion of "truth content" and of its measure is the tool Popper intends to be used for comparing two rival theories and establishing which is more "verisimilar" than the other (Popper 1972, pp. 47–53). In such a way the substantialist view of truth becomes very clear, since truth is gradually acquired along with the progress of theories. Though Popper often speaks of truth as a characteristic of propositions (adjectival sense), it is significant that he also stresses that truth must be understood as the class of all true statements: "If we speak about approach or approximation to truth, we mean 'the whole truth'; that is, the whole class of true statements" (Popper 1972, p. 55).

[2] If Plato is the founder of this substantialistic conception of truth, of this promotion of truth to the level of an ontologically subsistent entity instead of a property of the discourse, many others have followed him along this path, and not only those who, more or less explicitly or clearly, adhere to a Platonist philosophy (for instance, because they locate the seat of truth *in se* in the mind of God). We also find philosophers outside this school who have followed this idea even in

An interesting amplification of this attitude (and a witness to how widespread it is) is the concern we find in several domains to look for the 'true' something behind the alleged 'untrue' one. As we have already noted in Sect. 4.4.1, we are sometimes told that 'true' Christianity is not the Christianity we have believed till now; we are told that the 'true' Marx is not the Marx of the Marxists; we are invited to look for 'true' democracy, 'true' freedom, the 'true' man, and so on. What is typical of all these positions is the postulation of a discrepancy between what is manifest and what is hidden, with the implicit addition that the authentic, the paradigm, the model, is what is hidden and must be recovered in its purity by an effort of research which is already tacitly supposed to be endless (here we again find the notion of the regulative ideal).

This consideration has also shown how some part of the adjectival use of "truth" is the simple prolongation of the substantival one. Actually, when we use "true" not to qualify a sentence, but to qualify some other entity (as in the case of the expressions "true friendship," "true love," "true Christianity," and so on) we are implicitly, but sometimes even explicitly, hinting at an alleged authentic essence, or genuine paradigm, with respect to which the actual situation must be confronted and corrected. We shall now try to explain why this substantival conception of truth is untenable.

The reason might be summarised by saying that this conception is again a form of *epistemological dualism* in as much as it postulates (without producing either evidence or argument) a third ontological level between thought and reality. Simply labelling this intermediate level with the august name of "Truth" does not change the substance of the situation, nor does it confer on it greater plausibility. The form of epistemological dualism implied in this attitude concerns the 'semantic logos,' since we could explain it as a confusion of the intension and the reference of *concepts* (instead of saying that we know an object through an intension, the dualists say that we know the intension). We now simply have to repeat the same remark in the case of the 'apophantic logos.' Instead of saying that we know reality and express this knowledge apophantically *through true sentences*, the dualists say that we know truth.

Here we can repeat that the intentional nature of knowledge requires that our knowing activity terminate on *referents*, intensions being not referents themselves, but only the way referents are present to thought (or, if we prefer, intensions being the special kind of relation referents have to thought). If we break this trajectory,

---

(Footnote 2 continued)
quite recent times. Let us only mention Hegel's conception of truth as being identical to the 'totality,' and being therefore something hidden which must be brought to consciousness by a dialectical process involving the self-clarification of the Idea. Or let us mention Heidegger, who conceives of truth as a deposit which exists and must be uncovered (up to the point that he invents the well-known alleged etymology of the Greek *aletheia* as meaning somehow a 'dis-occultation'). But this conception is even wider, and finds expression in several branches of present day 'hermeneutic' philosophy, whose basic assumption is that our cognitive enterprise is in general a process of uncovering some self-subsistent concealed message that almost inevitably escapes our efforts to catch it completely or adequately.

not only have we no argument for doing so, but we immediately get into trouble. For let us assume for the sake of argument the strange claim that our efforts are not directed to knowing reality but to knowing truth. This would mean that in this case truth would become the *referent* of our knowing activity, and at this moment we should automatically be led into the situation envisaged by the definition of truth itself, and we should therefore say that we necessarily can or cannot attain 'true knowledge about truth' (or, equivalently, that we must provide a 'true sentence about truth'). This, besides being strange, is either circular if the *true knowledge* and the *truth* about which it is said to be true are identical, or implies a *regressum in infinitum* if they are distinct.

More could be said on this point, but we consider what we have presented here to be sufficient to show how the conception of truth as a subsistent deposit of eternal treasures is simply a pictorial way of giving shape to an unjustifiable aspect of epistemological dualism. It follows that, if this doctrine is untenable, there is no truth which we must 'approximate,' and thus no 'approximation to truth' takes place. We shall consider later, however, the legitimate requirements which are covered by this unhappy image.

Let us note, however, that there exists another domain of problems that is related to this notion of 'approximation to truth,' although only verbally. These problems are found in the technical treatment of what may be termed "the semantics of empirical theories," which studies the question of how to provide a rigorous semantics for formal systems that have been devised with the explicit purpose of formalising some empirical (e.g. physical) theory. Most of the research performed in this field involves the employment of the tools of the standard semantics one finds in mathematical logic, and specifically in model theory, which is essentially based on set-theoretic notions and procedures. According to this semantics, as is well known, variables are associated with specified sets as constituting their range, while extralogical constants are associated with or mapped onto set-theoretic entities such as elements of a particular universe or sets of ordered n-tuples of such elements in the case of monadic or n-place predicates. While in the case of formal systems devised for mathematical theories the above conventions are easily applied, this might not be the case as regards the languages of concrete empirical theories, which may contain 'vague' predicates, both of a qualitative and of a quantitative nature.

A qualitative example may be offered by the predicate 'bald' as used in everyday discourse, to which no precise indication is appended regarding how many hairs must be left on a person's head in order for the person to be considered non-bald. From this indeterminacy it follows that this predicate cannot receive an interpretation in the standard extensional sense just recalled, since there is no set (in this case, of persons) corresponding to it. This is so because the very definition of a set in standard set-theory requires that in order for S to be a set, it must be defined, for *every* x, whether $x \in S$ or $x \notin S$, and this condition cannot be fulfilled in the case of 'bald.' A quantitative example may be provided by some physical magnitude such as *mass* which, being intended in the formal language as a function, should correspond in the standard interpretation to some concrete procedure for assigning a *unique* real number as a value to every individual of the domain. Actual measurement procedures, however, only yield approximate mass measures, not single real numbers as their results. Thus in this case as well a straightforward application of the usual semantics is not at hand. (Note, by the way, that this applies to any formal approach, including that of the statement view with its presupposed bivalency.)

Several solutions have been proposed for overcoming this difficulty (one for example proposing to change the ontological basis of the semantics, introducing 'fuzzy sets' instead of sets proper for the interpretation of predicates). We are not going to present these proposed solutions, and shall only mention a solution which preserves the usual set-theoretic basis but, on the other hand, assigns to a predicate not a set, but a class of sets. This makes it possible to introduce the notion of the approximate truth of a sentence, requiring that the sentence be true in some 'structure' belonging to a certain class of structures, rather than in a single specified structure.

We shall not continue to give examples, but note rather that the attempted solutions to the problem of approximate truth (or 'partial truth' as it is sometimes called) that have been proposed are not at variance with the general notion of truth, but actually presuppose it and, moreover, try to make it work in those cases of semantic ambiguity in which it was believed not to apply.[3]

8.1.2 Scientific Sentences are Neither True nor False, but Only More or Less Probable

This claim has become popular after the already mentioned 'crisis' of scientific certainty which occurred at the beginning of the twentieth century, and it has been reinforced by the famous 'probabilistic' interpretation of quantum theory developed after 1927. As a consequence, it is frequently presented as expressing the quintessence of contemporary science (especially of physics) in popular and sometimes also in professional descriptions of it. Yet the two historical circumstances mentioned here lie at the root of two different notions of probability, neither of which, however, has to do with truth proper.[4]

The first notion of probability can be qualified as in a way contrary to that of *certainty*, rather than to that of truth, for what was challenged after the crisis of 'classical' (modern) science was not so much the possibility of our attaining true knowledge about nature. The idea was rather that, although we *de facto* may happen to gain such true knowledge, we can never be *sure* that this or that particular sentence or theory is true. Hence, all we can do is say that we (owing to both the theoretical and the practical performances of the particular sentence or

---

[3] A brief but adequate account of this problematic may be found in Przelecki (1976).

[4] For a survey of the different meanings of "probability," see Agazzi (1988c).

theory) have a more or less high degree of confidence in it, and estimate it to be, at best, highly 'probable'.

One cannot overlook, however, that this way of using the notion of probability is not the one adopted in the most exact contexts, where probability is usually attributed to an *event* and not to a sentence. Here, as we have just stressed, probability is interpreted as a *degree of confidence*. Hence it appears as an *epistemic* requirement, and not as a *semantic* requirement such as truth is. (According to our terminology, we should say that truth is an *apophantic* requirement, but the distinction also remains valid if one uses the most current terminology.) As a matter of fact, we might faithfully translate the alleged 'probabilistic' statement in this way, "The sentence in itself is either true or false, but our degree of confidence, or our certainty, with regard to its being true can only be more or less high but never total." It is therefore clear that this first sense of probability leaves the truth of scientific sentences and theories intact.[5]

Let us now consider the 'probabilistic' interpretation of quantum theory, first proposed by Max Born, that has become the 'official' way of considering this theory for the great majority of scientists. In order to condense the meaning of that interpretation into an immediately understandable (although somewhat oversimplified) statement, we might say that according to it we can never claim, for example, that an electron will be at a certain time in the position x_0, but that we can only give a certain probability that this event should happen (Or, to express it in a more rigorous way, we should say that this being the case is ontologically only probable to some degree).

We do not want to discuss whether this is the correct or the only possible interpretation of the equations of quantum theory. Rather, we shall assume that it is and note, first, that the notion of probability applies here in the ontological standard sense, that is, as the probability of the event that a certain electron occupies at time t_0 the position x_0. But now what can be said with regard to truth? In order to answer this question, we must consider which sentence it is whose truth is being questioned; and here we must be accurate and explicitly recognise that what comes into play is not the material sentence itself, but what the sentence says, that is, its *proposition*, as we have already stressed in a preceding section.

Which proposition a sentence expresses is not, so to speak, automatic, but depends on the *interpretation* we adopt. It follows that if we give to the certain quantum-theoretic sentences a probabilistic interpretation, what they 'speak about' are not events but probabilities of events, or events plus probabilities. Therefore, in

---

[5] This has always been clear in classical epistemology. In any textbook of scholastic epistemology (including modern ones), when the discussion regards the status of our intellectual acceptance of knowledge, certain standard steps are distinguished. We begin with *ignorance*, then we formulate certain conjectures and pass to a state of *doubt*, then, when we opt for one of such conjectures we pass to an *opinion* that we consider *probable*, and finally we may attain *certainty* (if we are lucky) when cogent reasons are established for the acceptance of our opinion. In contemporary analytic epistemology such traditional distinctions are often implicitly revisited, e.g., when knowledge is defined as 'justified belief'.

order to see whether this sentence is true, we have to check whether the event occurs with the stated probability; if it does, the sentence is true, otherwise it is false. As is well known, the ways of checking probabilistic sentences are normally bound to the consideration of relative frequencies, which can often be translated into the consideration of a suitable interval of a certain magnitude. In the case of our example, the probability p ($0 \leq p \leq 1$) of finding the electron at time t_0 in the position x_0 might be translated into the statement that we shall find at time t_0 the electron in the neighbourhood of x_0 of a given radius r_0. If we find the electron in this neighbourhood, we say that the probabilistic sentence was true, otherwise we say it was false.

The other way of speaking (which is unfortunately widespread) according to which the sentence that the electron is in x_0 at t_0 is not true, but only endowed with a probability p, is untenable because it is *inconsistent* with the interpretation of the sentence it presupposes. We cannot claim that quantum-theoretic sentences express probabilities and then evaluate their truth-values as though they were merely about events. This fact is even clearer when we consider that not only does the *non-occurrence* of an isolated event not disprove a probabilistic sentence about it, but neither does the *occurrence* of the event confirm the sentence. In fact, if this event occurs, for instance, only two times in one hundred tests, and the probability assigned to the event by the sentence was 90 %, we should say that the sentence was false, although the event has in fact occurred. We can therefore conclude that even this second interpretation of probability, according to which it is simply an additional feature predicated of the referents of sentences, leaves the truth of scientific sentences and theories intact, because it does not replace the requirement of "true" by that of "probable," but only indicates that we are invited to see whether "it is *true* that it is probable that" such and such an event occurs.

8.1.3 Theoretical Sentences in Science Cannot be Qualified as True but Only as Justified

The analysis proposed in the preceding pages may be almost completely accepted without necessarily leading people who share it to the conclusion that all scientific sentences contained in an accepted theory are true. Alwin Diemer, for example, who agrees almost entirely with the distinction between the substantival and adjectival uses of "truth," would also agree that only the adjectival use, applied to sentences, can be taken into consideration in science. Nevertheless he maintains that not all scientific sentences can be qualified as true according to this adjectival use. Following a claim already defended by Hans Reichenbach, he says that only sentences of an empirical character may be qualified as *true*, while theoretical sentences may, at best, be qualified only as *justified*.[6] This claim is understandable

[6] See Diemer (1964).

in the case of Reichenbach, who was a rather crude empiricist and identified truth with the neo-positivistic requirement of empirical verification, which certainly cannot be extended to theoretical sentences. It is not equally well understandable in the case of people (such as, e.g. Diemer) who do not share this empiricist tenet. We shall therefore explore the question in some detail.

Analysed to its decisive step, this issue must be reduced to the question of whether the definition of truth does or does not apply in the case of theoretical sentences. The answer seems to be the following: while it is rather uncontroversial that in the case of single empirical sentences the referent is available, this is not so for theoretical sentences. It follows that we are not in the position to provide a 'meaningful application' ("*sinnvolle Verwendung*," as Diemer says) of the concept of truth. Note that it is not being maintained that we cannot *know* whether or not theoretical sentences are true, or that they are neither true nor false in the sense of having some kind of intermediate status between truth and falsity, but precisely that they lie *outside the domain of application* of the concepts of truth and falsity. This is promptly confirmed by the fact that, on the same basis, it is also claimed that they cannot be considered false, for example, through a 'falsification' in Popper's sense. For the falsification would only amount to recognising the falsity of a basic sentence (i.e. of an empirical sentence) derived from the theoretical sentence under scrutiny. This is correct and possible, yet—Diemer says—it would not mean that the theoretical sentence had turned out to be false, but only that it was untenable, unjustified or something of the kind.

In the case of theoretical sentences the reason for this somewhat puzzling claim is, as we have just said, that they allegedly lack the indispensable property of having a referent, and this not so much because they contain no empirical terms, but because their structure is such that their meaning necessarily transcends any possible 'givenness'.

We must examine these two reasons separately. If we qualify a sentence as theoretical *because* it contains theoretical concepts, we would deny its having a reference *because* we claim that theoretical concepts have no reference, and this is the empiricist objection we have rejected in Sect. 4.5.6. (Do not forget that a positive answer to the question of whether scientific sentences are true implies attributing referents to theoretical concepts as well.) It would therefore be naive and question-begging to assume the very problem at issue as being solved in the negative.

We consider now those situations in which some scholars qualify a sentence as theoretical *because* it is not empirically testable although it contains only empirical predicates. The most typical (but not unique) case is that of empirical generalisations such as "all ravens are black," which are not testable because of the impossibility of checking *all* ravens. This is so because all ravens can under no circumstances be thought to be actually *given* to us to be checked, and this automatically means that they cannot constitute the *referents* (remember the identification of 'referent' and 'given' introduced in earlier discussions) necessary for determining our sentences to be either true or false. As can be seen, this has nothing to do with prejudices of an empiricist flavour against theoretical terms, but only with the intrinsic structure of one (very elementary) theoretical sentence.

Are there flaws in this argument? Not if one accepts two implicit, and engaging, presuppositions. The first is that of considering those sentences that express empirical generalisations to be theoretical, while, as we have discussed in Sect. 7.1, they are in fact *laws* that must be explained by *theories* in a proper sense. This is not, however, the most debatable presupposition. More debatable is the 'rough' and unjustified conception of the *correspondence theory* of truth that we criticised in Sect. 4.4. This theory, as we have pointed out, maintains that we should have at our disposal the entire explicit and fixed structure of reality displayed before us, in order to make sentences 'correspond' to it in a kind of point-to-point mirror image. Therefore it is only if we advocate *this version* of the correspondence theory that we can pretend to have examined all ravens in order to consider the sentence "'all ravens are black' is true" to be 'meaningful.' If, instead, we adopt the approach proposed earlier in this book, according to which, in the case of 'distributive universals,' "all" is equivalent to "any," we could very simply say that to claim that this sentence is true *means* that, *whatever* raven we should consider, it would turn out to be black. This allows for a *meaningful application* of the concept of truth, provides a testing procedure, and only leaves open the *epistemic* question (not the semantic, nor the apophantic question) of the degree of certainty we can have reached after a certain amount of successful tests involving ravens. An immediate advantage with this solution is, moreover, that it rehabilitates the epistemological (besides the apophantic) relevance of falsification, at least in those elementary cases in which it does not become involved in the well known difficulties which have nourished the discussion about different forms of falsificationism in recent years.

This question deserves further scrutiny because, independently of the implicit commitment to the 'rough' correspondence theory of truth, the thesis that the concept of truth cannot meaningfully apply to theoretical sentences has a far more important negative implication. As a matter of fact, as we have already remarked, this claim would amount to saying that those sentences *lie outside the domain of application of the concept of truth*. We shall therefore ask how such a claim can be tenable without a total subversion of the concept of truth itself.

We have already seen that truth is nothing but a property of sentences or, if one prefers, a relation between sentences and their referents. This is tantamount to saying that the *whole* domain of sentences is the *domain of application* of the concept of truth. The only question which remains is that of determining the domain of sentences in a proper sense; and, concerning this problem, there are some accepted views that we have analysed in Sect. 4.4 that make it clear that not every meaningful linguistic expression is a sentence (thus excluding, e.g. questions, exclamations, and imperatives). A sentence in a proper sense is what we have called an *apophantic* sentence, which can also be qualified as a *declarative* sentence, and which is characterisable, according to a criterion going back to Aristotle, as a linguistic expression which must be either true or false. If we should maintain that theoretical sentences are neither true nor false, we should exclude them from the domain of apophantic or declarative sentences altogether. But what kind of linguistic entity would they then be?

They would certainly not be imperatives, or questions, or invocations, and so on. But more than any linguistic or material feature, the *intention* of these statements is illuminating, and the intention of theoretical sentences in science is certainly a declarative one; or, if we prefer, they are uttered with the intention of saying something true.[7] It is undeniable that they enter into some process of 'justification,' but one must not overlook the fact that in this process they also play the role of sentences which *justify* rather than of sentences which *are justified*. The process of justification is, in empirical theories, basically that of deductive explanation in which theoretical sentences normally play the role of the *explanans* and not that of the *explanandum*. But they can play this role only inasmuch as the structure of formal logic is such as to provide a warranty of truth when we start from true sentences.

This shows again that if we leave truth out of consideration, the process of scientific explanation itself collapses. Indeed, we know that true consequences may be derived even from false assumptions. Hence, if the problem were *merely* that of obtaining a logical explanation, we should feel satisfied once such an explanation has been provided for our empirical sentences, *independently* of any further investigation about the truth of the assumptions. Yet we do not proceed in this way, but rather look for a variety of additional tools (functions, tests, and so on) for checking our hypotheses; and this is *only* meaningful if we require that, beside being formally able to provide explanations, our hypotheses must also be *true*. In this sense we cannot help agreeing with Popper when he says: "Our main concern in philosophy and in science should be the search for truth. Justification is not an aim; and brilliance and cleverness as such are boring".[8]

In conclusion, there is no possibility of denying that theoretical statements are apophantic sentences and, hence, that they necessarily fall within the domain of application of the concept of truth, and turn out to be either true or false. They are true if they "say that which is"; they are false if they "say what is not." Moreover, it is certain that theoretical sentences in science are retained (or even considered as 'justified') to the extent that they are believed "to say that which is" *about the domain of objects of their theory*. They are therefore claimed to be *true of* this domain. It should be superfluous to stress that this leaves the problem of the *certainty* of theoretical sentences open.

We do not discuss those positions that *identify* truth with justification. They have been authoritatively defended, for example by scholars such as Dummett and Putnam, with different degrees of force, but they do not constitute *rejections* of the notion of truth in science. They are rather proposals for a *definition* of truth that

---

[7] This corresponds to the deep aim of science (including modern science) that remains undeniable even by those who are sceptical about its capaqbility of attaining it, an aim that is well espressed in the title of Psillos (1999), "How science traks truth".

[8] Popper (1973), p. 44.

can reasonably be included among the forms of the 'coherence theory' of truth. We have already pointed out, however, that this theory actually proposes a *criterion for* rather than a *definition of* truth.[9]

8.1.4 Scientific Laws Lie

A position rather similar to that discussed above is advocated by those authors who admit truth for sentences that are very close to empirical evidence (such as empirical generalisations) but not for the most typical theoretical sentences in science (such as fundamental laws), and this not because truth does not apply to them in a pertinent way, but simply because they are declared *false*. The best known representative of this position is Nancy Cartwright, whose proposals are metaphorically expressed in the very title of her book, *How the Laws of Physics Lie*.[10] In the book she maintains that the laws of physics "lie" because they fail to satisfy the requirement of "facticity" that they are commonly believed to satisfy. This requirement amounts to the ability to "describe facts about reality" so that, to the extent that they are able to do this, they are true (p. 54). But, she says, the paradigmatic laws of physics, that is, its *fundamental laws*, "do not describe true facts about reality. Rendered as descriptions of facts, they are false; amended to be true, they lose their fundamental, explanatory force" (ibid.). The reason adduced for this claim is that, owing to their generality and to the fact that they concern only very restricted aspects of reality, they are unable to match any concrete situation in which several aspects, forces, or interactions, are involved. Even when one tries to explain a concrete fact through the "composition of several causes,"

> the force of these explanations comes from the presumption that the explanatory laws 'act' in combination just as they would 'act' separately. It is critical, then, that the laws cited have the same form, in or out of combination. But this is impossible if the laws are to describe the actual behaviour of objects. The actual behaviour is the resultant of simple laws in combination. The effect that occurs is not an effect dictated by any one of the laws separately. In order to be true in the composite case, the law must describe one effect (the effect that actually happens); but to be explanatory, it must describe another. There is a trade-off here between truth and explanatory power (p. 59).

---

[9] This is implicitly recognised by Putnam, who says:

> Whereas Dummett identifies truth with justification, I treat truth as an *idealization* of justification. Truth cannot simply be justification, I argue, for any number of reasons: truth is supposed to be a property of a statement that cannot be lost, whereas justification can be lost (in fact justification is both tensed and relative to a person), justification is a matter of degree, whereas truth is not (or not in the same way) etc. (Putnam 1983, p. 84).

Let us note, however, that it remains unclear how something that is *not* truth (i.e. justification) becomes identical with truth in the limit. It seems that a certain confusion between truth and certainty is implicit in Putnam's position.

[10] See Cartwright (1983).

It is clear from these statements (and from the general approach of her book) that the author not only is unaware of all the literature devoted to *idealisation* in science, but does not realise that fundamental laws *encode* certain features of natural phenomena that are only partially *exemplified* in any particular 'fact,' and precisely because this fact also exemplifies several other features that can be encoded by different laws. The privilege accorded to "facticity" brings the author to give full cognitive weight to what she terms *'phenomenological' laws*, that is, in practice, to empirical generalisations: "The great explanatory and predictive power of our theories lies in their fundamental laws. Nevertheless the *content* of our scientific knowledge is expressed in the phenomenological laws" (p. 100). This affirmation is debatable since (as we have maintained in the present work) fundamental laws also provide scientific knowledge, in the full sense that we have considered.

Cartwright's position, however, is different. She maintains that "the ultra-realist thinks that the phenomenological law is true *because of* the more fundamental laws. One elementary account of this is that the fundamental laws make the phenomenological laws true. The truth of the phenomenological laws derives from the truth of the fundamental laws in a quite literal sense—something like a causal relation exists between them" (ibid.).

We shall not contest whether such ultra-realists actually exist or have existed (the vague allusions made by the author are certainly insufficient as an historical backing for this claim), but shall rather note that in any sensible philosophy of science the truth of the phenomenological laws is not made "dependent" on the truth of the fundamental laws, and even less "caused" by it. The (admitted) truth of the fundamental laws provides *reasons* for the (independently ascertained) truth of the phenomenological laws and, in fact, in the case of a collision between the phenomenological and the fundamental laws, what is questioned, and must be "saved" (if possible) is the truth of the fundamental laws.

Cartwright, however, is too acute a scholar to remain fully prisoner of the logical-empiricist outlook that implicitly determines the position mentioned above (in particular, the last sections of her book contain interesting elements indicating an overcoming of this outlook). Therefore, it is not surprising that (accepting an explicit indication of Adolf Grünbaum) she advocates a view of the relationship between fundamental and phenomenological laws that is much in keeping with the relationship between encoding and exemplifying that we maintain in the present work:

> I call this kind of account of the relationship between fundamental and phenomenological laws a *generic-specific* account. It holds that in any particular set of circumstances the fundamental explanatory laws and the phenomenological laws that they explain both make the same claims. Phenomenological laws are what the fundamental laws *amount* to in the circumstances at hand. But the fundamental laws are superior because they state the facts in a more general way so as to make claims about a variety of different circumstances as well" (p. 103).

Here the difference is seen simply as a relation between generality and particularity. But what is interesting is that, in the description of how the fundamental laws apply to the particular situations expressed by the phenomenological laws, we find an excellent characterisation of what it means for a concept or a sentence to *encode* properties that are partially *exemplified* in concrete cases, and to be open to exemplification in other cases. But now it is also clear that one could hardly maintain that fundamental laws are not true in a given situation if they "make the same claims" as the phenomenological laws that are declared true in the same situation.

With similar arguments we can refute analogous objections against the capability of science to attain truth because its theoretical objects (such as material points, perfect gases, rigid bodies, and so on) do not exist. We have already discussed this objection and seen that it rests upon a failure to distinguish between encoding and exemplifying, and we shall not repeat those remarks here.

8.1.5 How can Scientific Theories be True if they are Usually Refuted After a More or Less Short Life?

This objection is based on the historical fact that no scientific theory is known to have survived indefinitely. Since the creation of modern science in the seventeenth century, the life of scientific theories has become increasingly shorter, and there is no plausible reason for claiming or even hoping that our present theories, which in the different domains of science have been proposed only recently, will last for the rest of human history. Because of this historical evidence Popper's falsificationist philosophy of science presents itself, on first impression, as a happy intuition regarding the genuine 'spirit' of science, for it may be seen as the expression of the awareness that the destiny of a scientific theory is that of being proved false, sooner or later. But if a theory, or even a simple sentence, is true, its truth must be eternal, since truth consists in the relation between the sentence (which does not change) and reality (which also does not change), and this relation cannot cease to hold at some particular moment. Or, if one prefers, we should say that if an (untensed) sentence is found to be false at a given time, this means that it has always been false, even before that time, and that it will continue to be false for the rest of time. No exception to this is actually provided even by tensed sentences, since they can be considered as having time itself among their referential attributes.

It follows that the admission that our theories will be proved false sometime in the future automatically implies that they are already false now. Therefore, science cannot help being a perpetual collection of falsities and not of truths. Because of his desire to avoid this pessimistic outlook on the history of science, Popper was led to propose his conception of verisimilitude or approximation to truth, which might be qualified as an effort to explain how it is possible to come *closer* to truth while remaining in falsity.

We can immediately reject at least one formulation of this objection by recalling that it presupposes the substantival notion of truth, which has already proved untenable at least in the case of scientific theories. This is so, in particular, with regard to the Popperian formulation we have just recalled. It is not said, however, that this objection necessarily presupposes such a conception of truth. It can be formulated in terms of the correct relation between sentences and reality, which is contained in the definition of truth we have adopted. It is in answer to this precisely formulated question that the approaches presented in this work will join and give what we believe to be the solution to the whole problem of the truth of scientific sentences, or, for the sake of brevity, of 'scientific truth'.

The central point of our argument will be a precise evaluation of the claim (for which we have provided the necessary foundation in previous sections) that scientific truth is always a *relative* truth, in the sense that every scientific sentence is always true (or false) 'of' the specific objects which constitute the particular domain of the theory in which the sentence occurs. This remark has already led us to specify that these objects are the *referents* of the sentence in question; and this fact has obliged us to modify the ambiguous statement which makes truth consist in a relation between a sentence and reality, by saying instead that truth consists in the relation between a sentence and its intended referents.

We have called the first statement ambiguous, not because it is wrong to say that truth consists in the relation between a sentence and reality, but because it is not sufficient. Though referents are surely real, not everything real is the referent of some given sentence. Once this fact is recognised, the eternity of truth will not be expressed in the form we (purposely) used above, but rather through this new one which, by the way, does not presuppose the correspondence theory of truth (which could be too easily accommodated in the above form), and which has the correct form of a *conditional* statement:

> if a sentence does not change, and its referents do not
> change, then its truth (or falsity) does not change.

Our problem has now received a precise formulation. In order to see whether the falsification of scientific theories really challenges their truth (which seems prima facie tautologically obvious) we must see how they behave with respect to the above condition. It is clear that a sentence S, which was claimed to be true in a given theory, cannot be proved false within the very same theory, at least in almost all the known cases. In other words, not only is it impossible that an 'immediately true' sentence be falsified in a theory where it is a protocol sentence, but it is also impossible that a theoretical sentence be proved to be at the same time both true and false within the same (consistent) theory. Furthermore, it is also difficult for a theoretical sentence to be falsified by an unexpected new empirical sentence incompatible with it. We shall admit this possibility because, after all, errors can be made in science just as everywhere else. But what is the *scientifically significant* effect of claiming that a *theoretical* sentence is false? It is simply that of claiming that its contradictory sentence is true. But this new sentence, being theoretical as well and not being compatible with *S* in the scrutinised theory, must necessarily be claimed in another theory.

Note that we are here envisaging the case in which *S* is declared false, and not the case in which it is simply, so to speak, 'problematic.' For imagine that *S* has been challenged by an empirical sentence *E* that contradicts it; from this moment on, judgement on *S* is suspended. It may happen that, by means of some addition, correction, or more or less *ad hoc* adjustments, we are able to remove the difficulty and continue to accept *S* in the theory. It may even happen that we leave the situation in a kind of limbo, hoping to get rid of the difficulty at some later time. It is only if we find *another* theory, which is held to be more satisfactory than the one which was previously accepted, that we actually say that *S* has been recognised as *false*, and reject it. In other words, we are not only claiming that truth, but that falsity as well must be accompanied in science, by *reasons*; and these can be provided only by theories. Let us note that we have spoken of the rejection of sentences and not of theories, considering what we said in the discussion concerning the hermeneutic dimension of science. This means, in particular, that the rejection of a single theoretical sentence could induce a readjustment of the theory that remains within the accepted 'model' or *Gestalt.*

We might face difficulties with regard to our condition of eternal truth, for let *S* have the form *Pa*; in order to say that *Pa* is true in theory *T* and false in theory *T′* we have to know that *Pa* is in agreement with its referent in *T* and is not in agreement with its referent in *T′*. But this challenges the eternal truth of *S* only if we can show that *S* is the same proposition (not only the same sentence!) in *T* and *T′*, and that the referent of *Pa* is the same referent in the two cases as well. It is exactly this condition which is hardly fulfilled in all known cases. As we have already explained on a preceding occasion, the *concepts* designated by *P* and by *a* are hardly the same in the two theories, and the *meaning* of *S* is therefore bound to change. This implies that the sentence is no longer the same either, and we must therefore conclude that a *new* sentence, so to speak 'resembling' *S*, has been proved false in *T′* while leaving S (eternally) true in *T*. This is what the supporters of 'theory-ladenness' affirm, since for them the (undeniable) meaning variance also applies wholly to referential concepts. For reasons already explained at length, we cannot subscribe to this view; and we can admit that truth is preserved in both theories only if the identity of the referents is preserved and the two theories simply express certain *complementary* views on them (a situation rather unknown in the past, but that has been imposed by quantum physics). Besides this possibility, we must also take into consideration the possibility in which the referent is no longer the same, though denoted by the same words.

We already know that this happens when the operational concepts receive a different operational definition, and we are not going to repeat here what has already been explained elsewhere. In this case it is even more obvious that *S*, being 'about' different objects in the two theories, may be true in one and false in the other, without violating the eternity of the truth of a *given* sentence about *given* objects. (To be exact, we should say that the sentence is no longer *pertinent* in the new referential context; but we need not be that rigorous for the sake of our argument.)

All we have been saying has accepted the unrealistic assumption that it is possible to 'falsify' an isolated sentence. But the discussions in the literature have abundantly shown that this is practically never the case, and that only theories as a whole can be accepted or rejected. Taking this into account, we must admit that the condition for the eternity of truth is always fulfilled for, two theories being necessarily different, either they give rise to the unproblematic situation comparable to that of two sentences being 'complementarily' true of the same referent, or to the even less problematic situation of two sentences being true of two different referents. Our puzzling conclusion is therefore that no falsification of a theory is properly possible, and in such a way the entire objection is met.

We shall consider, however, some obvious reactions to this conclusion, that apparently completely reverse Popper's falsificationism. For instance, do we really believe that Ptolemaic astronomy *is* still true, that the corpuscular theory of light was not disproved by experimental results on the velocity of light, that Newtonian mechanics was not disproved by relativistic and quantum mechanics, and so on?

Our answer is that these theories have been disproved to the extent that they are believed to speak about 'things'; but they are still true if they are correctly judged on the basis of what they say about their 'objects.' For example, Ptolemaic astronomy indeed contained much more than what was 'objectively' stateable on the basis of the criteria of referentiality available at the time it was accepted (which were those of naked eye observation and of certain geometrical-astronomical instruments). In particular, the most famous tenet of this astronomy, concerning the 'intrinsic' or 'absolute' state of motion of the sun and of the earth, contained a completely non-operational predicate in that only relative motions of the earth, the sun, the moon, and the stars were observable; and they were and *are* accounted for quite well by this astronomical theory. Actually, we still use it for the limited purposes in which only these aspects of astronomical reality matter (such as when we make calendars, or make appointments to meet, e.g. at sunset, and so on). This objective content remained unaltered in the subsequent astronomical theories, which only modified the general conceptual framework of the Ptolemaic theory according to a process which we considered when we treated the problem of theory change.[11]

Something similar may be said about the corpuscular and wave theories of light. The corpuscular theory gave a good account of many aspects of this 'thing,' and the theory was only rejected because it turned out that other aspects of light escaped its 'conceptual space' and required a new one. This new conceptual framework was mistakenly assumed to exhaust the 'thing,' and we know that both conceptual frameworks had to be recovered and combined in order to account for

[11] As a matter of fact, the 'scientific proofs' of the rotation of the earth on its axis were obtained only in the nineteenth century, and those proposed by Galileo and Newton were not correct. In the meanwhile the Copernican theory was preferred because of several 'gestaltic' advantages of the type we have already discussed.

the optical phenomena known today. We can therefore say both that the corpuscular theory was (and is) true of the corpuscular *aspects* of light (or of light seen under a corpuscular viewpoint or, even better, objectified by means of corpuscular predicates), and that the wave theory was (and is) true of the undulatory aspects of light, and finally that our present wave-particle theory of light is true of light as we objectify it in present-day physics. What is important in this comment is that we are hinting at an idea which we have already met in discussing the question of theory change, that is, the idea that after a certain time an objectification meets its 'limits,' and *without being proven false*, it is proven *partial*, that is, not such as to exhaust reality. Note that, by this, we do not maintain that each different theory has to do with a *different reality* (this is would be like Feyerabend's 'epistemological realism'), but that they have to do with *different aspects* or attributes of reality, which we express more precisely by saying that each theory has to do with different *referents* that result from different *objectifications* of reality. Let us say immediately, in any case, that this 'partial truth' has nothing to do with an approximation to truth or with an 'approximate truth' of the kind we have already discussed. Indeed, it has the meaning of a 'complete truth on a partial domain.' The mention of the problem of approximation invites us to discuss the third example mentioned above. It is actually often said that the transition from Newtonian to relativistic or quantum mechanics represents progress inasmuch as the latter two are better approximations than is the former (or, similarly, it is said that Newtonian mechanics constitutes only a first approximation with respect to the exact, or at least more exact, formulation of truth we reach in the other two theories). The arguments adduced in support of this claim are well known and need not be recalled here.

However, we cannot agree with this interpretation for two main reasons. The first is that it coincides more or less with such untenable doctrines as epistemological dualism (the reality 'hidden' behind the phenomena can only be indefinitely approached by endless chains of further approximations), or with the substantialist theory of truth (the unattainable truth is more closely approximated in the new theory than in the old). The second is that it completely misconceives what we have already explained when we said that these theories do not approximate more or less faithfully to the exact knowledge of the *same object*, but are concerned with *different* objects. (So e.g. van der Waals' theory of gases is concerned with gases consisting of atoms that have volume while the theory employing the ideal gas model concerns gases consisting of atoms that have no volume.) This second point will give us the opportunity to clarify how the order of approximation *enters into the definition of the object.*

This thesis is the straightforward consequence of our general point of view, according to which scientific objects are 'clipped out' of things by operational predicates which are defined on the basis of operational procedures. Every operational procedure is given (or, better, is characterised) by a certain order of approximation or margin of error. This is especially evident when operations are performed by means of concrete instruments, as in physics, where it is well known that it is incorrect to say that the value of a measured magnitude has been found, for example, to be equal to x, since we should always say $x \pm e$, where e is the margin of error of the applied measurement procedure. This means, in particular, that it makes no sense to carry out calculations leading, for example, to the expression of the length of a body as being equal to 5.00021 cm. if the instrument on which the length calculations are based in that context only admits of a margin of error of one millimetre. The alleged accuracy would simply lead to a *meaningless* statement. When we say "meaningless" we wish to be taken literally, because if the meaning of the operational predicate has been introduced by means of a measurement limited by a certain order of approximation, it is clear that we are not using this meaning (or we are misusing it) if we pretend that it is bound to a different order of approximation. In short: *the order of approximation is an integral component of the referential aspect of objectivity*.

What has been said here with respect to operational criteria may be extended to theoretical conditions as well. To the extent that two theories, be it for operational or for theoretical reasons, have to do with different orders of approximation, they already have to do with different objects. But then the immediate consequence is that it is incorrect to say, for example, that by using a balance of a certain type I can measure the mass of a particular body 'only within the range of 1 mg.' This way of speaking is adequate for everyday discourse having to do with things. But within a scientific context, in which we assume that this balance was accepted as the standard for measuring mass, we must say that the mass of that given body *is* $m \pm e$ mg.

Even common sense has, on occasion, the awareness that 'maximum exactness' may be identical with 'meaninglessness.' Let us imagine that one claims to use such a refined instrument as to be able to determine the height of a mountain with an accuracy of one millimetre. We shall immediately say that this claim is meaningless, not because we do not know which powerful instrument to use for such a determination, but simply because it is intrinsically impossible to fix with the precision of one millimetre where the base of the mountain actually begins. The same may be repeated for the breadth of a country road, which may be bordered by fields in such a way that it would be meaningless to say with accuracy that its side ends at one particular place rather than one millimetre further over. What is so obvious in the case of these 'things,' when they become 'objects' of some measurement in everyday life, should be even clearer for objects of specific scientific disciplines, which can *only* be considered *as* objects, and for which it makes absolutely no sense to imagine the epistemological 'thing in itself' which *has* the exact mass, the exact length, and so on, lying behind our 'approximate' measures.

Let us note, by the way, that these remarks are useful in the discussions concerning 'approximate truth,' 'semantic ambiguity,' and so on, which we briefly mentioned above. Several problems raised in such discussions will simply prove to be artificially posed, and to dissolve as meaningless (a conclusion which may already be drawn, perhaps, from consideration of the fact that actual science is never troubled with them).

Another remark: if questions of approximation are already that important in the domain of those sciences in which 'exact' tools for measurement are at one's disposal, they are of even more importance in sciences where these tools are of no use, and where, therefore, the claim of 'maximum precision' would really risk appearing as totally meaningless. This is the case, in general, for the 'human' sciences, where the methods of objectification usually have rather wide margins of error. It is therefore quite strange, in those cases, to read papers which present allegedly exact quantitative estimations when the operational procedures adopted are far from sufficient to allow such claims. The same may be said of sentences of an historical or philological character when the alleged precision is sometimes the result of ingenuity and speculation rather than being based on the reliability actually given by the existing documents or texts. On the other hand, we must not lay blame for this solely on scholars in the human sciences; natural scientists often do not behave much better. Think only of the bold self-confidence with which some authors fix times for the origin of life on earth, or of the universe itself.

In conclusion, let us stress how the consideration of the order of approximation is related to something more engaging than a pure discussion of an epistemological nature, as the present discussion may be thought to be. As a matter of fact, working scientists well know how 'orders of magnitude' are relevant to every question, and how they can often determine authentic clear-cut distinctions in the properties and behaviour of the entities they study; microphysics, biology, psychology, sociology, and so on are full of examples of this kind.

8.1.6 The Controversial Nature of Scientific Truth

The above discussion has completed our task of justifying the claim that scientific sentences and theories can be true with respect to their theoretical parts as well. We must be aware, however, that this conclusion is valid on the semantic (or rather apophantic) plane, but leaves the question open on the epistemic level. In other words, once a sentence is formulated and its domain of referents is fixed, it cannot escape having a meaning (semantic level), and being in a truth-or-falsity relation with its referents (apophantic level) which necessarily makes it either true or false about these referents. This is independent, however, of other kinds of *relations* the sentence may have under other aspects, which provide it with properties *other* than those of truth or falsity. Some of these properties express the relation the sentence may have to people uttering, or using, or considering it (both at an individual and a collective level). Such properties are usually qualified with terms such as "certainty," "reliability," and so on, that refer to attitudes subjects may have towards a sentence, such as 'knowing that S,' 'believing that S,' and such. For the sake of brevity we shall call this level of consideration *epistemic*, and speak therefore of an 'epistemic logos' near the 'semantic logos' and the 'apophantic logos.' Now, it is clear that just as the meaning of a sentence is compatible both with its truth and with its falsity, so is the truth of the sentence compatible both with its being known or believed to be true, and its not being known or believed to be true. Yet, in science (but not only in science) we want to know with certainty that some sentences are true or, if this optimal situation does not obtain, to have at least a justified degree of confidence, or of belief, in the truth of those sentences.

Justification, which we have shown to be mistakenly believed to be a replacement for truth in the case of theoretical sentences, actually has an important role to play on the epistemic level. In fact, on the apophantic level we are still left with the problem of 'evaluating the success' we have been able to reach in the application of the only two tools at our disposal for obtaining truth, that is, evidence and argument, which are applied in the different strategies we invent in our efforts to obtain true sentences. Therefore, it is clear that all the critical investigations suggested for discarding prejudices, for analysing the conditions of empirical inquiry, for taking disturbances into account, for checking the accuracy of instruments, and so on belong in a broader sense to the evaluation of the 'quality' of the evidential support we provide for claiming that some sentences are 'immediately true.' Besides this, we try to draw the maximum advantage from the argumentative tools. We do this not only by refining the deductive procedures, but also and especially by determining better strategies for increasing the 'inductive support' of those sentences which do not derive their strength immediately from evidence, or from being deduced from immediately (or otherwise established) true sentences, but rather by admitting true sentences as their logical consequents.

From what we have just said, it should be clear that neither the requirement of evidence nor that of argumentation are matters of course, but are rather the result of complex processes which have the character of 'performative' *activities* and are exposed to the risk of certain shortcomings. This is the reason why a scientific sentence is always controversial, at least in principle. For one may doubt that all the requirements have been adequately checked. Moreover, the intrinsic non-definitiveness of inductive procedures always leaves the questions *theoretically* open. (This is why the fact that a theoretical sentence has met with a great number of positive checks cannot give the total certainty that it would not fail the next test.)

We must distinguish, however, this property of being controversial in principle, or this theoretical non-definitiveness, from what we like to call the *relative practical definitiveness* scientific theories are actually able to attain. By this expression we mean the following. The adjective "relative" reminds us that the theory is being claimed or proposed as true only relatively, that is, 'with regard to its objects,' and that it is likely to be 'superseded' (not falsified) when other criteria of objectification are advanced. The adjective "practical" means that, after a 'sufficient' number of checks, favourable tests, successful predictions, useful applications, fruitful adoptions in the context of various problems, convincing logical connections with other fields of knowledge, and so on there is no *reasonable ground* for not being confident about the theory. In other words, we could say that our degree of confidence is very close to 1, if we want to express it in terms of probability. But in any case let us stress that it is a question of *practical* certainty, that is, of certainty which is not entirely founded on purely logical arguments, as is known to everyone who has some familiarity with inductive logic and related topics. One should not underestimate, on the other hand, this 'practical definitiveness,' because it justifies our conviction with regard to the 'cumulative' nature of science, which obstinately survives all efforts to persuade us that scientific theories are 'incommensurable,' and that science proceeds only through devastating 'revolutions'.

Let us also note that it is because of this possibility of reaching a practical definitiveness that we may conceive of the possibility of making errors and yet progressing in science. The conceptual situation regarding errors is quite vaguely determined in the existing research in the philosophy of science. All authors seem simply to ignore errors, and to concentrate solely on the ascertaining of 'positive' knowledge; and this is true not only of those scholars who like to look at the successful side of the scientific enterprise, but also of those who lay stress on falsification, for falsification always depends on ascertaining some 'positive' fact which speaks against an hypothesis. The typical problem of error, however, is different, and has something to do with 'not finding something,' rather than with 'finding that not-something'.

In order to appreciate this fact, we must try to imagine scientific research as research in the most usual and even trivial sense of this word, such as when we are looking for a book in a library without having had the chance to consult the library's catalogue. We formulate successive hypotheses about the book's location, and we may often be wrong. But when, at last, we find our book, we have no reason to be sceptical about the success of our search. The same happens in science. We first delimit our domain of inquiry (as when we limit our search to books, or even to a single book, and not to other things), and then we start formulating hypotheses and testing them. Most of these hypotheses may turn out to be false, that is, to be falsified by the 'data' which are fixed by our criteria of referentiality. But after a certain time something is found with certainty, in a sense rather analogous to that we employed in the case of our book in the library.

Our example cannot, of course, be taken too far, if only because in empirical science an absolute certainty is logically impossible for quite simple and frequently explained reasons. But, on the other hand, one is also not to value this fact too highly. Once a theory has been patiently constructed and received a sufficient amount of independent confirmations, it reaches a stage of relative truth and of practical definitiveness with regard to its objects, and it is correct to be confident that it is and will forever remain true about these objects. Indirect (but 'practically' decisive) evidence in favour of this is the fact that all problems we are able to formulate, for example, in terms of the Newtonian predicates proper, can be handled and solved by resorting to those same predicates and to the tools of the Newtonian theory in its present state (and we do not see why this should not continue to be the case tomorrow).

8.1.7 The Controvertibility and Non-absoluteness of Scientific Truth

The fact that a scientific sentence is always controversial in principle (i.e. that it can be challenged and become the object of controversy) has as a consequence (or, actually, presupposes) that it is *controvertible*, that is, that it might, at least in principle, be disproved sooner or later. This, as we have repeated several times, is the typical situation for theoretical sentences, which are claimed to be true only on the basis of their admitting true consequences, which is in turn a necessary but not a sufficient condition for truth. On the other hand, it is appropriate to say that scientific theories as such are constituted entirely of theoretical sentences, since single empirical sentences are rather the starting point from which theories are constructed (when they inductively give rise to empirical laws), or express the result of experiments designed to put theories to the test. (Actually, no scientific theory we find in textbooks, for example, speaks of single individual events proper, at least if we leave aside the historical sciences.) Now, this controvertibility of scientific knowledge is of an epistemic character which, though of no harm within science, may become of significance when the discourse involves human existential requirements regarding certainty; and it may therefore help us to understand the existence of other domains of human inquiry and activity, besides science, where man tries to satisfy these requirements. We shall return to this problem later.

A related problem, which also arises when one considers scientific truth as such, is its already stressed characteristic of relativity. We have considered this relativity, thus far, as expressing a 'restriction to a special domain,' but it also has another flavour, which corresponds to the classical absolute-relative dichotomy. According to this dichotomy, relative is that which is non-absolute, that is, which is conditioned or dependent on something else. It is clear from what has been said in the foregoing parts of this work that scientific truth is relative also in this second sense, for its being a truth 'relative to objects' makes it depend on the conditions in which such objects are 'built up,' and these conditions are constituted by the presence of many *data* which the theory cannot influence.

As a matter of fact, as already noted, in order to have the horizon of objectivity, some 'data' must be present in two distinct senses. In the first sense, data are the starting point of the process of objectification, because this must begin with the intention of certain subjects to seek the agreement of other subjects regarding a particular content of knowledge which is presented to them as a 'datum.' This datum, as we have already remarked several times, may be considered as a 'thing' which is present to the private consciousness of every subject, but which must be operationally surveyed in some of its aspects (or, if one prefers, from various viewpoints). Only after this can it become the 'object' of a certain intersubjective discourse.

In order to perform the objectification procedures effectively, the operations, i.e. the instruments and the ways of employing them, must also be considered as 'data' by all the subjects who try to communicate through them. This is so simply because it is unthinkable that two subjects may come to agree about anything without already having a common basis on which they are in agreement. Such a basis, as we have explained, is represented by a wide spectrum of background knowledge that must be taken for granted before the subjects begin to test their notions. This means that the instruments and the background knowledge are not objectified inside a certain science but rather lay outside that particular objectification, because they represent the *condition* for its being established. All this amounts to saying that objectivity operates between two poles, neither of which belongs to it. These poles are constituted by the two different kinds of 'data' that every objectification must *presuppose*. Only within these presuppositions can the *intention* of making an intersubjective investigation become effective.

The very fact that we have inevitably been led to speak of conditions and presuppositions already points to the non-absoluteness of the knowledge which is bound to these conditions or presuppositions. This non-absoluteness would be removed only if such conditions or presuppositions were given by a kind of intrinsic necessity, but we know that this is not the case, for even 'things,' as we have stressed, are not such 'in themselves,' but only relatively. Indeed, we have already had the opportunity of emphasising the 'contingency' of every scientific objectification, that is, of every such objectification's being something that has no intrinsic necessity but only the nature of an 'historical determinateness'.

As a consequence of this, two intellectual requirements may emerge. One is the need to search for the possibility of obtaining absolute knowledge proper. Though such a requirement is rather infrequently expressed today, it actually plays a significant role at an existential level. We shall return to it in the final chapter in connection with the question of the relationship between science and metaphysics. The other requirement is the 'questioning' of the conditions of a certain scientific objectification. This radical possibility (and legitimacy) of questioning may be expressed by saying that any scientific statement is *controvertible* and that, if a certain statement is proposed as absolutely *incontrovertible*, it is not scientific for this very reason. An analysis of this aspect of science will be our concern in the Chap. 9.

Chapter 9
The Context of Making Science

9.1 Science and Society

Starting in the 1960s, a 'sociological' trend in the philosophy of science began to develop, and since that time has become very influential, despite the strong opposition of several authoritative scholars. As with the 'linguistic turn,' this 'sociological turn' has its strong and its weak points. These depend on the interpretation given to its key thesis, that is, the *social dependence of science*.

For a better appreciation of the different facets of this issue, a brief description of the historical reasons for the ascendancy of this view may be useful. These are represented by the almost accidental convergence of two different cultural streams: neo-Marxism in continental Europe, and the sociology of knowledge in the United States. European neo-Marxists maintained (for reasons that will be explained later) that science belongs to the "ideology" of a given society, in the sense coined by Marx and Engels, that is, as the product of the economic structure of that society, in which the dominant class prompts intellectual and institutional means for the defence and legitimisation of its privileges. Since neo-Marxism has been quite influential in several countries in the western part of Europe for at least three decades, this doctrine of the social dependence of science has been widely advocated there as well.[1]

The second stream, i.e. the sociology of knowledge, was explicitly inaugurated by Karl Mannheim in 1929 in his *Ideology and Utopia*, in which he investigated the social dependence of the form and content of our cognitive activity in general.[2] He maintained, however, that mathematics and the natural sciences are not affected by this "existential determination." A contrary view arose, however, with Thomas Kuhn's *The Structure of Scientific Revolutions* of 1962, which in a certain sense offered a sociological view of science. More precisely, Kuhn's book was based on an historical-sociological approach and raised issues that were certainly

---

[1] We must mention the Frankfurt School in particular (M. Horkheimer, T. Adorno, H. Marcuse, J. Habermas and K.-O. Apel), but also other writers such as L. Althusser and L. Goldmann in France. In Agazzi (1992), especially Chap. 2, this issue is treated in detail.

[2] See Mannheim (1929).

E. Agazzi, *Scientific Objectivity and Its Contexts*, DOI: 10.1007/978-3-319-04660-0_9,

© Springer International Publishing Switzerland 2014

of relevance to the philosophy of science; however, the stress he laid on the dependence of "paradigms" on the acceptance of a given "scientific community" introduced a 'micro-sociological' dimension in the philosophy of science that was rather unusual at that time. Since the academic force and influence of sociology (in general, and of the sociology of knowledge in particular) were a well-established fact in the Anglo-American world, the optimal conditions were at hand for an increase in the influence of the sociological approach to science over the following decades,[3] at the same time that the neo-Marxist trends were advocating this approach for other reasons.

These are, very briefly, the facts explaining the 'fortune' of the sociological view of science. But they are certainly not sufficient for evaluating its merits and its possible limitations. Some of these merits are easy to admit and may be summarised in the remark that submitting science to a sociological study can open new vistas and lead to interesting results. In particular, what has been expressed in this book regarding the *historical determinateness* of scientific objectivity and the *historical dimension* of science clearly also includes *sociological* components, since "history" has been understood not as referring essentially to the past, but as meaning the whole context of human life and activities in all ages. Therefore, it is not sociological considerations in themselves, but the possible *consequences* of such considerations that can give rise to limitations and objections. Usually, these derive from 'absolutising' the sociological approach by attempting to *reduce* all forms of study of science to it, with the consequence of reducing all aspects of science to social factors. In particular, we must distinguish the consequences of this approach on the cognitive level and on the practical level, and also consider the relations between individual and collective contributions to the growth of science.

9.1.1 Is Science a Social Product?

The presently good fortunes enjoyed by the sociological view of science are the projection of a much more general attitude of our time that sees human achievements as the result of a collective effort, rather than as the contribution of exceptional individuals. The historical development of science—that traditionally was seen as the result of the titanic efforts of a few exceptional individuals—came to be considered instead as a kind of glorious monument which has grown thanks to the accumulation of new discoveries made by a legion of 'normally' gifted scientists in a nearly anonymous way. Both perspectives have been criticised for lack of a sufficient awareness of what the genuine *social* nature of science actually is, which cannot be equated with the simple fact of its being a 'collective'

[3] The best known works of this school are Bloor (1976), Barnes (1977), Latour and Woolgar (1979), and Knorr-Cetina (1981).

enterprise. Indeed, science is not so much the 'work' of many individuals as the 'result' of many cultural, economic, and political factors whose complex structure constitutes a *social* milieu determining the form and content of the scientific 'work' itself.[4] The adoption of this perspective has also condemned as dated and insufficient the logico-empirical analysis of scientific theories, which was blamed for having made of science a fictitious and abstract entity, with the result that it was unable to account for its real dynamic development.[5]

In the field of the history of science, the consideration of the 'giants' was superseded by an interest in the patient investigation of minor contributions, on the work done by less well known researchers belonging to the scientific community immediately preceding or contemporaneously with the great geniuses, on the preparatory steps and on the general ideas which the great discoveries could employ as a fertile growing ground. This trend led in a quite natural way to the conception of science as a *social product*. This expression is ambiguous because, taken literally, it should mean that society is able to *produce* something, whereas society is an abstract entity and only individuals actually produce things. A more reasonable meaning is that society provides the *preconditions* for science. Again, these preconditions cannot be understood as being necessary and sufficient for the existence and development of science. Not only can the single social factors not be considered sufficient (this is obvious), but they cannot be considered necessary in a strict sense either (this would mean that without a certain specific factor a particular development of science was impossible).

All this means that we cannot consider the influence of society on science according to a *deterministic* pattern. This influence has the characteristics of that *historical* determinateness of which we have spoken in Sect. 6.1, and which does not amount to an historical *determination*, as we have explained there.

---

[4] This claim has nourished a widespread discussion concerning the so-called 'internal' and 'external' history of science. The result of this discussion has been (if one can venture to draw such conclusions) that a correct history of science must maintain a balance between the recording of internal technical and disciplinary developments on the one hand, and the recording of external social and historical conditions of these developments on the other.

[5] What we are saying points out a positive side of this criticism. There is however also a negative side: the dichotomy that this criticism reinforced also meant that the logico-linguistic 'philosophical' approach (as opposed to the 'sociological' one) came to be seen as being the only philosophical approach, so that alternatives approaches were completely left out of consideration. Non-sociological philosophy of science has come to be thought to be the same thing as logico-linguistic philosophy of science. Thus the weaknesses of the logico-linguistic approach have come to be seen as weaknesses of the philosophical approach more generally. This remark concerns especially the American intellectual milieu, whereas a cross-fertilisation between history and philosophy of science has been typical of the European continental tradition since the beginning of the twentieth century (a tradition, by the way, in which logico-linguistic approaches had only a limited influence). This fault, however, cannot be charged only to the sociologist philosophers of science, since the idea that other approaches and even topics (such as e.g. ethical issues) did not belong to 'genuine' philosophy of science was also frequent among the philosophers of science of analytic inspiration, as we shall have the opportunity to consider in the sequel.

Consequently, no one of the single views regarding science is sufficient to account for the *whole nature* of science. Furthermore the conceptions now considered as dated contain correct aspects that cannot be neglected or discarded. Science certainly implies some kind of internal, cumulative acquisition of knowledge, and is organised according to certain logical and methodological rules; and it owes its growth to the extraordinary impulse given from time to time by some exceptionally talented individuals, although the contribution given to its progress by the scientific community at large, by the social and cultural environment, by the material conditions of society, and so on, do play a role that we can qualify as a set of *favourable preconditions*.

9.1.2 Cognitive Aspects of the Social Contextualisation of Science

One of the most criticised consequences of the sociologistic conception of science maintained by certain representatives of this doctrine was the rejection of scientific objectivity. The 'social dependence' of science has been interpreted as a *social relativism*, very similar to the *cultural relativism* advocated in other contexts. According to this view, a given social or cultural context is characterised, in particular, by its own intellectual *categories*, which include conceptual schemes as well as inference patterns, explanatory models, synthetic interpretations, overall worldviews, and so on. Therefore, in any society *reality* simply is the result of a *construction* determined by the intervention of these cognitive factors; and it would be naive to believe that this reality is something existing in itself and endowed with its own objective features.

This is, obviously, a crucial issue for the whole perspective advanced in this work, which has defended the notion of scientific objectivity as something that admittedly entails a relativity of scientific knowledge, but in the sense of relativity *to specific objects*, and not of relativity *to cognitive attitudes*. We have already discussed, however, a sufficient number of points that will enable us to recognise the *partial* legitimacy of the sociologistic claim, and the incorrectness of its extreme relativistic conclusions.[6]

---

[6] We have spoken of a "sociologistic" interpretation of science precisely to stress the extreme features of this view—as can be found in the works mentioned in note 3 and as have become the most fashionable in the past few decades, a view that has certain affinities, e.g., with some of Feyerabend's positions. It must be stressed, however, that not *every* sociological interpretation of science entails such extreme consequences. In particular, one of the most authoritative contemporary philosophers of science and technology, Paul Durbin, advocates a sociological interpretation of science grounded on the tradition of the American pragmatism of Peirce, Mead and Dewey. Durbin has offered a much more balanced perspective in which the features of objectivity, individual creativity and originality can be accommodated in a more satisfactory way. Some of Durbin's works are mentioned in the References.

We have maintained that every scientific discipline, every branch of a single discipline, and every line of research within the single branches are determined by the choice of some restricted and increasingly specialised 'points of view' under which reality is investigated, so that the domain of the discourse of every scientific study is wholly circumscribed by a certain set of predicates which explicitly express the said viewpoint. In our treatment of the historical determinateness and hermeneutic dimension of science we have also recognised that not only these points of view, but also the operational criteria of referentiality, are 'relativised' to the patrimony of ideas, the background knowledge, and the technological and material conditions that surround science in a given epoch. By doing this we have attributed the correct place to the *legitimate* claims of the sociological doctrine.

Nevertheless, we have also analysed in detail the *ontological independence* of the referents, and this entails that reality (even the 'referred' and intended reality) is not *constructed by* but *known through* the cognitive instruments available. In other words, it would be very naive, for example, to say that colours and forms of things are *constructed* by our sense of sight, rather than being *known* through this sense (so that blind people are unfortunately deprived of such an access to these attributes of things). Similarly, we can unproblematically admit that certain particular features of a given language, or certain abstract concepts included in the intellectual patrimony of a given culture, and so on, account for its forms of expressing knowledge and even of 'knowing' reality. But this only means that these cognitive tools enable people to investigate aspects and attributes of reality that are made accessible through such tools. This does not automatically entail that other cultures could not avail themselves of different tools for investigating other equally relevant aspects of reality, nor that they would be *totally* prevented from knowing the aspects we know by resorting (perhaps less efficaciously) to *partially* different tools.

The discourse we have developed about the comparability, complementarity, compatibility, and incompatibility of different scientific objectifications applies here in a natural way, with the realist consequences we have stressed. We would only like to add a *factual* consideration. Science has shown itself to be the most powerful *intercultural* discourse humankind has been able to produce, a discourse that can be *understood* and *tested* by people belonging to the spatially and temporally most distant of cultures and societies. This happens because *intersubjectivity* is the fundamental characteristic of science; and it holds not only among individual scientists, but also with regard to those super-individual entities that are societies and cultures. By the way, this explains why the founder of the sociology of knowledge (Mannheim) excluded the exact sciences from the domain of application of his theories.

If one considers the doctrines of the sociologists of science without prejudice, one can see that their epistemological relativism depends on an insufficient distinction between the cognitive and the practical aspects of the social involvement of science. We shall study this issue by considering the cultural context in which this confusion was generated.

9.1.3 The Question of the Neutrality of Science

The conception of science as a social product was significantly backed in the 1950s and 1960s by the disputes regarding the 'neutrality' of science, since this conception constituted the frame of reference for practically all the positions that denied the said neutrality. Until the middle of the twentieth century, science had been considered as a realm involving a disinterested, impartial and objective search for truth and reliable knowledge, unaffected by external influences and pressures, insensitive to ideological conflicts, ready to help humankind by offering it efficacious tools for solving all kinds of problems. Nor was science considered responsible for the morally negative (or positive) applications that people made of its discoveries.

After the Second World War, however, many people began to stress that morally negative *applications* of science were actually rather common, that the *conditioning* of scientific research by different kinds of 'power' was not negligible, and also that the self-propelling growth of science and technology was about to produce, more or less automatically, undesirable and even terrible *consequences*, of which pollution and certain ecological disasters were only preliminary signs. Therefore, it seemed obvious that science ought not remain impartial and 'neutral' given these facts.

A second step soon followed, when criticism of science did not concentrate on its possible uses and consequences, but on its cognitive performance, denying that it was that model of research providing impartial, public, testable and critical knowledge that it had long been believed to be. It was claimed, on the contrary, that science is always the product of some particular social community, that it develops from the basic worldview and tenets of this community, that it inevitably tends to serve the interests of its dominant class, and to offer intellectual support to that class's ideology. The alleged objectivity and testability of scientific doctrines was claimed to be purely fictitious, since the hierarchical organisation of the scientific community, the connections between its political and/or economic leaders, the control exerted over publications, the access to research funding, and the actual possibility of expressing dissident (scientific) opinions were all determined on the basis of extra-scientific criteria. In such a way science was condemned to be (willingly or unwillingly) a 'servant of the powers that be,' and to reflect (consciously or unconsciously) their ideology.

This wave of criticism was influenced by political and ideological goals and, as we have mentioned, was mainly developed by certain trends of Western 'unorthodox' Marxism, which sought to undermine one of the most solid pillars of the intellectual opposition to Marxism. This contraposition of science and ideology was particularly stressed, for instance, by Popper, who condemned Marxism (and every other ideology) as a dated, old fashioned, dogmatic, and irrational proposal for facing and solving social, economic and political problems. Whereas 'orthodox' Marxists belonging to the Soviet Union and other communist countries tried to reject this criticism by defending the old 'classical' tenet that Marxism is the only scientific approach to society, Western neo-Marxists (being aware of the extreme weakness of this claim) tried to maintain that there is no basic difference between science and ideology, for science itself is ideologically inspired and committed. Therefore, according to Western neo-Marxists, science cannot serve as a tool for criticising ideologies. This explains the particular position that came to be assigned to the thesis that science is a social product. This thesis was the pivotal argument used as a foundation for the claim that science is intrinsically ideological and deprived of any objectivity, a claim used as a tool for discrediting it as a model of an honest and independent search for truth.

The defenders of the neutrality of science did not deny that scientific applications can be dangerous, but maintained that this is not a fault of science or scientists but of those who *use* scientific knowledge. Scientific knowledge is 'neutral' since can be applied to attain both useful and dangerous goals, and must not be distracted from its search for truth by such extra-scientific preoccupations. The defenders of the neutrality of science also rejected the criticism addressed to the cognitive neutrality of science, considering it to be a deformation of the genuine nature of science, a deformation advocated for partisan political reasons.

The debate remained sterile not only because the rival arguments were often biased by contrary prejudices, but (more substantially) because the meaning of the very notions of neutrality and science were not clarified. Therefore, we shall propose such a clarification before continuing with our discussion. The meaning attributed to "neutrality" in the debate was not the most common one, according to which neutrality consists in an equidistance from two opposed parties, but rather that of an *independence* from motives, prejudices, interests, conditionings and ends.

As to science, we must recognise that we usually conceive of it in two different ways. On the one hand we consider it to be a *system of knowledge*, and identify it, for example, with the content of textbooks, journal articles, theories, and so on, which is arrived at according to certain criteria of *objectivity* and *rigour*. On the other hand we also conceive of science as a *system of human activities* of a professional nature; and, for example, we say that a person *does* mathematics or physics rather than *doing* (as her profession) e.g. music or carpentry.

This distinction is very relevant to the issue of neutrality. If we consider science as a human activity, it is obvious that it *cannot* and even *must not* be neutral. Like every human activity, it presupposes personal and collective motives; it serves certain purposes, as well as certain more or less legitimate interests; it depends on conditionings of various kinds; it is subjected to moral and political considerations; it receives philosophical and ideological interpretations and inspirations, and so on. However, if we consider science as a system of objective knowledge, we must recognise that it is and *must be* neutral with respect to all these elements. In other words, the *cognitive value* of a scientific statement or theory *must* rely upon objective scientific criteria only. For example, if a certain genuine discovery has been made within a military research programme that has morally unacceptable goals, it remains *scientifically valid* notwithstanding the negative intentions which were behind the programme. On the contrary, if an alleged discovery is actually mistaken, it remains *scientifically invalid* even though it might have been reached within a generous research effort made, for example, for the purpose of finding a remedy for a certain form of cancer. (People's intentions can determine *what* is investigated scientifically, but not the *results* of the investigation.) We can now easily understand, by the way, why 'socialist' and 'capitalist' scientists in fact develop the *same* mathematics, physics, chemistry, and so on. This is simply the consequence of the objective knowledge contained in these disciplines being totally independent of the social and ideological context where it is gained.[7]

The brief treatment we have devoted to the question of the neutrality of science has enabled us to appreciate the importance of considering the context of *making* science, even when one's interest is in studying the *cognitive* performance of science, as it is in this book. As a matter of fact, scientific *knowledge* is not a self-producing entity, but something that is gained and produced by a specific human *activity*—more precisely by that activity whose *defining goal* (as we have explained in Sect. 5.2.1) is the pursuit of rigorous and objective knowledge. This activity, however, is a *whole* in which it is possible to *distinguish* but not *separate* several concurring factors, such as intentions, proposals, interests, conditions, and so on. They are linked by a net of feedback loops; and our legitimate preoccupation is that of ensuring that the effect of such complex interaction, though leading to a certain 'shaping' of scientific knowledge, does not destroy its 'defining characteristics,' since this would amount to an elimination of science in its proper sense.

Therefore, it does not matter how many examples might be provided of concrete cases in which scientific research or its results can be shown to be more or less distant from the requirements of objectivity and methodological rigour they

[7] A delicate position in this respect is that of the social sciences. It would be naive to deny that these sciences are influenced by political, ideological, and social conditions. There are several reasons for this. The most important is that the set of 'viewpoints' which, as mentioned earlier, constitute the conceptual precondition for opening a scientific inquiry on society cannot help being the articulation of some 'global view' of society itself, a view which necessarily contains many philosophical components, but also political and ideological components, that determine the choice of the aspects of social life that are considered as especially 'relevant.' Moreover, everyone who firmly holds certain general conceptions (and this is always the case with ideological tenets) is psychologically inclined to support his or her 'certainties,' to look for solid arguments for defending them, and to fight opposing conceptions. In such a way it is natural that one try to develop a science (i.e., a 'social science') that could provide a factual basis or a rigorous system of arguments to serve as a kind of 'confirmation' of the correctness of the general views held. We know that certain ideologies, such as Marxism, pretended to be nothing but a 'scientific' interpretation of society and history. Given this, and given also the well-known tendency of ideologies to 'instrumentalise' everything for the sake of their victory, it is clear that the social sciences in particular (as well as to a lesser extent the natural sciences) are very exposed to the risk of manipulation and distortion, if there is some advantage from doing so for the dominant ideology. All this means, however, a *risk* of discarding objectivity in favour of partisanship. Nevertheless, though this risk is serious and has quite often led to unreliable social theories, this does not mean that objectivity is completely *impossible* in the social sciences. Max Weber has amply discussed this problem, which we cannot tackle here. For a detailed study of this question as well as of the problem of the neutrality of science, we refer to Agazzi (1992), Chaps. 3 and 8.

ought to respect. Such cases would simply be examples of *bad* science, or even of forgeries perpetrated under the guise of science, or of pure and simple betrayals of science committed with the view of attaining other goals, such as economic profit, prestige, power, and so on. All this, however, could not prevent science from being what it is, nor eliminate the fact that it can be performed (and is normally performed) in conformity with its genuine nature.

Let us note that this is an absolutely general situation. For example, we could agree that democracy is a very valuable form of political system; and this would be true despite the fact that in many concrete cases dictators have claimed that they were respecting democracy while actually carrying out an oppressive policy, or that dominant parties in democratically ruled countries preserve their power simply by organising electoral frauds in a formally democratic process of political elections. The same could be said of religion, which can be considered as a precious spiritual force for individuals and societies, but can also be used as an instrument for obtaining power or making money by many people. Moreover, the genuine spirit of religion would not be tarnished even if the highest religious authorities were to evince scandalous behaviour clearly in contrast with the precepts of their own religion. The examples could easily be multiplied.

In conclusion, giving consideration to the 'context of making science' is of great importance for a better appreciation of the 'complex reality' of science, provided we do not 'dissolve' science in this context. This consideration is essential not only in order to be aware that the context affects the cognitive *aspect* of science without undermining its cognitive *value*, but also in order to recognise the legitimacy of exploring *other* aspects of science related to its being a human activity. These aspects cannot be fully appreciated if one considers only the individual activity of scientists; a broadening of the perspective to the collective and social aspects of this activity is required. Thus we shall devote more consideration to this topic.

9.1.4 Individuals and Society in Scientific Work

One reason we cannot claim that science is 'nothing but' a social product is the role that individuals play in science. The traditional view that attributed to geniuses and intellectual giants the whole merit of scientific progress was certainly exaggerated, but it contained a precious grain of truth. Science, like the arts, philosophy, literature, and so on is man's creation and is grounded therefore on a *creative power*. Now creativity is an individual gift and not a social feature. Of course, personal creativity may remain unexpressed or frustrated if suitable social and cultural conditions do not support it, but this cannot mean that such conditions are able by themselves to *produce* the discovery or invention of what is *new*. Indeed, many examples exist of great persons (in all human civilisations and cultures) whose achievements enormously overstepped the humble and very unfavourable social and cultural conditions in which they were born and raised.

Moreover, legions of normally gifted people, with an excellent academic training, are only able to perform routine work of no exceptional value. This means that, even though several great scientific achievements were 'prepared' by many preconditions and knowledge that was accumulated piecemeal, it nevertheless remains true that an exceptional and unpredictable act of personal insight, of genuine individual creativity, was necessary to bring together the scattered useful elements, to overlook the redundant information, to organise in a coherent and testable picture the separate components which did not by themselves suggest (let alone impose) such a synthesis. Most revolutionary scientific discoveries or theories originated from a reflection that some particularly perspicacious spirit made on some small and neglected detail, or on some mistaken view that had been taken for granted by the scientific community of his or her time.

Having recognised the indispensable role of individuals in the progress of science, we can proceed to seeing why the historical development of science imposed a significant appreciation of collective work. We can begin with the presentation of the traditional conception of the natural sciences.

Until the end of the nineteenth century, science was perceived as an enterprise where man put questions to Nature and forced her to give him answers. The records of these answers constitute the growing richness of scientific knowledge. It was rather natural to imagine this 'dialogue' between man and Nature as a kind of confrontation between two partners, and think that the most important secrets jealously defended by this Sphinx might be wrenched from her only by some exceptionally genial Oedipus who appeared in history from time to time. A valid scientific statement was seen as an utterance, made once and for all by a single person, which was granted its truth-value by Nature herself and not by some consensus of other individuals. As we have seen, already at the end of the nineteenth century the traditional trust in the idea that science can come into direct contact with nature was lost, and the 'objective' validity of scientific statements was no longer considered as relying upon a 'correspondence' with the structure of nature, but rather upon the intersubjective consensus of the scientific community. This means that science could no longer be seen as an individual, but rather a necessarily collective enterprise.

The above considerations concern the cognitive structure of science. Another approach, concerning rather the current practice of contemporary science, also stresses its collective character. Indeed, contemporary scientific research increasingly implies the collaboration of several people, especially on its experimental side. No experiment of relevance in advanced physics, for example, can be performed by a single person. Such an experiment requires a team of well-harmonised specialists, who follow the course of the experiment day and night for several weeks, and are charged with the task of interpreting different features that arise, using their different specialised scientific competences. This state of affairs is mirrored in the scientific literature, where it is hardly possible to find a paper carrying the name of one single author; and, in any case, references to specialised literature always indicate that work in isolation is impossible in modern science. Such a fact makes every scientific result the product of a collective effort.

Another 'internal' aspect of modern scientific research points more directly to its social connections, that is, the increasing financial support needed. It is clear that money comes to science from the social structure in which science is embedded, and this rule finds no exception even in the case of 'self-financing' studies that are such because their results may be given in exchange for incoming funds proceeding from particular social agencies. At this stage we can already see how at least a feedback between the internal life of science and the external life of society is at work. But this will become more evident if we proceed to consider the 'external' side of the coin.

We need not develop this point in detail, as it corresponds to the core of the arguments proposed in the polemic against the neutrality of science. Let us therefore mention only those that are most relevant. The first concerns the constraints imposed on science by political decisions that can direct the course of scientific research by allocating or cutting funds for different research programmes according to a given 'scientific policy' (not to speak of political interference in the selection of people who head scientific institutions). Also economic power can act in a similar way in steering or limiting freedom of scientific research, especially because most such research is applied or goal-oriented, and such goals are determined outside science, which must 'serve' them.

This kind of argument has been exploited mainly to discredit science, as we have already said, and this because only the *negative* aspects of this situation have been stressed. But it also contains *positive* aspects. For instance, a certain dependence of scientific research on political decisions may also mean that science is put to the service of the community. The fact that military and industrial contractors hire scientists may also mean that science is led to contribute to the nation's defence (and not simply to *war* as such), and to economic development (and not simply to *profit*), while the large presence of applied research is simply an inevitable and not intrinsically negative consequence of the fact that a modern society tends to make use of the contents of scientific knowledge to improve the quality of life of its population in different areas. The negative aspects are certainly there (and it was opportune that a critical consciousness of them be stimulated by certain persons, independently of their further intentions), but we cannot be one-sided. A genuine wisdom, proportioned to our era, requests that we try to minimise the negative and maximise the positive, being aware that a purely positive development has never been possible in any human activity at any time in human history.

9.1.5 The Social Impact of Science

Our reflections have led us to insist on the influence of society on the 'making' of science, and this in a certain sense justifies a reasonable acceptance of the thesis that science is a 'social product.' We pass now to a symmetric kind of consideration, concerning the effects of science on the life of society, which could justify a reasonable acceptance of the thesis that society is a product of science.

Scientific progress has led to fundamental modifications in our social life. This is particularly evident if we consider the typical offspring of science—technological development. The products of technology have so deeply penetrated our everyday life, including the smallest details, that the real 'natural environment' of modern man is his artificial world, and any return to virgin nature is nothing but an illusion. This is too obvious to deserve a detailed clarification. Just a bit less obvious is how the presence of science and technology has modified our interior nature, bringing about new ways of looking at things, new worldviews, new interpersonal relations, new expectations and personal needs, new social hierarchies, new ethical situations and problems, new facilities, and new difficulties of every kind. But this aspect of science and technology has also received sufficient attention in recent decades, so we shall simply mention it now.

An equally well known aspect is that, with the dangers and risks of technological development becoming particularly evident, science and technology have ceased to be seen as a source of benefit for humankind, since it has also shown itself to be highly unlikely that such risks could be controlled as a kind of automatic result of the internal development of science. The possibility of controlling science through science was tacitly taken for granted. Now we are aware that such a conviction was too optimistic for various well-known *practical* reasons. There is, however, an additional reason that indicates how a *social* condition must also be satisfied. The appreciation of such dangers can only occur at a social level, since humankind or society at large seems to be exposed to these dangers, and even to the risk of a future annihilation, whereas the single individual is inclined to think that the tragedy would occur, if at all, after his death. This is why we find it so difficult to master such problems. To do so we need a 'social way of thinking,' a mentality in which society (extended also to future generations) is the point of reference. However, despite our presumption of being socially conscious and sensitive, we are bound to an individualistic experience of life, and are unable to think in terms of social dimensions or, therefore, to take those decisions which necessarily involve such a mental attitude.

Consideration of the various impacts science has on society (which shows that the effects of science are not only positive, but also constitute a burden and even a possible danger) spontaneously leads to a discussion of the 'social costs' of science. Already at the purely financial level it is clear that money allocated to scientific research is unavoidably subtracted from other social destinations such as hospitals, schools, public welfare, and so on. Therefore science becomes morally obliged to compensate for this. Several other 'costs' of a non-financial kind must be paid for the development of science, but we need not mention them since they have been discussed extensively in the literature. We also leave out of consideration those of these costs—such as species' extinctions—that cannot be repaid. What we would like to point out is that this consideration should not lead us to the conclusion that science ought to be totally engaged in this 'reimbursement,' since it lies in the interest of society to preserve a free space for the disinterested personal creativity that, as we have seen, is a precondition for science. This implies that if the whole of scientific research were goal-oriented, applied, or even directed towards 'useful' targets, we would seriously risk condemning science to its death. This, however, does not concern only the importance of avoiding an 'atrophying' of science which would itself be detrimental to society, it also concerns the fact that it is in the interest of society to preserve creativity, personal initiative, a critical attitude, and freedom of spirit, all of which are precious human qualities employed in several different fields. Moreover, the said duty of providing a return must be seen as a *responsibility* rather than an *obligation* of science and, as is well known, responsibility concerns in the last analysis individuals and, more precisely, free individuals. Responsibility can germinate in a reflecting and conscious free will, but not as a consequence of predetermined constraints. We shall return to this issue in the section devoted to the ethics of science.

9.1.6 A Systems-Theoretic Approach

The relations between science and society have appeared to be bi-directional, or better multidirectional. In fact, we can only roughly consider science as an entity standing before another entity which is society, since the interrelations exist between science and several 'subsystems' of society. Moreover science itself is 'embedded' in society and does not exist outside it. This suggests that the most suitable way to understand the said multidirectional relations is to adopt a *systems-theoretic* model, that is, a model inspired by systems-theory—a discipline which, by the way, we have mentioned on previous occasions.[8]

What distinguishes systems-theory from set theory is that sets are *collections of elements*, that is, of individual entities that 'belong' to the set (the relation is that of pure *membership*), whereas systems are global entities constituted by *subsystems*, and subsystems entertain *functional (and usually causal) relations* with other subsystems and with the global system itself. Moreover, each subsystem is characterised by the performance of certain *specific functions*. But at the same time it contributes to the functioning of other subsystems, while its own functioning depends on that of these subsystems and of the whole system as well.

All this can be expressed in terms of (functional) outputs, inputs and feedback loops among the subsystems and the global system, and entails that a proper functioning (or even the survival) of the global system is bound to the fact that every subsystem can perform its specific functions within a given margin of efficiency and, at the same time, that the proper functioning of a subsystem does not prevent other subsystems from operating sufficiently well for the whole system to exist. This 'dynamic equilibrium' is technically expressed by the notion of *optimisation*, which means that the most satisfactory functioning of the global system occurs when no subsystem 'maximises' its performance but only maintains

---

[8] The considerations that follow were developed at length in Agazzi (1987), which is reproduced as Chap. 12 of Agazzi (1992). An original re-elaboration of this perspective is offered in Marcos (2010).

it within the limits that allow it to perform adequately without damaging the adequate functioning of other subsystems (which constitute its *environment*).

Society can be considered (for the limited sake of the present discussion) as a global system (though it is clearly not such from a more comprehensive point of view, since it is a subsystem of the planet's ecosystem, which is not global either, being dependent on energy from the sun). In this system many subsystems interact while they are nevertheless autonomous in the sense that each of them is characterised by a specific function that it performs thanks to 'internal' dynamic mechanisms. These systems are *open* (because they can receive inputs from the environment in the form of matter, energy, information, and the like) and *adaptive* (because they are able to modify their internal functioning to cope with external inputs and, in such a way, safeguard that functioning). They are also *social* in the sense that their functioning and adaptation are the result of human *actions* intentionally tending to the pursuit of certain goals, including the possibility of modifying to a certain extent the goals and the functional structure of the subsystem itself. Examples of such social systems are the scientific, the technological, the economic, the industrial, the political, the military, the administrative, the educational, the religious, the legal, and so on.

What we have said above, speaking of the influence of society on science and of the impacts of science on society, can be very easily expressed in terms of inputs, outputs, and feedbacks. *Every* social subsystem establishes these relations with its social environment in the form of demands, stimulations, and facilitations from its own side, and constraints, competing options, and even oppositions from without. This continuous circulation of influences assures the satisfactory functioning of society as a whole *if* a dynamic equilibrium is maintained in the form of the *optimisation* of which we have spoken. This means that the *autonomy* of each subsystem must be respected (that is, that its 'internal' functioning must not be jeopardised) and that no 'maximisation' of the role of one or a few subsystems be allowed to result, which would be detrimental to the functioning of other subsystems and the functioning or even survival of society as a whole. Considering science, this amounts to recognising that there is a basic systemic reason for defending its 'internal' autonomy and, at the same time, for recognising that it is systemically bound to receiving inputs from the social environment and to adequately responding to them even by 'adapting' itself to such inputs, provided this does not entail a loss of its internal autonomy.

We can now venture to draw some conclusions from our reflections.

9.1.7 Conclusions and Broadening of the Discourse

We can summarise a couple of conclusions of our discourse:

A. The primary task of science must remain its *defining goal*, that is, the search for truth, i.e. *objective knowledge*. This goal is primary in the sense that no social reasons or imperatives could justify a deviation from this line and admit concealing or deforming the truth. On the other hand, this is the only guarantee given to society in order to rely on scientific objective knowledge for solving its problems. Even if in some very special case a manipulation of the truth might seem to obtain a particular social goal, it is certain that the price of this achievement (i.e., the loss of trust in the *reliability* of science) would be too high to be compensated by the momentary advantage. This stresses an ineliminable aspect of ensuring the autonomy and independence of science with respect to society.

B. This autonomy is fully compatible with the service that science has to provide to society. Concretely speaking, this service may be understood as being help that science brings to the solution of problems of social relevance; and it is clear that the best way of providing this service is to rely upon sound and objective knowledge. The disposition to try to solve these problems can be 'stimulated' by an appropriate scientific policy, but is best secured by the maturation of a sense of social responsibility in the consciousness of scientists.

C. All this may be expressed in different terms by saying that science has to be 'socially committed' but not 'socially dependent'; and this may also be seen as a consequence of its intrinsic nature. As we have already seen, science is both 'made by history' as well as being a 'factor in history.' In the first respect modern science is tied up with the whole historical environment. This environment stimulates it and provides it with its problems, background knowledge, technical tools, conceptual frameworks and categories, and so on. In the second respect, science has propelled history and made it depend on its development. This contribution to the unfolding of history is and must be 'specific,' and this means that only insofar as science preserves an 'identity,' and is not dissolved into a mere overall social context, may it make its full contribution to the material and spiritual growth of humankind.

The above conclusions were obtained (as we have noted at the beginning) by considering society as the *global* system, but we have also noted that it is not really such (even if we ideally consider it to coincide with the whole of humankind) since it is not *isolated* but is embedded into the planetary ecosystem. This entails that a consistent adoption of our systems-theoretic approach cannot avoid taking this further horizon into consideration. And this is not a purely intellectual requirement. Indeed it is well known that, according to several serious investigations and analyses of the present status of our world, precisely the growth of science and technology constitutes a serious danger for the survival of our species because such a growth is detrimental for the ecosystem. This means that we are certainly entitled to go on applying our idea of optimisation, but at the same time we must be aware that this optimisation is thoroughly dependent on the relation between the social system and the rest of the ecosystem. Concretely speaking this means that we must seriously consider the thesis that technological development, supported by scientific results, has led to a situation where the human species is on the verge of extinction. This is not the result of 'bad choices' in applying technology, but simply the result of applying it. The flaw, from an ecological point of view, is in the techno-scientific system itself, not in the particular things the system is used to do.

One might perhaps hope to escape this issue by pointing out that it does not concern science as such, which limits its task to offering objective and reliable knowledge. But this was precisely the position of the defenders of the absolute 'neutrality' of science, that we have shown to be untenable throughout our discussion: our *concrete* problems are due to technological development, which in the last 150 years has been directly supported by modern science; and, in particular, we have also stressed the 'consubstantiality' of science with technology that, especially in contemporary societies, justifies the adoption of the expression "techno-science" (which we have also used sporadically).

What are the solutions offered for such concrete problems that are proposed today? They are those that can be envisaged *within* the techno-scientific system itself, following the illusion that science and technology will be able to remedy the problems they produce. But the problems scientific knowledge is devoted to solving are short-term, and their solution only leads to greater long-term problems (pollution, bacterial resistance, species extinctions, etc.), and this is something intrinsic to the very nature of science and technology, whose optics and perspectives are always *partial and limited*, as we have abundantly shown in this work. This characteristic does not concern only the physical sciences and their derived technologies, but science in general, as we have maintained, so that we cannot hope to solve our problems by resorting to sciences different from natural sciences, such as, for example, social sciences. Indeed, any science is distinguished by its operations, and other 'sciences' would have different operations with respect to natural sciences. So, in this light, on the one hand modern natural science is no more objective than any other science but, on the other hand, any other science is no less partial and limited than natural science.

We are apparently in a deadlock, but this is the consequence of having implicitly admitted that there are no rational means for analysing and trying to solve human problems than those offered by science and technology; and this is, after all, the old and persistent ideology of positivism, that we can call with a more modern denomination *scientism*. Therefore, a possible solution to our problem might come from recognising that, besides undisputable *scientific rationality*, there are *other forms of rationality* that, in particular, have a broader *scope* than scientific rationality and may help in giving to the ideal of optimisation the amplitude of application needed for the really 'global' problems we are facing today, We are not obliged to go too far to find such forms of 'rationality outside science,[9] they have been present in the whole history of humankind and, among them, there is a specific field in which rational investigation has been devoted to problems that are of relevance to our issues, this field is *ethics*, to which we can attribute the status of a particular subsystem in our systems-theoretic approach.[10]

---

[9] See Agazzi (2012c).

[10] The sense in which ethics can be considered a subsystem of the global social system is clarified in Agazzi (1992, Ch. XIII).

9.2 Science and Ethics

The discussion and attempts at clarification of the foregoing section provide a framework for the treatment of an issue that used to be considered alien to the philosophy of science in the past, but which has gained increasing importance during the past decades, that is, the question of the relation between science and ethics. The reasons for the diffidence towards and even opposition to this kind of problem are mainly historical. Especially in the Anglo-American culture "philosophy of science" was almost a technical term used to denote that particular way of analysing science which was inaugurated by logical empiricism and was continued in the tradition of analytic philosophy. This approach—as we have often repeated—consisted in a logico-methodological analysis of science considered from an essentially linguistic point of view, and was concerned with the cognitive aspects of science. A general 'philosophical' reflection on science was neglected and even considered to be vacuous and sterile. Thus the only branches of philosophy that were allowed to contribute to this enterprise were logic, epistemology, and some fragments of ontology; ethics remained outside of this framework.[11]

Moreover, this approach was largely developed within an empiricist philosophical tradition in which the Human separation of the descriptive and the normative was a commonplace. Thus on this view science (which tries to state 'how things are') has nothing to do with ethics (that tries to say 'how things ought to be'). Finally, in the Continental tradition much stress had been laid (following a thesis advocated by Max Weber) on the fact that science is and must be 'value-free,' and that value-judgements are prohibited in science in order to safeguard its objectivity. Since moral judgements are paradigmatic value-judgements, any contact of science with ethics appeared as spurious and dangerous. At most one might be entitled to ask whether science is a 'good thing' or a 'bad thing' but, in this case, the obvious answer seemed to be that science is 'in itself' a good thing of

[11] Persons who have worked within the philosophical community can certainly confirm that, still in the 1980s, it was common to listen to very authoritative philosophers of science expressing severe judgments against those people who dared to introduce sociological or ethical considerations in the 'serious' domain of philosophy of science; and, as a matter of fact, it was only in 1987 that a section devoted to the Ethics of Science appeared in the official program of the International Congress of Logic, Methodology and Philosophy of Science. On the other hand, in the universities, several professors of Moral Philosophy jealously affirmed as belonging exclusively to their own disciplinary competence issues regarding the ethics of science and technology. In order to overcome the narrowness of this perspective a radical "rethinking of the philosophy of science" was needed, in which this discipline would no longer be equated with an 'epistemology of science' but considered as a full philosophical reflection upon science, to which several philosophical disciplines (including ethics, ontology, metaphysics, semiotics, phenomenology, hermeneutics, and possibly a few others as well) besides epistemology, logic and philosophy of language could be called upon to cooperate. This 'rethinking,' in particular, has inspired also the redaction of the present work, and was explicitly advocated by the present author in an invited lecture at a plenary session of the XXII World Congress of Philosophy (Seoul 2008; see Agazzi 2012d).

which bad people can at times make bad use. We have seen in the preceding section that by the end of the first half of the twentieth century this optimistic view of science began to be criticised; and since that time the legitimacy of moral, social, and political judgements on science (implying also controls and limitations) has been advocated with growing insistence (though probably with little effect). We shall limit ourselves to the consideration of the ethical aspect.

9.2.1 Does an Ethics of Science Make Sense?

Several people have reacted against the proposal of submitting science to moral judgement since they have seen such a proposal not only as the first step towards a *controlling* of science, a controlling that would not only jeopardise the *freedom* of scientific research, but as constituting an illegitimate *external influence* in the very structure and life of science. Even the idea of controlling science has an ethical flavour, since the freedom of science has historically been one of the most significant instances of *freedom of thought*, one of the most highly valued ideals of modern civilisation. The idea that submitting science to moral judgement would only jeopardise the freedom of scientific research, and constitute an illegitimate external influence on the structure and life of science, is an expression of the fear that admitting such an external influence would put scientific objectivity at risk, that is, jeopardise the defining characteristic of science itself. Allowing that science be submitted to moral judgement would amount to a regression to obscurantist attitudes, where certain persons were credited with an authority of censorship over intellectual productions, with the additional consequence that we should no longer have free access to the reliable objective knowledge of science. For these reasons many scholars have been led to defend the 'neutrality' of science in the sense of science's being independent and separate (in particular) from ethics. However, a particular distinction we have made in discussing the issue of the neutrality of science will help us see the distinction between science as a system of knowledge and science as a human activity.

Considered as a system of knowledge, science is and must be independent of ethics, since statements are admitted in science under the assumption that they are or could be *true*, and there are no 'morally acceptable' or 'morally prohibited' truths. This sounds very obvious today, but not so very long ago particular statements or theories (including scientific ones) were prohibited in Europe, such bans being pronounced and implemented in the name of various ethical, religious and ideological doctrines. The present situation can be considered to be (and as a matter of fact *is*) an historical conquest of freedom. But the banning of 'interferences' also corresponds to a more intrinsic feature of science: the *condition* for admitting in science a certain statement is that it be (at least putatively) true, and the *criteria* for ascertaining this truth are certain specialised forms of *empirical evidence* and *logical arguments*. These are also the criteria for rejecting a statement. Two consequences follow. First, that moral criteria cannot be admitted for acceptance or rejection of scientific statements, no matter whether they are at variance with or in addition to the specific criteria of science. This must be so because moral criteria serve to discriminate what is *right* (or *good*) from what is *wrong* (or *bad*), and not what is *true* from what is *false*. Therefore, moral judgements are *not pertinent* as criteria of intrascientific acceptability and, *in this sense*, science has nothing to do with ethics. To increase the intuitive accessibility of this claim, let us first imagine that someone says "this mathematical theorem is correct, but it contradicts my moral convictions," or, "this experimental result is scientifically sound, but we cannot accept it for moral reasons." We would certainly not try to convince the person making such a claim that she is 'wrong,' but simply say that her discourse is 'meaningless,' and perhaps try to convince her of the fact.

Second, some collisions may occur regarding the possibility of considering a certain statement *true*, owing to the different *criteria* accepted for such an evaluation. If these controversies occur *within* science, they can be settled by considering the different criteria of objectification adopted, criteria which normally entail a difference in the 'domain of objects,' as we have seen in discussing the issue of theory comparison. A more complex argument is required when the criteria admitted for ascertaining the truth of a *descriptive statement* are radically different, such as, for example, when a divine revelation instead of empirical evidence is accepted as a criterion of truth. In such cases we can repeat that the 'other' discourse is not *pertinent* to scientific statements whose *relative truth* (in the sense we have already explained) is established on the basis of certain specific criteria of referentiality. The legitimacy of the 'other' discourse, however, can be admitted if it can be shown that it has *non-descriptive goals*, but strives rather to make it easy for people to understand the *meaning* of a certain message by resorting to familiar images, without being 'literally true,' and this because the *intended reference* of that discourse does not concern empirically ascertainable 'states of affairs.' (The classical example is that of the admission or rejection of the Copernican theory at the time of Galileo.) In conclusion, we maintain that, as far as its *cognitive aspect* is concerned, science has the right to maintain a full autonomy and independence in the *acceptance* of its statements, and this without minimising either the inputs or the feedbacks it receives from its cultural and historical environment which, as we have repeated several times, constitute parts of its objectivity.

9.2.2 The Ethics of Science

We must conceive of the situation differently when we consider science as a human activity (or a complex system of human activities). On this level the competence of ethics to deal with science, and the *pertinence* of ethical judgements to scientific activity, cannot be discarded, and are actually unavoidable. In fact the defining goal of ethics is that of establishing when a human action is morally right (or good) and when it is morally wrong (or bad), and to elaborate *criteria* for the correct formulation of such *moral judgements*. In this respect ethics is similar to a science and, like science, presents different 'theories' regarding what is good or bad, and the various criteria for expressing correct ethical judgements. These theories are submitted to logical analysis, arguments are put forth defending or attacking them, concrete examples and counterexamples are presented with a role similar to that of experimental confirmation or refutation in science—even axiomatic presentations of ethics have been elaborated, and so on. For these reasons several scholars consider ethics to be a particular science, though we do not use this terminology because it doesn't rely upon those criteria of intersubjectivity that we have proposed as characteristic of science; rather, we simply qualify ethics as a 'rational investigation.' Apart from this admission of the 'seriousness' not only of moral concerns but also of the treatment they receive in ethics, our interest here is to point out that every human action belongs to the legitimate field of inquiry of ethics, including those actions that are performed in *doing* science. The acts of scientists are bound to respect the fundamental requirement of ethics, that is, the conformity with moral *duty*.

The admission of an ethical commitment on the part of science is not unusual, though it obviously does not concern science in a proper sense (science being an abstract entity that, as such, does not perform actions) but scientists (who do concretely act). Such a moral commitment is expressed, at a minimal level, by saying that the *duty* of scientists is to do their work with the most scrupulous adhesion to the methodological requirements of their science, resisting the temptation to stray from this duty for whatever reason. This has sometimes been called the *ethics of objectivity*,[12] and we gladly recognise that this is the primary duty of scientists. But we do not agree that 'the ethics of science' *reduces* to this.[13] Nor can we be content with the enlargement of this perspective that consists in the obvious admission that a scientist also has the duty of being a good father, a good citizen, not telling lies, respecting promises, and so on. We maintain that ethical problems exist also in his *scientific practice*, independently of this being 'scientifically correct.'

The moral judgement regarding any human action could be divided into four fundamental steps: the evaluation of the ends, the means, the circumstances or conditions, and the consequences. It is only if an action has passed the examination of all these aspects, and has been found morally legitimate from the point of view of each of them, that it can be considered morally licit. In the case of science, attention is normally limited to the first aspect, and it is easy to conclude that, the specific goal of science being the attainment of truth, this goal immediately entails the passing of a positive moral judgement on science. This argument is, at best, pertinent to *pure* science. In the case of *applied* science, on the other hand, it is

---

[12] See Monod (1971).

[13] For a detailed treatment of the issues discussed in this section, see Agazzi (1992), especially Chaps. 13 and 14.

obvious that certain of the scientific research goals could be morally objectionable; and this could entail a negative moral judgement being passed on this research.

But also in the case of the moral acceptability of the goals, objections could concern the *means* devised for the research in question. (Cf. the ethical debates regarding experimentation on human embryos for scientific or therapeutic purposes.) If the means can also be considered morally admissible, the *conditions* and *circumstances* can pose moral problems. (For example, the allocation of public funds for 'big science' has led people to ask whether it is morally right to devote such great quantities of money to fostering natural science, while many urgent social problems require funding.) Finally, the consideration of the *consequences* of scientific and technological advance is precisely the problem that started the ethical debates on science and technology (by 'consequences' we mean the *unintended* consequences, i e., those which were not among the goals pursued).

The reflections presented here in a synthetic form are sufficient to show that ethical concerns are far from irrelevant to the concrete practice of science, and that their consideration is by no means an *intrusion* of ethics into the domain of science. This can be easily understood if we refer to our systems-theoretic approach. Scientific activity is performed within a particular open and 'adaptive' social subsystem. In particular it entertains relations of input, output and feedback, including such relations with the 'ethical subsystem,' and this not for 'moralistic' reasons but simply for *systemic* reasons. Therefore, as we find it normal that certain constraints could be imposed on scientific activity for financial, energy-related, political, or technological reasons, we must admit that scientific activity could also be constrained for certain moral reasons.

We would like to point out, however, that the same situation also holds for the 'moral system.' This system has specific functions that it must perform in an 'autonomous' way, but at the same time it is open and adaptive, in the sense that its efficient functioning requires that it be sensitive to the inputs coming to it from other social subsystems, including in particular the scientific system. Such inputs have in general the form of *ethical problems* which originate in the context of scientific activity (as in the context of several other specialised activities) that require *ethical solutions* which cannot be found on the basis of scientific criteria. Ethics must provide these solutions, at least in the sense of elaborating principles, norms and criteria for finding such solutions. But it cannot do this by pretending to 'apply' some general and immutable principles and norms to these problems. Since these norms must be applied to concrete action situations, the real 'solution' must be tailored to the concrete situation which, very often, is totally *new* and even previously unimaginable. On the other hand, the correct interpretation of this situation, including the interpretation of the moral problems actually involved in it, is hardly possible without mastering the relevant scientific knowledge. Many people lament today over the 'crisis' of ethics and morality, and attribute it to a 'loss' of traditional values. This is only partially true; a deeper reason is that ethics has been considered to be a closed and static system of principles and norms. This has produced an atrophying of its force due to the fact that people of today often feel that ethics 'speaks' to a human being who belongs to other times, and does not offer a solution to moral problems that it is insufficiently prepared to treat. Therefore, the interrelation between ethics and science is far from expressing or entailing a 'supervisory right' of ethics over science, since ethics itself is pushed to accept an internal dynamisation in order to cope with the ethical problems created by science.

9.2.3 The Freedom and Responsibility of Science

The preoccupations of those who see a risk for the *freedom* of science in the acceptance of moral judgements on science can now be analysed under a suitable light. If this means freedom of investigation, of producing and exchanging knowledge, of completely autonomously choosing the criteria for admitting statements and theories (that is, if this freedom regards the cognitive side of science), it must be defended against the pretence of limiting it in the name of alleged moral imperatives. Since the cognitive goal of science is that of contributing to the search for truth, that is, of serving a very high *value*, it is a *moral duty* to respect and protect such a freedom.[14] However it is well known that *freedom of action* requires a more careful treatment. Indeed—as we have already noted—the progress of civilisation can be seen as a continuous enlargement of the domains of human 'freedoms,' accompanied by a suitable *regulation* of the *exercise* of such freedoms that usually implies a *limitation of the freedom of action*.

Such a limitation has never been found illegitimate in itself, since it simply corresponds to two obvious needs: (i) since actions can produce harm to other persons, restrictions are needed to avoid this; (ii) one person's freedom of action ends where the next person's begins, or, in other words, the *limitations* of one's freedom of action are imposed by the requirement not to hamper others' freedom of action. This second requirement is a consequence of accepting the *universality* of freedom (of action). If everyone has the right to this freedom, it cannot be unlimited for anyone. Let us note that this corresponds to the idea of *optimisation* presented above, which also contains the suggestion of a criterion for non-arbitrary limitation: the freedom of action of an individual (or of a subsystem) must be limited *only to the extent* that it would otherwise hamper the freedom of other individuals (or subsystems). In conclusion, freedom of science is not only compatible with, but necessarily bound to, a *regulation* of scientific activity.

Since we have here been concerned only with *moral* problems, the conclusion of our considerations is that people doing science *ought to* accept limitations in their practical activity. This is a *moral obligation* that, as such, concerns the *conscience* of individual scientists, and has the form of a *duty* or moral *imperative*

[14] We do not tackle the much more complex question of whether there is also a duty to fund science, especially because duties, in a moral sense, concern only individuals, and it is not simple to 'extend' them to society. Nevertheless we can say that a certain 'systemic commitment' of society to fund science emerges from our systemic reflections.

for them. For this reason it does not have the force of a *compulsion*, since every individual retains his freedom of choice and can act against his conscience with respect to his duty. This is, however, a significant obstacle when limitations of freedom of action must be imposed for social reasons. In this case, some *legal* instruments must be introduced that could *impose*, with the legitimate use of force, the respect of such limitations. This happens in every domain, and there is no reason why it could not be admitted in the case of scientific activity. Nevertheless, something that is common to every legal prescription is that it should avoid appearing as a pure compulsion, but should be compatible with a 'free acceptance' of the limitations it imposes. The solution to this delicate problem is offered by the concept of *responsibility*.[15] This certainly implies freedom (of choice and action), for only free persons can be considered responsible for their actions. But it also implies obligations, because a responsible person is one who freely accepts obligations that, in particular, can entail limitations to her freedom of action. This is the *ideal* situation for morally inspired conduct, since in it freedom is the attitude of conscience that respects the (moral) law, and obligation is the expression of a law that respects conscience.

If this situation can be extended to the laws understood in their technical *legal* sense, the problem of a conciliation between freedom and regulation is satisfactorily solved. Such a situation, however, is 'ideal' and difficult to satisfy in the case of 'collective' activities such as science. A first approximation to this situation is constituted by a *self-regulation* of the scientific community, within which certain ethically sensitive activities are carried on. Many people believe that this is the only form of regulation compatible with the freedom of science. This solution, however, is insufficient for at least two reasons. First, why should a particular scientific community have the right to decide what is good and what is bad for the whole of society? Moreover, scientists are no better endowed than anyone else when it comes to establishing norms of an essentially moral character. Second, simple self-regulation might not suffice to impose morally correct conduct on recalcitrant people in the case of particularly objectionable practices. Therefore some public binding regulation must also be envisaged in this case, with the force of a genuine law.

Here our problem returns, since laws capable of 'respecting one's conscience' should be the institutional expression of *moral imperatives*; that is, they should emanate from a *public ethics*. Unfortunately such an ethics is not at hand today, since in every society different ethical convictions exist, and there is no socially shared moral code. We live in *pluralistic* societies in which freedom of thinking and freedom of conscience are rightly recognised values. Therefore, what we can propose for a better approximation to our ideal situation is an honest confrontation of the different ethical approaches. This should be done with the view of obtaining the largest possible consensus on issues that ought to be regulated by

---

[15] For a detailed analysis, see Agazzi (1989c), where the question of a collective or, better, participant responsibility is also addressed.

democratically elaborated laws, leaving other less urgent and more detailed issues to the self-regulation of the scientific communities, and finally leaving to the free conscience of the individual scientist the decision on particular special cases.

One should not believe that the responsibility of which we have spoken only concerns scientists. It is an attitude that must be adopted by all members of society, and which corresponds, as said, to the fact that any citizen commits himself to acting in conformity with his *duty*. Only if this attitude becomes general can we expect the advent of that atmosphere of mutual *confidence* that will be the best guarantee also of a morally correct development of scientific activity. We must have or create a situation in which everyone feels engaged and respects his duty not to harm other people, to the extent that he or she is sure that other people too feel the same duty not to harm him/her. In the last analysis the problem of regulating a free science can find its real solution in the diffusion of a *sense of responsibility* and of a *sense of duty* in science, a diffusion which (sustained by the participation of all citizens in the decision-making process) would make this regulation at the same time acceptable and reasonable. But this is a problem of *public education* that we cannot tackle here.

Chapter 10
Science and Metaphysics

10.1 Criteria for a Distinction

In concluding this work we want to devote some consideration to the relations between science and metaphysics, an issue that has been constantly present in the history of Western philosophy and is far from having lost its intrinsic interest in our time as well, despite the fact that it is often considered with suspicion in the 'official' philosophy of science. Whereas in ancient and medieval philosophy these two notions were strictly related (indeed metaphysics was considered the best example of science) in modern times a process of separation has developed, which began as a distinction in the seventeenth century and terminated as an opposition in the twentieth century, when the view of science proposed by positivism became dominant even in the mind of general public. This was a consequence of a change occurred in the meaning of science itself, determined by the emergence of a new paradigm of knowledge, that is, of modern natural science. The acceptance of this paradigm led Kant to ask the question whether metaphysics can actually be a science, and he answered this question in an allegedly negative sense. The subsequent developments gradually transformed this difference into a real opposition: science is the only genuine form of knowledge that can supersede metaphysics only by fighting against it and overcoming it. Positivists and neo-positivists strongly advocated this position, and it became very influential. In fact also many people who do not adhere to an anti-metaphysical attitude admit that modern science could begin by 'freeing itself from metaphysics,' and that 'metaphysical interferences' are only detrimental to the progress of science. This has been, obviously, the position of those who gave to science the pre-eminent position among the human cultural and intellectual performances (being more or less 'tolerant' toward metaphysics). But also those who gave the privileged position to the humanities and philosophy on the cultural and intellectual plane (showing 'tolerance' toward the sciences on the practical plane) considered science and metaphysics as opposite approaches to reality. A minority of modern philosophers of science have adopted a more conciliatory position, typically expressed by Popper and his followers, who have recognised in metaphysics the source of several worldviews that can give rise to

E. Agazzi, *Scientific Objectivity and Its Contexts*, DOI: 10.1007/978-3-319-04660-0_10,
© Springer International Publishing Switzerland 2014

useful "conjectures" capable of evolving in truly scientific "falsifiable" hypotheses and theories. In this case, science remains the most genuine form of knowledge, and metaphysics is credited with a subsidiary role with respect to science, a role that the majority of 'genuine' metaphysicians, however, would hardly accept.[1] The considerations concisely summarised above have been already expressed on preceding occasions in this work. Our intention now is to see, after the detailed exploration of the nature of science we have proposed, whether a satisfactory relation between science and metaphysics can be envisaged in which the legitimate aims of both would be respected.

An obvious condition for a methodologically correct development of the proposed discourse is a definition of the two concepts of science and metaphysics. We shall satisfy this condition here only sketchily since we do not want to go into the analysis of the quite different meanings that have been given to the terms "science" and "metaphysics" in the philosophical tradition. Therefore, instead of a proper 'definition' of science, we will be content with indicating some 'paradigmatic examples' of it, that is, the natural sciences such as they are pursued nowadays; but at the same time we will refer to those general features of science, explored in this work, that have allowed us to elaborate a concept of science as 'rigorous and objective knowledge' that applies 'analogically' to many fields of investigation different from the natural sciences.

We cannot use the same procedure regarding metaphysics since we certainly cannot rely upon 'paradigmatic examples' of metaphysics sharing certain common features and having a recognised circulation in contemporary philosophy. Nor can we take seriously that nonprofessional meaning, according to which metaphysics is a kind of general and vague world-picture which does not actually say anything precise but is tolerable and even useful to the extent that it may be the fountainhead of several heuristic outlooks, the majority of which will probably remain at the stage of fantasies, but a few of which may give rise to more precise conjectures capable of inaugurating 'serious' scientific disciplines and theories. Indeed the 'serious' metaphysician is not interested in such an impoverished and 'ancillary' conception of metaphysics, and vindicates for it the status of respectable *knowledge*. If we take metaphysics in this more engaging sense, two basic meanings of it may be found already at the beginning of its official history (i.e., in the *Metaphysics* of Aristotle). On the one hand, metaphysics is conceived as the science of "reality as such," that is, of the most universal *features* of reality, whose aim can

[1] What we have just said must be taken only as a sketchy presentation of what we could call a 'prevailing atmosphere' in Western culture during the nineteenth and twentieth centuries. As a matter of fact, several important scholars have defended there being a more or less explicit and significant interconnection between science and metaphysics (though not always using this terminology), from Whewell to Duhem, Meyerson, Einstein, Campbell, Enriques, Northrop, Harré and many others. In Dilworth (2007), a work thematically devoted to the illustration and defense of such a position, abundant references and quotations are offered attesting to this trend in the philosophy of science. An author who has done important work along this trend is A. Chakrawartty. See especially Chakrawartty (2007).

also be qualified as the investigation of the highest "principles" of reality itself; on the other hand, it is conceived as the science of those *dimensions* of reality that overstep its empirically ascertainable level (or, to put it briefly, as the science of the 'supersensible').[2]

This double characterisation is already sufficient for establishing a *distinction* between science and metaphysics, since the natural sciences (but also other sciences in the *modern* sense of this concept) are not concerned with the study of the universal features of reality as such, and they are also restricted to those levels of reality which are empirically knowable. On the contrary, if one should try to *separate* science and metaphysics on the ground of some formal or methodological criteria (more or less in the sense of qualifying science as 'solid knowledge') it would prove hardly possible to succeed in this enterprise (as we will see in the sequel), and the ancient claim that metaphysics is a science could hardly be disproved.

It is interesting to remark that Kant had in mind both meanings of metaphysics in his works, and had, moreover, characterised science as 'solid knowledge.' As the result of his "critical" investigations, he was convinced that metaphysics was admissible as a "science" in its (suitably reformulated) sense of being the doctrine of the most universal features of *knowable* reality, while it could not be credited with the qualification of "science" in its second sense, that is, in the sense of being the knowledge of a supersensible dimension of reality. The suitable reformulation just mentioned was that metaphysics can be seen as the doctrine of the most universal *features* of our knowledge, that is, of those a priori features which make knowledge at all possible, and which determine in such a way the whole domain of the *objects* of knowledge (the world of *phenomena*). Instead of thinking of metaphysics as the science of the universal and necessary principles of *what exists*, Kant confined it to being the science of the universal and necessary principles of *what is knowable*. The second traditional part of metaphysics—that of being the science of the supersensible—was displaced from the status of "knowledge" to the status of a (rationally legitimate) "faith."[3]

---

[2] Sometimes metaphysics understood as an inquiry on the universal features of reality is called a *transcendental* inquiry, while when it is understood as an investigation into the supersensible dimensions of reality is called a doctrine of the *transcendent*. As we shall soon see, Kant admitted metaphysics in the first sense, but not in the second. These terms, however, have often been used with different meanings in modern philosophy, and we shall therefore avoid using them in the sequel.

[3] What we have said corresponds especially to the discourse developed by Kant in his *Prolegomena* (1783), but we cannot deny that Kant may even have engaged himself in proposing a genuine 'metaphysics of science.' This he did in the *Metaphysical Foundations of Natural* Science (1786), that is, in the period of his 'critical' thought, including the time during which he wrote the first edition of the *Critique of Pure Reason* (1781), the *Prolegomena* (1783), and the second edition of the *Critique* (1787). In this work he explores natural science (practically, Newtonian physics) based on the norms established in the *Critique*, realising in such a way a project that he had had in mind already in his 'pre-critical' period (as is explicitly attested in a letter to Lambert of 1765). The need to afford a natural philosophy that was 'exact', free of speculative adventures, and solidly anchored to mathematics, had been present to his mind as the need for a 'metaphysical foundation', but could be satisfied only after the deep reform of the

10.2 The Universal as a Precondition for Knowing the Individual: The Conceptual Roots of Metaphysics

Metaphysics, in the above sketched double sense, was not invented by Aristotle, but rather by Plato; he was the first to explicitly remark that we can "know" an individual entity only if we can "recognise" it as being a particular instantiation of a universal model, such that a pre-existing knowledge of these models is necessary if we want to *explain* how everyday knowledge is possible. These universal models are the well-known Platonic *Ideas* which in this way actually play the role of "a priori conditions for the possibility of knowledge" in a sense very close to that of Kant. We can also add that Plato did not maintain that "we know our mundane ideas," that is our representations (at variance with modern epistemological dualism), but rather that "we have known the eternal Ideas," the inborn souvenirs of which are the conditions of empirical knowledge, conditions which become active on the occasion of encountering the empirically given material objects. It is no wonder, therefore, that he was confronted with the problem of the *origin* of our ideas (in the modern sense of our representations), a question which he answered through the mythical doctrine of "reminiscence." This doctrine led him to admit the supersensible world of the Ideas—which had to be immaterial because of their universality—and this was one of the main reasons (not the only one, however) for his being the founder of metaphysics also in the second sense, that is, as the doctrine of a supersensible dimension of reality. In order for his move to be really efficient, he had also to claim that empirical reality is actually 'modelled' according to the supersensible Ideas, and he therefore developed his doctrine of the *mímesis* and *méthexis* (things are "copies" of the Ideas and "participate" of their nature).

A possible way of preserving the unavoidable function of the universal as a precondition for the knowledge of the individual—without resorting to a supersensible world—was that of making universals inborn structures or *functions* (and not inborn *contents*) of our intellect; and this was practically the Kantian solution. A different way was that of maintaining that the universal dimensions are built-in features of reality which our intellect can *abstract* from it; this is Aristotle's

(Footnote 3 continued)
notion of metaphysics he had attained in the *Critique*. Indeed, in this work of 1786 he shows the passage from the "pure" concepts and principles of the understanding presented in the "Analytics" of the *Critique of Pure Reason* to genuine natural science, that is, to the study of the motion of material points (i.e. mechanics). One must note, however, that the *Metaphysical Foundations* did not attract much attention when it was published (in a 1795 letter to Kant by Kiesewetter this fact is deplored). But the work was often discussed later by the idealists, and influenced their often arbitrary metaphysical interpretations of natural science, so that the final outcome was a discredited mixture of metaphysics and science that favoured the positivist hostility to metaphysics in general.

solution, which Kant could not adopt because he was sharing the 'epistemological dualism' of his age, according to which we do not know "things in themselves," but only our representations.

Modern science has for a long time overlooked the fact that the individual can be known only within the framework of a universal model. This was due to the fact that scientific inquiry does not start from nothing, but from everyday knowledge, which already singles out individual objects and events thanks to the intervention of certain universals. Therefore, empirical science could take these individual objects and events as starting points (the so-called 'empirical data') remaining unaware of the preliminary work of unification required for them to be 'given' to it. In fact, no single item of knowledge consists of an isolated sensation, or of a scattered multiplicity of sensations, but always of *organised units* of such multiplicities which we usually call individuals. Hence, individuals are not 'atoms,' but units in which the atoms are organised according to some structure, form or *Gestalt*, so that 'knowing something' always amounts to knowing it 'as something' (as we have already discussed within a different context).[4] This 'knowing as' points precisely to the presence of that unity which Plato and Aristotle called *eidos* (or *form*) and modern psychology calls *Gestalt*. But between Plato and modern psychology this fundamental fact has been very often considered (and differently interpreted) by a great deal of philosophers. In particular, Kant was well aware that knowledge is necessarily a *synthesis*, not only at its most advanced level of the "a priori synthesis," but already at the more elementary level of the "empirical synthesis."

But even this distinction between 'atoms' and 'units' is in need of clarification. First of all, the atoms are not given prior to the unit, but may be singled out by an *analysis* of the whole *Gestalt* of which they appear as constituents. Second, this *Gestalt* may serve to organise other and different atoms, and in *this* sense it is universal. Third, the atoms themselves may be 'identified' because they have in turn a certain *Gestalt* (which enables us to say that they are *the same* atoms—the Platonic "recognising"—when they are organised in different structures and units). In conclusion, there is no moment in which our knowledge can dispense with the universal, be it because we need the 'unity of the multiplicity,' or because we must be able to grasp 'the permanent under the mutable.'

If we understand this priority of the universal over the individual, of the whole over the parts, we can also easily understand the fundamental inspiration of metaphysics: if the most penetrating knowledge is that which uncovers and makes explicit the universal features of things, the regulative ideal becomes that of scrutinising the most universal features of reality "as such," as Aristotle put it (and

[4] These unifications occur within the commonsense apprehension of reality, and constitute, essentially, that which Sellars has called the "manifest image" of the world, an image that, in such a way, is the indispensable precondition and permanent framework for the construction of the "scientific image." This is why (as we have already explained) we cannot share Sellar's thesis that the manifest image is intrinsically wrong and must be superseded (at least as a regulative ideal) by the scientific image, which is right.

as Kant recognised, though in the form of his "transcendental" restructuring of metaphysics).

Once this step is taken, it may become tempting also to take another one: if we succeed in uncovering the most universal features of reality, why not use them to account for the particular aspects or components of reality? But this program becomes misleading if it is understood as the possibility of *deducing* by purely logical means the particular from the universal. This often happened with the ancient "philosophy of nature" which, as a consequence, remained very poor as far as actual knowledge of the particular features of nature was concerned, and mainly for this reason was replaced by modern natural science with Galileo and his followers. The reason for this fallacy is not difficult to see: it is hardly possible to *deduce* from a *Gestalt* the actual details of its components. They are not given without the *Gestalt* but they are not logically *entailed* by it, they must be *ascertained.* Kant was conscious of this fact. In his work *Metaphysical Foundations of Natural Science* he deduced a priori many principles of mathematical physics, but he always remained aware that "the particular natural laws, since they concern empirically determined phenomena, cannot be exclusively deduced from the categories, though they are all submitted to the categories" (*Critique of Pure Reason*, B 165). But several other modern philosophers, from Descartes to Hegel, could not resist this temptation of deducing physics from metaphysics, and it was because of their bad examples that a drastic 'liberation from metaphysics' came to be seen as a necessary condition for the development of science.[5]

10.3 Modern Science as a Non-metaphysical Kind of Knowledge

The core of the Galilean revolution—which mainly for this reason deserves being considered as the starting point of 'modern' science, as we have discussed in Sect. 1.4—consisted precisely in the breaking of the said illusion or, if one prefers, in rejecting the deductive link from metaphysics to science. When Galileo proposed the grasping of the "intimate essence" of things to be a "desperate enterprise," and that we should rather content ourselves with becoming acquainted with "some of [the thing's] affections," he was rejecting the tenet that knowledge of the universal (the essence) is a necessary precondition for knowing the

[5] It is this kind of aprioristic dependence of science on metaphysics maintained by several even famous philosophers (from Descartes to Hegel) that produced the anti-metaphysical reaction of the neo-positivists, as can be seen, for instance, in this declaration of Hans Reichenbach: "[Modern scientists] refuse to recognize the authority of the philosopher who claims to know the truth from intuition, from insight into a world of ideas or into the nature of reason or the principles of being, or from whatever super-empirical source. There is no separate entrance to truth for philosophers. The path of the philosopher is indicated by that of the scientist." (Reichenbach 1949, p. 310).

particular, i.e. that one must know the whole before descending to the details. This attitude was anti-metaphysical not only because it discarded the possibility of uncovering the features of 'reality as such' but because, even in front of a single thing or process, it discarded the possibility of knowing it 'as such' (i.e., in its 'whole') but only under a few very restricted aspects. Coming then to those aspects which it is hopeful to know, he confined them to those which are empirically describable and testable, and also mathematically expressible. In this way also the second aspect of metaphysics (the investigation of supersensible dimensions of reality) was discarded, not as something which is absurd, meaningless, or non-existent, but as something which is not relevant for the knowledge of nature (and that might even overstep the capabilities of our knowledge).

Newtonian science fully accepted this approach, and applied it with dramatic success, providing it with a kind of irresistible practical confirmation. In this way universality in a proper sense was dismissed from science and replaced by the less committed notion of "generality" which corresponds to the idea of a progressive enlargement of the evidence proposed by experience. In this sense Newton says that natural laws must be uniquely "deduced from the phenomena," but this deduction (which subsequently was qualified rather as "induction") can never grant the certainty of universality, as Hume's criticism soon proved.

We can now easily see the itinerary leading from Galileo to Kant. From Galileo, Kant accepted the inscrutability of the essence and the impossibility of knowing the supersensible, with Galileo (and Newton) he shared the thesis that only experience provides the content of our knowledge, with Hume he agreed that experience cannot grant universality, but only a practically reliable generality. Nevertheless, he wanted to restate universality (also in natural science) and was led to reintroduce the universal as a precondition for empirical knowledge, in the form of an a priori of reason, as we have seen.

10.3.1 Does Science Need a Priori Universals?

Let us now set aside what we have remarked concerning the fact that scientific inquiry actually starts from a certain commonsense apprehension of the world, which is already organised according to several *Gestalten*, and ask whether in its *specific* work science can dispense with certain preconceived universal unifications. This is certainly not the case. Taking, for example, Newtonian physics (allegedly "deduced from experience" alone), it is easy to see that even the simple law $f = ma$ does not result from pure experience, not only in the Popperian sense that it was 'conjectured' before being tested, but in the sense that it presupposes a certain way of 'looking at things.' For example, force, which replaces the old notion of cause of acceleration, has all the features of the traditional "efficient cause"; moreover, it is thought of as a cause acting upon bodies 'from the outside.' (This was a possibly unconscious but profound step, since traditional causes were thought most often to be teleological, and to move beings 'from within' towards their "natural goal.") Mass plays at the same time the ancient role of "matter" (it is actually defined as "quantity of matter") and that of "substance" (being what remains permanent under all changes). This not to mention other concepts such as those of space, time, and action at a distance, which were 'metaphysically' discussed and challenged from the time of Newton up to the present. Now it is certainly true that the law in question was discovered through the contribution of experience, but it is unquestionable that this was possible because it could be *conceived* within a pre-existing conceptual framework, that furnished the universal elements for its formulation, some of which we have pointed out.[6]

What we have said concerning this elementary example may be repeated for dozens of cases in the history of the sciences, and we need not dwell on this point, which has been sufficiently illustrated in the pertinent literature of the last decades. We want rather to remark that this phenomenon is even more impressive if we consider scientific *theories*, instead of scientific *laws*. Unfortunately (as we have already stressed) the distinction between laws and theories has been too often overlooked in recent philosophy of science. While laws may be seen as gestaltisations which try to describe '*how* things are,' theories are systems of *hypotheses* which again express a *Gestalt*, but with the purpose of explaining '*why* things are so.' There is certainly no discontinuity between laws and theories, and they even share some common characteristics. But it is important not to ignore this specific difference.

[6] We would add too much to the already respectable size of this work if we should embark on the illustration of *how* metaphysical considerations constitute the prerequisite framework of science. Let us simply say that certain metaphysical principles belonging to *general ontology* (e.g., the principle of the permanence of substance, or the principle of causality) receive a 'specialization' when they are 'applied' to the specific ontology of a certain science, that is, as we have explained in preceding chapters, when the interest of the inquiry focuses on certain restricted 'attributes' of reality. So, for example, in Newtonian mechanics mass, motion, space and time play the role of substances, while force plays the role of cause (the cause of the *change* of motion that must be compatible with the 'conservation' of the quantity of motion). An additional metaphysical principle that does not belong to general ontology, but only to the *special ontology* of Nature is that of the "uniformity of nature." This last principle is the rational prerequisite for looking for natural *laws*, as well as for planning experiments and making predictions. The general causality principle is the rational prerequisite for constructing *theories* that should show *why* certain empirical laws are so, as a consequence of the basic specific properties and laws of the substances involved. Therefore, the general metaphysical principles are specialized in principles, laws and theories of a particular science. In Dilworth (2007) a detailed and convincing presentation is offered of this process and particular stress is laid upon the role of *principles* in science, an aspect that we did not explicitly treat in this book considering that it is included in the idea of a "hermeneutic framework" we have presented. This topic certainly deserves the deeper analysis provided by Dilworth.

10.3.2 Metaphysics as a Prerequisite for Science?

Can the universal background or conceptual framework of which we have spoken be called "metaphysical"? We certainly feel allergic towards using this qualification, especially because we have seen that natural science could take its start with a certain 'liberation' from metaphysics, and because positivistic no less than anti-positivistic philosophies have nearly convinced us that science and metaphysics are irreconcilable enemies. (The vindication of the rights of the one seemed to necessarily demand the negation of the rights of the other.) However, if we take metaphysics in the first of its two basic meanings—that is, as the exploration of the most universal features of reality—the issue may be seen under a much less polemic light. Under this aspect, metaphysics appears as the unrolling of the general conditions for the *intelligibility* of reality, and in this sense it is unavoidable. Everyone who starts speaking about something must have *understood* it in some way, this way reflecting in its turn his understanding of other more general features of reality. In this sense it is impossible not to have an implicit (and often unconscious) metaphysics, articulated into several levels. Science is no exception to this, since it cannot exist without using certain criteria of intelligibility which are prior to its specific work.

As a matter of fact, the detailed presentation of the philosophical and historical background against which the Scientific Revolution took place, presented in the first chapter of this work, not only makes clear in what precise sense modern natural science started with a 'liberation from metaphysics.' (This liberation amounts to abandoning the pretention that the grasping of the "essence" of natural bodies is the prerequisite for knowing their particular behavior as a *deductively necessary consequence* of the essence.) It also shows that the determination of the actual subject matter of natural science amounts to the choice of a particular *ontological domain*, that is, the domain of the *affections*. And such a determination was made by explicit reference to *metaphysical* doctrines that had been elaborated by scholastic philosophy, and which continued to concern the most prominent philosophers and scientists of the seventeenth century. In this sense it is not correct to say that the Scientific Revolution constituted a dismissal of metaphysics altogether. It constituted rather the provision of the *general metaphysical framework* for the new science. The precisation of which would include the provision of standards of *intelligibility*.

The discourse developed in this section would appear much more peaceful and acceptable if we had spoken of *ontology* instead of *metaphysics*, since the first term does not produce the unconscious negative reactions that the term "metaphysics" often produces. One should be aware, however, that the term "ontology" was coined only in the seventeenth century, and precisely to indicate a subdomain of metaphysics, that is, what was called "general metaphysics" understood essentially in the classical sense of a doctrine of reality as such. This general metaphysics had to be distinguished from the "special metaphysics" that concerned certain great subdomains of reality, such as the physical world, the human soul, and God. In more recent times the situation has to a certain extent changed, in the sense that ontology has received full recognition as a philosophically respected discipline that can be articulated into a general ontology and particular ontologies. Along this line one can say that ontology is concerned with analysing the different "kinds of reality" (something that was also present in the classical tradition, as we have already noted). In particular this view is reflected in the notion of *regional ontologies* that we have also adopted in this book in order to denote the *domains of objects* of the different sciences. Until now we have characterised such domains of objects or regional ontologies through the *criteria of reference* used by any particular science, mentioning also occasionally that the specificity of the domain of objects also entails differences in the kind of arguments and the "criteria of rigor" used in the different sciences. Now we can add something that should be totally evident, that is, that each ontological region is also characterised by its *ontological principles*, which are in part refinements or specialisations of the most 'general ontological' (i.e. metaphysical) principles, and in part have the status of hypothetical *presuppositions* for which neither cogent rational arguments nor empirical evidence is provided. For example, absolute time and absolute space are such ontological principles of Newtonian mechanics, but are not assumed in relativity theory. It is straightforward that when we pass from physics to biology, psychology, sociology and other sciences, the respective ontological principles must differ to a larger extent. This is a strong argument against any form of reductionism; it should also help us understand the sense in which there are discrepancies among different scientific theories, and the reasons for these discrepancies. For example, we have already noted that quantum mechanics does not falsify classical mechanics because the operational criteria of the two theories are different. Now we could add, for example, that certain 'astonishing' or 'surprising' statements of special relativity that have puzzled many scientists and have been tried to be accounted for by means of complicated 'interpretations' may lose their paradoxical appearance once one is aware that they were paradoxical only because such statements were judged within the ontology of classical mechanics, whereas they are natural and can be literally accepted once they are properly encompassed within the ontology of special relativity, in which, in particular, absoluteness of time and space are no longer presupposed ontological principles.[7]

In fact every time the advancement of some science has been presented as a 'liberation from metaphysics,' it has actually been tantamount to discarding *a particular* metaphysical framework and accepting (often unconsciously) a different one. For example, discarding determinism in quantum physics did not mean eliminating all metaphysical views from microphysics, but meant simply replacing the "classical" deterministic metaphysics of nature with a new indeterministic one. It is much more reasonable to be aware of the metaphysics one has, rather than having one without knowing it.

---

[7] For an excellent discussion of this point see Mittelstaedt (2011).

10.4 The Mutual Dynamics of Metaphysics and Science

Two major obstacles have made a better understanding of the relation between metaphysics and science difficult. The first is the idea that metaphysics is a purely a priori speculation which dogmatically pretends to impose its eternal, unchangeable and absolute 'principles' on other forms of knowledge, and on science in particular. The second is that metaphysics would consider science as a kind of corollary or application of its tenets. Both these views are wrong, and we will begin to show this starting from the second.

The relationship between science and metaphysics is analogous to the relation between experiments and theories in science. As we have already discussed, experiments presuppose a theory, since they are designed and performed by using the concepts, laws, methods of a certain theory, and with the view of answering 'questions' asked within it. In this sense they 'depend' on the theory. However, their outcome does not depend on, and inevitably introduces a modification in, the theory. As we have already stressed in Sect. 7.1, if an experiment is successful, it not only 'confirms' or 'corroborates' the theory, but actually *enriches* it, by adding an additional detail to the *Gestalt* of the domain of objects which the theory is about. If an experiment shows a 'negative' result, the theory must be modified, its proposed *Gestalt* proves not to be fully adequate, and it may have to be abandoned and replaced by another. What we have said of experiments may be repeated with some modifications of the 'data' of a theory. In short (as we have seen in detail in Sect. 6.3), there is a continuous *feedback* between the *Gestalt* and its components; the global view provided by the original *Gestalt* is put to test through the analysis of its details, several of which were not scrutinised at first, and the verdict of this scrutiny is fully open. If I am shown a photo, I may at a first glance 'recognise' it as being one of an old friend, and after careful examination I may be led either to confirm this first judgment (and even to uncover some previously unnoticed details in the face of my friend), or to 'recognise' that this was the photo of another person altogether.

What theories are with regard to experiments and empirical data, metaphysical frameworks are with respect to scientific theories. They are *Gestalten* of a higher order within which theories take shape. Therefore, theories 'depend' on these more general criteria of intelligibility but are not 'deduced' from them and interact with them in a *feedback* loop which, in any case, produces modifications (of different importance) in the metaphysical background. For example, classical mechanics, with its implicit equating of the concept of cause with that of force, gradually induced a restriction of the concept of causality to that of "efficient" causality, and even of lawfulness (as is patent both in Hume and in the *Critique of Pure Reason*). This in turn led to an unconscious identification of causality with determinism so that, when determinism was challenged by quantum physics, the "principle of causality" seemed to be disproved. But this stimulated a critical revision of the said principle, leading to distinguishing it from determinism, and to adding further precisions and distinctions, in addition to the many that had already been introduced in the history of philosophy. It would be unfair to say that metaphysicians have tried all possible means for 'saving' their old 'eternal' principle of causality, that affirms that every change has a cause. It is correct to say that many philosophers have tried to see how the criterion of intelligibility provided by the causality principle could be reshaped in order to cope with relativity and quantum physics. This actually means a real 'regestaltisation' of this principle (which, by the way, has been reshaped hundreds of times in the history of philosophy, so that there is no immutable and untouchable formulation of it). To sum up, not only is there an influence of metaphysical frameworks on scientific theories (as nowadays several scholars have shown) but there is also the no less significant influence of scientific theories on metaphysics, and this issue might deserve more philosophical attention than it has received up to now. All this becomes easier to express if we consider the said 'regestaltisations' not as reformulations of the metaphysical principle as such, but rather as modulations of the principle into different *ontological principles* of the different sciences in the sense discussed in Sect. 10.3.[8]

The above reflections have paved the way to the discussion of the first tenet we have mentioned, that is, that metaphysics is dogmatic a priori speculation. Only ignorance of the history of philosophy may allow a defense of this tenet. Metaphysics has always been an effort to deeply understand reality, to make it *intelligible*; and in this sense its attitude does not differ from that which science adopts in its different specific fields. In this sense metaphysics has been characterised by the careful elaboration of *concepts*, much more than by the formulation of tenets. These concepts have been used for answering fundamental *questions* or problems (history of philosophy contains much more questions than answers), and these problems were also very 'concrete' (of course, if we are able to see them in their historical context). Obviously, only few of them were related to the understanding of the physical world, and this explains why many metaphysical doctrines which give rise to epistemological, logical, moral, existential, and political problems, may give the impression of being remote from the intellectual style of science; but it would be naive to pretend that everything interesting and important has to be related to science.

In order to avoid misunderstanding, a last remark may be appropriate. By saying that metaphysical frameworks are preconditions of scientific inquiry, we are not maintaining that the elaboration or *study* of metaphysics is a necessary prerequisite for doing science. We simply claim that it is not possible to *do* science

[8] This feedback from scientific knowledge to metaphysics can have even more significant impacts, in the sense of implying, for example, the rejection of general metaphysical models of physical reality. For instance, Massimo Pauri maintains that "The ontological breakthrough implied by the discovery of the *atomization of action* is so radical that quantum theory represents the death certificate of atomism in a very deep sense. In this sense, I believe, it historically represents the most conspicuous empirical disproof of a general philosophical thesis about the world" (Pauri 1997, p. 175). Developing his approach, the same author maintains later that "Planck's discovery of the *atomization of action* leads to the fundamental recognition of an ontology of *non-spatial abstract entities* (Quine) for the quantum level of reality (QT) as distinguished from the necessarily *spatio-temporal* experimental revelations (*measurements*)" (Pauri 2011, p. 1677).

without *having* some kind of metaphysical background and *using* it, but we do not claim that it is impossible to do science without explicitly and consciously *knowing* metaphysics. The situation is similar to that of a native speaker of a certain language: He or she may speak correctly and fluently despite being unaware of the grammatical rules of that language (as well as of its complete dictionary), which he or she simply *has* and *uses* in an unconscious way. But this does not mean that it makes no sense to devote time and skill to the *study* and *explicitation* of all this, and, for instance, write grammars and dictionaries of that language. Moreover, it is true that even a native speaker often refines the use of his language by studying its grammar in a textbook, or by consulting dictionaries—which amounts to *reflecting* on one's language, becoming *aware* of its structures and richness, and in this way also *availing* oneself of it more adequately. The attitude of a *reflecting* scientist with regard to metaphysics should be essentially the same: He or she should have for it the same respect as a writer has for linguistics. A novelist will probably never engage himself in preparing a dictionary or grammar of his native language, he also probably knows that grammars and dictionaries are not dogmatically fixed codes of correct linguistic behavior, but rather inventories which aim at recording the most general features of a living language to which they adapt in the course of time. However, he will never claim that these works are abstract and useless speculations, and from time to time he might consult them profitably. Moreover, he will be aware that these works do not only take into account the particular sectorial language he professionally works with, but many other sectors as well: poetic, juridical, technical, philosophical uses must be recorded and accounted for, which may be alien to his normal use of the language but have right of citizenship in a study devoted to language *as a whole*. A scientist scorning metaphysics would be similar to a native speaker who scorns grammars and dictionaries with the excuse that "he knows how to use his own language." Perhaps he sometimes makes mistakes of orthography, uses syntactically or grammatically incorrect expressions, or misunderstands the meanings of certain words. Of course, he cannot help having 'his own' implicit grammar and dictionary, but these are not necessarily all correct. Quite different, obviously, is the situation in those cases in which new expressions are created, new meanings are introduced, or certain deviations from standard rules are implemented, not just out of ignorance, but for some more or less clear and conscious purpose.

10.5 Metaphysics as an Approach to the Supersensible

We come now to the second aspect of metaphysics, that which makes of it a discourse concerning supersensible levels of reality. Since I have discussed this issue elsewhere,[9] I will limit myself to only a few remarks. We have seen in the

[9] See especially Agazzi 1977, 1981c, 1988d.

above that the *meta-empirical* is already present in science, since the universal background, the universal *Gestalt*—which constitutes the conditions of intelligibility for any scientific domain, and the preconditions for describing data and proposing theories—are not given in experience but make this very experience possible. This is, however, a meta-empirical framework which *applies* to experience, and is taken into consideration only *as far as* it applies to experience. Metaphysics, in its second step, goes further and tries to see whether the meta-empirical may be endowed with a much more engaging *ontological status*, that is, whether there *exist* entities which are not empirically ascertainable. Of course, in several contexts it is spoken, for example, of a "religious experience" in the sense of an immediate acquaintance with the divine (which is supersensible), but in those cases one could not speak of metaphysics proper. Metaphysics, at least in the sense envisaged here, is characterised as an effort of *rationally* reaching the supersensible, starting from experience in the more usual meaning.

In this enterprise the meta-empirical works no longer as a framework for *intelligibility*, or comprehension, or understanding, but as a means of *explanation*. This is already the case with science. For example, when in physics elementary particles are introduced, they are not admitted because they are *seen* or *observed*, but because they are needed in order to *explain* what is seen or observed. This is true in general for all 'theoretical constructs' of science (and not only of physics). They are neither induced nor deduced from sense experience, but rather *inferred* from it, essentially as *causes* of the observable phenomena (which one tries to understand and explain within the general *Gestalt* initially adopted). This fact is of great significance, since it shows that for the construction of science as a *cognitive* enterprise, a *synthetic use of reason* must be admitted; that is, a use in which reason is credited with the ability to afford new knowledge, and not just with the function of transforming, without change, the knowledge provided by the senses. (See Sect. 4.5.6 on this issue.) In the case of the sciences, however, these unobservable entities are conceived of and described by using the same conceptual tools of this *Gestalt*, so that they belong to the same *domain of objects* as the one the *Gestalt* aims to organise. They are, so to speak, those parts of the domain which one cannot know *by acquaintance*, but only *by argument*. In this sense they are 'meta-empirically known,' but they are not 'metaphysical entities'; they still belong to the *whole of experience*. The features which we use for characterising them are still those which we use for the observable entities (possibly arranged in some 'artificial' new way). We may even hope to observe them one day, and sometimes may actually succeed. They are characterised, for instance, through reference to physical magnitudes that we can measure in the experiments for which they are said to account.

With metaphysics things are different. The *metaphysical entities* are characterised by properties which are not found in the empirical world (e.g., they are not located in space and time, they do not possess mass, energy and so on). Even when they are claimed to be the causes of empirically ascertainable entities or events, the features of these cannot be deduced as a consequence of their features. But is this the fault of metaphysics? Certainly not. Every particular science works inside its specific *Gestalt* and cannot overstep it without ceasing to be *that* science. Similarly, *all* empirical sciences necessarily refer to some empirically determinable domain of objects so that empirical science *as a whole* cannot overstep the *whole of experience*. But metaphysics, being concerned with the most universal characteristics of reality, or with reality 'as such,' cannot take as a *precondition* of its discourse (as empirical science *must* do) that non-empirical features be a priori *excluded* from reality.

These statements can be even better understood by considering the relations between truth, reference, ontology and meaning that we discussed at length in other sections of this work. If we admit experiment as the *only* criterion for truth, it follows that we cannot have truth outside of this criterion; but if truth entails reference to reality, it also follows that truth can only be 'about' such types of reality as are accessible by means of that criterion. We have also seen that, if one admits a certain truth, one must also admit the existence of those realities that are accessible through the admitted criterion, and that are of the *kind* (that is, characterised by the properties) that this criterion is able to capture. In conclusion, if experiment is admitted as the only criterion for truth, the *ontology* of the discourse relying solely on this criterion is drastically restricted to experimentally directly accessible entities only. A more relaxed (and more vague) criterion for truth can be observability, and, by applying the same reasoning as before, we must say that, if observability is the *only* criterion for truth, the ontology of any discourse relying solely on this criterion will include exclusively observable entities. van Fraasen's position (as we have discussed at length) claims that science aims at establishing truth only regarding observable features of the world, and this amounts to admitting observability as the only criterion for *truth* (though other "virtues" can lead to the "acceptance" of a theory). This is why van Fraassen can consistently maintain that the ontology of science reduces to the domain of observable entities. There is, however, another possibility, that of admitting that observations, operations, experiments and the like, that is, *empirical tools* in a broad sense, be admitted as the *only* sources of meaning, but not as the *only* criteria for truth. This means that entities considered in a discourse adopting these sources of meaning must be characterised only through properties that are definable in terms of the *empirical attributes*, though truth about them could be attained also by means of not-exclusively-empirical criteria (e.g., by means of logical arguments). In this case, which we maintain to be that of the empirical sciences, the ontology remains restricted to *empirical entities*, in the sense of entities characterised through empirically grounded attributes, but not to empirically accessible entities only. This is the sense of our claim that the ontology of science remains within "the whole of experience," and that science proceeds through meta-empirical inferences that are not metaphysical in a proper sense. (By the way, this fact should convince people such as van Fraassen that one does not open the door to metaphysics by accepting the existence of unobservables; theirs still remains a *physical* existence.)

Doing metaphysics also includes not putting a priori delimitations on the meaning of basic concepts. Therefore, its task must be that of using such universal criteria of intelligibility for reality that are not bound to apply to empirically accessible entities or facts *only,* though obviously applying *also* to them. If, by carefully making *inferences* starting from empirical evidence, it should happen that these inferences compel us to admit non-empirical entities, we *must* admit them. Refusing to do so *because* they are not empirical would be equivalent to refusing to admit elementary particles in physics because they are not observable. Of course, there may be reasons in physics for being hesitant about the existence of certain particles, but these reasons are *not* that they are unobservable (but, let us say, that we are not fully convinced by the scientific *arguments* proposed for accepting them). For metaphysics we must adopt the same attitude; metaphysical conclusions concerning the existence of supersensible entities cannot be rejected because these entities are supersensible, but only if we can show that the metaphysical *arguments* produced for claiming their existence are insufficient or even wrong. To say that such entities *cannot exist* would not only be a dogmatic tenet, but even a *metaphysical* dogmatic one, because it would be in any case a statement concerning *reality as such.*

10.6 Metaphysics as Cognitive Enterprise

In science two kinds of knowledge are predominant, knowledge by acquaintance (empirical knowledge) and knowledge by argument (theoretical knowledge), but a non-negligible role is also played by *knowledge by reflection*, that is, by a critical reflection on tacitly accepted intellectual frameworks, concepts, principles or, to put it better, *conditions of intelligibility*. Relativity theory is probably the most eloquent example of this scientific knowledge by reflection since the critical investigation, criticism, and reshaping of accepted views concerning space, time, simultaneity, and other conditions of intelligibility of physical phenomena were in this theory the breakthrough that subsequently led to the empirical tests and theoretical developments made possible by the new *Gestalt*. This was not the only example in the history of science.

Metaphysics, conceived as the study of the most universal features of reality, may be seen as a great enterprise of reflecting knowledge, since it amounts to digging out the most general criteria of intelligibility of what we know. The Kantian 'acceptable' sense of metaphysics is fully in keeping with its being knowledge by reflection (on the contrary, if one were to strictly adhere to the Kantian claim that knowledge is possible *only* if referred to sense perceptions, his *Critique of Pure Reason*, and in particular his "transcendental deduction," could not be taken as expressing any knowledge at all). Metaphysics is also to a certain extent knowledge by acquaintance, since its aim is that of making intelligible reality as it is actually experienced in its most different manifestations, including those that are discovered by the sciences (as we have noted when we spoke of the feedback loop between metaphysics and science). Finally, metaphysics is knowledge by argument, and under this respect it develops its most *specific* task, that is, the investigation of the supersensible. This is why all efforts to find a sharp *demarcation* criterion of a *methodological* nature between science and metaphysics have been doomed to failure; there is no such criterion, because both are forms of knowledge, and share all the features of the cognitive enterprise. The difference can be found only in their respective *conceptual domain* and *intellectual interest*. For metaphysics, it is the domain of reality in its *whole* and the interest is that of finding its *ultimate explanation*. For science, the domain is circumscribed to some aspects of the empirically ascertainable features of reality, and the interest is that of explaining them within a preliminarily circumscribed framework of conceptual and operational tools.

The issue of *intellectual interest* deserves a few additional considerations. Humans have been 'interested in the supersensible' not just for *intellectual* reasons, but especially for *existential* reasons, because admitting or not admitting the existence of dimensions of reality that *transcend* the material world of our mundane existence can have a considerable impact on one's *sense of life*. The fact that God may exist and be the creator and regulator of the universe, as well as of human existence; the fact that human nature might be of an ontological level higher than pure animal life; the fact that there might be a continuation of our existence after our biological death are, for example, questions that have been answered in the affirmative by all cultures we know of in all historical times. From the admission of such supersensible realities several consequences have been derived regarding the *right way of living* that humans have to follow if they want genuinely to *save themselves* in the radical sense of not wasting their own lives. Religions have typically taken up the task of providing a kind of 'description' of the supersensible, and of formulating rites, rules and prescriptions for humans to follow in order to put themselves in the right relation with this superior domain. Both the descriptions and the prescriptions presented by religions are not based on intellectual *arguments*, but on narrations and revelations that are accepted by *faith*. The advantage of a faith (for those who have it) is that it provides *certainty*, and this is of paramount importance in all *vital questions*. No one is really ready to 'play one's life under hypothesis,' but everyone wants to be absolutely right when their own life is at stake. Adhering to a faith (be it religious, political or simply spontaneously human) is therefore a practically efficacious way of attaining what we could call "existential security." Doubt, however, can attack faith, and induce people to submit it to a more or less critical scrutiny. This scrutiny may regard not only the plausibility of the narrations or revelations the faith contains, but even the general views according to which the existence of the supersensible and its fundamental characteristics are proposed. The effort of clarifying and eliminating such doubts requires an *intellectual enterprise* based on evidence and arguments specifically concerned with the issue of the supersensible, and this is precisely the deep existential motivation that sustains *metaphysics* in its second and more specific sense (a sense that corresponds to the function of providing a 'conceptual space' and a 'logical justification' for faith in the supersensible).

Considering this motivation, we can easily understand certain differences with respect to science. Whereas we can qualify the intellectual attitude of science as *curiosity* in the high sense of open inquisitiveness, we should rather qualify the research on the supersensible as an *existential preoccupation.* Indeed every human feels that his life 'is not at stake' if he happens to be wrong in whatever scientific issue, and will not feel existentially insecure by admitting that the validity of any scientific statement, as we have seen, is 'relative' and 'controvertible.' But many feel that being right or wrong on the question of the supersensible could entail the gain or the loss of the fundamental value of their existence. This is why metaphysics is characterised by a fundamental striving for certainty and absoluteness while science (in its modern sense) has renounced this aspiration. This is the major reason for which science is not in the position of offering the *kind of knowledge* suitable for handling existential problems (something expressed also in the famous statement of Wittgenstein, "We feel that even if all scientific problems were solved, nothing would be made for the fundamental problems of man"). If things are so, it would certainly be arbitrary not only to declare 'pseudo-problems' those that science cannot face, but also to dogmatically exclude that other *kinds of cognition* different from science (and in particular not sharing the characteristic of its objectivity) could be viable for handling such problems. Kant himself, after all, elaborated a complex *rational* discourse for legitimating the metaphysical claims regarding the supersensible on the ground of moral arguments.

One might think that all the above simply concerns those people who have a 'faith' in the supersensible and try, therefore, to 'justify' it or even to give it a rational 'foundation.' This impression is wrong. Also those who have the 'faith' that the supersensible does not exist remain with the problem of providing some rational justification for this claim and, especially, of proposing a *sense of life* in accordance with this perspective. As a matter of fact, materialistic philosophies have taken up this task in the course of human history, but this does not mean they have escaped the issue of the supersensible; unless the positions taken were simply dogmatic, they necessarily amounted to entering this issue and answering *in the negative* its fundamental questions. In conclusion, metaphysics can perhaps be 'ignored,' but not 'eliminated,' also in its second sense.

Other alleged differences between science and metaphysics (which should demonstrate that the latter does not deserve serious consideration as a cognitive enterprise) are actually very questionable. Let us simply mention a couple of them. One concerns the controversial character, the lack of intersubjective agreement, and the refusal of "falsification" possibilities which is said to exist in metaphysics and not in science. But living science, that is, that which is not crystallised in textbooks as representing, so to speak, the 'accepted inheritance' of the past, is no less open to controversies than metaphysics (and this is far from being a weakness, but constitutes rather the spring of its progress), while in metaphysics itself there are 'accepted traditions' which pass away peacefully and become obsolete much like old scientific theories. On the other hand, scientific theories may often be protected against falsification by their supporters with no less obstinacy than metaphysical tenets (think, e.g., of Darwinism and neo-Darwinism in biology). Recent philosophy of science has stressed (even to an excessive degree) this fact. Of course, one would not deny that it is usually easier to refute a wrong scientific thesis than a metaphysical one, but this has to do with the higher degree of generality of metaphysics (also in science general laws or principles are much more difficult to disprove than particular statements).

A common reproach to metaphysics is that it is always concerned with the same 'eternal' problems without attaining any stable solution of them, so that it is compelled always to start again from scratch. This too is an oversimplified view. Metaphysics is perpetually changing because knowledge of reality (i.e., reality such as it has to be accounted for, which cannot help being reality such as it is actually known) evolves with time; and new dimensions of it are uncovered. In a similar way man also changes, since he elaborates new concepts, new attitudes, new ideals, is submitted to new historical conditions; and for this reason he problematises reality with changing interests and attitudes. This *must* lead to an evolution of metaphysics which does not exclude the permanence of certain basic features, reshaped in new ways. But science too is subject to similar conditions. One could say that science is still inquiring into the ultimate constituents of matter, that it is still trying to satisfactorily understand the mathematical and the physical continuum, that it is perpetually asking the question about the origin of the universe. Why should these facts be considered as a labour of Sisyphus in the case of metaphysics and not of science? In fact they show that a rough 'cumulative' picture of the progress of science is not less arbitrary than an 'anarchist' picture of metaphysics. The correct view is not that which introduces anarchism in science as well, but that which recognises that both fields are subject to change, and that rational inquiry may preserve truth without making it static, and recognise the limitations of our cognitive successes without denying them.

Appendix
The Semantics of Empirical Theories[1]

A.1 The Concept of Empirical Data

A.1.1 Proposals for an Intensional Semantics of Empirical Theories

1. The necessity of empirical theories' including 'data' may be considered the fundamental difference between them and formal theories. From a methodological point of view, this fact may be seen as constituting a rather radical difference in the way the two different types of theory fulfil the essential condition of possessing 'immediate truth'. *Formal* theories are characterised by the fact that the immediate truth of some of their sentences is 'stated' by the theories themselves, whereas in the case of *empirical* theories such a truth is considered to be something *discovered*, which comes from outside the theory. Moreover, the theory is thought to construct its own *internal truth* while at the same time maintaining this *external truth* and, in a way, to include it. In other words: every scientific (empirical) theory has the problem of ascertaining the truth of its accepted sentences, and this may rather often be accomplished by generating it out of the truth of previously accepted statements; but this in turn is only possible if there are sentences which possess their own truth intrinsically. While formal theories may be qualified as

[1] What follows is (with only a few stylistic corrections) a paper presented at a conference entitled "Formal Methods in the Methodology of the Empirical Sciences" held at Jablonna (Warsaw) in June 1974, and published in the proceedings of that meeting (see Agazzi 1976). The reason I reproduce the paper here is that I have often expressed, in the present work, remarks regarding the inadequacy of the standard *extensional semantics* (that is adopted in mathematical logic and, more specifically, in its 'model theory') when it is applied to empirical rather than formal theories. I have maintained that, in order to cope with such issues, an *intensional semantics* should be adopted. In order not to increase the complexity and the length of this book, however, I did not develop the reasons for the said inadequacy, nor give a detailed presentation of the nature of this intensional semantics. This second task in particular would require several technical elaborations (to which I have had the opportunity of devoting some attention in the course of the many years that have elapsed since the publication of the 1976 paper, but without their attaining a completely satisfactory form). For this reason I decided that at least a first glance into this thematic could be afforded by reproducing that paper, whose basic views I still consider tenable despite their brief presentation.

E. Agazzi, *Scientific Objectivity and Its Contexts*, DOI: 10.1007/978-3-319-04660-0,
© Springer International Publishing Switzerland 2014

those which simply 'single out' some of their sentences as being endowed with such a truth, empirical theories must learn from outside which their immediately true sentences are.

This way of considering 'data' is slightly different from the more usual one, which conceives of them as *events*, as *facts*, as structures or properties of the 'external world'. Here, on the other hand, we have directly connected them to the problem of *immediate truth*. Nevertheless, it will be seen how this choice makes a rather precise treatment of certain questions possible, which would hardly have been the case in the usual context, for the concept of truth applies properly to sentences (statements) and for this reason the reference to scientific theories becomes straightforward, as they are but particular systems of sentences. In this way the problem of characterising the concept of empirical data loses its common-sense vagueness, and becomes the particular problem of characterising the concept of an immediately true sentence in an empirical theory.

We must now make more precise the above-sketched distinction between formal and empirical theories, which was expressed by saying that the former find their immediate truth 'inside' themselves, while the latter find it 'outside' themselves. If we look at this problem from a modestly formal point of view we must admit, first of all, that both kinds of theories, if properly conceived, turn out to be sets of sentences which cannot possess as such any truth-value, but must always receive it 'from the outside'. This occurs after their having been interpreted on some suitable domain of individuals in which the language of the theory receives a *model*. This is true and obvious, but the different behaviour of the empirical and the formal theories comes out when one looks at the way they behave towards the models of their languages. In the case of a formal theory, FT, which has some set of postulates (let us call it P), if it turns out that a certain model of its language L is not a model of P, we do not worry but simply look for another model of the language, hoping to discover, eventually, a model of L which would also be a model of P. But even if after a certain time our efforts are not successful, we do not discard our formal theory. It is in this sense that we can say that the truth of P, although it cannot be concretely shown unless a model of P is found, is nevertheless not considered essential in order to retain P, which can also be expressed by saying that P is considered to be true 'inside' the theory.

In the case of an empirical theory we behave differently. If a model of its language L turns out not to be a model of one of its sentences S, this sentence is immediately discarded; and the same is the case for every set of sentences which cannot have as its model the model of the language. The difference appears now quite patently: in the case of *formal* theories the sentences of the theory are stable, this stability determining the choice of the acceptable models of the language; in the case of *empirical* theories, we have a stability of the model of the language, which determines the selection of the acceptable sentences of the theory.

Better understood, this fact must be expressed by saying that, while in the case of formal theories the language is commonly supposed to have many *possible models*, in the case of empirical theories it is supposed to have (at least theoretically, if not practically) one single model, i.e., its 'intended' model. There is surely nothing new in this remark, but its consequences do not seem to have been fully investigated. As a matter of fact, such a uniqueness of the model is in such deep contrast with the current way of thinking in model theory that one must expect, in a way, to find serious difficulties in applying its tools to the semantics of empirical theories. More explicitly, one must be prepared to find difficulties when, after having applied the usual devices of the logico-mathematical semantics (which are all conceived in order to relate a formal language to 'arbitrary' universes) one is faced with the problem of ensuring the uniqueness of the interpretation that makes the formal sentences true of the intended model. Let us also note that this model-theoretic problem is strictly bound to the exact characterisation of the concept of empirical data, because there could be no data without the possibility of ensuring that a language has its *intended* meaning.

The purely theoretical reflections made above about the difficulty of taking full advantage of model-theoretic tools in the semantics of empirical theories are promptly confirmed by the actual efforts made to provide such an application of methods. It is well known, e.g., that the uniqueness of the model cannot be guaranteed by linguistic devices, such as that of adding to the set of sentences of an empirical theory other sentences of the kind termed "meaning postulates": model theory teaches us that, also in this case, we should not be able to distinguish the universe of our objects from other universes isomorphic with it (*isomorphism theorem*) and that, moreover, this would already be an exceptionally lucky case, for it is usually not possible to distinguish it even from non-isomorphic universes; this always happens, in particular, in the case of infinite universes (results on categoricity).

As a consequence of these well known facts, the *semantic determinateness* of the language of an empirical theory has been investigated along certain 'non-verbal' paths, e.g. by resorting to what have been termed *ostensive definitions.* But these too revealed particular weak points of their own. In fact, to define a predicate ostensively was meant to 'point out', one after the other, a certain number of concrete objects for which the predicate is to hold, and a certain number of objects for which it does not hold. This implies, obviously, that only a finite and even a rather small set of 'positive standards' and of 'negative standards' can be concretely put forth; and at this point semantic ambiguity appears inevitable. For, taking an object **x** which belongs neither to the first nor to the second set (and this must necessarily be the case since both sets are finite), we should not know whether our predicate holds or does not hold for **x**. It would seem that a possible way out might be afforded by saying that, after having ostensively provided a certain amount of positive and negative standards, this very fact should put everyone in the position of considering them as 'instantiations' of a certain predicate P, as 'examples' taken out of the 'class' that constitutes the proper denotation of P and which will contain every future object of which P is true.

But this solution too is weak for at least two reasons: first, because the objects which were 'pointed out' and sampled together as positive standards may well have more than one feature in common, which means that they could appear as instantiations of more than one predicate, and this would immediately imply a *semantic ambiguity* for P, which would denote all these different classes. Second, even if the selected positive standards had only one feature in common, and so uniquely determined a class, the problem of accepting a new object x in this class should be solved on the basis of a certain ‘criterion’, which can only be a ‘similarity’ criterion. But this acceptance procedure is necessarily based on a judgment and no longer on an ostension. Moreover, in order for this judgment to be secure, it should be supported by the knowledge not of the positive standards, but of their unique common feature, which cannot be pointed out and, thus, goes beyond the ostensive procedure. On the other hand, without such knowledge, the acceptance judgment would remain subjective and vague.

Both verbal and non-verbal devices have thus proved rather ineffective in overcoming semantic ambiguity. On the other hand, this overcoming appears to be of decisive importance with regard to the ‘data’ inside an empirical theory, for such data are the touchstone for deciding on the acceptability not only of single statements, but of the entire theory. They are, in a way, the only unshakeable part of the theory, and they could not play this role unless they were free from ambiguity.

If we turn our attention to the two types of failures we considered above, we should admit that one of them seems hardly avoidable: it is the one bound to the possibility of a ‘verbal’ characterisation of the ‘datum’. Against this possibility stand not only the very detailed and exact results obtained in model theory (such as the isomorphism theorem and the categoricity results), but also a general epistemological consideration, i.e. the awareness that data are never, in science, the result of linguistic activities, but rather of ‘non-verbal’ activities, such as observation, instrument manipulation, the modification of concrete situations, etc. It seems therefore advisable to look for a better specification of the possibilities hidden in the non-verbal devices, which would be able to avoid the limitations that appeared in the case of the ostensive definitions.

The main source of inadequacy as regards ostensive definitions seems to be the fact that they can reveal the objects to us, but not the predicates. If, for example, I want to provide a small child with knowledge of the notion ‘red’, I might point out to it, e.g., a red ball, the red hood of its doll, a red skittle; but I could not be sure that, in place of the concept of red, it is not starting to form in its mind the concept of toy, for example. Moreover, while it is quite possible that, from a psychological point of view, concept formation linked to everyday terminology follows a path similar to that just sketched, it is rather obvious that this is not the case with respect to scientific languages. For these languages the question is not that of becoming gradually trained in the more or less correct use of some vocabulary, but rather that of becoming acquainted with its exact meaning. As a consequence, while certain ostensive procedures might prove useful as mental training for the employment of a language according to the accepted standards of the community of speakers of that language, it has not the ‘logical’ force sufficient to provide this language with the definiteness of meaning that is required in science.

But this argument, though of a certain value from a very general point of view, does not touch the very reason for the inadequacy of ostensive definitions when it comes to scientific assignments of meaning. Such a reason may be briefly indicated as follows: ostensive definitions can only show us 'things' of everyday experience, but not 'objects' of any specific science. This statement must be clarified somewhat, especially since we too have used the term "object" previously, when speaking of the kind of concrete things which are 'pointed out' by ostensive definitions. We employed such a term then because at that stage our discourse was still rather informal; but starting now we must sharply distinguish between a 'thing' and a scientific 'object'. The reason for this distinction is quite natural: no exact science is concerned with *generic* 'things', but always with 'things considered from a certain viewpoint,' and what is important actually turns out to be this 'viewpoint'. So, e.g., a sheet of paper on which a red drawing is traced can be considered from the viewpoint of its weight, thereby becoming an 'object' of physics, but it can also be considered from the viewpoint of the composition of the red ink with which the drawing was traced, and it thus becomes an 'object' of chemistry; and if it is considered from the viewpoint of those spatial properties of the drawing which remain invariant under certain deformations of the sheet itself, it becomes an 'object' of topology, and so on. In other words, our sheet of paper, though being but a single 'thing', can become a very large group of 'objects', depending on the different sciences that may be concerned with it.

Now, how can one clarify this notion of 'viewpoint' which makes a scientific 'object' out of an everyday 'thing'? The answer can be given by considering what the different sciences do in order to treat 'things' from their 'viewpoints': they submit them to certain specific manipulations of an *operational* character, which put the scientist in the position of being able to answer certain specific questions he can formulate about these things. Such operational procedures may be the use of a ruler, of a balance, of a dynamometer, in order to establish some physical properties of the 'thing' such as its length, its weight or the strength of some force exerted on it; they may be the employment of some reagents to determine its chemical composition, etc.

At this point, the whole situation becomes a bit clearer. The true issue, in the case of those empirical predicates that may be called *observational*, does not concern pointing out their 'empirical' or 'factual' denotations (which would mean, if properly understood, providing a complete ostensive enumeration of the members of the class denoted by the predicate—which is impossible—or pointing out only a finite number of them—which would fall short of the goal, as already remarked), nor pointing out their 'abstract' denotation (that is, their 'intension', which cannot be pointed out because it is a mental entity). Rather, the true issue concerns the provision of a positive or negative answer regarding the truth of certain *sentences*. As a consequence, if certain operational criteria are at hand which prove sufficient for that purpose, we must say that these very criteria are able to 'operationally' define our observational predicates (i.e., the predicates used in these sentences).

We shall now sketch how in such a way the difficulties that were met in the case of ostensive definitions disappear. Let us consider, e.g., the predicate 'flammable' and assign to it, as an operational criterion for testing whether it holds true of a given object **x**, that of putting **x** over a flame: the answer will be positive if that flame spreads on it, otherwise it will be negative. Such an operational procedure constituting a 'criterion', it may be applied a potentially infinite number of times and, in such a way, the class of the objects which would be assigned to the denotation of this predicate will also be potentially infinite, thus eliminating a first weak point of the ostensive definition. Second, this criterion being univocally determined, it will be affected neither by some difference, nor by any casual resemblance, between the objects which have actually been grouped together up to a certain moment (this means that if they were, by chance, all red things, there would be no risk of taking redness to be the quality of being flammable, for it was not mentioned in the description of the operation, and as a consequence does not concern the 'objects'). The example shows, further, how the criterion of operationality allows one also to consider as observational many predicates which are not such from the viewpoint of the ostensive definition. In fact, while it is conceivable that one can construct a set of red objects by ostension, it is not conceivable that one can do the same with flammable objects (or with objects endowed with any other 'dispositional' property), on the basis of simply perceiving them.

In an operational definition there is, to be precise, an ostensive aspect (one must point out the different 'instruments' to be used in the relevant operation(s), and also how to employ them). But this aspect only concerns a finite and normally small number of ostensions, which concern the *predicate* and not the objects it refers to, and which allow for an unambiguous definition of the predicate together with an indefinite possibility of applying it to objects.

After this brief explanation, the concept of an *empirical datum* may be clarified in the following way: it is a sentence (statement) which proves to be true according to the direct and immediate application of some of the operational criteria that have been accepted for defining the *ground-predicates* of that particular empirical science. Such ground-predicates must be still better determined: we shall see that they are the ones that directly enter the definition of the *objects* of an empirical science. In order to see this we shall now leave this informal discourse and enter a more formal treatment of the subject.

2. According to the oversimplifications which are currently accepted in the literature, we can suppose an empirical theory T to be expressed in a first-order language L, which must contain, among its descriptive constants, some observational predicates $O_1, \ldots, O_n$, as well as theoretical predicates $T_1, \ldots, T_p$. What makes this theory empirical rather than formal is the existence of a model $\boldsymbol{M}$ of its language, which may be identified with a structure of the following kind:

$$\boldsymbol{M} = <\mathrm{U}, \mathbf{R}_1, \ldots, \mathbf{R}_s>$$

where U is a non-empty set of 'individuals', and $\mathbf{R}_1, \ldots, \mathbf{R}_s$ are relations in U, the total number of which should be $s = n + p$, so that every O-predicate and every T-predicate can be interpreted on one of these relations (or, to put it differently, may be considered as the name of that relation in L). The set U provides the range of the individual variables of L. Once the model of L is fixed, it is straightforward to define the model of every sentence α of L: unary relations are identified with subsets of U, n-ary relations with sets of ordered n-tuples of elements of U and then, given a sentence $\alpha \equiv$ Px we say that is true in $\boldsymbol{M}$ (or that $\boldsymbol{M}$ is a model of α) if the individual **x** of U 'named' by x belongs to the subset **P** of U 'named' by P; if $\alpha \equiv Rx_1, \ldots, x_n$ we say that $\boldsymbol{M}$ is a *model* of α if the ordered n-tuple $<\mathbf{x}_1, \ldots, \mathbf{x}_n>$ of individuals of U named by $x_1, \ldots, x_n$ belongs to the set **R** of ordered n-tuples 'named' by R. The way of defining α's being true in $\boldsymbol{M}$ for every non-atomic α is well known.

There are two more or less explicit assumptions lying at the basis of this discourse: (i) that the individuals of U and the relations on U must be conceived as 'given'; (ii) that the set U must be decidable (i.e., given an individual **x**, it is always possible to decide whether $\mathbf{x} \in U$ or $\mathbf{x} \notin U$), while the relations on U are not necessarily decidable (i.e., given an n-tuple $<\mathbf{x}_1, \ldots, \mathbf{x}_n>$ it is not always decidable whether $<\mathbf{x}_1, \ldots, \mathbf{x}_n> \in \mathbf{R}$ or $<\mathbf{x}_1, \ldots, \mathbf{x}_n> \notin \mathbf{R}$). This essential undecidability of the relations on U is the reason for the semantic ambiguity we have spoken about in the preceding section, for it means that, given a certain **x**, we are sometimes unable, e.g., to state whether or not it belongs to **P**, which amounts to saying that we cannot decide whether or not $\boldsymbol{M}$ is a model of Px. The above-mentioned failures of the verbal and non-verbal devices to avoid semantic ambiguity express the impossibility of making the relations on U decidable by means of those devices. We shall explicitly remark that, owing to the ineffectiveness of the ostensive definitions, this ambiguity also holds if we restrict ourselves to what we could call the observational sub-model $\boldsymbol{M}^{o}$ of our language (i.e., the model of its O-predicates).

We shall now try to explain what the semantics of an empirical theory should look like in order to fit the methodological approach of the operational criteria of definition for predicates we advanced in the preceding section. Given the language L of an empirical theory, we shall still distinguish among its descriptive constants the O-predicates $O_1,, O_n$ and the T-predicates $T_1, \ldots, T_p$. But now the O-predicates will be considered as *operational* and not as *observational* (remember that dispositional predicates may turn out to be operationally definable, while not being observational in a strict sense). Our first problem (and actually the only one which will be discussed in this paper) concerns the semantic definiteness of the operational predicates. We shall therefore confine our treatment to the operational sub-model $\boldsymbol{M}^{o}$ of L or, in other words, to the model $\boldsymbol{M}$ of the operational sub-language L_o of language L. Our model will be something like the following:

$$\boldsymbol{M}^{o} = <U, \Omega, O, R, \mathbf{P}^{o}_{1} \ldots \mathbf{P}^{o}_{n}>$$

where Ω is a finite set of instruments, O is a finite set of operations, R is a finite set of results (i.e. of observational outcomes of concrete operations), while every $\mathbf{P}^{o}_{i}$ is an element of the Cartesian product $\{\Omega \times O \times R\}$. To make this statement clear through an example, let Ω contain a gold-leaf electroscope ω_1, let O contain the operation o_1: "to put **x** in contact with the free plate of ω_1"; let R contain the result r_1: "the gold-leaf of ω_1 is repelled". In this case, $\mathbf{P}_i^o$ could be for instance, $<\omega_1, o_{1,}, r_1>$ that is, intuitively, 'operation o_1 is performed on **x** and the gold-leaf of the instrument ω_1 is repelled', which could be seen as an operational definition of the unary predicate of 'being electrically charged'.

The most peculiar feature of our definition of $\boldsymbol{M}^o$ is that no explicit mention of any universe U is made in it, contrary to what is the case in 'extensional' semantics, while a clear 'intensional' character is expressed by the fact that relations are effectively 'given' by reference not to set-theoretical entities, but to certain meaningful *conditions*. On the other hand, in our example we have spoken of an '**x**' to be put in contact with the electroscope. This could sound strange, but it is in agreement with our previous distinction between 'things' and 'objects': **x** is here an indefinite 'thing', which becomes an 'object' of the theory T only at the moment that *all* the operational procedures accepted in T (i.e., explicitly codified in Ω and O) prove applicable to it. Then, in our semantics the individuals of the universe surely must appear, but they are not 'given': they are 'singled out' step by step via the application of the *operational criteria*. The set of individuals is thus 'constructed' and remains always 'open', exactly as every empirical science requires. For example, a book, though not normally considered to be an object of electrical science, can nevertheless be studied by this science, if somebody were interested in its electric properties.

If one should find it too embarrassing to accept that objects are constructed by predicates, we could make the innocent admission that there exists an 'overall universe of discourse' to which all the individual variables of every language may be referred, provided it is understood that a theory T is solely concerned with that subset of the overall universe to which all the operational criteria explicitly stated by the semantics of T do actually apply.

This methodological choice, besides being rather close to the actual practice of scientific inquiry, has many advantages. First of all, as already remarked, it leaves the universe of the objects of a theory 'open' and potentially infinite; second, it leaves similarly open and potentially infinite, for quite analogous reasons, the subset of objects (or the set of n-tuples of objects) which corresponds to every predicate. Moreover, every O-relation is decidable, for, in order to be accepted as an 'object' of the theory, a particular **x** must have proven itself to be manipulable by all the prescribed operations, while every such manipulation always leads to a result which is selected as a kind of defining 'clause' of a certain P. This implies that an **x** enters as an object of T and, at the same time, is effectively decided upon as far as its belonging (alone, or as inserted in an n-tuple with other objects) to every O-relation concerned. This implies, of course, that no semantic ambiguity is possible here, as is easily seen when we proceed to explaining the concept of the model of a sentence α.

Let us consider, for the sake of brevity, only the simple case of an atomic sentence O $x_1, \ldots, x_n$. O is interpreted on a certain $\mathbf{P}_i^o$, which is supposed to hold for an n-tuple of objects $<\mathbf{x}_1, \ldots, \mathbf{x}_n>$ if, and only if, submitting them to certain manipulations by means of some ω_i belonging to Ω, according to a given operation o_i, belonging to O, there will be a certain result r_i stated in R. As a consequence, when an assignment is made which maps the individual variables of a sentence onto some generic 'things' of the 'overall universe,' it must first of all be apparent whether these 'things' can also be admitted as belonging to the universe of T; and in this case it is automatically decidable whether $\mathbf{P}_i^o$ is true of them or not. Indicating by *Ver* ($\boldsymbol{M}^o$) the set of atomic sentences which are true in $\boldsymbol{M}^o$ (or the set of atomic sentences of which $\boldsymbol{M}^o$ is a model), we shall say, for an $\alpha \equiv O_i\, x_1 \ldots, x_n$:

$$\alpha \in Ver(\boldsymbol{M}^o) \leftrightarrow <\omega_i, o_i>, \text{ applied to } <\mathbf{x_1}, \ldots, \mathbf{x_n}> \text{ gives as a result } r_i.$$

Let us remark how suitable it is to have operational criteria 'singling out' objects instead of having them as 'given'. Suppose we have $\alpha \equiv$ Px and that x has been interpreted on the 'thing' **x** which is a toothache, while P has been interpreted on our previously described predicate as meaning 'being electrically charged'. If we were in the traditional situation of considering the objects as 'given', we should conscientiously say that α is false in $\boldsymbol{M}$, as the predicate of being electrically charged is not true of the toothache. But this conclusion would puzzle many people, who would rightly point out that Px turns out to be 'meaningless' more than 'false' in $\boldsymbol{M}$. If we adopt, instead, the viewpoint of our intensional semantics, we should immediately see that the operational criteria attached to P (i.e., the employment of an electroscope, etc.) cannot be used with such an **x**, and by this simple fact **x** does not belong to our universe, and hence α could be neither true nor false, but would simply be meaningless in our theory, exactly as every man on the street would maintain.

But what would be said if, e.g., we take **x** to be the moon? It certainly does not sound meaningless to ask whether the moon is electrically charged. But, on the other hand, it is surely not possible to test such a predicate using an electroscope, as is prescribed by our operational definition. Should we then discard the moon from the objects of our theory? The answer to this question involves some additional considerations. First of all, we must remember that our discourse was restricted to the operational predicates, and the fact that, in common scientific practice predicates which have been originally defined in an operational way are also applied to 'inaccessible' objects already suggests that this might be possible thanks to the 'mediation' of the theory, i.e., thanks to the presence of some T-predicates in it.

From this point of view we can say that the inclusion of something in the universe of the objects of a theory may happen either directly, as a consequence of the application of the operational criteria, or indirectly, through the employment of theoretical tools. But here we have a slightly different question: the problem is not so much that of having T-predicates which might refer to operationally inaccessible objects, as that of having an O-predicate (such as that of being electrically charged) which seems to apply outside the domain of its defining operations. This problem is actually not easy, and I have tried to treat it elsewhere, suggesting that an operational concept be defined not by a single operation but by an 'equivalence class' of operations, two operations being called equivalent

(i) if there is a certain set of objects to which both can apply, and (ii) if the results of their applications to these objects is the same.[2]

This can only happen if at least some fragment of the theory is employed; and the consequence is an enlargement of the universe. In fact, the objects of the theory are obliged to be possible arguments of *all* the predicates of the theory; and this means that, if two operations o_1, and o_2 of the theory T can be applied to two different sets of objects, only the intersection of these sets is included in the universe of T. But if we accept the defining of the predicates not by single operations but by equivalence classes of operations, it follows in our example that, if operations o_1 and o_2 are equivalent, the union and not the intersection of their sets of objects is included in the universe of T. The theory thus allows for a first enlargement of its universe by stating the 'equivalence' of certain different operations; but it can also ensure a 'connection' between predicates which can allow the *inference* that one O-predicate is true of **x** from the fact that a particular other O-predicate is true of **x**, this inference being testable by the actual performance of the operations involved. Once the validity of this inference is tested, it becomes the basis for admitting its validity also with respect to those cases in which it cannot be directly tested, i.e., when the first O-predicate can be operationally tested on a particular **y**, while the second cannot. In this case we may say that the second predicate is also true of **y**, though we cannot test it. In such a way we have actually an 'extension' of the model $\boldsymbol{M}^{o}$, which comes to include objects that are still characterised by O-predicates without actually being manipulable by *all* the operations of the theory.

If we now take all the above sketched remarks together (their formal treatment is not problematic, and we omit it for brevity) and conceive of O-predicates, besides being defined through equivalence classes of operations, as 'extendable' thanks to the theory, we can qualify such predicates as ground-predicates, and require that every object of the theory be characterised with reference to *all* of them. The reason for privileging them to such an extent as to consider them as the 'makers of the objects' is strictly bound to what has been said in the first section of this paper about the clipping out of scientific 'objects' from everyday 'things': we remarked then that an object results when a thing is investigated from certain points of view and there are tools for answering immediate questions about it. The operational criteria are such tools; they are the effective incarnation of such points of view, and it is thus quite legitimate to assume the O-predicates related to them to be basic predicates of the empirical theory concerned with the 'objects' that have emerged. Note, furthermore, that when an empirical theory has to put its sentences to the test, this cannot be done unless one comes step by step down to these operational procedures, which receive a confirmation of their foundational character from this fact as well.

---

[2] This problem was first treated in my book, *Temi e problemi di filosofia della fisica*, Milan 1969 (pp. 128–130) without any formal apparatus. It was later investigated formally by M. L. Dalla Chiara Scabia and G. Toraldo di Francia in their paper, 'A Logical Analysis of Physical Theories', in *Rivista del Nuovo Cimento*, 3 (1973), pp. 1–20.

After the above considerations it appears quite obvious to qualify as the 'empirical data,' or simply 'data', of an empirical theory all the atomic sentences which are true in $\boldsymbol{M}^{o}$ and all the negations of the atomic sentences which are false in $\boldsymbol{M}^{o}$, i.e. all the atomic sentences (possibly negated) which are built up exclusively by resorting to O-predicates.

Without entering the complex questions which arise when T-terms come into play, we shall briefly hint at some points which seem worth mentioning. First, it is perhaps worthwhile to point out that the kind of 'intensional' semantics proposed in this paper is not, after all, so complicated and cumbersome as may be thought at first sight. It is in fact a rather naive belief to think that it would be easy to actually 'give' a universe U of individuals, as is presupposed in the current extensional semantics: it must be found much easier, from a concrete point of view, to 'give' three finite sets of 'instruments', 'operations' and 'results', which can rather easily be described in the metalanguage and even be practically 'pointed out' if necessary. When we pass to the interpretation of the predicates, the current extensional semantics assigns to them some particular set-theoretical entities which are very easy to speak of but practically impossible to demonstrate, and this is immediately reflected in the concept of the model of a sentence: here again it is easily said that α is true in $\boldsymbol{M}^{o}$ if the relation P is true of the objects $<\mathbf{x}_1, \ldots, \mathbf{x}_n>$; but how such a fact may be determined remains a rather enigmatic affair. On the other hand, the operational definition of the predicates makes such a crucial step quite manageable, as was shown above.

It is also interesting to note that no isomorphism theorem holds in our semantics. The reason is simple: in the extensional semantics, if two universes U and U′ have the same cardinality and a certain relation **R** is 'given' on U, a corresponding relation **R**′ can be easily 'induced' on U′ by simply stating:

$$<f(\mathbf{x}_1),\ldots,f(\mathbf{x}_n)> \in \mathbf{R}' \leftrightarrow <\mathbf{x}_1,\ldots,\mathbf{x_n}> \in \mathrm{R}$$

$f(\mathbf{x}_1), \ldots, f(\mathbf{x}_n)$ being the images of $\mathbf{x}_1, \ldots, \mathbf{x}_n$ under the one-one correspondence f, which must exist in order to insure that the two universes have the same cardinality. In the case of our semantics, nothing of the kind is possible, for no one can be sure that, 'given' two operationally defined predicates P_1 and P_2, every time P_1 holds true of its objects, P_2 holds true of certain 'corresponding' objects. Because of this, no such correspondence between objects can be established as a rule. On top of this, relations cannot be 'induced' from one model to another for, if they are obtained by 'copying' the operational definition of the first model, they simply turn out to coincide with those from which they were supposed to be derived, and the two models thus coincide. If, on the other hand, they are characterised by different operations, there is no *a priori* warranty that they will remain 'parallel' in their behaviour. This fact holds even if P_1 and P_2 are referred to the same 'universe' (i.e. when they belong to the same theory T). In fact, it is quite possible sometimes to prove something like:

$$\forall \mathrm{x}\ (\mathrm{P}_1\mathrm{x} \leftrightarrow \mathrm{P}_2\mathrm{x})$$

but this simply means that we have found an *empirical law* connecting two different properties of our objects.

If P_1 and P_2 are both O-predicates, we could take advantage of this law and declare 'equivalent' the two operational criteria on which the predicates are founded, putting them in the same 'equivalence class.' But we are not compelled to do this (think of the predicate 'magnetic' defined by the operational criterion of attracting iron filings, or by the criterion of inducing an electric current by motion near a circuit). In this last case we would prefer to say that we have discovered a new empirically testable property of our objects. This can be generalised to the case of non-operational predicates, and it expresses the fact that in the empirical sciences (though in mathematics as well) we frequently arrive at establishing the 'equivalence' of certain properties without meaning by that that they are one and the same property; and this is perhaps one of the ways of giving an exact characterisation of the fact that science always proceeds by 'synthetic' and 'synthetic a priori' judgements or, if we prefer, that there cannot properly be scientific inquiry without 'data' (empirical and otherwise).

References

Adickes, E: Kant und das Ding an sich. Pan Verlag Rolf Heisse, Berlin (1924). Reprint Olms, Hildesheim/New York (1977)
Agassi, J.: Science in Flux. Reidel, Dordrecht (1975)
Agassi, J.: Science and Society. Reidel, Dordrecht, 1981 (1976)
Agazzi, E.: Introduzione ai problemi dell'assiomatica. Vita e Pensiero, Milano (1961)
Agazzi, E.: Riflessioni su alcuni nuovi orizzonti della logica matematica. In: Logica e analisi, special issue of Archivio di Filosofia, pp. 47–69 (1966)
Agazzi, E.: Il "Saggiatore". Barbera, Firenze (1967)
Agazzi, E.: Temi e problemi di filosofia della fisica. Manfredi, Milano (1969), 2nd edn. Abete, Roma (1974)
Agazzi, E: The concept of empirical data. Proposals for an intensional semantics of empirical theories. In: Przelecki, M. et al. (eds.) Formal Methods in the Methodology of Empirical Sciences, pp. 153–167. Reidel, Dordrecht (1976)
Agazzi, E.: The Role of Metaphysics in Contemporary Philosophy. Ratio **19**(2), 162–169 (1977)
Agazzi, E.: From Newton to Kant: The impact of physics on the paradigm of philosophy, In: Fuyas M (ed.) Christian Theology in the Context of the Scientific Revolution. Gutenberg, Athens (1978), Report from Abba Salama IX, pp. 59–76 (1978a)
Agazzi, E.: Eine Deutung der wissenschaftlichen Objektivität. Allgemeine Zeitschrift für Philosophie **3**(1978), 20–47 (1978b)
Agazzi, E.: Le matematiche come teorie e come linguaggio. Epistemologia **1**, 165–182 (1978c)
Agazzi, E.: Non-contradiction et existence en mathématiques. Logique et Analyse **21**, 459–481 (1978d)
Agazzi, E.: Is scientific objectivity possible without measurement? Diogenes **104**, 93–111 (1978e)
Agazzi, E. (ed.): Modern Logic. A Survey. Reidel, Dordrecht (1981a)
Agazzi, E.: Logic and Methodology of Empirical Sciences. In: Agazzi (1981a), pp. 255–282 (1981b)
Agazzi, E.: Considerazioni epistemologiche su scienza e metafisica. In: Huber, C. (ed.), Teoria e metodo delle scienze, pp. 311–340. Università Gregoriana, Roma (1981c)
Agazzi, E.: La natura del modello. In Dalle Fratte, G. (ed.) Teoria dei modelli in pedagogia, pp. 31–64, 234–242. Federaz. Prov. Scuole Materne, Trento (1984)
Agazzi, E.: Commensurability, incommensurability, and cumulativity in scientific knowledge. Erkenntnis **22**, 51–77 (1985)
Agazzi, E.: The historical dimensions of science and its philosophy. Diogène **132**, 60–79 (1985)
Agazzi, E.: La funzione del modello nella scienza. Il Quadrante scolastico **31**, 52–62 (1986)
Agazzi, E.: A systems-theoretic approach to the problem of the responsibility of science. Zeitschrift für allgemeine Wissenschaftstheorie **18**(1–2), 30–49 (1987)
Agazzi, E.: Wave, particles and complementarity. In: Tarozzi G., van der Merwe A. (eds.) The Nature of Quantum Paradoxes, pp. 53–74. Kluwer Academic Publishers, Dordrecht (1988a)

E. Agazzi, *Scientific Objectivity and Its Contexts*, DOI: 10.1007/978-3-319-04660-0,
© Springer International Publishing Switzerland 2014

Agazzi, E.: On the different kinds of truth. In: Incardona N. (ed.) Les formes actuelles du vrai, pp. 11–38. Enchiridion, Palermo (1988b)
Agazzi, E.: Probability: a composite concept. In: Agazzi E. (ed.) Probability in the Sciences, pp. 3–26. Kluwer, Dordrecht (1988c)
Agazzi, E.: Science and metaphysics before nature. In: McLean. G. (ed.) Person and Nature, vol. 1, pp. 3–13. The International Society for Metaphysics, Studies in Metaphysics, Washington (1988d)
Agazzi, E.: Do experiments depend on theories or theories on experiments? In: Batens, D., van Bendegem J.P. (eds.) Theory and Experiment, pp. 3–13. Reidel, Dordrecht (1988e)
Agazzi, E.: Logo semantico e logo apofantico. Epistemologia **12**(Special Issue), 17–47 (1989a)
Agazzi, E. (ed.): Naive realism and naive anti-realism. Dialectica **43**, 83–98 (1989b)
Agazzi, E.: Responsibility: the genuine ground for the regulation of a free science. In: Shea, W.R., Sitter B. (eds.) Scientists and their Responsibility, pp. 203–219. Watson Publishing International, Canton (1989c)
Agazzi, E.: Il bene, il male e la scienza. Le dimensioni etiche dell'impresa scientifico-tecnologica. Rusconi, Milanno (1992)
Agazzi, E.: On formalism. In Floistad G. (ed.) Philosophical Problems Today, vol. 1, pp. 74–137. Kluwer Academic Publishers, Dordrecht (1994a)
Agazzi, E.: Was Galileo a realist? Physis **31**(1), 273–296 (1994b)
Agazzi, E. (ed.): Realism and Quantum Physics. Rodopi, Amsterdam/Atlanta (1997a)
Agazzi, E.: On the criteria for establishing the ontological status of different entities. In: Agazzi (1997a), pp. 40–73 (1997b)
Agazzi, E.: The relation of mathematics to other sciences. In: Agazzi, E., Darvas, G. (eds) Philosophy of Mathematics Today, pp. 235–259. Kluwer Academic Publishers, Dordrecht (1997c)
Agazzi, E.: From technique to technology: the role of modern science. In: Agazzi, E., Lenk, H. (eds.) Advances in the Philosophy of Technology, pp. 135–144. Delaware Society for Philosophy and Technology, Newark (1999)
Agazzi, E.: Observability and referentiality. In: Agazzi, E., Pauri, M. (eds.) The Reality of the Unobservable, pp. 45–57. Kluwer Academic Publishers, Dordrecht, Boston and London (2000)
Agazzi, E.: Metaphysical and scientific realism. In: Marsonet, M. (ed.) The Problem of Realism, pp. 35–63. Ashgate Publishing, Aldershot (2001)
Agazzi, E.: Continuità e discontinuità fra scienza e senso comune. In: Agazzi, E. (ed.) Valore e limiti del senso comune, pp. 341–355. Angeli, Milano (2004)
Agazzi, E.: Meaning between sense and reference. Impacts of semiotics on philosophy of science. Semiotica **188**(1/4), 29–50 (2012a)
Agazzi, E.: Ragioni e limiti del formalismo. Saggi di filosofia della logca e della matematica, Angeli, Milano (2012b)
Agazzi, E.: Rationality outside science. In: Lektorsky, V., Guseynov, A. (eds.) Rationality and its Limits, pp. 32–53. Institute of Philosophy of the Russian Academy of Sciences, Moscow (2012c)
Agazzi, E.: Rethinking philosophy of science today, In: Selected papers from the XXII World Congress of Philosophy. J. Philos. Res. **37**, 85–101 (2012d)
Alai, M.: van Fraassen, Observation and Belief. In: D'Agostino et al., pp. 663–675 (2010)
Alai, M.: Levin and Ghins on the 'No-Miracle' argument and naturalism. Eur. J. Philos. Sci. **2**(1), 85–110 (2012)
Bacon, F.: Novum Organum. Recent edition in: Encyclopaedia Britannica Inc. Great Books of the Western World, Chicago, London, etc., vol. 30 (1952) (22nd repr. 1978) (quotations are from this edition, 1620)
Balzer, W., Moulines, U., Sneed, J.: An Architectonic for Science. The Structuralist Approach. Reidel, Dordrecht (1987)
Barnes, B.: Interests and the Growth of Knowledge. Routledge and Kegan Paul, London (1977)

Black, M.: Language and Philosophy. Cornell University Press, Ithaca (1949)
Black, M.: Models and Metaphors. Cornell University Press, Ithaca (1962)
Block, N.: Advertisments for Semantics for Psychology. In: French, P., Uheling, T., Wettstein, H. (eds.) Midwest Studies in Philosophy. University of Minnesota Press, Minnesota (1986)
Bloor, D.: Knowledge and Social Imagery. Routledge and Kegan Paul, London (1976)
Boas Hall, M.: Robert Boyle on Natural Philosophy. An Essay with Selections from his Writings. Indiana University Press, Bloomington (1965)
Bochenski, J.: Formale Logik, Alber, Freiburg and München (1956). English transl. by Thomas, I., as A History of Formal Logic. Chelsea Publishing, New York (1970)
Boehner, P.: Medieval Logic. University Press, Manchester (1952)
Bohr, N.: Atomic Theory and the Description of Nature. Cambridge University Press, Cambridge (1934)
Bohr, N.: Atomic Physics and Human Knowledge. Wiley, New York (1958)
Bontadini, G.: Studi sulla filosofia dell'etá cartesiana. La Scuola, Brescia (1947)
Bontadini, G.: Indagini di struttura sul gnoseologismo moderno. La Scuola, Brescia (1952)
Born, M.: Physics in My Generation. Pergamon Press, London/New York (1956)
Born, M.: Natural Philosophy of Cause and Chance. Dover, New York (1964)
Boyd, R.: The current status of scientific realism. In: Leplin, J. (ed.) Scientific Realism, pp. 41–82. University of California Press, Berkeley (1984)
Boyle, R.: Works. Rivington et al., London, 1772, 6 vols. (1672)
Braithwaite, R.B.: Scientific Explanation. Cambridge University Press, Cambridge (1953)
Brentano, F.: Psychologie vom empirischen Standpunkt, Wien (1874) (3rd edn. Leipzig 1924, 2 vols: reprint. Meiner, Hamburg 1955–1959). English transl. Routledge and Kegan Paul, London; Humanities Press, New York (1973)
Bridgman, P.W.: The Logic of Modern Physics. MacMillan, New York (1927)
Bridgman, P.W.: The Nature of Physical Theory, Princeton University Press, Princeton (1936); Reprint Dover, New York, 1936 and later.
Bridgman, P.W.: Reflections of a Physicist. Philosophical Library, New York (1950)
Bunge, M.: Foundations of Physics. Springer, Berlin (1967a)
Bunge, M. (ed.): Delaware Seminar in the Foundations of Physics. Springer, Berlin (1967b)
Bunge, M.: Quantum Theory and Reality. Springer, Berlin (1967c)
Bunge, M.: The Bell inequalities and all that. Philosophia Naturalis **26**, 121–134 (1989)
Butts, R., Pitt, J. (eds.): New Perspectives on Galileo. Reidel, Dordrecht and Boston (1978)
Buzzoni, M.: Conoscenza e realtà in K.R. Popper. Angeli, Milano (1982)
Buzzoni, M.: Scienza e tecnica. Teoria ed esperienza nelle scienze della natura. Studium, Roma, (1994)
Buzzoni, M.: Erkenntnistheoretische und ontologische Probleme der theoretischen Begriffe. J. Gen. Philos. Sci. **28**, 19–53 (1997)
Campbell, N.R.: Physics: The Elements. Cambridge University Press, Cambridge (1920); Reprinted as Foundations of Science. Dover, New York (1957)
Carl, W.: Sinn und Bedeutung. Studien zu Frege und Wittgenstein. Hain, Königstein/Ts (1982)
Carnap, R.: Der logische Aufbau der Welt, Weltkreis-Verlag, Berlin (1928); English transl. by George, R.A.: The Logical Structure of the World. Routledge and Kegan Paul, London (1967)
Carnap, R.: Logische Syntax der Sprache. Springer, Wien (1934). English trans. with additions: Kegan Paul and Trubner, London (1937). Several reprints, e.g. Humanities Press, New York (1951)
Carnap, R.: Testability and meaning. Philos. Sci. **3,** 419–471 (1936); and **4,** 1–46 (1937)
Carnap, R.: Introduction to Semantics. Harvard University Press, Cambridge (1942); reprint with Formalization of Logic in one volume in 1959
Carnap, R.: Meaning and Necessity. Chicago University Press, Chicago (1947); 2nd edn. 1956
Carnap, R.: Meaning Postulates. Philos. Stud. **3**, 65–73 (1952)

Carnap, R.: The Methodological Character of Theoretical Concepts. In: Feigl, H., Scriven, M., Maxwell, G. (eds.) Minnesota Studies in the Philosophy of Science, vol. 1, pp. 38–76. University of Minnesota Press, Minneapolis (1956)

Cartwright, N.: How the Laws of Physics Lie. Clarendon Press, Oxford (1983)

Cassirer, E.: Substanzbegriff und Funktionsbegriff, B. Cassirer Verlag, Berlin (1923). Engl. transl. (Substance and Function, Chicago, Open Court, 1923 (quotations are taken from this edition))

Chakrawartty, A.: A Metaphysics for Scientific Realism: Knowing the Unobservable. Cambridge University Press, Cambridge (2007)

Chisholm, R. (ed.): Realism and the Background of Phenomenology. The Free Press, Glencoe (1960)

Churchland, P., Hooker, C. (eds.): Images of Science. Essays on Realism and Empiricism, with a Reply from Bas C. van Fraassen. University of Chicago Press, Chicago (1985)

Cohen, L.J.: The Diversity of Meaning. Methuen, London (1962)

D'Agostino, M., et al. (eds.): New Essays in Logic and Philosophy of Science. London College Publications, London (2010)

Davidson, D.: True to the facts. J. Philos. **66**, 748–764 (1969)

Davidson, D.: Knowing one's mind. Proc. Addresses A.P.A. **60**, 441–458 (1986)

Descartes, R.: Oeuvres. In: Adam, C., Tannery, P. (eds.), 13 vols. Cerf, Paris (1897–1913)

Devitt, M.: Realism and Truth. Princeton University Press, Princeton, 1984 (1981) (third edition, with an afterword, 1997)

Diemer, A.: Was heisst Wissenschaft? Hain, Meisenheim (1964)

Diez, J., Lorenzano, P.: La concepción estructuralista en el contexto de la ilosofía de la ciencia del siglo XX. In: Diez, J., Lorenzano, P. (eds.) Desarrollos actuales de la metateoría estructuralista, pp. 13–78. Universidad Nacional de Quilmes, Bernal (Buenos Aires) (2002)

Dilworth, C.: The Metaphysics of Science, 2nd edn. Springer, Dordrecht (2007)

Dilworth, C.: Scientific Progress, 4th edn. Springer, Dordrecht (2008)

Drake, S.: Discoveries and Opinions of Galileo. Doubleday, New York (1957)

Drake, S.: Galileo's new science of motion. In: Bonelli, R., Shea, R. (eds.), pp. 131–156. New York: Science History (1975)

Drake, S., O'Malley, C.D.: The Controversy on the Comets of 1618. University of Pennsylvania Press, Philadelphia (1966)

Dreyfus, H. (ed.): Husserl: Intentionality and Cognitive Science. MIT Press, Cambridge (1982)

Duhem, P.: Sozein ta phainomena. Essai sur la notion de théorie physique de Platon à Galilée. Hermann, Paris (1908). English trans.: University of Chicago Press, Chicago (1969)

Duhem, P.: Le système du monde. Histoire des doctrines cosmologiques de Platon à Copernic, 5 vols. Hermann, Paris (1913–1917). Complete edition in 10 vols.: ibid. (1954–1958). Partial English trans.: Ariew, R.: Medieval Cosmology. University of Chicago Press, Chicago (1985)

Dummett, M.: Frege. Philosophy of Language. Duckworth, London (1973)

Durbin, P.T. (ed.): Technology and Contemporary Life. Reidel, Dordrecht (1979)

Durbin, P.T. (ed.): Philosophy of Technology. Kluwer, Dordrecht (1989)

Durbin, P.T.: Social Responsibility in Science, Technology and Medicine. Lehigh University Press, Bethlehem (1992)

Edwards, P. (ed.): The Encyclopedia of Philosophy. Macmillan, New York (1967)

Eisler, R.: Wörterbuch der philosophischen Begriffe, 3rd edn. Mittler, Berlin (1927)

Ellis, B.: What science aims to do. In: Churchland, P.M., Hooker, C.A. (eds.) Images of Science, pp. 48–74. University of Chicago Press, Chicago (1985)

Etchemendy, J.: Tarski on Truth and Logical Consequence. J. Symb. Logic **53**, 51–79 (1988)

Etchemendy, J.: The Concept of Logical Consequence. Harvard University Press, Cambridge (1990)

Faye, J.: Niels Bohr: His Heritage and Legacy. Kluwer, Dordrecht (1991)

Feigl, H., Maxwell, G. (eds.): Current Issues in the Philosophy of Science. Holt, Rinehart and Winston, New York (1961)

Feigl, H., Scriven, M., Maxwell, G. (eds.): Minnesota Studies in the Philosophy of Science. University of Minnesota Press, Minneapolis (1956–60) (vol. I (1956), vol. II (1958), vol. III (1962)
Feyerabend, P.K.: How to be good empiricist. A plea for tolerance in matters epistemological. In: Nidditch, P.H. (ed.) The Philosophy of Science. Oxford University Press, London (1963)
Feyerabend, P.K.: Science in a Free Society. New Left Books, London (1978)
Field, H.: Tarski's theory of truth. J. Philos. **69**, 347−375 (1972)
Fine, A.: The natural ontological attitude, in Leplin, pp. 83−107 (1984)
Fine, A.: Unnatural attitudes: realist and instrumentalist attachments to science. Mind **95**(378), 149–179 (1986)
Frede, M., Patzig, G.: Aristotele's 'Metaphysik Z,' Text, Uebersetzung und Kommentar, 2 vols. Beck, München (1988)
Frege, G.: Begriffschrift, eine der arithmetischen nachgebildete Formelsprache des reinen Denkens. Nebert, Halle (1879). Repr. Olms, Hildesheim 1964. English trans. in van Heijenoort, J.: Frege and Gödel: Two Fundamental Texts in Mathematical Logic. Harvard University Press, Cambridge, 1970
Frege, G.: Die Grundlagen der Arithmetik. Eine logisch mathematische Untersuchung über den Begriff der Zahl. Marcus, Breslau (1884). 2nd ed. 1934. Repr. Oxford, 1950 (German-English, revised ed., 1953) and Olms, Hildesheim, 1961
Frege, G.: Über das Trägheitsgesetz. Zeitschrift für Philosophie und philosophische Kritik **98**, 145−161 (1891). Repr. in Frege (1967)
Frege, G.: Über Sinn und Bedeutung. Zeitschrift für Philosophie und philosophische Kritik **100**, 25−50 (1892). Repr. in Frege (1967)
Frege, G.: Der Gedanke. Eine logische Untersuchung. Beiträge zur Philosophie des Deutschen Idealismus **1**, 58−77 (1918–1919). Repr. in Frege (1967)
Frege, G.: Kleine Schriften (miscellanea). In: Angelelli, I. (ed.) Wissenschaftliche Buchgesellschaft, Darmstadt (1967). English edition: Blackwell, Oxford (1984)
French, P., Uheling, T., Wettstein, H. (eds.): Midwest Studies in Philosophy. University of Minnesota Press, Minneapolis (1986)
Galilei, G.: Opere: Le opere di Galileo Galilei, Edizione Nazionale a cura di A. Favaro, Barbera, Firenze (1929−1939). Repr. 1964−66, 20 vols
Galilei, G.: Il Saggiatore. In: Opere, cit., vol. VI, pp. 197−372 (1623). Partial English trans.: S. Drake in Drake and O'Malley (1966): The Assayer, pp. 151−336
Galilei, G.: Dialogo sopra i due massimi sistemi del mondo. In: Opere, cit., vol. VII (1632). English transl.: Dialogue Concerning the Two Chief World Systems, 2nd edn., transl. by Drake, S., forward by Einstein, A., University of California Press, Berkeley (1967)
Galilei, G.: Discorsi e dimostrazioni matematiche intorno a due nuove scienze. In: Opere, cit., vol. VIII (1638). English trans.: Two New Sciences, trans. by Drake, S., University of Wisconsin Press, Madison (1974)
Geymonat, L.: Galileo Galilei, Einaudi, Torino (1957). Engl. trans. by Drake, S., McGraw-Hill, New York (1965)
Giancotti Boscherini, E.: Lexicon Spinozianum, 2 vols. Nijhoff, The Hague (1970)
Glymour, C.: Explanation and Realism, in Leplin, pp. 173−192 (1984)
Goclenius, R.: Lexicon philosophicum. Frankfurt (1613); Marburg, 1615. Repr. Olms, Hildesheim and New York, 1980
Goodman, N.: Languages of Arts. An Approach to a Theory of Symbols, Hackett Publishng Company, Indiana (1968) (2nd edn., 1976)
Hacking, I.: Representing and Intervening: Introductory Topics in the Philosophy of Natural Science. Cambridge University Press, Cambridge (1983)
Hacking, I.: Experimentation and Scientific Realism, in Leplin, pp. 154−172 (1984)
Hall, R.A.: The Scientific Revolution, 1500−1800. The Formation of Modern Scientific Attitude. Longmans, London (1954). Bacon Press, Boston, 1954, 3rd ed. 1966

Hamilton, W.: In: Hamilton, W., Mansel, H.L., Veitch, J. (eds.) Lectures on Logic. Edinburgh and London (1860). 2nd revised edn., 1866 (they constitute vols. 3–4 of Lectures on Metaphysics and Logic). Repr. Fromann-Holzboog, Stuttgart, 1969

Hanson, N.R.: Patterns of Discovery. Cambridge University Press, Cambridge (1958)

Harré, R.: Matter and Method. MacMillan, London (1964)

Harré, R.: The Principles of Scientific Thinking. MacMillan, London (1970)

Harré, R.: Varieties of Realism. Blackwell, Oxford (1986)

Hegel, G.W.F.: Pänomenologie des Geistes, Würzburg/Bamberg (1807). Recent edition: Meiner, Hamburg, 1952. English trans. by Billie, J.B.: The Phenomenology of Mind, 2nd ed. Allen and Unwin, London (1931)

Hegel, G.W.F.: Wissenschaft der Logik, Nürnberg, 2 vols (1812–1816). Recent edition: Suhrkamp, Frankfurt a. M., 1969, 2 vols. English trans. by Johnstout, W.H., Struthers, L.G., London, 1929, 2 vols.

Heisenberg, W.: The Physicist's Conception of Nature. Hutchinson, London (1958a) (original German edition, 1955)

Heisenberg, W.: Physics and Philosophy. Harper and Brothers, New York (1958b)

Helmholtz, H. v.: Ueber die Erhaltung der Kraft, Reimer, Berlin (1847)

Hempel, C.: Fundamentals of Concept Formation in Empirical Science. Chicago University Press, Chicago (1952)

Hempel, C.: The theoretician's dilemma. In: Feigl, H., Scriven, M., Maxwell, G. (eds.), vol. II, pp. 37–98. University of Minnesota Press, Minneapolis (1958)

Hempel, C.: Aspects of Scientific Explanation. Free Press, New York (1965)

Hertz, H.: The Principles of Mechanics (1899). English trans. by Jones, D.E., Walley, J.T. (eds.) MacMillan, London. (Orig. German edn. 1894)

Hesse, M.: Models and Analogies in Science. University of Notre Dame Press, Notre Dame (1966)

Hübner, K.: Kritik der wissenschaftlichen Vernunft. Alber, Freiburg and München (1978), 2nd edn. 1979. English trans. University of Chicago Press, Chicago (1983)

Husserl, E.: Logische Untersuchungen, 2 vols. Niemeyer, Halle (1901). 4th repr. 1968; English trans. by Findlay, J.N., Logical Investigations. Routledge, London (1970)

Husserl, E.: Ideen zu einer reinen Phanomenologie und phanomenologischen Philosophie. I, Niemeyer, Halle (1913) (almost unchanged editions 1922, 1929). New edition by Biemel, W., with inclusion of later Husserl annotations and considerations, in the collection Husserliana: Nijoff, The Hague, 1950. New edition by Schuhmann, K., in the collection Husserliana, 1976 (the original text is separated from Husserl's subsequent unpublished elaborations). English trans. by Boyce Gibson, W.R. (with Husserl's Preface for this translation), Ideas: General Introduction to Pure Phenomenology, Allen and Unwin, London; McMillan, New York, 1931 (repr. 1952)

Husserl, E.: Formale und transzendentale Logik. Niemeyer, Halle (1929); new edn. by Jenssen, P., In: Husserliana: Nijoff, The Hague (1974)

Husserl, E.: Zur Phenomenologie der Intersubjektivität (1929–1935). In: Husserliana, Band XV, Nijoff, The Hague (1973)

Ingarden, R.: Der Streit um die Existenz der Welt, 2 vols. Niemeyer, Tübingen (1964–1965)

Jacobi, F.H.: David Hume über den Glauben, oder Idealismus und Realismus. Ein Gespräch. G. Loewe, Breslau (1787)

Kant, I.: Critique of Pure Reason, trans.: Kemp Smith, N., McMillan, London, 1929 (1781)

Kelvin (see Thomson, W.): Notes of Lectures on Molecular Dynamics and the Wave Theory of Light, stenographically reported by A.S. Hathaway, Baltimore (1884)

Keuth, H.: Realität und Wahrheit. Mohr (Siebeck), Tübingen (1978)

Kneale, W., Kneale, M.: The Development of Logic. Clarendon Press, Oxford (1962)

Knorr-Cetina, K.: The Manufacture of Knowledge: An Essay on the Constructivist and Contextual Nature of Science. Pergamon Press, Oxford (1981)

Kokoszynska, M.: Über den absoluten Wahrheitsbegriff und einige andere semantische Begriffe. Erkenntnis **6**, 143–165 (1936)
Krajewski, W.: Correspondence Principle and the Growth of Science. Reidel, Dordrecht (1977)
Kripke, S.: Naming and Necessity. Harvard University Press, Cambridge (1980)
Kuhn, T.: The Structure of Scientific Revolutions. Chicago University Press, Chicago (1962)
Kuhn, T.: Reflections on my critics. In: Lakatos, I., Musgrave, A. (eds.) Criticism and the Growth of Knowledge. Cambridge University Press, Cambridge (1970)
Kuhn, T.: Second thoughts on paradigms. In: Suppe, F. (ed.) The Structure of Scientific Theories. University of Illinois Press, Urbana (1974)
Küng, G.: Ontologie und logische Analyse der Sprache. Springer, Wien (1963). English trans.: Reidel, Dordrecht (1967)
Küng, G.: The world as noema and referent. J. British Soc. Phenom. **3**(1), 15–26 (1972)
Küng, G.: Husserl on pictures and intentional objects. Rev. Metaphys. **26**, 670–680 (1973)
Lahti, P., Mittelstaedt, P. (eds.): Symposium on the Foundations of Modern Physics. World Scientific Publication, Singapore (1987)
Latour, B., Woolgar, S.: Laboratory Life: the Social Construction of Scientific Facts. Sage, London (1979)
Laudan, L.: Progress and its Problems. University of California Press, Berkeley (1977)
Laudan, L.: A Confutation of Convergent Realism, in Leplin, pp. 218–249 (1984)
Laudan, L.: Science and Relativism. University of Chicago Press, Chicago (1990)
Leinfellner, W., Kraemer, E., Schark, J.: (eds.) Language and Ontology. In: Proceedings of the 6th International Wittgestein Symposium, 1981, Hoelder, Pickler and Tempsky, Vienna (1982)
Lenk, H.: Interpretationskonstrukte: Zur Kritik der interpretatorischen Vernunft. Suhrkamp, Frankfurt a.M. (1993)
Lenk, H.: Interpretation und Realität. Suhrkamp, Frankfurt a.M. (1995a)
Lenk, H.: Schemaspiele. Ueber Schemainterpretationen und Interpretationskonstrukte. Suhrkamp. Frankfurt a.M. (1995b)
Lenk, H.: Erfassung der Wirklichkeit. Eine interpretationsrealistische Erkenntnistheorie. Könighausen und Neumann, Würzburg (2000)
Leplin, J. (ed.): Scientific Realism. University of California Press, Berkeley (1984a)
Leplin, J.: Truth and Scientific Progress. In: Leplin (1984a), pp. 193–217 (1984b)
Le Pore, E. (ed.): New Directions in Semantics. Academic Press, London (1987)
Le Pore, E., Loewe, J.: Dual Aspect Semantics. In: Lepore (1987)
Locke, J.: An Essay Concerning Human Understanding (1690). Recent edition by Nidditch, P.H., Clarendon Press, Oxford (1975) (quotations are made from this edition)
Longino, H.: Science as Social Knowledge. Princeton University Press, Princeton (1990)
Mach, E.: Compendium der Physik für Mediciner. Braumüller, Wien (1863)
Machamer, P.: Galileo and the Causes. In: Butts and Pitt (eds.), pp. 161–180 (1978)
Mackie, J.L.: Problems from Locke. Clarendon Press, Oxford (1976)
McGinn, C.: The Structure of Content. In: Woodfield (ed.) (1982)
McMullin, E.: Newton on Matter and Activity. University of Notre Dame Press, Notre Dame (1978)
McMullin, E.: A Case for Scientific Realism. In: Leplin (1984), pp. 8–40 (1984a)
McMullin, E.: Two Ideals of Explanation in Natural Science. In: French, P.A., et al. (eds.) Causation and Causal Theories. University of Minnesota Press, Minneapolis, pp. 205–220 (1984b)
Mally, E.: Gegenstandstheoretische Grundlagen der Logik und Logistik. Barth, Leipzig (1912)
Marcos, A.: Ciencia y acción. Una filosofía práctica de la ciencia. Fondo de Cultura Económica, Mexico (2010)
Margenau, H.: The Nature of Physical Reality. McGraw-Hill, New York (1950)
Margenau, H., Park, J.L.: Objectivity in Quantum Mechanics. In: Bunge (1967b), pp. 161–187

Marty, A.: Untersuchungen zur Grundlegung der allgemeinen Grammatik und Sprachphilosophie. Niemeyer, Halle (1908)
Mathieu, V.: L'oggettivitá nella scienza e nella filosofia moderna e contemporanea. Accademia delle Scienze, Torino (1960)
Maxwell, G.: The Ontological Status of Theoretical Entities. Minn. Stud. Philos. Sci. **3**, 3–27 (1962)
Maxwell, J.C.: Treatise on Electricity and Magnetism, 2 vols, 2nd edn. Clarendon Press, Oxford (1881).
Meinong, A.: Ueber Annahmen. Barth, Leipzig (1902)
Meinong, A.: Ueber Gegenstandstheorie. In: Meinong (1904b). English transl. On the Theory of Objects. In: Chisholm (1960), pp. 76–117 (1904a)
Meinong, A. (ed.): Untersuchungen zur Gegenstandstheorie und Psychologie. Barth, Leipzig (1904b)
Minazzi, F.: Galileo "filosofo geometra". Rusconi, Milano (1994)
Mittelstaedt, P.: The problem of interpretation of modern physics. Found. Phys. (Special Issue: Festschrift in Honor of Evandro Agazzi) **41**(11), 1667–1676 (2011)
Moody, E.A.: Truth and Consequence in Medieval Logic. North Holland, Amsterdam (1953)
Morris, C.: Foundations of the Theory of Signs. Chicago University Press, Chicago (1938) (vol. 1, no. 2 of the International Encyclopedia of Unified Science). Reprinted in Charles Morris, Writings on the General Theory of Signs, Mouton, The Hague, 1971
Mulligan, K.: Husserl on States of Affairs in the Logical Investigations, Epistemologia, XII/ Special number, pp. 207–234 (1990)
Mulligan, K., Simons, P., Smith, B.: Truth-makers. Philos. Phenom. Res. **44**(3), 287–321 (1984)
Musgrave, A.: The ultimate argument for scientific realism. In: Nola (1988), pp. 229–252 (1988)
Nagel, E.: The Structure of Science. Routledge & Kegan Paul, London (1961)
Niiniluoto, I.: Truthlikeness. Reidel, Dordrecht (1987a)
Niiniluoto, I.: Varieties of Realism. In: Lahti-Mittelstaedt (1987), pp. 459–483 (1987b)
Niiniluoto, I.: Critical Scientific Realism. Oxford University Press, Oxford (1999)
Nola, R.: Realism and Relativism in Science. Kluwer, Dordrecht (1988)
Nowak, L.: The Structure of Idealization. Reidel, Dordrecht (1980)
Parsons, T.: Nonexistent Objects. Yale University Press, New Haven (1980)
Pasch, M.: Vorlesungen über neuere Geometrie. Teubner, Leipzig (1882)
Pauri, M.: The Quantum Space-Time and Observability. In: Agazzi (1997a), pp. 175–195 (1997)
Pauri, M.: Epistemic *primacy*, ontological *elusiveness* of spatial *extension*: is there an evolutionary role for the *quantum*? Found. Phys. (Special Issue: Festschrift in Honor of Evandro Agazzi) **41**(11), 1677–1702 (2011)
Pieri, M.: Della geometria elementare come sistema ipotetico-deduttivo: monografia del punto e del moto, Memorie della Reale Accademia delle Scienze di Torino. Classe di Sc. Fisiche, Matematiche e Naturali, Serie 2, 49, pp. 173–222 (1899)
Pieri, M.: Sur la géométrie envisagée comme un système purement logique. Bibliothèque du Congrès International de Philosophie, pp. 367–404. Colin, Paris (1901)
Planck, M.: Where is Science Going?, with a preface by Einstein, A., trans. of Planck's selected essays by J. Murphy, Norton, New York (1932), repr. Ox Bow Press, Woodbridge (1981)
Planck, M.: Wege zur physikalischen Erkenntnis. Hirzel, Leipzig (1933)
Plato: The Collected Dialogues of Plato. In: Hamilton, E., Cairns, H. (eds.) Bollingen Series LXXI. Random House, New York (1961)
Poincaré, H.: La science et l'hypothèse. Flammarion, Paris (1902). English trans. Science and Hypothesis. Scribner, New York (1907) (quotations are taken from the latter edition)
Poincaré, H.: La valeur de la science. Flammarion, Paris (1904). English trans. The Value of Science. Dover, New York, Repr. 1958 (quotations are made from the latter edition)
Popper, K.R.: The Logic of Scientific Discovery (6th reviewed impression, 1972). Hutchinson, London (1959)

Popper, K.R.: Conjectures and Refutations (3rd reviewed edition, 1969). Routledge & Kegan Paul, London (1963).
Popper, K.R.: Objective Knowledge. Clarendon Press, Oxford (Repr. with corrections) (1973)
Popper, K.R.: Realism and the Aim of Science. Rowman and Littlefield, Totowa (1983)
Prior, A.N.: Correspondence theory of truth. In: Edwards, P. (ed.) Encyclopedia of Philosophy, I, pp. 223–232. MacMillan, London (1967)
Prior, A.N.: Objects of Thought. In: Geach, P.T., Kenny, A.J.P. (eds.) Clarendon Press, Oxford (1971)
Przelecki, M.: The Logic of Empirical Theories. Routledge and kegan Paul, London (1969)
Przelecki, M.: Fuzziness as multiplicity. Erkenntnis **10**, 371–380 (1976)
Psillos, S.: Scientific Realims. How Science Traks Ttruth. Routledge, New York (1999)
Putnam, H.: The meaning of meaning. Minn. Stud. Philos. Sci. **7**(1975), 131–193 (1975a)
Putnam, H.: Mathematics, Matter and Method (Philosophical Papers, vol. 1); Mind, Language and Reality (Philosophical Papers, vol. 2). Cambridge University Press, Cambridge (1975b)
Putnam, H.: Reason, Truth and History. Cambridge University Press, Cambridge (1981)
Putnam, H.: What is Realism? In: Leplin (1984), pp. 140–153 (1982)
Putnam, H.: Realism and Reason (Philosophical Papers, vol. 3). Cambridge University Press, Cambridge (1983)
Putnam, H.: The many faces of realism. Open Court, La Salle (1987)
Queraltó, R.: Technology as a New Condition of the Possibility of Scientific Knowledge. In: Agazzi and Lenk (eds.) (1999)
Reichenbach, H.: The Philosophical Significance of the Theory of Relativity. In: Schilpp (1949), pp. 287–311 (1949)
Rescher, N.: The Coherence Theory of Truth. Oxford University Press, Oxford (1973). Repr. University Press of America, Washington (1982) (quotations are from this edition)
Rescher, N.: Empirical Inquiry. The Athlone Press, London (1982)
Rescher, N.: Scientific Realism: A Critical Reappraisal. Reidel, Dordrecht (1987)
Righini Bonelli, M.L., William, S. (eds.): Reason, Experiment and Mysticism. Science History Publications, New York (1975)
Ronchi, V.: Il cannocchiole di Galileo e la scienza del Seicento. Einaudi, Torino (1959)
Russell, B.: On denoting. Mind **XIV**, 479–493 (1905). Repr. In: Marsh, R.C. (ed.) Logic and Knowledge. Allen & Unwin, London (1956)
Russell, B.: The philosophy of logical atomism. The Monist **28** (1918) and **29** (1919). Repr. in vol. 8 of The Collected Papers of B. Russell, Allen and Unwin, London (1986). Repr. Pears, D.F. (ed.) Russell's Logical Atomism. Fontana, London (1972) (quotations are made from this edition)
Schilpp, P.A.: Einstein Philosopher-Scientist. Open Court, La Salle (1949)
Searle, J.: Intentionality: An Essay in the Philosophy of Mind. Cambridge University Press, Cambridge (1983)
Sellars, W.: The Language of Theories. In: Feigl and Maxwell (eds.) (1961), pp. 57–77 (1961)
Sellars, W.: Science, Perception and Reality. Routledge and Kegan Paul/ Humanity Press, London/New York (1963)
Sellars, W.: Science and Metaphysics. Variations on Kantian Themes. Routledge and Kegan Paul, London (1968)
Sen, P.K.: Reference and Truth. Indian Council for Philosophical Research with Allied Publishers Limited, New Delhi (1991)
Shapere, D.: The concept of observation in science and philosophy. Philos. Sci. **49**(1982), 485–525 (1982)
Shapere, D.: Reason and the Search for Knowledge. Reidel, Dordrecht (1984)
Shapin, S.: The Scientific Revolution. University of Chicago Press, Chicago, 1966 (1994)
Simons, P.M.: Moments as Truth-Makers. In: Leinfellner, W., Kraemer, E., Schark, J. (eds.), pp. 159–161 (1982)
Smart, J.J.: Between Science and Philosophy. Random House, New York (1968)

Smith, B. (ed.): Parts and Moments. Studies in Logic and Formal Ontology. Philosophia, Munich (1982)
Smith, N.K.: A Commentary on Kant's Critique of Pure Reason. MacMillan, London (1923)
Smith, P.: Realism and the Progress of Science. Cambridge University Press, Cambridge (1981)
Sneed, J.: The Logical Structure of Mathematical Physics (2nd edn. 1979). Reidel, Dordrecht (1971)
Spinoza, B.: Cogitata Metaphysica. English trans. available in: Spinoza, B., The Principles of Descartes' Philosophy, trans. with introduction by Britan, H.H., Open Court, Chicago, 1905 (quotations are taken from this edition) (1663)
Stegmüller, W.: The Structuralist View of Theories. Springer, Berlin, 1979 (1977)
Stepin, V.: Theoretical Knowledge. Springer, Dordrecth (2005)
Strawson, P.: On referring. Mind **59**, 320–344 (1950). Repr. in Logico-Linguistic Papers. Mehuen, London (1971)
Tarozzi, G.: Science, metaphysics and meaningful philosophical principles. Epistemologia **11** (1988), Special issue "Science et métaphysique", pp. 97–104
Tarski, A.: Pojecie prawdy w jezykach nauk dedukcyjnych. In: 'Travaux de la Société des Sciences et des Lettres de Varsovie, Classe III Sciences Mathémathiques et Physiques,' no 34, Warsaw (1933). Better known after the German translation: Der Wahrheitsbegriff in den formalisierten Sprachen. In: Studia Philosophica **1**, 261–405 (1935). English trans. by Woodger, J.H.: In: Tarski, A., Logic, Semantics, Metamathematics. Papers from 1923 to 1938, pp. 152–278. Clarendon Press, Oxford (1956)
Tarski, A.: Grundlegung der wissenschaftlichen Semantik. In: Actes du Congrès International de Philosophie Scientifique, Paris, vol. III, pp. 1–8 (1936)
Tarski, A.: The semantic conception of truth and the foundations of semantics. Pilos. Phenom. Res. **4**, 341–375 (1944)
Thiel, C.: Sinn und Bedeutung in der Logik Gottlob Freges. Hain, Meisenheim am Glan (1965). English trans. Reidel, Dordrecht (1968)
Thomson, W.: (Lord Kelvin) Notes of Lectures on Molecular Dynamics and the Wave Theory of Light, stenographically reported by A. S. Hathaway, Baltimore (1884)
Trendelenburg, A.: Geschichte der Kategorienlehre, Berlin (reprint G. Olms), Hildesheim, 1963 (1846)
van Fraassen, B.C.: The Scientific Image. Clarendon Press, Oxford (1980)
van Fraassen, B.C.: To Save the Phenomena. In: Leplin, pp. 250–259 (1984)
van Fraassen, B.C.: Empiricism and the Philosophy of Science. In: Churchland and Hooker (eds.) (1985), pp. 245–308 (1986)
van Fraassen, B.C.: Scientific Representation. Paradoxes of Perspective. Oxford University Press, Oxford (2008)
von Wright, G.H.: Explanation and Understanding. Routledge & Kegan Paul, London (1971)
Wettstein, H.K.: Has Semantics Rested on a Mistake?. Stanford University Press, Stanford (1991)
Whewell, W.: The Philosophy of the Inductive Sciences, 2nd edn. Franck Cass, London, 1967 (1847)
Woodfield, A. (ed.): Thought and Object: Essays in Intentionality. Clarendon Press, Oxford (1982)
Zalta, E.: Intensional Logic and the Metaphysics of Intentionality. MIT Press, Cambridge (1988)

Index of Names

E. Agazzi, *Scientific Objectivity and Its Contexts*, DOI: 10.1007/978-3-319-04660-0,
© Springer International Publishing Switzerland 2014

MIX
Papier aus verantwortungsvollen Quellen
Paper from responsible sources
FSC® C105338

If you have any concerns about our products,
you can contact us on
ProductSafety@springernature.com

In case Publisher is established outside the EU,
the EU authorized representative is:
Springer Nature Customer Service Center GmbH
Europaplatz 3, 69115 Heidelberg, Germany

Printed by Libri Plureos GmbH
in Hamburg, Germany